Chemical and Biochemical Processes for Energy Sources

Chemical and Biochemical Processes for Energy Sources

Editors

Venko N. Beschkov
Konstantin Petrov

Basel • Beijing • Wuhan • Barcelona • Belgrade • Novi Sad • Cluj • Manchester

Editors

Venko N. Beschkov
Institute of Chemical
Engineering
Bulgarian Academy
of Sciences
Sofia
Bulgaria

Konstantin Petrov
Institute of Electrochemistry
and Energy Systems
Bulgarian Academy
of Sciences
Sofia
Bulgaria

Editorial Office
MDPI
St. Alban-Anlage 66
4052 Basel, Switzerland

This is a reprint of articles from the Topic published online in the open access journals *Processes* (ISSN 2227-9717), *Energies* (ISSN 1996-1073), and *ChemEngineering* (ISSN 2305-7084) (available at: https://www.mdpi.com/topics/chemical).

For citation purposes, cite each article independently as indicated on the article page online and as indicated below:

Lastname, A.A.; Lastname, B.B. Article Title. *Journal Name* **Year**, *Volume Number*, Page Range.

ISBN 978-3-7258-1357-5 (Hbk)
ISBN 978-3-7258-1358-2 (PDF)
doi.org/10.3390/books978-3-7258-1358-2

Contents

About the Editors

Venko N. Beschkov

Venko N. Beschkov, PhD, DSc, was born in 1946 in Sofia, Bulgaria. He achieved his PhD in 1978 and his DSc degree in 1996 from the Bulgarian Academy of Sciences. He was appointed as an Associate Professor in 1984 at the Institute of Chemical Engineering at the Bulgarian Academy of Sciences and he was appointed as a full Professor in 1997 at the same institute. His present interests are chemical and biochemical processes for environmental protection and for the utilization of renewable energy sources. He has published over 200 scientific papers, 2 monographs, and 14 chapters in selected issues. He has been the editor of two books on selected topics. He was the Head of the Institute of Chemical Engineering at the Bulgarian Academy of Sciences (1993–2014) and the Deputy Minister of Environment (1991–1992).

Konstantin Petrov

Konstantin Petrov, PhD, was born in 1955 in Shabla, Bulgaria. He graduated in Electrochemical Engineering from the University of Chemical Technology and Metallurgy, Sofia, Bulgaria. He defended his PhD thesis in 1994 at the Institute of Electrochemistry and Energy Systems, Bulgarian Academy of Sciences. He completed his post-doc at the Texas A&M University, U.S.A., in 1995–1996. He was appointed as an Associate Professor in 1998 and as a full Professor in 2011 at the Institute of Electrochemistry and Energy Systems. His research interests lie in the fields of materials chemistry, catalysis, and physical chemistry, namely electrocatalysis and fuel cell applications. He has also been the Deputy Director of the same institute mentioned above (2010 2015)

Preface

The interest in chemical and biochemical methods for the production of renewable energy sources is governed by the depletion of fossil fuel resources on a global scale and the threatening pollution of the atmosphere by greenhouse gases, namely carbon dioxide, methane, and nitrogen oxides.

The biofuels can successfully replace, at least partially, fossil fuels, enabling the closure of the carbon cycle within the near future, i.e., the released carbon dioxide after use of these biofuels can be captured and recycled by chemical or biochemical methods. On the other hand, these biofuels (ethanol, biogas, and biodiesel) can be utilized for value-added chemical production, thus partially replacing crude oil as a feedstock for industrial organic syntheses, involving the so-called "biorefinery concept".

This reprint illustrates these options for saving fossil fuels through the use of renewable biological resources converted into energy and chemical products. These goals are attained by the methods of chemical catalytic processes and biochemical processes, based on microbiology and bioengineering methods.

Venko N. Beschkov and Konstantin Petrov
Editors

Article

Coprocessing Corn Germ Meal for Oil Recovery and Ethanol Production: A Process Model for Lipid-Producing Energy Crops

Yuyao Jia [1,2]**, Deepak Kumar** [2,3]**, Jill K. Winkler-Moser** [4]**, Bruce Dien** [2,4]**, Kent Rausch** [1]**, Mike E. Tumbleson** [1] **and Vijay Singh** [1,2,*]

[1] Department of Agricultural and Biological Engineering, University of Illinois, 1304 W. Pennsylvania Avenue, Urbana, IL 61801, USA; yjia17@illinois.edu (Y.J.); krausch@illinois.edu (K.R.); mtumbles@illinois.edu (M.E.T.)
[2] DOE Center for Advanced Bioenergy and Bioproducts Innovation, University of Illinois, 1206 W. Gregory Drive, Urbana, IL 61801, USA; dkumar02@esf.edu (D.K.); bruce.dien@usda.gov (B.D.)
[3] Department of Chemical Engineering, State University of New York College of Environmental Science and Forestry, Syracuse, NY 13210, USA
[4] USDA, Agricultural Research Service, National Center for Agricultural Utilization Research, 1815 N University St, Peoria, IL 61604, USA; jill.moser@usda.gov
* Correspondence: vsingh@illinois.edu

Abstract: Efforts to engineer high-productivity crops to accumulate oils in their vegetative tissue present the possibility of expanding biodiesel production. However, processing the new crops for lipid recovery and ethanol production from cell wall saccharides is challenging and expensive. In a previous study using corn germ meal as a model substrate, we reported that liquid hot water (LHW) pretreatment enriched the lipid concentration by 2.2 to 4.2 fold. This study investigated combining oil recovery with ethanol production by extracting oil following LHW and simultaneous saccharification and co-fermentation (SSCF) of the biomass. Corn germ meal was again used to model the oil-bearing energy crops. Pretreated germ meal hydrolysate or solids (160 and 180 °C for 10 min) were fermented, and lipids were extracted from both the spent fermentation whole broth and fermentation solids, which were recovered by centrifugation and convective drying. Lipid contents in spent fermentation solids increased 3.7 to 5.7 fold compared to the beginning germ meal. The highest lipid yield achieved after fermentation was 36.0 mg lipid g^{-1} raw biomass; the maximum relative amount of triacylglycerol (TAG) was 50.9% of extracted oil. Although the fermentation step increased the lipid concentration of the recovered solids, it did not improve the lipid yields of pretreated biomass and detrimentally affected oil compositions by increasing the relative concentrations of free fatty acids.

Keywords: oil-bearing energy crops; corn germ meal; oil recovery; cellulosic ethanol

Citation: Jia, Y.; Kumar, D.; Winkler-Moser, J.K.; Dien, B.; Rausch, K.; Tumbleson, M.E.; Singh, V. Coprocessing Corn Germ Meal for Oil Recovery and Ethanol Production: A Process Model for Lipid-Producing Energy Crops. *Processes* **2022**, *10*, 661. https://doi.org/10.3390/pr10040661

Academic Editor: Giuseppe Toscano

Received: 2 March 2022
Accepted: 28 March 2022
Published: 29 March 2022

Publisher's Note: MDPI stays neutral with regard to jurisdictional claims in published maps and institutional affiliations.

1. Introduction

The need to reduce greenhouse gas emissions provides a strong impetus for the production of renewable fuels. Biofuels produced from dedicated biomass crops and agricultural residues are considered as future sustainable alternatives to petroleum-based fuels. Biodiesel, currently produced mainly from vegetable oils, is one of the most widely used biofuels. Its production results in the release of fewer greenhouse gases and pollutants into the atmosphere than fossil fuels while yielding comparable engine performance [1,2]. It can be integrated into the existing infrastructure for fueling diesel vehicles. In addition, its production process has been improved technically and economically via the development of bio-based catalysts for the transesterification reaction [3] and application of silica in the purification of crude biodiesel [4]. However, soybean yields, the major crop for biodiesel, constrain production and add costs to biodiesel. In the US, 57.5 billion gal of distillate fuel was consumed in 2020, while only 1.82 billion gal of biodiesel was produced—61% using soybean oil [5,6]. Biodiesel retailed at a national average price of 2.36 to 3.51 $ gal^{-1} during the first quarter of 2020, which is generally higher than the average retail price of diesel fuel at 2.43 $ gal^{-1} [7,8]. The manufacturing cost of biodiesel is determined by the cost of bio-oil [2,9].

The development of energy crops that accumulate lipids in their vegetative tissues can overcome the oil supply barrier because they can be grown on marginal lands and/or provide higher production yields compared to traditional oil seed crops (e.g., soybean). Recent advances in plant biotechnology have led to the successful engineering of lipid production in tobacco, sugarcane and potato plants, albeit at low lipid levels [10–16]. This allows for the simultaneous production of biodiesel from plant lipids and high-value fuels or chemicals from cellulosic residues. A coproduction scheme for oil-bearing energy crops affords an opportunity to boost production and lower the cost for biodiesel. Technoeconomic analysis [17] predicted that biodiesel production costs from lipidcane containing 2% to 20% lipid content in plant stems (0.59 to 0.89 $ L^{-1}) were less than soy oil (1.08 $ L^{-1}). A major technical challenge to utilizing an oil-bearing energy crop is developing an efficient process to recover oil from vegetative tissues and convert recalcitrant lignocellulosic biomass into fermentable sugars [18,19].

Due to the insufficient availability of engineered oil-bearing crop samples, corn germ meal, with 2.3% oil on a dry basis [20], was used as a model lignocellulosic energy crop. Corn germ meal, the solid residue of corn germ after oil extraction from wet-milled corn, is a low-value product rationed in ruminant animal diets [21,22]. Liquid hot water (LHW) pretreatment, an essential step to enhance the enzymatic recovery of fermentable sugars, solubilizes hemicellulose and partially removes lignin. Previously, corn germ meal was used as a model feedstock to optimize LHW pretreatment conditions for the lipid enrichment of solids [20]. At the optimal pretreatment severity, lipid content in residual solids increased 2.2 to 4.2 fold.

Having established the compatibility of LHW for oil recovery [20], the goal of this study was to ferment the pretreated germ meal and investigate the fate of oil during hydrolysis and fermentation. Simultaneous saccharification and co-fermentation (SSCF) is a method that combines the enzymatic hydrolysis of carbohydrates to monosaccharides and their fermentation to ethanol by yeast; metabolically modified yeast *Saccharomyces cerevisiae* [23–26] can be used to ferment glucose (from the hydrolysis of cellulose) and xylose (from the hydrolysis of hemicellulose) together. SSCF is expected to increase ethanol conversion efficiencies and final ethanol titers compared to other common fermentation schemes (e.g., separate enzymatic hydrolysis and fermentation, and simultaneous saccharification and fermentation) [14,27,28]. A higher final ethanol titer reduces distillation costs and energy consumption [29]. Removal of lipids post-fermentation would be advantageous because it would streamline the entire process. This study also examined the fractionation of lipids between the liquid and residual solids following fermentation.

2. Materials and Methods

2.1. Corn Germ Meal Samples

Corn germ meal samples were collected from a commercial wet milling facility and dried in an oven maintained at 45 °C for 24 h to reduce moisture content to less than 5% on a wet basis. Dried samples were stored at 4 °C. Corn germ meal is composed of 13.6% water/ethanol extractives, 31.0% glucan, 22.4% xylan, 5.1% acid-insoluble lignin, 8.4% acid-soluble lignin, and 0.03% ash, on a dry basis [20].

2.2. Hot Water Pretreatment

Corn germ meal samples and deionized water were mixed to form a slurry at 20% (w w^{-1}) solid content; slurry was transferred into 50 mL stainless steel tube reactors (316 stainless steel tubing, 1.905 cm outer diameter × 0.165 cm wall × 10.478 cm length, SS-T12-S-065–20, Swagelok, Chicago Fluid System Technologies, Chicago, IL, USA). Reactors were capped with stainless steel caps (316 stainless steel, SS-1210-C, Swagelok, Chicago Fluid System Technologies, Chicago, IL, USA). Reactors were immersed in a fluidized sand bath (IFB-51 Industrial Fluidized Bath, Techne Inc., Burlington, NJ, USA) and heated and held at 160 °C or 180 °C for 10 min (Figure 1). The reaction temperature was monitored by a thermocouple (Penetration/Immersion Thermocouple Probe Mini Conn (temperature ranges from −250 °C to 900 °C), McMaster-Carr, Robbinsville, NJ, USA) inserted into one tube reactor and connected to a data logger (HH306/306A, Datalogger Thermometer, Omega, Stamford, CT, USA). After pretreatment, tube reactors were submerged in cold water to quench the reaction.

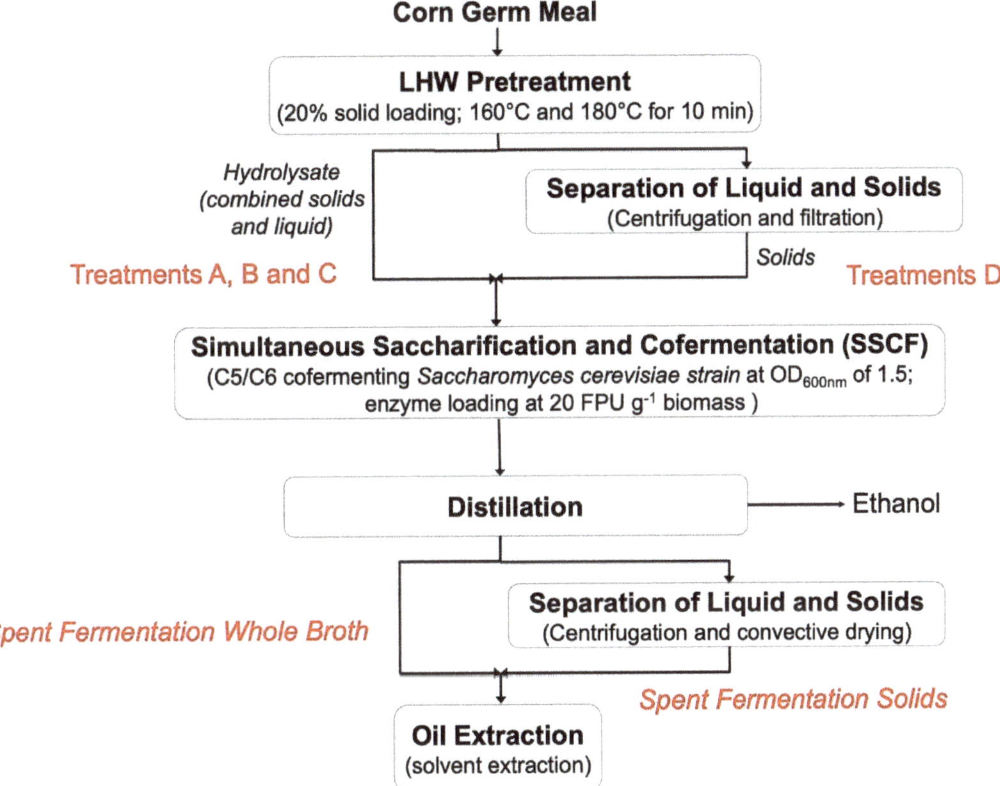

Figure 1. Flow diagram of experimental process to produce lipids and ethanol from corn germ meal. Corn germ meal was pretreated. Pretreated hydrolysate or solids were fermented for 96 h to produce ethanol. After distillation, lipids were extracted from spent fermentation whole broth or solids.

The severity parameter R_o is defined by Equation (1) [18,30–32]. Severity of LHW pretreatment was measured by the log severity factor, represented by Log R_o. The log severity factor is a common method for describing pretreatment conditions [18,32–36].

$$R_o = t \times \exp[(T - 100)/14.75] \tag{1}$$

where R_o is the severity parameter, t is the residence time (min), and T is the pretreatment temperature (°C).

For the fermentation of pretreated solids only, pretreated hydrolysate was centrifuged at $15,000 \times g$ for 15 min to collect pretreated solids. Liquid was separated immediately and collected for HPLC (Bio-Rad Aminex HPX 87H, Biorad, Hercules, CA, USA) to measure levels of inhibitors and acids generated from pretreatment.

2.3. Yeast Culture Preparation

A co-fermenting *Saccharomyces cerevisiae* yeast strain provided by DSM (Heerlen, The Netherlands) was used. The yeast was prepared according to Wang et al. [32]. Seed culture was obtained by inoculating a single colony of the yeast strain in 3 mL YPD media (2% $w\,v^{-1}$ glucose, 2% $w\,v^{-1}$ Bacto™ peptone and 1% $w\,v^{-1}$ yeast extract) at 32 °C and 225 rpm for 24 h. One mL of seed culture was mixed with 25 mL of YPD media, and the yeast culture was incubated at 32 °C and mixed at 225 rpm; the optical density (OD_{600nm}) of the culture was measured using a spectrophotometer set to 600 nm (Evolution 60S, Thermo Scientific, Waltham, MA, USA). When the culture reached an OD_{600nm} of 4, it was centrifuged at $2305\times g$ and the cell pellet resuspended in phosphate-buffered saline to an OD_{600nm} of 50. Resuspended yeast culture was preserved for later use during fermentation.

2.4. Simultaneous Saccharification and Co-Fermentation (SSCF)

Pretreated biomass was fermented using procedures adapted from Wang et al. [32]. Biomass samples were mixed with deionized water to achieve desired solid content, 1 M citrate buffer (pH 6.0) to achieve a final buffer concentration of 0.05 M and YP medium (2% $w\,v^{-1}$ bacto peptone and 1% $w\,v^{-1}$ yeast extract). The pH was adjusted to pH at 6.0 by adding calcium hydroxide solution. A mixture of cellulases and hemicellulases (NS22,257, Novozymes North America, Inc., Franklinton, NC, USA) with an enzymatic activity of 231 FPU mL^{-1} was added at 20 FPU g^{-1} dry substrate and the culture was inoculated to a beginning OD_{600nm} of 1.5 in the final mixture. For raw or pretreated solids (treatment A and D), SSCF was conducted at 20% $w\,w^{-1}$ solid content. For LHW pretreated hydrolysate at 160 °C or 180 °C at 10 min (treatment B and C), SSCF was conducted at 15% and 14% ($w\,w^{-1}$), respectively. These are the solid contents of LHW pretreated hydrolysate after addition of YP media, buffer, yeast culture, enzyme, and carbon hydroxide solution [37]. Enzyme blanks and samples in duplicate were incubated at 32 °C and 150 rpm. Samples (0.5 mL) taken at 0, 3, 6, 9, 24, 48, 72 and 96 h were analyzed using HPLC (Bio-Rad Aminex HPX-87H, Biorad, Hercules, CA, USA) for sugar and ethanol concentrations. Three treatments (A, B, C) of SSCF for LHW pretreated corn germ meal were first conducted (Table 1, Figure 1): in treatment A, which served as the control, raw germ meal was fermented at 20% ($w\,w^{-1}$) solid content; in treatment B, 160 °C LHW/10 min pretreated germ meal hydrolysate (combined solid and liquid fractions) was fermented at 15% ($w\,w^{-1}$) solid content; in treatment C, 180 °C LHW/10 min pretreated hydrolysate (combined solid and liquid fractions) was fermented at 14% ($w\,w^{-1}$) solid content. Treatment C was expected to have higher ethanol conversion efficiency than treatment B because of its higher pretreatment temperature [18,20,38]. However, as discussed in detail in Section 3.1, this expectation was not met, which could be due to inhibitors released into the liquid portion of pretreated germ meal mixture during pretreatment. Therefore, treatment D was conducted to improve treatment C, whereby the liquid proportion of the pretreated hydrolysate (180 °C LHW/10 min) was separated, and the solid fraction was fermented at 20% ($w\,w^{-1}$) solid content. After fermentation, ethanol was removed from the spent fermentation whole broth by heating the broth at 90 °C for 1 h [39]. Ethanol conversion efficiency (%, $w\,w^{-1}$) was calculated based on beginning glucan and xylan contents in untreated germ meal. Beginning glucan and xylan contents were calculated based on compositions of untreated germ meal, including 31.0% glucan and 22.4% xylan, obtained from Jia et al. [20]. Solid content of pretreated germ meal hydrolysate was determined to calculate amounts of beginning untreated germ meal used for treatments B and C. Solid recovery rate (36.8%) of 180 °C LHW/10 min [20] was used to calculate amounts of beginning untreated germ meal used for treatment D.

Table 1. Treatments of SSCF for LHW [1] pretreated corn germ meal.

Treatment	Pretreatment Condition	Log Severity Factor (Log R_0)	Solid Content (%, $w\,w^{-1}$)
A [2]	Untreated	0	20
B [3]	160 °C LHW/10 min	2.8	15 [4]
C [5]	180 °C LHW/10 min	3.4	14 [4]
D [6]	180 °C LHW/10 min	3.4	20

[1] LHW = liquid hot water pretreatment. [2] A: Raw germ meal was fermented at 20% $w\,w^{-1}$ solid content. [3] B: 160 °C LHW/10 min pretreated germ meal hydrolysate (solid and liquid fraction) fermented at 15% $w\,w^{-1}$. [4] Highest possible solid content was calculated when no additional deionized water was added with addition of YP medium, buffer, resuspended yeast culture, enzyme, and carbon hydroxide solution. [5] C: 180 °C LHW/10 min pretreated germ meal hydrolysate fermented at 14% $w\,w^{-1}$. [6] D: Liquor fraction of 180 °C LHW/10 min pretreated germ meal hydrolysate was removed prior to fermentation. Solid fraction was separated and fermented at 20% $w\,w^{-1}$.

2.5. Lipid Extraction

Spent fermentation whole broth was centrifuged at 15,000× g for 15 min to remove supernatant, and the residual spent fermentation solids were dried for 24 h at 45 °C using a convection oven. Lipids were extracted from spent fermentation whole broth and solids using extraction procedures outlined by Huang et al. [39]. Briefly, spent fermentation whole broth or solids were mixed with hexane (15 mL) and isopropanol (10 mL). The mixture was homogenized twice (1 min each time) using a homogenizer (LabGen 700, Cole Parmer, Vernon Hills, IL, USA) and mixed for 10 min using a wrist action shaker (HB-1000 Hybridizer, UVP LLC, Upland, CA, USA). Next, 16 mL of sodium sulfate solution (6.7% $w\,v^{-1}$) was added to the sample and was mixed for another 10 min. The mixture was centrifuged at 5× g for 20 min and supernatant was transferred to a tared centrifuge tube for evaporation under nitrogen. The tube was reweighed after lipids were dried. Extractable lipid yield (mg lipid g^{-1} raw biomass) was calculated as the ratio of extracted lipids and beginning untreated germ meal. Lipid concentrations of spent fermentation solids were calculated as the ratio of extracted lipids and weight of solid samples from which lipids were extracted.

2.6. Lipid Analyses

For lipid composition analysis, dried oil extracts were redissolved in hexane to a concentration of 10 mg mL^{-1}. They were filtered (0.45 μm PTFE) into HPLC vials. Samples were analyzed using a HPLC system (LC-20AT, Shimadzu, Columbia, MD, USA) equipped with a photodiode array (PDA) (SPD-M20A, Shimadzu, Columbia, MD, USA) and evaporative light scattering defectors (ELSD) (Model-LTII, Shimadzu, Columbia, MD, USA) [20]. External standard curves were developed with the ELSD for triacontane (hydrocarbon, HC), behenyl behenate (wax ester, WE), cholesteryl oleate (steryl ester, SE), monoolein (monoacylglycerol, MG), diolein (diacylglycerol, DG), triolein (triacylglycerol, TAG), oleic acid (free fatty acid, FFA) and cholesterol (sterol, ST), and the PDA set at 320 nm for oryzanol (steryl ferulate, SF). Calibrations were linear for all components except for SE, which was fitted a with second-order polynomial, all with $R^2 \geq 0.99$. Method limits of quantitation were 0.5% ($w\,w^{-1}$) for HC, WE, SE, DG, MG, TAG and ST and 0.05% ($w\,w^{-1}$) for SF.

Fatty acid profiling of extracted lipids was performed using procedures outlined by Quarterman et al. [40]. Dried oil extracts were trans-esterified using 2 mL of hexane and 0.2 mL of 2M KOH in methanol. The resulted fatty acid methyl esters (FAME) were analyzed by gas chromatography with flame ionization detection (GC-FID) (Agilent Technologies, Santa Clara, CA, USA) using an Agilent HP-88 capacity column (30 m × 0.25 mm). A commercial standard with 31 known FAMEs (Nu-Chek Prep, Inc., Elysian, MN, USA, cat. No. GLC-411) was used to identify retention times of FAMEs in the samples. Corn oil was analyzed as a control. Percent of total fatty acids was calculated based on the sum of all FAME peak areas identified by GC-FID.

2.7. HPLC Analysis of Sugars, Organic Acids, and Ethanol

Pretreated hydrolysate from LHW and samples from SSCF were centrifuged at $15,000 \times g$ for 15 min. Supernatants were filtered through a 0.2 μm PTFE filter and analyzed by HPLC (Bio-Rad Aminex HPX-87H, Biorad, Hercules, CA, USA) to determine concentrations of monosaccharides, organic acids, common inhibitors, and ethanol.

2.8. Statistical Analysis

Analysis of variance (ANOVA) and Tukey's honest significant difference (HSD) test were performed using R software (V.3.6.1, R Foundation for Statistical Computing, Vienna, Austria) to compare means of final ethanol titers, ethanol conversion efficiencies, inhibitor concentrations, lipid yields, and lipid concentrations. All experiments were analyzed using two replicates. The significance level for difference was chosen as 5% ($p < 0.05$).

3. Results and Discussion

3.1. Ethanol and Sugar Profile during SSCF

The corn germ meal was pretreated at three severities (treatment A, B, and C; Table 1). For treatments A, B, and C, glucose and xylose were consumed as fast as they were released after 24 h of fermentation (Figure 2). The glucose and xylose concentrations peaked at 3 h into the fermentations. Ethanol was produced rapidly as glucose and xylose were consumed. Treatments A, B, and C resulted in final ethanol titers of 29.3, 32.5, and 27.9 g L^{-1} and ethanol conversion efficiencies of 38.9%, 60.1%, and 49.9%, w w^{-1}, respectively (Figure 3). Therefore, treatment B resulted in the greatest ethanol titer and ethanol conversion efficiency among these three treatments ($p < 0.05$).

Although more severe pretreatment conditions were previously observed to aid the release of monosaccharides [18,38], ethanol conversion efficiency did not increase uniformly with pretreatment severity in this study. This can be explained by the inhibitor concentrations in the LHW pretreated hydrolysates (Table 2): LHW at 160 °C for 10 min had lower amounts of 5-hydroxymethylfurfural (HMF) and furfural in the pretreated liquid (0.16 and 0.70 g L^{-1}, respectively) compared to LHW at 180 °C for 10 min (0.62 and 2.31 g L^{-1} HMF and furfural, respectively). Other inhibitors in the pretreated liquid, including lactic acid (formation of lactic acid is explained in Appendix A), formic acid, and acetic acid, were observed at such low concentrations that they could not cause inhibition [41–44]. The HMF and furfural concentrations were higher than the inhibitor concentrations observed in pretreated hydrolysates of sugarcane bagasse and biomass sorghum [18,32,34]. It is possible that the inhibitors present in the hydrolysate used in treatment C led to the lower ethanol titer. Microbial inhibitors are soluble; therefore, treatment C was modified (in treatment D) by removing the pretreatment liquor. The solid content was increased to 20% in SSCF. These changes led to an increased final ethanol titer of 32.4 g L^{-1}, which demonstrated that inhibitors were responsible for the poorer fermentation results for treatment C. It might be supposed that removing the liquid prior to fermentation (treatment D) would be preferable; however, this led to a decreased ethanol conversion efficiency of 15.7%, w w^{-1}, which was caused by the significant loss of fermentable sugars released during pretreatment by discarding pretreated liquor [20]. Thus, fermenting pretreated solids instead of the whole hydrolysate (combined solid and liquid portion) was disadvantageous for fermentation. The ethanol titer for treatment A was surprisingly high given that it had received no pretreatment. This is due to the lower recalcitrance of seed cell walls compared to stems and leaves [45–47]. Therefore, it is anticipated that pretreatment will be essential for SSCF of plants engineered for oil production stored in stems.

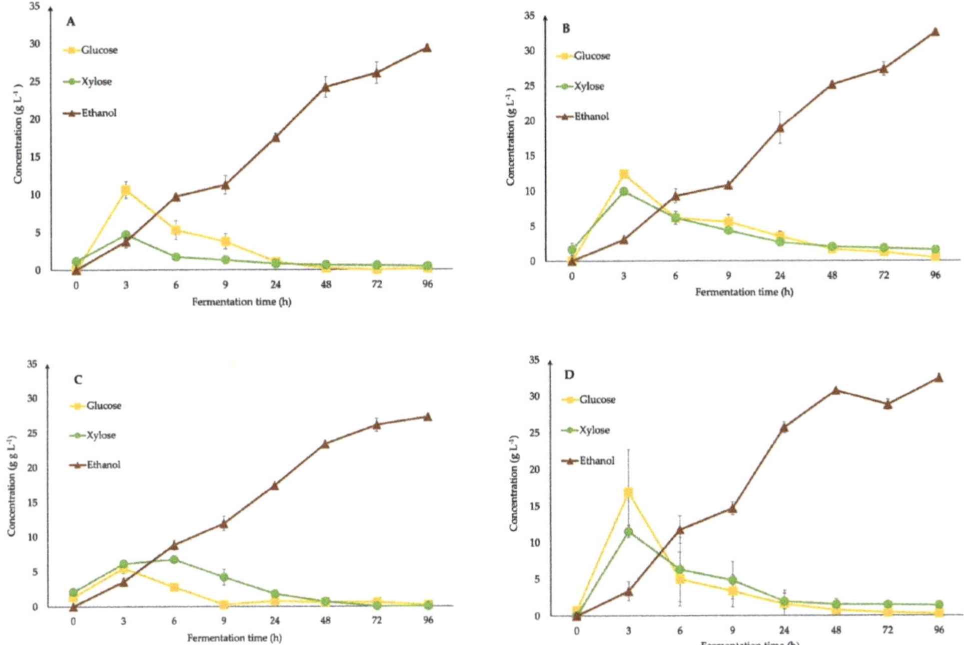

Figure 2. Glucose, xylose, and ethanol concentrations during 96 h of SSCF in treatments A, B, C, and D. Error bars represent ± one standard deviation. Two replicates per mean were used. (**A**) Treatment A: raw germ meal fermented at 20% $w\,w^{-1}$ solid content. (**B**) Treatment B: 160 °C LHW/10 min pretreated germ meal hydrolysate (solid and liquid fraction) fermented at 15% $w\,w^{-1}$. (**C**) Treatment C: 180 °C LHW/10 min pretreated germ meal hydrolysate fermented at 14% $w\,w^{-1}$. (**D**) Treatment D: solid fraction from 180 °C LHW/10 min pretreated germ meal hydrolysate was separated and fermented at 20% $w\,w^{-1}$.

It is notable that treatment B had the highest ethanol conversion efficiency (Figure 3) and its final ethanol concentration was comparable to that for treatment D, where the liquid/inhibitors had been removed prior to fermentation. The final ethanol titers for treatments B and D were 32.5 and 32.4 g L^{-1}, respectively. These ethanol titer results compare favorably to past results: [48] reported an ethanol concentration of 29.5 g L^{-1} after SSCF of dilute acid pretreated corn stover; [49] reported an ethanol concentration of 36.5 g L^{-1} after SSCF of aqueous ammonia pretreated corn stover; [50] observed an ethanol concentration of 32.6 g L^{-1} after SSCF of alkali pretreated wheat straw. The final ethanol titers of treatments B and D were in the same range of reported values from the co-fermentation of lignocellulosic feedstocks, although further improvement is desired [29,48–51].

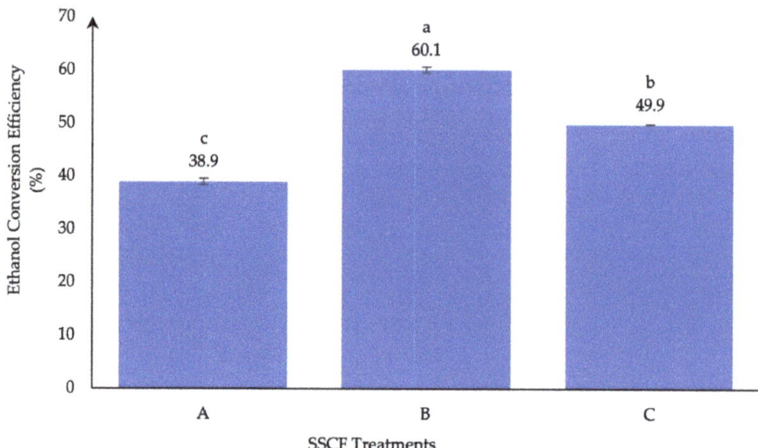

Figure 3. Mean ethanol conversion efficiency (%, $w\,w^{-1}$) at end of 96 h of SSCF in treatments A, B, and C. Error bars represent ± one standard deviation. Means without same letters were different. Two replicates per mean were used. Treatment A: raw germ meal fermented at 20% $w\,w^{-1}$ solid content. Treatment B: 160°C LHW/10 min pretreated germ meal hydrolysate (solid and liquid fraction) fermented at 15% $w\,w^{-1}$. Treatment C: 180 °C LHW/10 min pretreated germ meal hydrolysate fermented at 14% $w\,w^{-1}$.

Table 2. Inhibitor concentrations of LHW [1] pretreated liquid sample [2].

Inhibitors (g/L)	Treatments	
	B (160 °C LHW/10 min)	C (180 °C LHW/10 min)
Lactic acid	0.39 ± 0.04 a	0.36 ± 0.01 a
Formic acid	0.43 ± 0.52 a	0.86 ± 0.01 a
Acetic acid	0.38 ± 0.37 a	1.19 [3] a
Levulinic acid	BDL [4]	BDL
HMF [5]	0.16 ± 0.04 b	0.62 ± 0.01 a
Furfural	0.70 ± 0.16 b	2.31 ± 0.02 a

[1] LHW = liquid hot water pretreatment. [2] Results are represented as mean ± 1 standard deviation. Two replicates per mean were used. Means without same letters within one row were different. [3] Standard deviations below 0.01 were not displayed. [4] BDL = below the detectable limit (0.01 g/L). [5] HMF = 5-hydroxymethylfurfural.

The ethanol conversion efficiency of pretreated corn germ meal was relatively low compared to typical lignocellulosic feedstocks that are plant stems or leaves, such as sugarcane bagasse, hemp, corn stover, reed, and switchgrass. Ethanol conversion efficiencies of fermentation using LHW-pretreated typical cellulosic biomass ranged from 75% to 94% (Table A1 in Appendix B) [32,52–55]. Typical cellulosic feedstocks contain cell walls with greater recalcitrance, explaining why their glucan content was higher than germ meal and increased or remained the same after LHW, and they were pretreated with more severe conditions without significant loss of solids or carbohydrates [32,52–56]. Germ meal, however, contains cell walls with lower recalcitrance, and the recalcitrance was reduced during the wet milling process as well [22,45–47,57]. Significant loss of solids and sugars after 160 °C LHW/10 min was observed for germ meal [20], which could be related to its low ethanol conversion efficiency.

3.2. Lipid Recovery after SSCF

SSCF of LHW pretreated biomass was expected to extract cell wall carbohydrates during enzymatic hydrolysis and promote lipid recovery from the spent fermentation whole broth (both solids and liquid) and spent fermentation solids. Treatment C resulted in the highest lipid yield of 36.0 mg lipid g^{-1} raw biomass (Figure 4). Extractable lipid yields after fermentation increased with increased pretreatment severity. However, following pretreatment and before fermentation, the highest lipid yield achieved by extraction from pretreated germ meal was 34.0 mg lipid g^{-1} raw biomass [20]. No difference was detected between the highest lipid yield after fermentation and the highest lipid yield from pretreated germ meal ($p > 0.05$), indicating that SSCF did not further improve the lipid yields of pretreated biomass.

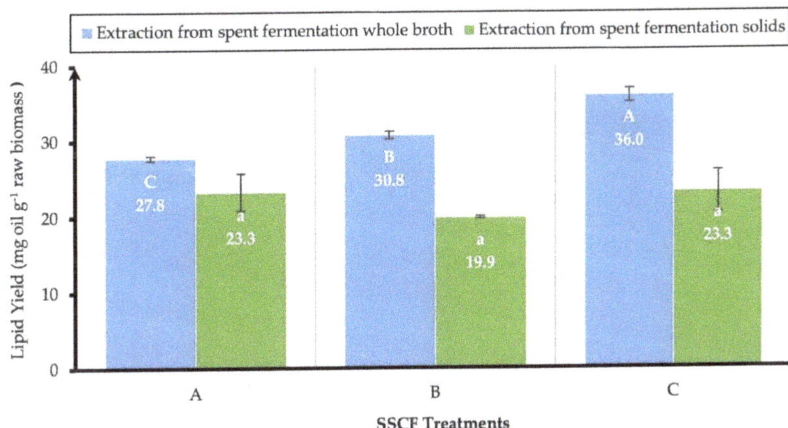

Figure 4. Extractable lipid yields (mg lipid g^{-1} raw biomass) for treatments A, B, and C using SSCF and LHW pretreated germ meal. Error bars represent ± one standard deviation. Two replicates per mean were used. Data within one series without same letters were significantly different ($p < 0.05$): the uppercase letters were designated for extraction from spent fermentation whole broth; the lowercase letters were designated for extraction from spent fermentation solids. Treatment A: raw germ meal fermented at 20% $w\ w^{-1}$ solid content. Treatment B: 160°C LHW/10 min pretreated germ meal hydrolysate (solid and liquid fraction) fermented at 15% $w\ w^{-1}$. Treatment C: 180 °C LHW/10 min pretreated germ meal hydrolysate fermented at 14% $w\ w^{-1}$.

While the extractable lipid yield was always greater for spent whole broth versus solids, the difference increased with pretreatment severity for A, B, and C (4.5, 10.9, and 12.7 mg lipid g^{-1} raw biomass, respectively), which largely reflects lipids/fatty acids that migrated to the liquid section. This means that more lipids were released to the fermentation liquor during SSCF with increased severity. In previous studies [58–60] using corn germ, 90% of lipids were stored in the scutellum of the germ, a storage organ mainly composed of tissues, in the form of lipid droplets (LD), in which they are protected by a phospholipid layer and oleosin molecules. To extract lipids within LDs, LDs need to be dissociated from the storage tissue matrix and then chemically or physically disrupted to release the lipids [61–63]. SSCF following pretreatment theoretically could liberate LD to release lipids into the fermentation broth. Due to the absence of pretreatment in treatment A, fewer cell wall carbohydrates were hydrolyzed, and thus fewer LD were released from the tissues into the fermentation broth compared to treatments B and C. As expected, lipid released from the solids increased with increased pretreatment severity. However, the observed trend could be also partially related to the convective drying of spent fermentation solids. Some lipids are sensitive to thermal decomposition [64–67] and thus might have partially degraded in spent fermentation solids.

Lipid concentrations of spent fermentation solids from treatments A, B, and C were 9.7%, 12.0%, and 13.2%, $w\ w^{-1}$, respectively (Figure 5). Lipid concentrations of spent fermentation solids increased 4.2 to 5.7 fold compared to original germ meal (2.3%) [20]. Compared to a 2.2- to 4.2-fold increase in the lipid content of germ meal after LHW pretreatment reported in [20], SSCF further improved the lipid concentrations. Treatments B and C both had the highest lipid concentrations of spent fermentation solids (Figure 5).

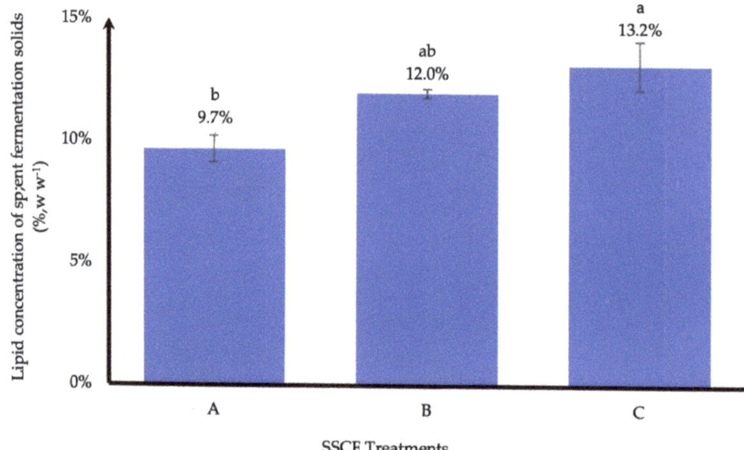

Figure 5. Lipid concentrations of spent fermentation solids (%, $w\ w^{-1}$) from treatments A, B, and C. Error bars represent ± one standard deviation. Means denoted with different letters are different. Two replicates per mean were used. Treatment A: raw germ meal fermented at 20% $w\ w^{-1}$ solid content. Treatment B: 160 °C LHW/10 min pretreated germ meal hydrolysate (solid and liquid fraction) fermented at 15% $w\ w^{-1}$. Treatment C: 180°C LHW/10 min pretreated germ meal hydrolysate fermented at 14% $w\ w^{-1}$.

For treatment D, lipid yields from spent fermentation whole broth and solids were 15.2 and 12.9 mg lipid g^{-1} raw biomass, respectively; the lipid concentration of spent fermentation solids was 12.9%. Lipid yields decreased compared to treatment C because of the significant loss of biomass by removing the pretreated liquor prior to fermentation. Therefore, removing pretreated liquor prior to fermentation (treatment D) was disadvantageous for lipid recovery.

3.3. Compositions of Oil and FAME from Oil

The recovered lipids were examined for composition (Figure 6). The largest fraction of lipids was TAGs, constituting 31.7% to 50.9% of extracted oil (Figure 6). Removing the liquor prior to fermentation (treatment D) lowered the TAG content of the spent whole fermentation broth from 41.3% to 31.7% (Figure 6), indicating another disadvantage of removing the liquor prior to fermentation. The maximum TAG content (50.9%) was observed in spent fermentation solids for treatment B. DG contents were up to four-fold higher in oil from treatments A and C compared to oil from treatments B and D (Figure 6). FFA contents constituted 8.1% to 16.1%, $w\ w^{-1}$ oil (Figure 6), so a neutralization step would be required to prevent FFA from hindering the transesterification reaction if the oil was used for biodiesel production [68–70]. MG, SE, SF, and HC contents were minor constituents (Figure 6), together constituting up to 18.0%, $w\ w^{-1}$ oil.

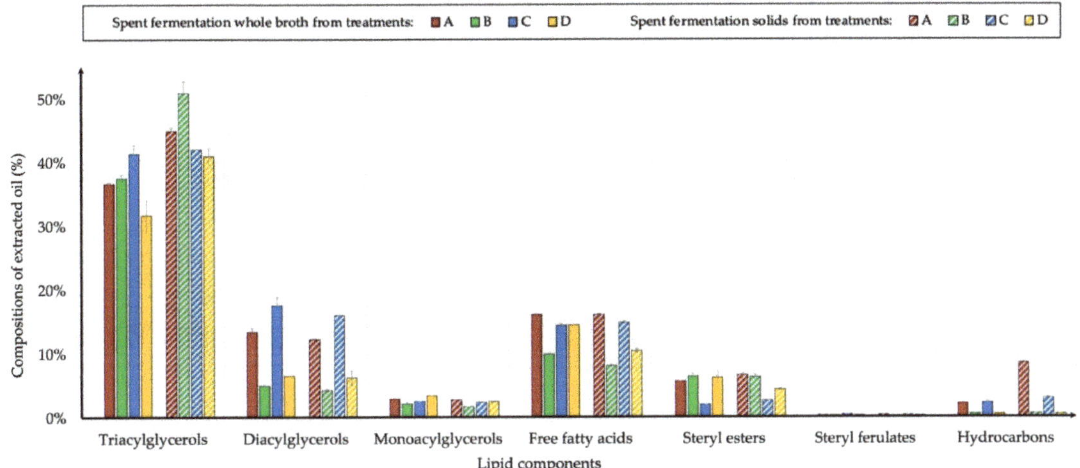

Figure 6. Compositions (%, $w\,w^{-1}$ oil) of extracted oil from spent fermentation whole broth and spent fermentation solids from treatments A, B, C, and D. Error bars represent $\pm$ one standard deviation. Two replicates per mean were used. Treatment A: raw germ meal fermented at 20% $w\,w^{-1}$ solid content. Treatment B: 160 °C LHW/10 min pretreated germ meal hydrolysate (solid and liquid fraction) fermented at 15% $w\,w^{-1}$. Treatment C: 180 °C LHW/10 min pretreated germ meal hydrolysate fermented at 14% $w\,w^{-1}$. Treatment D: solid fraction from 180 °C LHW/10 min pretreated germ meal hydrolysate was separated and fermented at 20% $w\,w^{-1}$.

Following pretreatment with LHW and before SSCF, the LHW pretreated corn germ contained 71.6% of TAG, 0.3% of FFA, and 6.2% of minor constituents [20]. Therefore, SSCF led to a substantial increase in the amount of relative FFAs (Figure 6), presumably because of the saponification of TAGS during SSCF. Increased FFA content is undesirable [65,66].

In the fatty acids analysis, palmitic acid (C16:0), palmitoleic acid (C16:1), stearic acid (C18:0), oleic acid (C18:1), linoleic acid (C18:2), alpha-linoleic acid (C18:3), and arachidic acid (C20:2) were identified. Linoleic acid (C18:2) was the major component of all recovered oil extracts, constituting 53.1% to 54.4% of total fatty acids (Figure 7). Compositions of total FAME obtained from the extracted oil across treatments were similar, indicating that SSCF did not affect the fatty acid composition of TAGs. Moreover, the fatty acid compositions were similar to the corn oil control sample, including 54.4% linoleic acid, 29.2% oleic acid, and 12.7% palmitic acid.

Overall, SSCF detrimentally affected the compositions of oil from LHW pretreated germ meal.

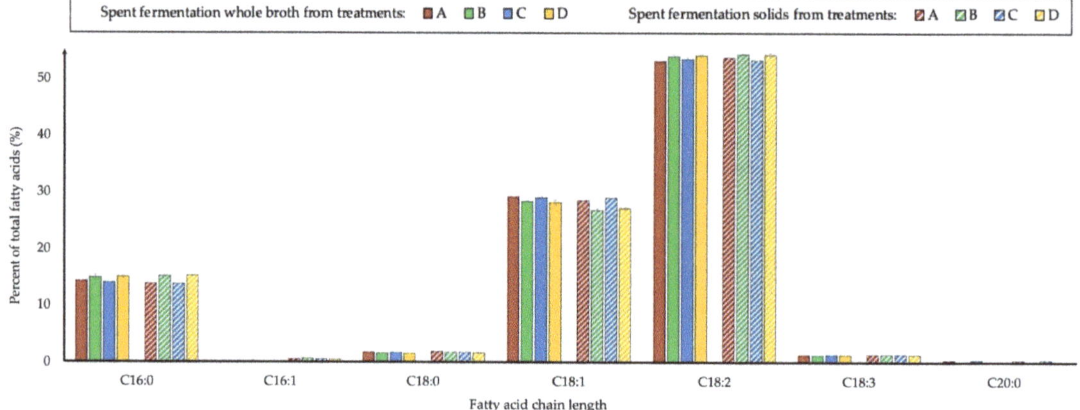

Figure 7. Compositions (%) of total FAME obtained from extracted oil from spent fermentation whole broth and spent fermentation solids from treatments A, B, C, and D. Error bars represent ± one standard deviation. Two replicates per mean were used. Treatment A: raw germ meal fermented at 20% $w\,w^{-1}$ solid content. Treatment B: 160 °C LHW/10 min pretreated germ meal hydrolysate (solid and liquid fraction) fermented at 15% $w\,w^{-1}$. Treatment C: 180 °C LHW/10 min pretreated germ meal hydrolysate fermented at 14% $w\,w^{-1}$. Treatment D: solid fraction from 180 °C LHW/10 min pretreated germ meal hydrolysate was separated and fermented at 20% $w\,w^{-1}$.

4. Conclusions

After fermentation, the highest lipid yield was 36.0 mg lipid g^{-1} raw biomass, the maximum relative amount of TAG was 50.9% of extracted oil, and the highest lipid concentration of spent fermentation solids was 12.0%. Compared to lipid recovery and compositions following pretreatment and before fermentation, the fermentation step did not improve the lipid yield of pretreated biomass and detrimentally affected oil compositions by increasing the relative concentrations of free fatty acids.

Glucose and xylose fermentation profiles during SSCF and compositions of FAME were similar across all treatments. Treatment B appears to be the most favorable as it resulted in the maximal ethanol titer (32.5 g L^{-1}), maximal ethanol conversion efficiency (60.1%), the highest lipid concentration (12.0%) of spent fermentation solids, and maximum relative amounts of TAG (50.9% of extracted oil).

5. Patents

A provisional patent application has been filed from this work (U.S. Patent Application No.: 62/945,438).

Author Contributions: Y.J. conducted all the experiments, analyzed the data, and prepared the manuscript; D.K. helped formulate the study and edited the manuscript; J.K.W.-M. conducted oil composition analyses and edited the manuscript; B.D. conducted FAME analyses and edited the manuscript; K.R. and M.E.T. helped formulate the study and edited manuscript; V.S. is the PI of this project, helped formulate the study, and edited the manuscript. All authors have read and agreed to the published version of the manuscript.

Funding: This work was funded by the DOE Center for Advanced Bioenergy and Bioproducts Innovation (U.S. Department of Energy, Office of Science, Office of Biological and Environmental Research under Award Number DE-SC0018420). Any opinions, findings, and conclusions or recommendations expressed in this publication are those of the author(s) and do not necessarily reflect the views of the U.S. Department of Energy. This work was supported in part by the U.S. Department of Agriculture, Agricultural Research Service. Mention of trade names or commercial products in this publication is solely for the purpose of providing specific information and does not imply recommendation or endorsement by the U.S. Department of Agriculture. USDA is an equal opportunity provider and employer.

Institutional Review Board Statement: Not applicable.

Informed Consent Statement: Not applicable.

Data Availability Statement: The data presented in this study are available on request from the corresponding author.

Acknowledgments: DSM provided the co-fermenting yeast strain and Novozymes North America the enzymes. Stephanie Thompson conducted the FAME analysis. Julie A. Anderson assisted in lipid analysis.

Conflicts of Interest: The authors declare no conflict of interest. The funders had no role in the design, execution, interpretation, or writing of the study.

Appendix A

Although the formation of lactic acid during pretreatment has been observed for various lignocellulosic feedstocks, including sorghum, corn stover, and sugarcane bagasse [19,32,71], it is unclear how lactic acid is formed during pretreatment. Possible reasons suggested by researchers [44,71–73] include wet oxidation of sugar and lignin compounds, endwise truncation of polysaccharides, and contamination of lactic-acid-producing bacteria in biomass.

Appendix B

Table A1. Ethanol conversion efficiencies of fermentation using LHW [1] pretreated lignocellulosic feedstocks [2].

Feedstock	Pretreatment Conditions	Fermentation	Ethanol Conversion Efficiency (%)	Reference
Sugarcane Bagasse	180 °C LHW/10 min	SSCF [2]	94%	[32]
Hemp	170 °C LHW/30 min	SSF [3]	75%	[53]
Corn stover	190 °C LHW/15 min	SSCF [2]	88%	[52]
Reed	210 °C LHW/20 min	SSF [3]	86%	[55]
Switchgrass	200 °C LHW/10 min	SSF [3]	86%	[54]

[1] LHW = liquid hot water pretreatment. [2] SSCF = simultaneous saccharification and co-fermentation. [3] SSF = simultaneous saccharification and fermentation.

References

1. Graboski, M.S.; McCormick, R.L. Combustion of Fat and Vegetable Oil Derived Fuels in Diesel Engines. *Prog. Energy Combust. Sci.* **1998**, *24*, 125–164. [CrossRef]
2. Haas, M.J.; McAloon, A.J.; Yee, W.C.; Foglia, T.A. A Process Model to Estimate Biodiesel Production Costs. *Bioresour. Technol.* **2006**, *97*, 671–678. [CrossRef] [PubMed]
3. Santos, S.; Puna, J.; Gomes, J. A Review on Bio-Based Catalysts (Immobilized Enzymes) Used for Biodiesel Production. *Energies* **2020**, *13*, 3013. [CrossRef]
4. Catarino, M.; Ferreira, E.; Soares Dias, A.P.; Gomes, J. Dry Washing Biodiesel Purification Using Fumed Silica Sorbent. *Chem. Eng. J.* **2020**, *386*, 123930. [CrossRef]
5. EIA. *January 2021 Monthly Energy Review*; U.S. Energy Information Administration: Washington, DC, USA, 2021.
6. EIA. *Monthly Biodiesel Production Report with Data for November 2020*; U.S. Energy Information Administration: Washington, DC, USA, 2020.
7. U.S. Department of Energy. *DOE Clean Cities Alternative Fuel Price Report*; U.S. Department of Energy: Washington, DC, USA, 2020.
8. EIA Gasoline and Diesel Fuel Update. Available online: https://www.eia.gov/petroleum/gasdiesel/index.php (accessed on 29 December 2020).
9. Apostolakou, A.A.; Kookos, I.K.; Marazioti, C.; Angelopoulos, K.C. Techno-Economic Analysis of a Biodiesel Production Process from Vegetable Oils. *Fuel Process. Technol.* **2009**, *90*, 1023–1031. [CrossRef]
10. Parajuli, S.; Kannan, B.; Karan, R.; Sanahuja, G.; Liu, H.; Garcia-Ruiz, E.; Kumar, D.; Singh, V.; Zhao, H.; Long, S.; et al. Towards Oilcane: Engineering Hyperaccumulation of Triacylglycerol into Sugarcane Stems. *GCB Bioenergy* **2020**, *12*, 476–490. [CrossRef]

11. Vanhercke, T.; El Tahchy, A.; Liu, Q.; Zhou, X.-R.; Shrestha, P.; Divi, U.K.; Ral, J.-P.; Mansour, M.P.; Nichols, P.D.; James, C.N.; et al. Metabolic Engineering of Biomass for High Energy Density: Oilseed-like Triacylglycerol Yields from Plant Leaves. *Plant Biotechnol. J.* **2014**, *12*, 231–239. [CrossRef] [PubMed]

12. Vanhercke, T.; Belide, S.; Taylor, M.C.; El Tahchy, A.; Okada, S.; Rolland, V.; Liu, Q.; Mitchell, M.; Shrestha, P.; Venables, I.; et al. Up-Regulation of Lipid Biosynthesis Increases the Oil Content in Leaves of *Sorghum Bicolor*. *Plant Biotechnol. J.* **2019**, *17*, 220–232. [CrossRef] [PubMed]

13. Vanhercke, T.; Dyer, J.M.; Mullen, R.T.; Kilaru, A.; Rahman, M.M.; Petrie, J.R.; Green, A.G.; Yurchenko, O.; Singh, S.P. Metabolic Engineering for Enhanced Oil in Biomass. *Prog. Lipid Res.* **2019**, *74*, 103–129. [CrossRef] [PubMed]

14. Xu, X.-Y.; Yang, H.-K.; Singh, S.P.; Sharp, P.J.; Liu, Q. Genetic Manipulation of Non-Classic Oilseed Plants for Enhancement of Their Potential as a Biofactory for Triacylglycerol Production. *Engineering* **2018**, *4*, 523–533. [CrossRef]

15. Zale, J.; Jung, J.H.; Kim, J.Y.; Pathak, B.; Karan, R.; Liu, H.; Chen, X.; Wu, H.; Candreva, J.; Zhai, Z.; et al. Metabolic Engineering of Sugarcane to Accumulate Energy-Dense Triacylglycerols in Vegetative Biomass. *Plant Biotechnol. J.* **2016**, *14*, 661–669. [CrossRef] [PubMed]

16. Liu, Q.; Guo, Q.; Akbar, S.; Zhi, Y.; El Tahchy, A.; Mitchell, M.; Li, Z.; Shrestha, P.; Vanhercke, T.; Ral, J.-P.; et al. Genetic Enhancement of Oil Content in Potato Tuber (*Solanum Tuberosum* L.) through an Integrated Metabolic Engineering Strategy. *Plant Biotechnol. J.* **2017**, *15*, 56–67. [CrossRef] [PubMed]

17. Huang, H.; Long, S.; Singh, V. Techno-economic Analysis of Biodiesel and Ethanol Co-production from Lipid-producing Sugarcane. *Biofuels Bioprod. Biorefining* **2016**, *10*, 299–315. [CrossRef]

18. Cheng, M.-H.; Dien, B.S.; Lee, D.K.; Singh, V. Sugar Production from Bioenergy Sorghum by Using Pilot Scale Continuous Hydrothermal Pretreatment Combined with Disk Refining. *Bioresour. Technol.* **2019**, *289*, 121663. [CrossRef]

19. Kumar, D.; Murthy, G.S. Impact of Pretreatment and Downstream Processing Technologies on Economics and Energy in Cellulosic Ethanol Production. *Biotechnol. Biofuels* **2011**, *4*, 27. [CrossRef]

20. Jia, Y.; Kumar, D.; Winkler-Moser, J.K.; Dien, B.; Singh, V. Recoveries of Oil and Hydrolyzed Sugars from Corn Germ Meal by Hydrothermal Pretreatment: A Model Feedstock for Lipid-Producing Energy Crops. *Energies* **2020**, *13*, 6022. [CrossRef]

21. Johnston, D.B.; McAloon, A.J.; Moreau, R.A.; Hicks, K.B.; Singh, V. Composition and Economic Comparison of Germ Fractions from Modified Corn Processing Technologies. *J. Am. Oil Chem. Soc.* **2005**, *82*, 603–608. [CrossRef]

22. Rausch, K.D.; Belyea, R.L. The Future of Coproducts from Corn Processing. *Appl. Biochem. Biotechnol.* **2006**, *128*, 047–086. [CrossRef]

23. Hoang Nguyen Tran, P.; Ko, J.K.; Gong, G.; Um, Y.; Lee, S.-M. Improved Simultaneous Co-Fermentation of Glucose and Xylose by Saccharomyces Cerevisiae for Efficient Lignocellulosic Biorefinery. *Biotechnol. Biofuels* **2020**, *13*, 12. [CrossRef]

24. Katahira, S.; Muramoto, N.; Moriya, S.; Nagura, R.; Tada, N.; Yasutani, N.; Ohkuma, M.; Onishi, T.; Tokuhiro, K. Screening and Evolution of a Novel Protist Xylose Isomerase from the Termite Reticulitermes Speratus for Efficient Xylose Fermentation in Saccharomyces Cerevisiae. *Biotechnol. Biofuels* **2017**, *10*, 203. [CrossRef]

25. Ko, J.K.; Lee, S.-M. Advances in Cellulosic Conversion to Fuels: Engineering Yeasts for Cellulosic Bioethanol and Biodiesel Production. *Curr. Opin. Biotechnol.* **2018**, *50*, 72–80. [CrossRef] [PubMed]

26. Tran Nguyen Hoang, P.; Ko, J.K.; Gong, G.; Um, Y.; Lee, S.-M. Genomic and Phenotypic Characterization of a Refactored Xylose-Utilizing Saccharomyces Cerevisiae Strain for Lignocellulosic Biofuel Production. *Biotechnol. Biofuels* **2018**, *11*, 268. [CrossRef] [PubMed]

27. Katahira, S.; Ito, M.; Takema, H.; Fujita, Y.; Tanino, T.; Tanaka, T.; Fukuda, H.; Kondo, A. Improvement of Ethanol Productivity during Xylose and Glucose Co-Fermentation by Xylose-Assimilating S. Cerevisiae via Expression of Glucose Transporter Sut1. *Enzym. Microb. Technol.* **2008**, *43*, 115–119. [CrossRef]

28. Wang, R.; Unrean, P.; Franzén, C.J. Model-Based Optimization and Scale-up of Multi-Feed Simultaneous Saccharification and Co-Fermentation of Steam Pre-Treated Lignocellulose Enables High Gravity Ethanol Production. *Biotechnol. Biofuels* **2016**, *9*, 88. [CrossRef]

29. Sassner, P.; Galbe, M.; Zacchi, G. Techno-Economic Evaluation of Bioethanol Production from Three Different Lignocellulosic Materials. *Biomass Bioenergy* **2008**, *32*, 422–430. [CrossRef]

30. Kim, S.M.; Dien, B.S.; Tumbleson, M.E.; Rausch, K.D.; Singh, V. Improvement of Sugar Yields from Corn Stover Using Sequential Hot Water Pretreatment and Disk Milling. *Bioresour. Technol.* **2016**, *216*, 706–713. [CrossRef]

31. Overend, R.P.; Chornet, E. Fractionation of Lignocellulosics by Steam-Aqueous Pretreatments. *Philos. Trans. R. Soc. Lond.* **1987**, *321*, 523–536. [CrossRef]

32. Wang, Z.; Dien, B.S.; Rausch, K.D.; Tumbleson, M.E.; Singh, V. Fermentation of Undetoxified Sugarcane Bagasse Hydrolyzates Using a Two Stage Hydrothermal and Mechanical Refining Pretreatment. *Bioresour. Technol.* **2018**, *261*, 313–321. [CrossRef]

33. Heitz, M.; Carrasco, F.; Rubio, M.; Chauvette, G.; Chornet, E.; Jaulin, L.; Overend, R.P. Generalized Correlations for the Aqueous Liquefaction of Lignocellulosics. *Can. J. Chem. Eng.* **1986**, *64*, 647–650. [CrossRef]

34. Ilanidis, D.; Stagge, S.; Jönsson, L.J.; Martín, C. Effects of Operational Conditions on Auto-Catalyzed and Sulfuric-Acid-Catalyzed Hydrothermal Pretreatment of Sugarcane Bagasse at Different Severity Factor. *Ind. Crops Prod.* **2021**, *159*, 113077. [CrossRef]

35. Iroba, K.L.; Tabil, L.G.; Sokhansanj, S.; Dumonceaux, T. Pretreatment and Fractionation of Barley Straw Using Steam Explosion at Low Severity Factor. *Biomass Bioenergy* **2014**, *66*, 286–300. [CrossRef]

36. Kim, Y.; Kreke, T.; Mosier, N.S.; Ladisch, M.R. Severity Factor Coefficients for Subcritical Liquid Hot Water Pretreatment of Hardwood Chips: Severity Factor for Liquid Hot Water Pretreatment. *Biotechnol. Bioeng.* **2014**, *111*, 254–263. [CrossRef] [PubMed]
37. Bondesson, P.-M.; Galbe, M. Process Design of SSCF for Ethanol Production from Steam-Pretreated, Acetic-Acid-Impregnated Wheat Straw. *Biotechnol. Biofuels* **2016**, *9*, 222. [CrossRef] [PubMed]
38. Binod, P.; Kuttiraja, M.; Archana, M.; Janu, K.U.; Sindhu, R.; Sukumaran, R.K.; Pandey, A. High Temperature Pretreatment and Hydrolysis of Cotton Stalk for Producing Sugars for Bioethanol Production. *Fuel* **2012**, *92*, 340–345. [CrossRef]
39. Huang, H.; Moreau, R.A.; Powell, M.J.; Wang, Z.; Kannan, B.; Altpeter, F.; Grennan, A.K.; Long, S.P.; Singh, V. Evaluation of the Quantity and Composition of Sugars and Lipid in the Juice and Bagasse of Lipid Producing Sugarcane. *Biocatal. Agric. Biotechnol.* **2017**, *10*, 148–155. [CrossRef]
40. Quarterman, J.; Slininger, P.J.; Kurtzman, C.P.; Thompson, S.R.; Dien, B.S. A Survey of Yeast from the Yarrowia Clade for Lipid Production in Dilute Acid Pretreated Lignocellulosic Biomass Hydrolysate. *Appl. Microbiol. Biotechnol.* **2017**, *101*, 3319–3334. [CrossRef] [PubMed]
41. Phowchinda, O.; Délia-Dupuy, M.L.; Strehaiano, P. Effects of Acetic Acid on Growth and Fermentative Activity of Saccharomyces Cerevisiae. *Biotechnol. Lett.* **1995**, *17*, 237–242. [CrossRef]
42. Narendranath, N.V.; Thomas, K.C.; Ingledew, W.M. Effects of Acetic Acid and Lactic Acid on the Growth of Saccharomyces Cerevisiae in a Minimal Medium. *J. Ind. Microbiol. Biotechnol.* **2001**, *26*, 171–177. [CrossRef] [PubMed]
43. Hasunuma, T.; Sung, K.; Sanda, T.; Yoshimura, K.; Matsuda, F.; Kondo, A. Efficient Fermentation of Xylose to Ethanol at High Formic Acid Concentrations by Metabolically Engineered Saccharomyces Cerevisiae. *Appl. Microbiol. Biotechnol.* **2011**, *90*, 997–1004. [CrossRef] [PubMed]
44. Graves, T.; Narendranath, N.V.; Dawson, K.; Power, R. Effect of PH and Lactic or Acetic Acid on Ethanol Productivity by Saccharomyces Cerevisiae in Corn Mash. *J. Ind. Microbiol. Biotechnol.* **2006**, *33*, 469–474. [CrossRef] [PubMed]
45. Fincher, G.B. Morphology and chemical composition of barley endosperm cell walls. *J. Inst. Brew.* **1975**, *81*, 116–122. [CrossRef]
46. Gibeaut, D.M.; Pauly, M.; Bacic, A.; Fincher, G.B. Changes in Cell Wall Polysaccharides in Developing Barley (Hordeum Vulgare) Coleoptiles. *Planta* **2005**, *221*, 729–738. [CrossRef]
47. Burton, R.A.; Fincher, G.B. Evolution and Development of Cell Walls in Cereal Grains. *Front. Plant Sci.* **2014**, *5*, 456. [CrossRef] [PubMed]
48. Zhu, J.-Q.; Qin, L.; Li, W.-C.; Zhang, J.; Bao, J.; Huang, Y.-D.; Li, B.-Z.; Yuan, Y.-J. Simultaneous Saccharification and Co-Fermentation of Dry Diluted Acid Pretreated Corn Stover at High Dry Matter Loading: Overcoming the Inhibitors by Non-Tolerant Yeast. *Bioresour. Technol.* **2015**, *198*, 39–46. [CrossRef] [PubMed]
49. Zhu, J.-Q.; Qin, L.; Li, B.-Z.; Yuan, Y. J. Simultaneous Saccharification and Co-Fermentation of Aqueous Ammonia Pretreated Corn Stover with an Engineered Saccharomyces Cerevisiae SyBE005. *Bioresour. Technol.* **2014**, *169*, 9–18. [CrossRef]
50. Zhang, Y.; Wang, C.; Wang, L.; Yang, R.; Hou, P.; Liu, J. Direct Bioethanol Production from Wheat Straw Using Xylose/Glucose Co-Fermentation by Co-Culture of Two Recombinant Yeasts. *J. Ind. Microbiol. Biotechnol.* **2017**, *44*, 453–464. [CrossRef] [PubMed]
51. Qin, L.; Zhao, X.; Li, W.-C.; Zhu, J.-Q.; Liu, L.; Li, B.-Z.; Yuan, Y.-J. Process Analysis and Optimization of Simultaneous Saccharification and Co-Fermentation of Ethylenediamine-Pretreated Corn Stover for Ethanol Production. *Biotechnol. Biofuels* **2018**, *11*, 118. [CrossRef]
52. Mosier, N.; Hendrickson, R.; Ho, N.; Sedlak, M.; Ladisch, M.R. Optimization of PH Controlled Liquid Hot Water Pretreatment of Corn Stover. *Bioresour. Technol.* **2005**, *96*, 1986–1993. [CrossRef]
53. Zhao, J.; Xu, Y.; Wang, W.; Griffin, J.; Wang, D. Conversion of Liquid Hot Water, Acid and Alkali Pretreated Industrial Hemp Biomasses to Bioethanol. *Bioresour. Technol.* **2020**, *309*, 123383. [CrossRef] [PubMed]
54. Faga, B.A.; Wilkins, M.R.; Banat, I.M. Ethanol Production through Simultaneous Saccharification and Fermentation of Switchgrass Using Saccharomyces Cerevisiae D5A and Thermotolerant Kluyveromyces Marxianus IMB Strains. *Bioresour. Technol.* **2010**, *101*, 2273–2279. [CrossRef] [PubMed]
55. Lu, J.; Li, X.; Zhao, J.; Qu, Y. Enzymatic Saccharification and Ethanol Fermentation of Reed Pretreated with Liquid Hot Water. *J. Biomed. Biotechnol.* **2012**, *2012*, 276278. [CrossRef] [PubMed]
56. Öhgren, K.; Bura, R.; Lesnicki, G.; Saddler, J.; Zacchi, G. A Comparison between Simultaneous Saccharification and Fermentation and Separate Hydrolysis and Fermentation Using Steam-Pretreated Corn Stover. *Process Biochem.* **2007**, *42*, 834–839. [CrossRef]
57. Himmel, M.E.; Ding, S.-Y.; Johnson, D.K.; Adney, W.S.; Nimlos, M.R.; Brady, J.W.; Foust, T.D. Biomass Recalcitrance: Engineering Plants and Enzymes for Biofuels Production. *Science* **2007**, *315*, 804–807. [CrossRef] [PubMed]
58. Huang, A.H.C. Oil Bodies and Oleosins in Seeds. *Annu. Rev. Plant Physiol. Plant Mol. Biol.* **1992**, *43*, 177–200. [CrossRef]
59. Tzen, J.T.C.; Cao, Y.Z.; Laurent, P.; Ratnayake, C.; Huang, A.H.C. Lipids, Proteins, and Structure of Seed Oil Bodies from Diverse Species. *Plant Physiol.* **1993**, *101*, 267–276. [CrossRef]
60. Swift, J.G.; O'Brien, T.P. Vascularization of the Scutellum of Wheat. *Aust. J. Bot.* **1970**, *18*, 45–53. [CrossRef]
61. Majoni, S.; Wang, T.; Johnson, L.A. Enzyme Treatments to Enhance Oil Recovery from Condensed Corn Distillers Solubles. *J. Am. Oil Chem. Soc.* **2011**, *88*, 523–532. [CrossRef]
62. Wang, T.; White, P.J. Lipids of the Kernel. In *Corn*; Elsevier: Amsterdam, The Netherlands, 2019; pp. 337–368. ISBN 978-0-12-811971-6.
63. Korn, E.D. (Ed.) *Methods in Membrane Biology*; Springer US: Boston, MA, USA, 1977; Volume 8, ISBN 978-1-4684-2912-1.

64. Lorenz, M.A.; Burant, C.F.; Kennedy, R.T. Reducing Time and Increasing Sensitivity in Sample Preparation for Adherent Mammalian Cell Metabolomics. *Anal. Chem.* **2011**, *83*, 3406–3414. [CrossRef]
65. Nawar, W.W. Thermal Degradation of Lipids. *J. Agric. Food Chem.* **1969**, *17*, 18–21. [CrossRef]
66. Nawar, W.W. Chemical Changes in Lipids Produced by Thermal Processing. *J. Chem. Educ.* **1984**, *61*, 299. [CrossRef]
67. Ulmer, C.Z.; Koelmel, J.P.; Jones, C.M.; Garrett, T.J.; Aristizabal-Henao, J.J.; Vesper, H.W.; Bowden, J.A. A Review of Efforts to Improve Lipid Stability during Sample Preparation and Standardization Efforts to Ensure Accuracy in the Reporting of Lipid Measurements. *Lipids* **2021**, *56*, 3–16. [CrossRef] [PubMed]
68. FAO. *Biofuels: Prospects, Risks and Opportunities*; The state of food and agriculture; Food and Agriculture Organization of the United Nations: Rome, Italy, 2008; ISBN 978-92-5-105980-7.
69. Hood, E.E. Plant-Based Biofuels. *F1000Research* **2016**, *5*, 185. [CrossRef] [PubMed]
70. Ribeiro, A.; Castro, F.; Carvalho, J. *Influence of Free Fatty Acid Content in Biodiesel Production on Non-Edible Oils*; Wastes: Solutions; Treatments and Opportunites: Guimarães, Portugal, 2011.
71. Du, B.; Sharma, L.N.; Becker, C.; Chen, S.-F.; Mowery, R.A.; van Walsum, G.P.; Chambliss, C.K. Effect of Varying Feedstock-Pretreatment Chemistry Combinations on the Formation and Accumulation of Potentially Inhibitory Degradation Products in Biomass Hydrolysates. *Biotechnol. Bioeng.* **2010**, *107*, 430–440. [CrossRef] [PubMed]
72. Jönsson, L.J.; Martín, C. Pretreatment of Lignocellulose: Formation of Inhibitory by-Products and Strategies for Minimizing Their Effects. *Bioresour. Technol.* **2016**, *199*, 103–112. [CrossRef]
73. Maitra, S.; Singh, V. Balancing Sugar Recovery and Inhibitor Generation during Energycane Processing: Coupling Cryogenic Grinding with Hydrothermal Pretreatment at Low Temperatures. *Bioresour. Technol.* **2021**, *321*, 124424. [CrossRef]

Article

Study of Methane Fermentation of Cattle Manure in the Mesophilic Regime with the Addition of Crude Glycerine

Wacław Romaniuk [1], Ivan Rogovskii [2], Victor Polishchuk [2], Liudmyla Titova [2], Kinga Borek [1], Witold Jan Wardal [3,*], Serhiy Shvorov [2], Yevgen Dvornyk [2], Ihor Sivak [2], Semen Drahniev [4], Dmytro Derevjanko [5] and Kamil Roman [3]

[1] Department of Rural Technical Infrastructure Systems, Institute of Technology and Life Sciences, National Research Institute, 05-090 Raszyn, Poland; w.romaniuk@itp.edu.pl (W.R.); k.borek@itp.edu.pl (K.B.)
[2] Department of Mechanics, National University of Life and Environmental Sciences of Ukraine, 03041 Kyiv, Ukraine; rogovskii@nubip.edu.ua (I.R.); polischuk.v.m@gmail.com (V.P.); l_titova@nubip.edu.ua (L.T.); sosdok@i.ua (S.S.); dvornykevgen@gmail.com (Y.D.); sivakim@ukr.net (I.S.)
[3] Institute of Wood Sciences and Furniture, Warsaw University of Life Sciences, 02-787 Warsaw, Poland; kamil_roman@sggw.edu.pl
[4] Department of Mechanics, Institute of Engineering Thermophysics of National Academy of Science of Ukraine, 03057 Kyiv, Ukraine; dragnev@secbiomass.com
[5] Department of Mechanics, Polissia National University, 10008 Zhytomyr, Ukraine; derevyanko.dmutro@gmail.com
* Correspondence: witold_wardal@sggw.edu.pl

Abstract: The urgency of the study is due to the need to increase the productivity of biogas plants by intensifying the process of methane fermentation of cattle manure in mesophilic mode by adding to it the waste from biodiesel production: crude glycerine. To substantiate the rational amount of crude glycerine in the substrate, the following tasks were performed: determination of dry matter, dry organic matter, and moisture of the substrate from cattle manure with the addition of crude glycerine; conducting experimental studies on biogas yield during fermentation of cattle manure with the addition of crude glycerine with periodic loading of the substrate; and development of a biogas yield model and determination of the rational composition of crude glycerine with its gradual loading into biogas plants with cattle manure. The article presents the results of research on fermentation of substrates in a laboratory biogas plant with a useful volume of 30 L, which fermented different proportions of crude glycerine with cattle manure at a temperature of 30 °C, 35 °C, and 40 °C. The scientific novelty of the work is to determine the patterns of intensification of the process of methane fermentation of cattle manure with the addition of different portions of crude glycerine. A rapid increase in biogas yield is observed when the glycerol content is up to 0.75%. With the addition of more glycerine, the growth of biogas yield slows down. The digester of the biogas plant, where experimental studies were conducted on the fermentation of substrates based on cattle manure with the addition of co-substrates, is suitable for periodic loading of the substrate. As a rule, existing biogas plants use a gradual mode of loading the digester. Conducting experimental studies on biogas yield during fermentation of cattle manure with the addition of crude glycerine with periodic loading of the substrate makes it possible to build a mathematical model of biogas yield and determine the rational composition (up to 0.75%) of crude glycerine with its gradual loading in biogas plants. Adding 0.75% of crude glycerine to the substrate at a fermentation temperature of 30 °C allows to increase the biogas yield by 2.5 times and proportionally increase the production of heat and electricity. The practical application of this knowledge allows the design of an appropriate capacity of the biogas storage tank (gasholder).

Keywords: biogas plant; substrate; crude glycerine; cattle manure; biosludge

Citation: Romaniuk, W.; Rogovskii, I.; Polishchuk, V.; Titova, L.; Borek, K.; Wardal, W.J.; Shvorov, S.; Dvornyk, Y.; Sivak, I.; Drahniev, S.; et al. Study of Methane Fermentation of Cattle Manure in the Mesophilic Regime with the Addition of Crude Glycerine. *Energies* 2022, 15, 3439. https://doi.org/10.3390/en15093439

Academic Editor: Attilio Converti

Received: 11 April 2022
Accepted: 6 May 2022
Published: 8 May 2022

Publisher's Note: MDPI stays neutral with regard to jurisdictional claims in published maps and institutional affiliations.

1. Introduction

The lack of energy resources in the world determines the intensification of activities in the field of adaptation and operation of existing or new technologies for the production of clean and cheap energy from renewable sources [1,2]. Biofuels are used to replace petroleum fuels: biomass (wood, agricultural and processing waste, energy crops), fuel pellets, and biomass briquettes [3], including torrefied fuel pellets and briquettes [4], biodiesel, and biohydrogen. Biogas technologies are being actively introduced into production. The urgency of the study is due to the need to increase the productivity of biogas plants by intensifying the process of methane fermentation of cattle manure. Such manure is characterized by rather low productivity of biogas output, due to which modern biogas plants have low profitability [5]. In order to increase the yield of biogas from cattle manure, it is advisable to use as co-substrates—waste from agricultural and processing production, at the same time eliminating the need for their disposal. Such co-substrates include sediments of vegetable oils [6], soapstock [7], substandard flour [8], and extruded straw [9,10]. One of such co-substrates is crude glycerine, which is formed as a by-product of biodiesel production [11,12].

In [13–21], the compatible methane fermentation of cattle manure with crude glycerine was studied, while in [22–34] also the fermentation of other substrates (sewage, maize silage, algae, etc.) with crude glycerine. The papers note that the use of crude glycerine as a co-substrate generally increases the yield of biogas. At the same time, the rate of biogas yield increased with the addition of 2% of crude glycerine [35], while the addition of 10% of crude glycerine to the substrate inhibited the rate of biogas yield [35,36].

Thus, there is a need for further research to substantiate the concentration of crude glycerine in the substrate based on cattle manure. The aim of this study is to increase the yield of biogas from cattle manure by adding crude glycerine in mesophilic mode.

2. Materials and Methods

An experimental study of the methods and regimes of cattle manure methane fermentation with the addition of crude glycerine for biogas production was carried out at a lab-scale biogas plant (Figure 1a) in the educational scientific laboratory of bioconversion in the agro-industrial complex of the National University of Life and Environmental Sciences of Ukraine. The biogas plant includes a digester with a usable volume of 30 L and a gasholder (Figure 1b).

Biogas from the methane tank through the pipeline enters the gasholder, where it is stored. This installation uses a "wet" gasholder, which consists of two hollow cylindrical containers (the body and a cylinder-level gauge) as well as a guide. The body of the gasholder is filled with water, in which the hollow cylinder-level gauge floats like a boat. Biogas enters the inner cavity of the cylinder-level gauge, which rises above the body of the gasholder along with the guide that allows determining the presence and volume of gas in the gasholder. From the gasholder, biogas enters the gas stove by squeezing the mass of the cylinder-level gauge.

The digester (fermentation chamber) works with periodic substrate loading. To the substrate that consisted of 3.5 kg of cattle manure and 5 kg of water was added crude glycerine weighing 50, 100, 200, and 300 mL, or 0.6%, 1.2%, 2.3%, and 3.4% of the substrate mass, respectively. The 30 L digester was half loaded with the substrate (loading factor 0.5, loaded volume 15 L). When adding a new portion of the substrate, the fermented substrate was changed by half (emptying coefficient −0.5). The temperature regime of the digester during the study was 30 °C, 35 °C, and 40 °C (mesophilic mode). For control, experiments were performed without the addition of crude glycerine.

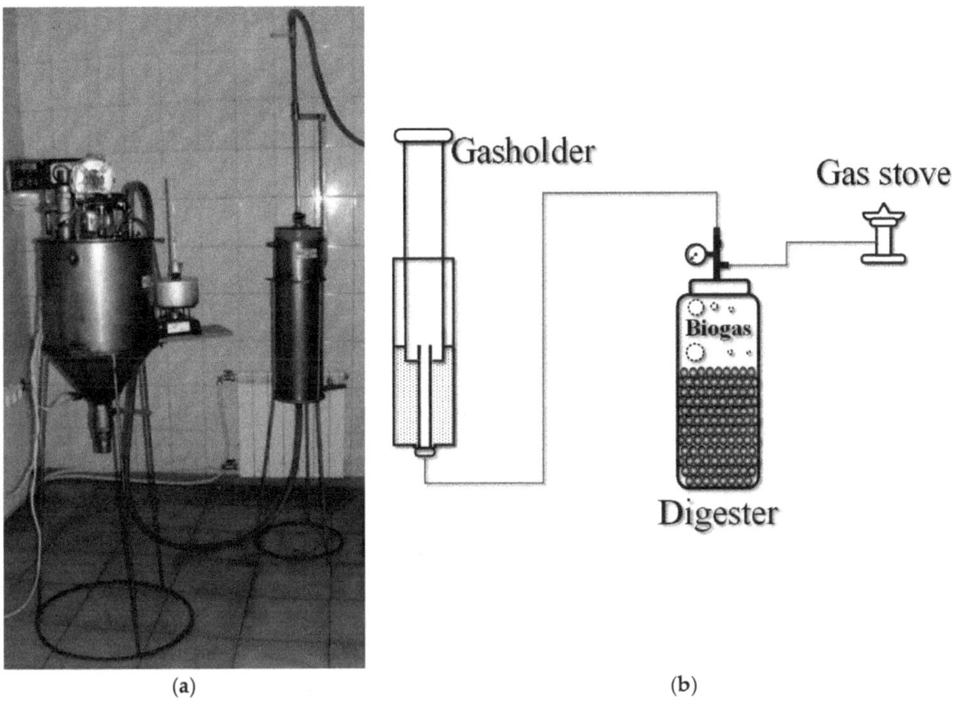

(a)   (b)

Figure 1. Lab-scale biogas plant: (a) general view and (b) principal scheme

The output of biogas was recorded once a day after raising the cylinder-level meter of the gasholder (on the guide of the cylinder-level-meter-attached scale, calibrated in cubic centimeters).

The volume of biogas generated in the digester in the interval between readings of the gasholder–cylinder lifting is determined by the formula:

$$V_b = \frac{\pi \cdot D_g}{4} \cdot H_g \tag{1}$$

where V_b is the volume of biogas generated in the digester in the interval between the readings of the gasholder–cylinder lifting (cm^3), D_g is the inner diameter of the cylinder-level gauge gasholder (cm), and H_g is the height of the cylinder-level gauge of the gasholder (cm).

The productivity of the digester on biogas is calculated by the formula:

$$Q_b = \frac{V_b}{\Delta t} \tag{2}$$

where Q_b is the productivity of the biogas (cm^3/h) and Δt is the time interval between readings of raising the cylinder-level gauge of the gasholder (h).

The time interval Δt between readings of raising the cylinder-level gauge of the gasholder is determined by the formula:

$$\Delta t = 24 - t_2 + t_1 \tag{3}$$

where t_2 is the reading time of the previous indicator of the cylinder-level gauge (h) and t_1 is the reading time of the current level of the cylinder-level gauge (h).

The total amount of biogas obtained during the experiment was specified by the formula:

$$V_{Genb} = \sum V_b \tag{4}$$

where V_{Genb} is the total amount of biogas obtained during the experiment (cm^3).

The average yield of biogas during the experiment was calculated by the formula:

$$V_{Avb} = \frac{V_{Gen\,b}}{t_{Gen}} \tag{5}$$

where V_{Avb} is the average biogas yield during the experiment (cm^3/day) and t_{Gen} is the total time of the experiment (days).

In most literature sources, the yield of biogas from the substrate is estimated in L/kg COP. Therefore, to be able to compare the obtained results with the data from the literature, we converted the obtained results into the dimension of L/(h·kg COP). To do this, we determined the dry weight of the substrate, which is loaded into the digester, according to the formula:

$$M_{DMs} = \frac{M_s \cdot DM}{100} \tag{6}$$

where M_{DMs} is the dry matter mass of the substrate (kg), M_s is the substrate mass (kg), and DM is the dry matter content in the substrate (%).

After that, the mass of dry organic matter of the substrate, which is loaded into the digester, is determined by the formula:

$$M_{DOMs} = \frac{M_{DMs} \cdot DOM_{DM}}{100} \tag{7}$$

where M_{DOMs} is the mass of dry organic matter of the substrate (kg) and DOM_{DM} is the dry organic matter content in the dry matter of the substrate (%).

If the percentage of dry organic matter in the substrate is known, the mass of dry organic matter of the substrate, which is loaded into the digester, is determined by the formula:

$$M_{DOMs} = \frac{M_{DMs} \cdot DOM}{100} \tag{8}$$

where M_{DOMs} is the mass of dry organic matter of the substrate (kg) and DOM is the content of dry organic matter in the substrate (%).

The hourly productivity of the biogas, referred to as the dry organic matter content in the substrate, is determined by the formula:

$$Q_{b/DOMhr} = 10^{-3} \cdot \frac{Q_b}{M_{DOMs}} \tag{9}$$

where $Q_{b/DOMhr}$ is the hourly productivity of the biogas, referred to as the content in the substrate of dry organic matter (L/(h·kg DOM)) and Q_b is biogas productivity of the digester (cm^3/h).

If the hourly productivity of the biogas, related to the dry organic matter content in the substrate, is multiplied by the time between readings of the cylinder-level gauge of the gasholder, the daily biogas productivity of the digester, related to the dry organic matter content in the substrate, is determined by the equation:

$$Q_{b/DOMd} = Q_{b/DOMhr} \cdot \Delta t \tag{10}$$

where $Q_{b/DOMhr}$ is the daily biogas productivity of the methane tank, referred to as the dry organic matter content in the substrate (L/(h·kg DOM)) and Δt is the time interval between readings of raising the cylinder-level gauge of the gasholder (h).

The accumulated biogas yield is determined by the formula:

$$Q_{b\,ac} = \sum_{Q=1}^{n} Q_{b/DOMhr} \tag{11}$$

where $Q_{b\,ac}$ is the accumulated biogas yield (L/(kg DOM)), i is summation index, and n is upper summation limit (days).

3. Results and Discussion

3.1. Experimental Study of Biogas Yield during Fermentation of Cattle Manure with the Addition of Crude Glycerine

The results of the experimental study of the dynamics of the rate of biogas output over time are shown in Figure 2.

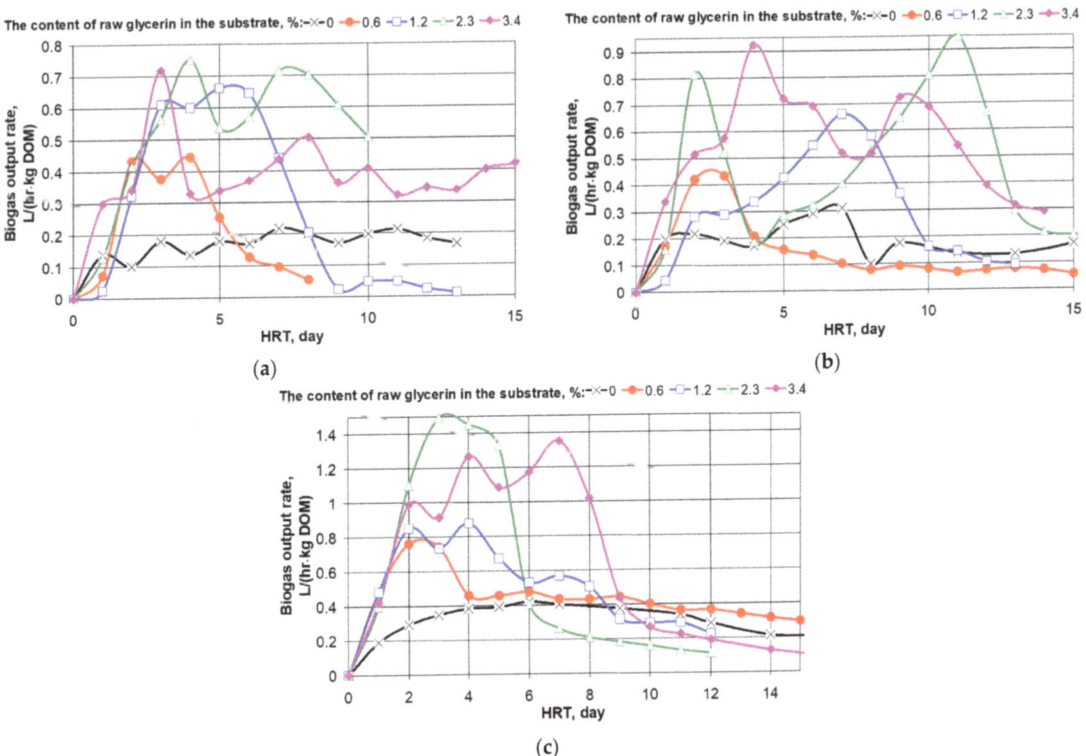

Figure 2. The rate of biogas yield during fermentation of cattle manure with the addition of crude glycerine at the fermentation temperature: (**a**) 30 °C, (**b**) 35 °C, and (**c**) 40 °C.

A characteristic feature of fermentation of cattle manure with the addition of crude glycerine is a short fermentation period, which is 6–12 days. Moreover, the lower the fermentation temperature, the shorter the fermentation period. If the volume of glycerine is higher in the substrate, the fermentation is more intense. There have been cases of clogging of the digester exhaust pipe with manure due to intense foaming. To avoid this phenomenon, the digester was only half filled with the substrate. Diauxia is also a characteristic phenomenon in the fermentation of cattle manure with the addition of crude glycerine.

The values of the maximum yield of biogas of cattle manure fermentation with the addition of crude glycerine in the periodic system of the digester loading are given in Table 1. The conversion factor of biogas yield from the dimension L/(h·kg DOM) to the dimension L/(h·kg) when adding 0.6% of glycerol to the substrate was 17.032 kg/(kg DOM); 1.2% glycerol, 15.867 kg/(kg DOM); 2.3% glycerol, 13.987 kg/(kg DOM); and 3.4% glycerol, 12.536 kg/(kg DOM).

Table 1. Maximum biogas yield of cattle manure fermentation with the addition of crude glycerine for the periodic loading system (L/(h·kg DOM)).

Glycerine Content in the Substrate, %	Maximum Biogas Yield, L/(h·kg DOM), at the Methane Fermentation Temperature		
	30 °C	35 °C	40 °C
0	0.217	0.311	0.402
0.6	0.445	0.604	0.761
1.2	0.662	0.792	0.879
2.3	0.752	0.949	1.494
3.4	0.931	1.200	1.352

The accumulated biogas yield of cattle manure fermentation with the addition of glycerine is shown in Figure 3.

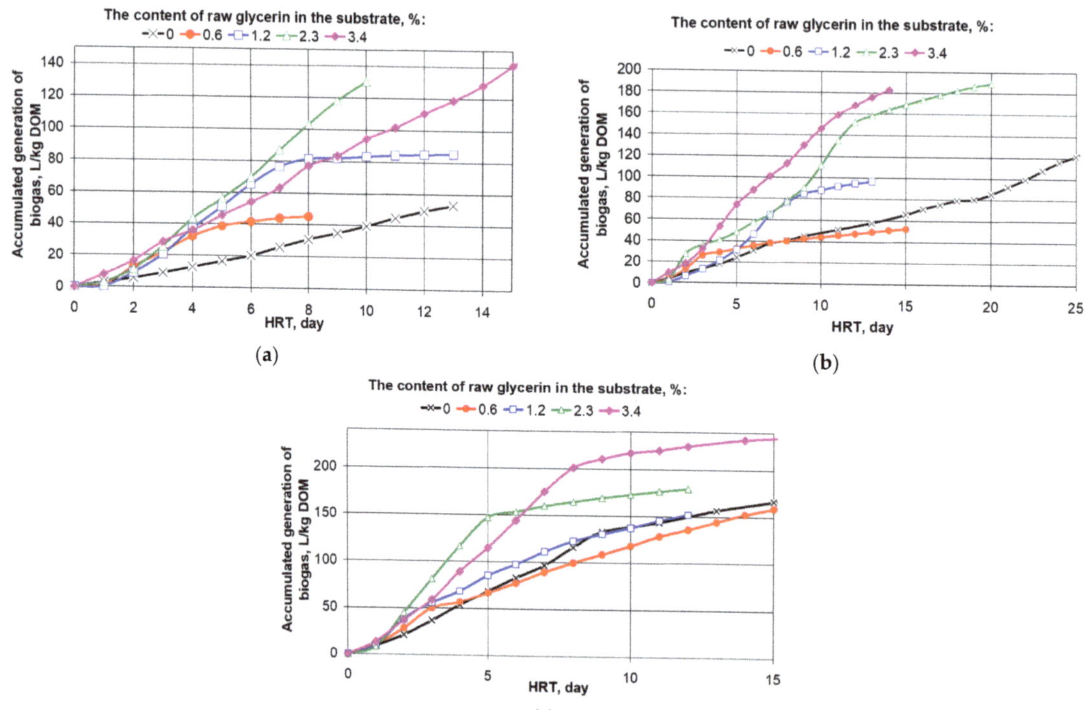

Figure 3. Accumulated biogas yield of cattle manure fermentation with the addition of crude glycerine at the fermentation temperature: (**a**) 30 °C, (**b**) 35 °C, and (**c**) 40 °C.

At low temperatures, the accumulated biogas yield of cattle manure fermentation with the addition of glycerol was higher in comparison with the accumulated biogas yield during cattle manure fermentation only. Thus, at a fermentation temperature of 30 °C, the accumulated biogas yield was about 47 L/(kg DOM) during 10 days of cattle manure fermentation with the addition of 8% DOM glycerol, it was 83.4 L/(kg DOM) in the case of 14.8% DOM glycerol, it was 130.2 L/(kg DOM) with 25.7% DOM glycerol, it was 94.3 L/(kg DOM) with 34.2% DOM glycerol, and it was 39.8 L/(kg DOM) with fermentation of cattle manure alone. However, as the fermentation temperature increased, the accumulated biogas yield of cattle manure fermentation only gradually exceeded the accumulated biogas yield of the fermentation of cattle manure with the addition of glycerol. Thus, at a fermentation temperature of 35 °C, the accumulated biogas yield was about 44.4 L/(kg DOM) during 10 days of cattle manure fermentation with the addition of 8% DOM glycerol, it was 88.3 L/(kg DOM) with 14.8% DOM glycerol, it was 111.7 L/(kg DOM) with 25.7% DOM glycerol, it was 146.5 L/(kg DOM) with 34.2% DOM glycerol, it was 85.2 L/(kg DOM) with 51% DOM glycerol, and it was about 50 L/(kg DOM) when fermenting only cattle manure. At a fermentation temperature of 40 °C, the accumulated biogas yield was 118.5 L/(kg DOM) during 10 days of cattle manure fermentation with the addition of 8% DOM glycerol, it was 137.9 L/(kg DOM) with 14.8% DOM glycerol, it was 173.2 L/(kg DOM) with 25.7% DOM glycerol, it was 217.4 L/(kg DOM) with 34.2% DOM glycerol, and it was about 140 L/(kg DOM) when fermenting only cattle manure. This phenomenon is explained not by the best biogas yield of cattle manure fermentation, but by the fact that this yield is stretched for a long time, and during cattle manure fermentation with glycerol, there is an intensive biogas yield for 3–5 days, after which it quickly stops.

The accumulated yield of biogas during the fermentation of cattle manure with the addition of crude glycerine is approximated by a polynomial.

$$Q_{accumul} = b_n \cdot t^n + b_{n-1} \cdot t^{n-1} + b_1 \cdot t + b_0 \tag{12}$$

where $Q_{accumul}$ is the accumulated biogas yield (L/(kg DOM)), t is the fermentation time (days), and n is the degree of the polynomial.

The coefficients of the polynomial (Equation (12)) are given in Table 2.

Coefficients of determination of the approximated curves (Equation (12)) are given with the coefficients of the polynomial in Table 2. They are approaching 1, which suggests that the obtained regression equations fairly accurately reflect the experimental data.

When tested by Fisher's test [37] (p. 201 (8.52)), the significance of the coefficients of determination was established. When tested by Student's criterion [37] (p. 201 (8.55)), it was established that all coefficients b_1 and b_2 were significant. The coefficients b_3 and b_4 were unreliable (except for coefficient b_3 for the regression equation, which describes the accumulated biogas yield at a fermentation temperature of 35 °C for the glycerol content in the substrate of 1.2%, which was significant). However, the results obtained by regression equation (Equation (12)), Table 2, without these unreliable coefficients, differed significantly from the experimental data, so these coefficients must be present in the regression equations. As a result of checking the significance of the coefficients of determination of the approximated curves (Equation (12)), Table 2, it was determined that they were significant.

Biogas obtained by fermentation of cattle manure with the addition of crude glycerine in the first few days of fermentation did not burn. Moreover, at lower fermentation temperature, increased the time in which no biogas combustion was observed. Thus, at a fermentation temperature of 30–35 °C, the absence of combustion in most cases was observed for 2–3 days, and at a temperature of 40 °C, it was for 1 day. In all cases, the next day after the start of combustion, biogas burned poorly, emitting less heat. Subsequently, the combustion of biogas stabilized with a combustion heat of 18–20 MJ/m^3 for a fermentation temperature of 30–35 °C and 14–18 MJ/m^3 for a fermentation temperature of 40 °C.

Table 2. Polynomial coefficients describing the accumulated biogas yield during the fermentation of cattle manure with the addition of crude glycerine.

No.	Glycerine Content in the Substrate, %	Polynomial Coefficients					R^2
		b_4	b_3	b_2	b_1	b_0	
1	2	3	4	5	6	7	8
Fermentation temperature 30 °C							
1	0	-	-	0.08	3.2	−0.62	0.9982
2	0.6	-	-	−0.57	11	−5	0.9759
3	1.2	-	-	−0.67	16.4	−13	0.9693
4	2.3	-	-	0.36	10.3	−5	0.9955
5	3.4	-	-	0.1	8.9	−4	0.9857
Fermentation temperature 35 °C							
6	0	-	-	-	5.77	−14.5	0.861
7	0.6	-	-	−0.26	7.2	1.3	0.9625
8	1.2	-	−0.128	2.32	−1.26	−0.12	0.9982
9	2.3	-	−0.043	1.1	4.36	4.7	0.988
10	3.4	-	−0.06	1.04	10.4	2	0.9978
11	6.6	-	-	0.09	9.4	−0.6	0.974
Fermentation temperature 40 °C							
12	0	-	−0.026	0.33	14.3	−7.7	0.9991
13	0.6	-	-	−0.26	14.5	0.9	0.9981
14	1.2	-	-	−0.69	21.2	−3.6	0.997
15	2.3	-	0.034	−2.5	41	−15	0.9721
16	3.4	-	-	−1.2	35	−20	0.9835

3.2. Simulation of Biogas Yield with Gradual Loading of the Digester Based on the Results of Experimental Studies of Biogas Yield with Periodic Loading

The digester of the biogas plant, where experimental studies were carried out on the fermentation of substrates based on cattle manure with the addition of co-substrates, is suitable for periodic loading of the substrate. The mode of the substrate's gradual loading is quite difficult to realize. However, in practice, on existing biogas plants, the periodic loading mode of the digester is rarely used; more often, a gradual loading is used, when the substrate is loaded into the digester in small portions after a certain time (usually about 1 h). The biogas yield reaches the maximum value that can be achieved with a periodic loading system, and it is maintained at this level throughout the operation of the biogas plant. Therefore, based on experiments with the periodic loading system of the digester, it is possible to model the biogas yield with a gradual loading system. The biogas yield in a gradual loading system is close to the maximum biogas yield in a periodic loading system.

The simulated yield of biogas during the fermentation of cattle manure with the addition of crude glycerine for the gradual loading of the digester is shown in Figure 4, and the values of the maximum biogas yield that were used to model the process are presented in Table 3.

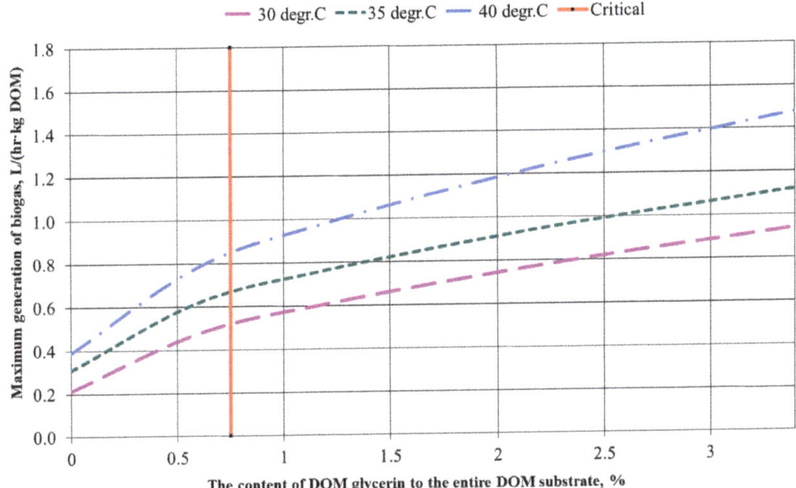

Figure 4. Simulated biogas yield during fermentation of cattle manure with the addition of crude glycerine for gradual loading of the digester.

Table 3. Power function coefficients (Equation (13)) describing the simulated biogas yield during the fermentation of cattle manure with the addition of crude glycerine for gradual loading of the digester.

Fermentation Temperature, °C	Power Function Coefficients			R^2
	b_2	b_1	b_0	
30	0.354	0.568	0.214	0.9826
35	0.412	0.553	0.308	0.9802
40	0.535	0.581	0.388	0.8821
45	0.377	0.59	0.859	0.8575
50	0.504	0.336	0.976	0.9316

The simulated yield of biogas during the fermentation of cattle manure with the addition of crude glycerine for the gradual loading of the digester is approximated by the power function:

$$Q_{b\,\mathrm{mod}} = b_2 \cdot G^{b_1} + b_0 \tag{13}$$

where $Q_{b\mathrm{mod}}$ is simulated biogas output for the gradual loading of the digester (L/(kg DOM)), G is the content of glycerol in the substrate (%), and b_0, b_1, and b_2 are coefficients of the power function.

The coefficients of the power function (Equation (13)) are given in Table 3.

Coefficients of determination of approximate functions (Equation (13)) describing the simulated biogas yield during the fermentation of cattle manure with the addition of crude glycerine for the gradual loading of the digester at all investigated fermentation temperatures approached 1, which suggests that the obtained regression equations fairly accurately reflect the experimental data. When tested by Fisher's test [37] (p. 201 (8.52)), the significance of the coefficients of determination was established for $\alpha = 5\%$ for all investigated regimes and also for $\alpha = 1\%$ for temperature regimes of the digester at 30 °C, 35 °C, and 40 °C. Student's test [37] (p. 201 (8.55)) showed that the coefficients of the power function are significant.

Since we considered the simulated biogas yield of cattle manure fermentation with the addition of crude glycerine for gradual loading of the digester depending on two variables, namely the glycerol content in the substrate, G, and the fermentation temperature

T, Equation (13) with the coefficients, given in Table 3, can be approximated by the function of two variables:

$$Q_{b\,mod} = \left[6.08 \times 10^{-5} \cdot G^4 - 9.58 \times 10^{-3} \cdot G^3 + 0.55894 \cdot G^2 - 14.29677 \cdot G + 135.659\right]$$
$$\times G^{\left(-1.21 \times 10^{-5} \cdot G^4 + 1.737 \times 10^{-3} G^3 - 9.208 \times 10^{-2} G^2 + 2.14187 \cdot G - 17.899\right)} + 4.15 \times 10^{-6} \cdot G^{3.1649} \qquad (14)$$

where Q_{bmod} is the simulated biogas output for gradual loading of the digester (L/(h·kg DOM)), G is the glycerol content in the substrate (%), and G is the fermentation temperature (°C).

The coefficient of determination was $R^2 = 0.8654$ of the regression equation, determined by Equation (14). It is not so close to unity to claim that the obtained regression equation accurately reflects the experimental data for the fermentation temperature of 30–40 °C and the glycerol content in the substrate is not more than 3.4%. However, greater accuracy was not achieved. When it was tested by Fisher's test [37] (p. 201 (8.52)), the significance of the coefficient of determination was established. Student's test [37] (p. 201 (8.55)) showed that all the coefficients of the approximation (Equation (15)) were significant, except for the coefficients 1.121×10^{-5} and 4.15×10^{-6}, which were unreliable, but their removal can cause a significant difference between the experimental and calculated values. The response surface, built according to the regression equation (Equation (14)), is shown in Figure 5.

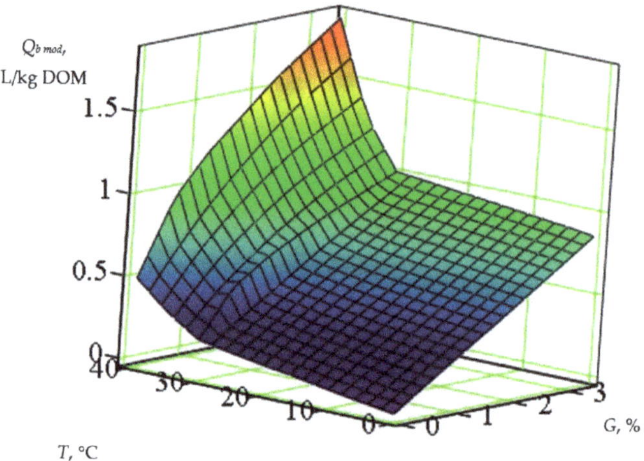

Figure 5. Response surface describing the simulated biogas yield for gradual loading of the digester during fermentation of cattle manure with the addition of crude glycerine.

As can be seen from Figures 4 and 5, with the addition of more crude glycerine, the biogas yield increased. However, it grew unevenly. When up to 0.75% of glycerine was added to the substrate, there was an intensive increase in biogas yield. With the addition of more glycerine, the growth of biogas yield slowed down. Thus, if at a fermentation temperature of 30 °C, an increase in the content of crude glycerine in the substrate from 0% to 0.75% causes an increase in biogas yield by 2.5 times, the increase in the content of crude glycerine in the substrate from 0.75% to 3.4%, i.e., by 25%, is only by 1.7 times. In addition, if the increase in crude glycerine in the substrate from 0% to 0.75% at a fermentation temperature of 40 °C causes an increase in biogas yield by 1.5 times, the increase in crude glycerine in the substrate from 0.75% to 3.4%, that is, by 25%, is only by 1.2 times.

Crude glycerine is a waste from biodiesel production. Currently, for several reasons, biodiesel production in Ukraine is developing at a slow pace. Therefore, to reduce the cost of valuable co-substrates, biogas producers, who have access to crude glycerine as a waste from biodiesel production, the resources of which are limited, are recommended to prepare a substrate with a content of not more than 0.75% of crude glycerine.

4. Conclusions

For increasing the efficiency of biogas production in the mesophilic regime, it is advisable to use crude glycerine, which is the waste from biodiesel production, as a co-substrate. Recycling crude glycerine causes problems for the biodiesel industry.

The scientific novelty of the work is the determination of the intensification patterns of the methane fermentation process of cattle manure based on the addition of different proportions of crude glycerine. An intensive increase in the biogas yield is observed if the glycerin content is 0.75% of the substrate. With the addition of more glycerine, the biogas yield decreases. Thus, at a fermentation temperature of 30 °C, the increase in the crude glycerine content in the substrate from 0% to 0.75% causes the biogas yield to increase by 2.5 times, while the increase in the crude glycerine content in the substrate from 0.75% to 3.4% causes the biogas yield to increase by only 1.7 times.

The increase in the crude glycerine content in the substrate from 0% to 0.75% at a fermentation temperature of 40 °C causes an increase in biogas yield by 1.5 times. In the case of the increase in the crude glycerine content in the substrate from 0.75% to 3.4%, the biogas yield grows by only 1.2 times.

Author Contributions: Conceptualization, W.R., I.R., V.P., L.T., K.B., W.J.W., S.S., Y.D., I.S., S.D., D.D. and K.R.; methodology, W.R., I.R., V.P., L.T., K.B., W.J.W., S.S., Y.D., I.S., S.D., D.D. and K.R.; software, W.R., I.R., V.P., L.T., K.B., I.S., S.D., D.D. and K.R.; validation, W.R., I.R., V.P., L.T., K.B., W.J.W., S.S., Y.D., I.S., S.D., D.D. and K.R.; formal analysis, W.R., I.R., V.P., L.T., K.B., W.J.W., S.S., Y.D., I.S., S.D., D.D. and K.R.; investigation, W.R., I.R., V.P., L.T., K.B., W.J.W., S.S., Y.D., I.S., S.D., D.D. and K.R.; resources, W.R., I.R., V.P., L.T., K.B., W.J.W., S.S., Y.D., I.S., S.D., D.D. and K.R.; data curation, W.R., I.R., V.P., L.T., K.B., I.S., S.D., D.D. and K.R.; writing—original draft preparation, W.R., I.R., V.P., L.T., K.B., W.J.W., S.S., Y.D., I.S., S.D., D.D. and K.R.; writing—review and editing, W.R., I.R., V.P., L.T., K.B., W.J.W., S.S., Y.D., I.S., S.D., D.D. and K.R.; visualization, W.R., I.R., V.P., L.T., K.B., W.J.W., S.S., Y.D., I.S., S.D., D.D. and K.R.; supervision, W.R., I.R., K.B., W.J.W., S.S., S.D., D.D. and K.R.; project administration, W.R., I.R., V.P., L.T., K.B., W.J.W., S.S., Y.D., I.S., S.D., D.D. and K.R.; funding acquisition, W.R., I.R., V.P., L.T., K.B., W.J.W., S.S., Y.D., I.S., S.D., D.D. and K.R. All authors have read and agreed to the published version of the manuscript.

Funding: This research received no external funding.

Institutional Review Board Statement: Not applicable.

Informed Consent Statement: Not applicable.

Data Availability Statement: Not applicable.

Conflicts of Interest: The authors declare no conflict of interest.

Abbreviations

COI	chemical oxygen index
DOM	dry organic matter
VS	volatile solids
DM	dry matter
HRT	hydraulic retention time

References

1. Hryniewicz, M.; Roman, K. Simulations of fuels consumption in the CHP system based on modernised GTD-350 turbine engine. *J. Water Land Dev.* **2021**, *51*, 250–255. [CrossRef]
2. Hryniewicz, M.; Strzelczyk, M.; Helis, M.; Paszkiewicz-Jasińska, A.; Steinhoff-Wrzesniewska, A.; Roman, K. Mathematical models use to yield prognosis of perennials on marginal land according to fertilisers doses. *J. Water Land Dev.* **2021**, *51*, 233–242. [CrossRef]

3.   Olugbade, T.; Ojo, O.; Mohammed, T. Influence of binders on combustion properties of biomass briquettes: A recent review. *Bioenergy Res.* **2019**, *12*, 241–259. [CrossRef]

4.   Olugbade, T.O.; Ojo, O.T. Biomass torrefaction for the production of high-grade solid biofuels: A Review. *Bioenergy Res.* **2020**, *13*, 1–17. [CrossRef]

5.   Lijo, L.; Gonzalez-Garcia, S.; Bacenetti, J.; Moreira, M.T. The environmental effect of substituting energy crops for food waste as feedstock for biogasproduction. *Energy* **2017**, *137*, 1130–1143. [CrossRef]

6.   Rogovskii, I.L.; Polishchuk, V.M.; Titova, L.L.; Sivak, I.M.; Vyhovskyi AYu Drahnev, S.V.; Voinash, S.A. Study of Biogas During Fermentation of Cattle Manure Using a Stimulating Additive in Form of Vegetable Oil Sediment. *ARPN J. Eng. Appl. Sci.* **2020**, *15*, 2652–2663.

7.   Polishchuk, V.M.; Shvorov, S.A.; Krusir, G.V.; Didur, V.V.; Witaszek, K.; Pasichnyk, N.A.; Dvornyk, Y.e.O.; Davidenko, T.S. Using soap waste from biodiesel production to intensify biogas generation during anaerobic digestion of cow dung. *Probl. Energeticii Reg.* **2022**, *1*, 97–107. [CrossRef]

8.   Polishchuk, V.M.; Shvorov, S.A.; Tarasenko, S.Y.E.; Antypov, I.O. Increasing the Biogas Release During the Cattle Manure Fermentation by Means of Rational Addition of Substandard Flour as a Cosubstrate. *Sci. Innov.* **2020**, *16*, 25–35. [CrossRef]

9.   Polishchuk, V.M.; Shvorov, S.A.; Zablodskiy, M.M.; Kucheruk, P.P.; Davidenko, T.S.; Dvornyk, Y.e.O. Effectiveness of Adding Extruded Wheat Straw to Poultry Manure to Increase the Rate of Biogas Yield. *Probl. Energeticii Reg.* **2021**, *3*, 111–124. [CrossRef]

10.   Polishchuk, V.M.; Shvorov, S.A.; Flonts, I.V.; Davidenko, T.S.; Dvornyk, Y.e.O. Increasing the Yield of Biogas and Electricity during Manure Fermentation Cattle by Optimally Adding Lime to Extruded Straw. *Probl. Energeticii Reg.* **2021**, *1*, 73–85. [CrossRef]

11.   Polishchuk, V.; Tarasenko, S.; Antypov Je Kozak, N.; Zhyltsov, A.; Okushko, O. Study of Methods of Biodiesel Neutralization with Aqueous Solution of Lymonic Acid. In Proceedings of the E3S Web of Conferences, 6th International Conference—Renewable Energy Sources (ICoRES 2019), Krynica, Poland, 12–14 June 2019; Volume 154, p. 02007. [CrossRef]

12.   Polishchuk, V.; Tarasenko, S.; Antypov Je Kozak, N.; Zhyltsov, A.; Bereziuk, A. Investigation of the Efficiency of Wet Biodiesel Purification. In Proceedings of the E3S Web of Conferences, 6th International Conference—Renewable Energy Sources (ICoRES 2019), Krynica, Poland, 12–14 June 2019; Volume 154, p. 02006. [CrossRef]

13.   Andriamanohiarisoamanana, F.J.; Saikawa, A.; Kan, T.; Qi, G.D.; Pan, Z.F.; Yamashiro, T.; Iwasaki, M.; Ihara, I.; Nishida, T.; Umetsu, K. Semi-continuous anaerobic co-digestion of dairy manure, meat and bone meal and crude glycerol: Process performance and digestate valorization. *Renew. Energy* **2018**, *128*, 1–8. [CrossRef]

14.   Simm, S.; Orrico, A.C.A.; Orrico, M.A.P.; Sunada, N.D.; Schwingel, A.W.; Lopes, W.R.T.; Lima Whittinghill, K.; Miranda de Vargas, F.; Sarolli Silva de Mendonça Costa, M. Contribute of crude glycerin to increase the efficiency of anaerobic digestion process of dairy cattle manure. *Environ. Prog. Sustain. Energy* **2018**, *37*, 1305–1311. [CrossRef]

15.   Simm, S.; Orrico, A.C.A.; Orrico, M.A.P.; Sunada, N.D.; Schwingel, A.W.; Costa, M.S.S.D. Crude glycerin in anaerobic co-digestion of dairy cattle manure increases methane production. *Sci. Agric.* **2017**, *74*, 175–179. [CrossRef]

16.   Pazuch, F.A.; Siqueira, J.; Friedrich, L.; Lenz, A.M.; Nogueira, C.E.C.; de Souza, S.N.M. Co-digestion of crude glycerin associated with cattle manure in biogas production in the State of Parana, Brazil. *Acta Sci. Technol.* **2017**, *39*, 149–159. [CrossRef]

17.   Regueiro, L.; Carballa, M.; Alvarez, J.A.; Lema, J.M. Enhanced methane production from pig manure anaerobic digestion using fish and biodiesel wastes as co-substrates. *Bioresour. Technol.* **2012**, *123*, 507–513. [CrossRef]

18.   Cremonez, P.A.; Feiden, A.; Teleken, J.G.; de Souza, S.N.M.; Feroldi, M.; Meier, T.W.; Teleken, J.T.; Dieter, J. Comparison between biodegradable polymers from cassava starch and glycerol as additives to biogas production. *Semin. Cienc. Agrar.* **2016**, *37*, 1827–1843. [CrossRef]

19.   Andriamanohiarisoamanana, F.J.; Yamashiro, T.; Ihara, I.; Iwasaki, M.; Nishida, T.; Umetsu, K. Farm-scale thermophilic co-digestion of dairy manure with a biodiesel byproduct in cold regions. *Energy Convers. Manag.* **2016**, *128*, 273–280. [CrossRef]

20.   Aguilar, F.A.A.; Nelson, D.L.; Pantoja, L.D.; dos Santos, A.S. Study of Anaerobic Co-digestion of Crude Glycerol and Swine Manure for the Production of Biogas. *Rev. Virtual Quim.* **2017**, *9*, 2383–2403. [CrossRef]

21.   Alvarez, J.A.; Otero, L.; Lema, J.M. A methodology for optimising feed composition for anaerobic co-digestion of agro-industrial wastes. *Bioresour. Technol.* **2010**, *101*, 1153–1158. [CrossRef]

22.   Andriamanohiarisoamanana, F.J.; Saikawa, A.; Tarukawa, K.; Qi, G.D.; Pan, Z.F.; Yamashiro, T.; Iwasaki, M.; Ihara, I.; Nishida, T.; Umetsu, K. Anaerobic co-digestion of dairy manure, meat and bone meal, and crude glycerol under mesophilic conditions: Synergistic effect and kinetic studies. *Energy Sustain. Dev.* **2017**, *40*, 11–18. [CrossRef]

23.   Khuntia, H.K.; Chanakya, H.N.; Siddiqha, A.; Thomas, C.; Mukherjee, N.; Janardhana, N. Anaerobic digestion of the inedible oil biodiesel residues for value addition. *Sustain. Energy Technol. Assess.* **2017**, *22*, 9–17. [CrossRef]

24.   Ferreira, J.D.; Volschan, I.; Cammarota, M.C. Co-digestion of sewage sludge with crude or pretreated glycerol to increase biogas production. *Environ. Sci. Pollut. Res.* **2018**, *25*, 21811–21821. [CrossRef]

25.   Ferreira, J.S.; Volschan, I.; Cammarota, M.C. Enhanced biogas production in pilot digesters treating a mixture of sewage sludge, glycerol, and food waste. *Energy Fuels* **2018**, *32*, 6839–6846. [CrossRef]

26.   Maragkaki, A.E.; Fountoulakis, M.; Kyriakou, A.; Lasaridi, K.; Manios, T. Boosting biogas production from sewage sludge by adding small amount of agro-industrial by-products and food waste residues. For results 6th International Symposium on Energy from Biomass and Waste. *Waste Manag.* **2018**, *71*, 605–611. [CrossRef] [PubMed]

27. Maragkaki, A.E.; Fountoulakis, M.; Gypakis, A.; Kyriakou, A.; Lasaridi, K.; Manios, T. Pilot-scale anaerobic co-digestion of sewage sludge with agro-industrial by-products for increased biogas production of existing digesters at wastewater treatment plants. *Waste Manag.* **2017**, *59*, 362–370. [CrossRef] [PubMed]

28. Nghiem, L.D.; Nguyen, T.T.; Manassa, P.; Fitzgerald, S.K.; Dawson, M.; Vierboom, S. Co-digestion of sewage sludge and crude glycerol for on-demand biogas production. *Int. Biodeterior. Biodegrad.* **2016**, *95*, 160–166. [CrossRef]

29. Athanasoulia, E.; Melidis, P.; Aivasidis, A. Co-digestion of sewage sludge and crude glycerol from biodiesel production. *Renew. Energy* **2014**, *62*, 73–78. [CrossRef]

30. Panpong, K.; Srisuwan, G.; O-Thong, S.; Kongjan, P. Anaerobic Co-digestion of canned seafood wastewater with glycerol waste for enhanced biogas production. *Energy Procedia* **2014**, *52*, 328–336. [CrossRef]

31. Panpong, K.; Srisuwan, G.; O-Thong, S.; Kongjan, P. Enhanced biogas production from canned seafood wastewater by co-digestion with glycerol waste and wolffia arrhiza. *Energy Procedia* **2014**, *52*, 337–351. [CrossRef]

32. Chavalparit, O.; Sasananan, S.; Kullavanijaya, P.; Charoenwuttichai, C. Anaerobic co-digestion of hydrolysate from alkali pre-treated oil palm empty fruit bunches with biodiesel waste glycerol. *J. Mater. Cycles Waste Manag.* **2018**, *20*, 336–344. [CrossRef]

33. Kacprzak, A.; Krzystek, L.; Ledakowicz, S. Co-digestion of agricultural and industrial wastes. For results 36-th International Conference of the Slovak-Society-of-Chemical-Engineering. *Chem. Pap.* **2010**, *64*, 127–131. [CrossRef]

34. Oliveira, J.V.; Alves, M.M.; Costa, J.C. Design of experiments to assess pre-treatment and co-digestion strategies that optimize biogasproduction from macroalgae Gracilaria vermiculophylla. *Bioresour. Technol.* **2014**, *162*, 323–330. [CrossRef] [PubMed]

35. Larsen, A.C.; Gomes, B.M.; Gomes, S.D.; Zenatti, D.C.; Torres, D.G.B. Anaerobic co-digestion of crude glycerin and starch industry effluent. *Eng. Agric.* **2013**, *33*, 341–352. [CrossRef]

36. Pokoj, T.; Gusiatin, Z.M.; Bulkowska, K.; Dubis, B. Production of biogas using maize silage supplemented with residual glycerine from biodiesel manufacturing. *Arch. Environ. Prot.* **2014**, *40*, 17–29. [CrossRef]

37. Förster, E.; Rönz, B. *Methoden der Korrelations und Regressionsanalise*; Verlag die Wirstschaft: Berlin, Germany, 1979; p. 302.

Article

Economic Analysis of Biogas Production via Biogas Digester Made from Composite Material

KeChrist Obileke [1,*], Golden Makaka [1], Nwabunwanne Nwokolo [1], Edson L. Meyer [2] and Patrick Mukumba [1]

[1] Department of Physics, University of Fort Hare, Private Bag X1314, Alice 5700, South Africa
[2] Fort Hare Institute of Technology, University of Fort Hare, Private Bag X1314, Alice 5700, South Africa
* Correspondence: kobileke@ufh.ac.za

Abstract: This study seeks to evaluate the economic implication of a biogas digester built from composite material to ascertain its cost effectiveness. The feasibility study conducted indicates that a brick made only of fixed dome digester costs between USD 3193.99 and USD 4471.59. This high cost is attributed to the construction material, thus prompting the need to use materials of lower cost for affordability and sustainability. Hence, the digester under study was made from composite material comprising high-density polyethylene (HDPE), bricks and cement. The inlet and outlet chambers were built using bricks and cement, while the digestion chamber was made from HDPE material. From the economic analysis conducted, the total initial investment cost of the biogas digester was reported to be USD 1623.41 with an internal rate of return (IRR) of 8.5%, discount payback period (DPP) of 2 years and net present value (NPV) of USD 1783.10. The findings equally revealed that the estimated quantity of biogas could replace 33.2% of liquefied petroleum gas (LPG) cooking gas. Moreover, the biogas daily yield of 1.57 m^3 generates approximately 9.42 kWh of electricity, which costs about USD 1.54. Thus, the study recommends the use of composite material of plastics and bricks in constructing the biogas digester, as it is cost effective and sustainable.

Keywords: composite materials; biogas; biogas digester; economic analysis; economic indicator

Citation: Obileke, K.; Makaka, G.; Nwokolo, N.; Meyer, E.L.; Mukumba, P. Economic Analysis of Biogas Production via Biogas Digester Made from Composite Material. *ChemEngineering* **2022**, *6*, 67. https://doi.org/10.3390/ chemengineering6050067

Academic Editors: Luis M. Gandía and Alírio E. Rodrigues

Received: 30 April 2022
Accepted: 15 July 2022
Published: 2 September 2022

Publisher's Note: MDPI stays neutral with regard to jurisdictional claims in published maps and institutional affiliations.

1. Introduction

Anaerobic digestion of waste is contributing significantly to solving energy, environmental and agricultural-related problems. This has encouraged the development of biogas technology globally as well as the need to study its economic viability [1]. Given the limited supply of fossil fuels and their negative impact to the environment, studies have been completed to find out environmentally friendly and renewable alternative fuels [2]. Bhatt and Tao [3] mentioned that current and future research in renewable energy has contributed to the rapid increase in investment and implementation of clean energy technologies around the world. To this development, the conversion of waste to energy through anaerobic digestion is a promising option. Moreover, the growing global concerns on sustainable waste management bring AD technology to light. It can promote sustainability and meet the world's renewable energy needs. In this regard, energy economist, industries and agencies are seeking low-cost technologies such as biogas digesters for the generation of energy. Several studies have looked at the economic feasibility of the biogas digester and its gas yield using different materials and substrates.

Kozlowski [1] economically evaluated the possibility of using dairy waste for the production of electricity and heat. The study reported that the generated waste from the dairy could produce approximately 14.785 MWh electricity and 57.815 GJ of heat. This supports the construction of biogas plants that can generate electrical power of 1.72 MW. Ogrodowczyk [4] studied the economic analysis of a biogas digester at a sugar factory. In the economic part of the study, the following economic indicators were determined: net present value (NPV) of 14,089.57 PLN (USD 3,294,844.78), internal rate of return (IRR) of 12.48 PLN (USD 2.92) and discounted payback period (DPP) of 8 years. The initial cost of investment of the biogas digester was 2,446.000 PLN, which is equivalent to USD 572,283.28. From the economic perspective, the study affirms that the higher NPV indicates an economic benefit of the bio-digester. On the other hand, the calorific value of methane from biogas was reported to be 9.17 kWh/m^3. A total of 13,104 MWh/year of energy was generated as electricity, costing 16,358 PLN/MWh.

Tufaner and Avsar [5] studied the economic analysis of biogas production from an anaerobic digestion system for cattle manure. The study aimed at determining the economic viability of an anaerobic digester system, and it focused on the domestic production conditions. In this study, it was reported that the total investment cost (fixed cost) for the 3 m^3 underground bio-digester was 4433 TL (Turkish lira), which is equivalent to USD 270.52. Looking at the economic perspective of the study, the liquefied petroleum gas (LPG) value of the biogas generated was 157 kg, and it had an annual biogas production of 365 m^3 and biogas annual turnover of 1210 TL (USD 683.66). Unfortunately, the study did not focus on the economic indicators and parameters. The findings revealed that the biogas equivalent of 4.4 LPG cooking gas of 12 kg per year can be produced from cow dung.

The techno-economic evaluation of biogas production from waste in a biogas digester was conducted by Al-Wahaibi [6]. The study aimed at assessing the economic feasibility because of the fluctuating value of biogas from food waste. Focusing on the economic analysis of the study, this revealed that cat USD 0.2944 m^3, breaking even occurred. Hence, any prices above this rate yield a positive net present value (NPV). The study reported the DPP and NPV of 6 years and USD 3108.00, respectively.

To analyze the economic performance of anaerobic digestion of a biogas digester in terms of its NPV and IRR concept, Gebrazgabher [7] conducted the study, using the green biogas plant in Netherlands as a case study. The total investment cost of the bio-digester was reported to be €675,000.00 (USD 7,245,990.00), with NPV and IRR as €400,000.00 (USD 4,293,600.00) and 21%, respectively. These economic indicators were necessary to measure the cost-effectiveness of the biogas digester. As seen in the previous studies, the study revealed a higher NPV value, which shows the greater economic benefit of the project. In the study, an electricity yield of 222.30 kWh ton^{-1} of feedstock was digested. However, electricity production of a total of 2 MW/year was obtained in the study.

A feasibility study on the anaerobic digestion of food waste was conducted by the National Institute of Renewable Energy [8]. This study focuses on assessing the feasibility of developing an anaerobic digestion tank for biogas production. According to the study, the cost of the biogas digester was USD 561.00 per ton, while the operation and maintenance costs and NPV were USD 77 to USD 140.00 and −6,762,992.00, respectively. The NPV results predicted that the project would lose money, despite reasonable food waste and locations that could support the biomass plant. This loss of money from the project can be due to the negative present value reported in the investigation. However, a biogas production rate of 15 ft 3 biogas/IbVS was produced, and the electricity cost was USD 0.078/kWh.

The previous studies mentioned above focus on the economic analysis of biogas digesters, assessing the energy content of the biogas yield and cost of electricity. During the literature review, the authors observed that the nature of the design in terms of type of the digester and materials as well as the design orientation (aboveground or underground) were not considered. This is necessary, as it also contribute to the feasibility of the project economically. The present study will fill this gap, thereby contributing to existing knowledge. Therefore, the aim of the study is to determine the economic feasibility of generating biogas in a biogas digester built from composite materials. The study will also address and calculate how much methane is required to replace the LPG gas used for cooking in the study site. The findings from this study will provide relevant information and serve as a guide to mostly energy economists and consultants regarding the level of economic feasibility to take up a biogas digester project. In this regard, it will provide energy savings and the amount of electricity required through financial and economic benefits. A holistic analysis of the variables and indicators or parameters that satisfy the feasibility of biogas digesters will also be presented. These will help to advise accordingly.

2. Materials and Methods

2.1. Materials

The Fort Hare Dairy Trust Production provided the organic waste (cow dung) used in the study. The inoculum used was taken from an existing working biogas digester located in one of the community engagement projects at Melani (Eastern Cape Province of South Africa). Thereafter, the collected inoculum was kept at room temperature of anaerobic condition.

2.2. Physiochemical Properties of the Substrate

The following parameters were examined and reported as follows: pH: 7.83 at 30 °C; total solids: 130,800 g/L; volatile solids: 110,467 g/L; chemical oxygen demand: 42,583 g/L [9].

2.3. The Economic Analysis

To evaluate the feasibility of installing a biogas digester, a preliminary economic study was conducted. The biogas digester was fed with cow dung collected from the Fort Hare Dairy trust farm. The produced biogas aimed to substitute the conventional LPG gas used for cooking in a residential building at the study site. This is to save cost. The reduction in cost through the economies of scale can be effective if the cow dung is processed at a higher plant scale. This is possible or obtainable through the collection and processing and development of a centralized animal waste (cow dung). For the sake of the economic evaluation, which is focus of our anaerobic digestion experiment, a scale of 9000 kg/month of cow dung was reported (Table 1).

Table 1. Calculation of estimated biogas production (per month).

Parameters Reported	Values
Amount of fresh cow dung (24% dry matter)	9000 kg/month
Amount of cow dung (dry matter)	6589.2 kg/month
Quantity of biogas produced	10,500 L/kg/month
Methane content present in the biogas	60%
Amount of methane produced	630 mL/g
Total amount of methane from the cow dung	4151.2 m^3/month

The economic parameters analyzed include total initial investment cost (fixed cost), total income, cost of maintenance, payback period, profitability, internal rate of return, cost annuity and cost of energy. The importance of these indicators or parameters is that it helps determine the extent of the feasibility in carrying out the biogas digester installation. According to Kabyanga [10], the financial assessment investment includes the following.

2.3.1. Net Present Value (NPV)

The net present value (NPV) is the sum of the present value of the money moved into and out of an investment project [11]. According to Ogrodowczyk [4], NPV determines the present rate of the total investment cost, considering the changes in the value of capital over time. For any project to be feasible or profitable, the NPV should be higher or equal to zero [4]. However, a discount rate of 3.6% was used for the project's study. Relating the NPV to the study, it focuses on the sum of the present value of all the cash inflow and outflow that is linked to the investment of the biogas digester project at time t = 0. This was calculated using Equation (1):

$$NPV = -I_0 + \sum_{t=0}^{T}(R_t - I_t)q^{-t} + L_T q^{-T} \tag{1}$$

Equation (1) can also be written as:

$$NPV = -I_0 = (R \times PF) + L_T \times q^{-T} \tag{2}$$

where I_0 is the cost of investment (USD) at the start of the project, T is the lifetime of the project in years, R or R_t is the annual returns/return in time period t, I_t is the investment in time period t, PF is known as the value factor present in years, L_T is the yield of liquidation or value of salvage and q^{-t} is the discount factor, which is calculated as

$$q^{-t} = \left(1 + \frac{i}{100}\right)^{-t} \tag{3}$$

2.3.2. Internal Rate of Return (IRR)

This is the obtainable interest tied up in a project of a particular investment. It computes at what interest the NPV will be zero. The internal rate of return can be expressed as

$$0 = -I_0 + \sum_{t=0}^{T} R_t\left(1 + \frac{IRR}{100}\right) + L_T\left(1 + \frac{IRR}{100}\right)^{-T} \tag{4}$$

Equation (3) can be written as

$$IRR = I_i - NPV\left(\frac{i_2 - i_2}{NPV_2 - NPV_1}\right) \tag{5}$$

2.3.3. Profitability or Return on Investment (ROI)

This measures the behavior of a project's average profit per time interval. This is calculated by dividing the net profit by the net worth. A high return on investment (ROI) favors the cost of the investment. The ROI is a parameter used to relate the profit and capital used in an investment. However, the profitability of ROI is expressed as:

$$ROI = \frac{Net\ profit}{Total\ investment} \times 100 \tag{6}$$

2.3.4. Annuity (A)

This is the fixed amount of money paid on an annual basis for an investment or project. Annuity can be either fixed or variable. The annuity is calculated as:

$$A = NPV \times RF\ (i, T) \tag{7}$$

where RF is the capital recovery factor, which is calculated as:

$$RF = \frac{q^t(q-1)}{q^t - 1} \tag{8}$$

In addition, i and T are the discount rate and period in years, respectively.

2.3.5. Cost Annuity (A_k)

The cost annuity denoted as A_k is the annual cost of a project [11]. The cost annuity is important, as it helps evaluate the favorability of an investment project based on cost per annum [12]. Cost annuity is calculated as

$$A_k = K_0 + (I_0 - L) \times R_F\,(i, t) + L \times I \tag{9}$$

where K_0 is the cost of operation per unit time (USD), I_0 is the cost of investment (USD), L is the liquidation time (years), R_F is the recovery periods (years), while i and t are the interest rate (%) based on assumption and project duration, respectively.

2.4. Cost Analysis Study

The study considered the employment of a fixed dome biogas digester design made from a high-density polyethylene (HDPE) and bricks/cement. Other materials such as Teflon, acrylonitrile butadiene styrene (ABS) and fiber-reinforced plastic can be used for the fabrication/construction of a digester chamber. However, the choice of HDPE in the present study was based on its durability as well as quick biogas production within a 3 to 4-day retention time [13], This is attributed to the nature or type of the material, which easily allows enhancement of the diversity of microbial communities (degradation of microbes), thereby increasing their synergistic activity and playing a vital role during acidogenic fermentation (one of the stages of anaerobic condition). During this process, it increases the total volatile fatty acid yield responsible for biogas production [14]. Another reason for the use of HDPE for the digester chamber was because of its characteristic to withstand the moisture exposure, corrosive gases and digestion of feedstock as well as prevent the emission of odor and gases (H_2S, CH_4, CO_2 and NH_4). In addition, it is a good insulator and can be warm, producing biogas at a lower temperature. HDPE possesses high corrosion resistance, high strength, no leakage, and good airtightness as well as fast and easy fabrication and installation [15]. Above all, it can withstand harsh environmental conditions and still maintain anaerobic conditions. This is different from other designs where bricks are used for the construction of the digester chamber. The use of bricks for the digester chamber results in defects such as cracks. Hence, this necessitated the use of an alternative material for fabrication/construction of the digester chamber, as shown in Figure 1.

Figure 1. Fabricated high-density polyethylene biogas digester.

In the study, brick/cements were used for the construction of an outlet and inlet chamber and not for the digester chamber because of the disadvantages mentioned earlier. In addition, HDPE was able to withstand the pressure within the entire biogas digester. This type of design is recommended for rural settings due to the ease of installation. The biogas digester chamber was built underground for thermal insulation, as shown in Figure 2. In so doing, the soil provides insulation for the system.

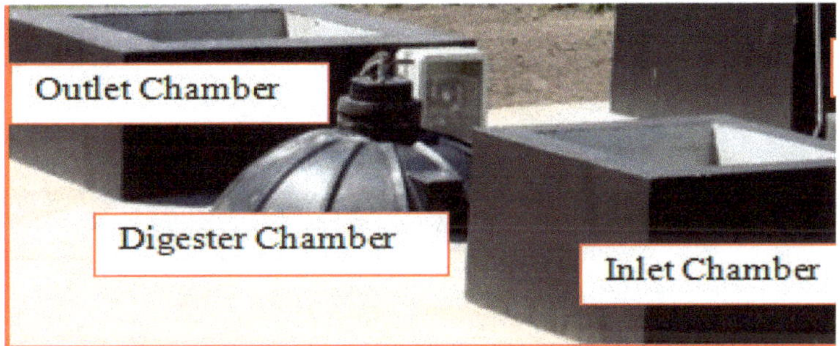

Figure 2. The biogas digester.

One major factor to be considered when choosing the volume of the biogas digester is the number of people that will benefit from the biogas when produced. According to Tufaner and Avsar [5], one person requires approximately 0.34–0.42 m^3 of biogas to cook a daily meal. Therefore, we assume that the quantity of biogas that a person would use for cooking daily is 0.5 m^3 day^{-1} on average and consider that a digester volume of 6 m^3 can produce 1 m^3 of biogas per day. The cost–benefit analysis was conducted on the biogas digester of 2.15 m^3 volume (small family size digester), which produced biogas of about 0.35 m^3 per day, as shown in Table 2. In terms of the animal dung, Tufaner and Avsar [5] assumed that a cow produced 10 kg of dung per day.

Table 2. Summary of cost analysis of the study.

Number of Cows Considered	Quantity of Cow Dung per Day (kg)	Volume of Biogas Digester (m^3)	Estimated Daily Cooking Demand (m^2/day)	Amount of Biogas per Day (m^3)	Amount of Electricity Generated (kWh)	Quantity of Cow Dung (kg/day)	Quantity of Water Added (L/day)
16	300	2.15	0.2	0.35	0.324	26	52

In the fabrication/construction of a biogas plant, the construction of cost items considered as fixed-cost items includes the following: a PVC pipe (110 mm thickness), concrete reinforcing mesh, concrete stone, all-purpose sand, UG rodding eye, and sealant. The variable costs considered include the labor costs, costs of fabrication of the digester chamber, and the cost of digging the pit. For the operating cost components, the maintenance and repair cost, water cost and labor cost were considered. The value of the biogas produced was calculated based on the equivalent price of a 5 kg cooking gas cylinder cost at approximately R 500.00–R 900.00 (USD 31.94–57.00) on average. However, the cost of cow dung, benefit of the fertilizer (digestate), and social and environmental benefit were not considered in the calculation.

Table 2 shows that the 300 kg of cow dung was produced from 16 cows, which was used to feed the 2.15 m^3 biogas digester. This generated 0.35 m^3 (0.324 kWh) of biogas per day for 0.2 m^2/day cooking demand daily for 5–8 households.

Therefore, the rate of cow dung required to be fed to the biogas digester daily can be calculated using the following:

$$m_s = \frac{E}{0.036} \times 0.1 \qquad (10)$$

where E refers to electrical energy (kWh/day).

Calculations:

- Amount of methane produced (mL/g) = amount of biogas produced $\times$ methane content produced;
- Total amount of methane from the cow dung = amount of cow dung $\times$ amount of methane produced $\times (10^3)/(10^6)$;
- LPG volume in one cylinder (L/cylinder) = Amount of LPG in a cylinder/Density of LPG;
- LPG volume in one cylinder (m^3/cylinder) = Volume of LPG (L/cylinder)/1000;
- Amount of gas after expansion = Volume of LPG (m^3/cylinder) $\times$ LPG expansion rate;
- Number of cylinders that can replaced by biogas per month = Total amount of methane based on the quantity of cow dung/amount of gas after expansion;
- Consumption reduction rate = Number of cylinders replaced by biogas per month/ estimated rate of LPG consumption $\times$ 100.

3. Results and Discussion

3.1. Estimated Gas Yield from Cow Dung Slurry

According to Nijaguna [16], a cow is assumed to produce 10 kg of dung per day, and 1 kg of cow dung produces 0.036 m^3 of biogas. Table 3 presents the estimated gas yield and energy rate of the biogas digester.

Table 3. Energy cost, yield, and rate of the feeding of the biogas digester.

Volume of Biogas Digester (m^3)	Amount of Dung Fed Daily (kg)	Expected Gas Yield per Day (m^3)	Expected Energy Generated (kWh)	Cost of Energy Expected (USD)
2.15	26	1.57	9.42	1.54

From the calculation, 26 kg of wet dung was fed daily to the digester, which produced on average, 1.57 m^3 of biogas per day. Hence, this generates 9.42 kWh of electricity per day, costing about USD 1.54. Recall that the 0.35 m^3 biogas per day can generate 0.324 kWh of electricity, as seen in Table 2. On assumption, 6 kWh of electricity per day is equivalent to 1 m^3 of biogas on average [16]. Regarding the mixing ratio, the rate of water to dilute the cow dung to form a slurry for biogas production is calculated in a mixing ratio of 1:1 waste to water [17].

3.2. Economic Analysis

3.2.1. Initial Total Investment Cost (Fixed Cost)

This involves the summation of the component of investment relating to building, equipment, land and working capital [18]. It is also known as the investment value, which represents the total amount of money spent in the biogas digester. Here, the cost of land was not considered, since the digester was installed in the research center of the FHIT, and land was not bought. Table 4 provides the initial total investment cost (fixed cost) of the fixed dome biogas digester, which involves various items that constitute the digester.

$$\text{Investment cost (IC)} = \sum \text{(investment cost of building, land, equipment and working capital)}$$

Table 4. Total investment cost of the biogas digester.

Material Used	Quantity	Unit Price (USD)	Total Cost (USD)
Cost of fabrication of the biogas digester	1	208.90	208.90
Portland cement (50 kg)	9	5.42	48.83
13 mm concrete stone	1.2 m^3	271.34	325.61
Clinker sp bricks	1850	159.70	295.44
All-purpose sand	1.84 m^3	19.99	36.79
UG rodding eye	1	4.52	4.52
110 mm PVC pipe	1	9.05	9.05
Polyfilla exterior crack filler	1	5.43	5.43
Concrete reinforcing mesh	1	22.61	22.61
Sealant	1	3.62	3.62
Supa lay hold	1	19.90	19.90
Brick force rolls (m^2)	8	1.25	9.99
Gas pipe	1	18.10	18.10
Thermal wool fiber	1	211.93	211.93
Others	1	55.21	55.21
Labor	2	173.74	347.48
Total			**1623.41**

From Table 4, it is revealed that total investment (fixed cost) to install the HDPE biogas digester was USD 1623.41. Although, in the present study, the slurry chamber of the bio-digester was buried underground to utilize the earth's thermal-insulating property and to minimize heat losses (Figure 2). Hence, it provides thermal insulation to the bio-digester. Considering a scenario where thermal insulating material such as the thermal wool fiber (RS PRO super wool fiber) was used, then, the additional cost of USD 211.93 will be part of the investment cost, as presented in Table 4. The thermal wool fiber is known for its excellent insulation from cold to heat. Table 5 presents the economic indicators or parameters used in the study.

Table 5. Study's economic analysis.

Economic Indicators	Values	Units
Total initial investment cost	1623.41	USD
Total income	1160.87	USD
Cost of energy	0.27	USD
Cost of maintenance	324.68	USD
Profitability (ROI)	67.7	%
Payback period	2	years
Net present value	1783.10	USD
Internal rate of return	8.5	%
Discount rate	3.6	%
Annuity	444.49	USD
Cost of annuity	587.77	USD

From Table 5, the project is worth embarking on and is financially feasible based on the data calculated as regards IRR, NPV and ROI. Hence, the high ROI, positive NPV and the above value of IRR against the discount rate shows that the project is accepted. The total revenue or income generated by the project was USD 1160.87, while the total investment cost was USD 1623.41. Looking at the rate of return on investment or return on investment (ROI), it is reported to be 67.7%, which suggests that the project is acceptable and feasible due to its higher cost. In addition, a two-year payback period was calculated as the recovery period; this means the time frame and duration required for the costs used in the project to be reimbursed. From the electricity point of view, the project is expected to generate 0.50 kWh of power that can operate 24 h a day. This produces approximately 9.42 kWh of electrical energy per day.

3.2.2. Annual Operating Cost (APC)

Operating cost refers to expenses in relation to the operation of the project. In this case, it includes the maintenance cost, cost of electricity, labor cost for feeding the digester, and any additional waste cost. It is usually calculated annually following the method in [18]. The costs of energy used in the process are not considered. It is assumed that the energy required for heating the digester is covered by the biogas energy by the digester. The cost of water was also not included, because the water used for the mixing of the substrate was from the tap water supplied by university. Regarding the waste (cow dung), the chemical oxygen demand (COD) value of the cow dung reported was 42,583 g/L. Maize silage is the main animal feed for cows and is known for being rich in energy and supporting higher dry matter intake (DMI) and milk yield. It contains about 30–35% dry matter [19]. The total solid of the cow dung obtained in the study was 13–15%, which is almost half of the maize silage total solids. Hence, the cost of maize silage, inoculum, and other items such as the feed related to the substrate were considered as an operational cost in relation to additional cost of the waste. The pump, which falls under the electricity cost, was also considered as an operational cost.

We considered the cost of maintenance (CM) as an operational cost, which is mostly needed. The cost of maintenance is calculated as 2% of the total initial investment cost divided by the lifetime of the biogas digester [20,21].

$$Cm = \frac{2\% \, of \, total \, cost \, of \, investment \, (USD)}{No \, of \, years \, of \, the \, system} \tag{11}$$

The cost of maintenance (Cm) is in USD/Year.

The operating and maintenance cost deals with the expenditures spent on labor on an annual basis. For instance, one unskilled laborer is responsible for the feeding and cleaning of the biogas digester and a skilled laborer has the responsibility of managing and performing advanced tasks regarding the system. In addition, the cost at which the gas valve and the pipe are replaced are considered in the cost of maintenance.

Operational cost = $\sum$ (building and equipment repair, labor, raw material, electricity, and water supply). The annual operational costs for the biogas digester are presented in Table 6.

Table 6. Annual operating cost for the biogas digester.

Cost Components	Cost (USD)
Additional cost of waste (maize silage and inoculum etc.)	63.88
Cost of water	N/A
Cost of electricity	35.13
Cost of maintenance and repair	324.68
Total	**423.70**

Considering the cost calculation of the biogas digester, the cost of 5 kg LPG cooking gas in South Africa is approximately R500.00–R900.00 (USD 31.94–57.00) on average. According to Tufaner and Avsar [5], 0.43 kg LPG is equivalent to 1 m^3 of biogas; also, 3–4 cows produced 1 m^3 day^{-1} of biogas. In a year, about 60 kg (12 LPG cooking gas cylinder of 5 kg) will be produced as 140 m^3 (60 ÷ 0.43) biogas equivalent. Table 7 shows the summary value of the cost price of the 2.15 m^3 biogas digester.

Table 7. Cost calculation of 2.15 m^3 biogas digester.

Parameters	Values	Units
Volume of biogas digester	2.15	m^3
Annual biogas production	140	m^3
LPG equivalent value of biogas	60	kg
Investment/fixed cost	1623.41	USD
Annual operating cost	423.70	USD

Focusing on the economic indicator of the project, Table 8 presents some of these indicators in relation to the methane gas content and project lifetime.

Table 8. Discounted payback period with the NPV and IRR on methane price.

Indicators	Values	Unit/Time
Methane gas content	0.085	(USD/m^3)
Discounted payback period (DPP)	2	years
NPV	1783.10	(USD)
Internal rate of return (IRR)	8.5	%
Project lifetime	10	years

The payback periods and net present value (NPV) consider the time value of money for the project lifetime. The NPV calculation is the sum of the discounted future cash flow less the total investment cost (fixed cost). Usually, this is made in the expenditure and working capital of the project. NPV could be positive or negative. Positive NPV indicates that the installation of the biogas digester is possible; hence, the value is created. On the other hand, negative NPV shows that the installation of the biogas digester should not go ahead; thus, it should be disregarded. Zero result of NPV means that no value is created, and therefore, no loss incurred during the project. As the value of the gas price is at USD 0.085/m^3, this yields a positive NPV, with a calculated discounted payback period rate of >10 years, as shown in Table 5. This implies that the installation of the biogas digester is financially feasible. This agrees with Al-Maghalseh [12].

Hence, it is recommended that investing in the installation of a biogas digester project should proceed carefully, considering the current and anticipated future gas rate, according to Al-Wahaibi [6]. On the other hand, internal rate of return (IRR) was employed as a second determinant of profitability. This came to 8%, as shown in Table 8. Assuming a scenario where the NPV is greater than zero and the IRR (8.5%) is greater than the discount rate (3.6%), the realization of the project is profitable to embark on. That means that the installation of the biogas digester will add value. Therefore, the study concludes that NPV and IRR make this possible regarding the benefits of the investments/projects. At this point, the internal rate of return of 8.5% calculated was said to be the rate at which the NPV was generated. Therefore, the rate of return on investment DPP is 2 years. On the other hand, if the IRR is less than the discount rate, then the essence of the project is defeated/destroyed and should not be embarked on. Interestingly, NPV and IRR are two determinants used to evaluate an investment or project.

In Table 9, the calculated amount of biogas can replace approximately 33.2% of LPG gas, which is currently consumed at Solar Watt FHIT for the purpose of cooking only. This value is different from the study conducted by Al-Wahaibi [6], where 28.6% of biogas replaced LPG gas. The discrepancy might be a result of the difference substrate and rate consumption.

Table 9. Calculation on the LPG replacement for cooking purposes.

Amount of LPG in one cylinder (constant)	5 kg/cylinder
Density of LPG	0.54 kg/L at 15 °C
LPG volume in one cylinder (L/cylinder)	9.26 L/cylinder
LPG volume in one cylinder (m^3/cylinder)	0.00926 m^3/cylinder
Expansion rate of LPG (constant)	270
Amount of gas after expansion	2.500 m^3/cylinder
Number of cylinders replaced by biogas per month	1.66 cylinders/month
Estimated rate of consumption of LPG per month for cooking only	5 cylinder/month
Consumption reduction rate	33.2%

4. Conclusions

The essence of this study was to analyze or evaluate the economic feasibility of the HDPE biogas digester, which was installed at the solar watt park of the University of Fort Hare, South Africa. The study was completed because of high cost of construction and installation of biogas digesters. The calculated net present value, profitability, and the annuity was high, which agrees with the literature that the study is favorable, profitable, feasible and acceptable to embark on. The calculated DPP is 2 years, with an NPV of USD 1783.10 and IRR of 8.5%. From the study, it was observed that 33.2% of LPG gas, currently consumed at the study site for cooking only, was replaced with biogas.

Author Contributions: Conceptualization, K.O.; formal analysis, K.O. and N.N.; supervision, G.M. and E.L.M.; writing—original drafting of the article, K.O. and N.N.; writing, reviewing, and editing, K.O., N.N. and P.M.; methodology, G.M., E.L.M. and P.M.; writing and proof reading of manuscript, K.O., N.N., G.M., E.L.M. and P.M. All authors have read and agreed to the published version of the manuscript.

Funding: This research received no external funding.

Institutional Review Board Statement: Not applicable.

Informed Consent Statement: Not applicable.

Data Availability Statement: The data presented in this study are available on the request from the corresponding author.

Acknowledgments: We acknowledged the Govan Mbeki Research Development Centre (GMRDC) University of Fort Hare, South Africa.

Conflicts of Interest: The authors declare no conflict of interest.

References

1. Kozlowski, K.; Pietrzykowski, M.; Czekala, W.; Dach, J.; Kowalczyk-Jusko, A.; Jozwiakowski, K.; Brzoski, M. Energetic and economic analysis of biogas plant with using the dairy industry waste. *Energy Rep.* **2019**, *183*, 1023–1031. [CrossRef]
2. Koberg, M.; Gedanken, A. Optimization of bio-diesel production from oils, cooking oils, microalgae, and castor and jatropha seeds: Probing various heating sources and catalysts. *Energy Environ. Sci.* **2012**, *5*, 7460–7469. [CrossRef]
3. Bhatt, A.; Tao, L. Economic perspective of biogas production via anaerobic digestion. *Bioengineering* **2020**, *7*, 74. [CrossRef] [PubMed]
4. Ogrodowczyk, D.; Olejnik, T.; Kazmierczak, M.; Brzezinski, S.; Baryga, A. Economic analysis for biogas plant working at sugar factory. *Biotechnol. Food Sci.* **2016**, *80*, 129–136.
5. Tufaner, F.; Avsar, Y. Economic analysis of biogas production from small anaerobic digestion systems for cattle manure. *Environ. Res. Technol. J. Park Acad.* **2019**, *2*, 6–12.
6. Al-Wahaibi, A.; Osman, A.; Al-Muhtaseb, A.; Alqaisi, O.; Baawain, M.; Fawzy, S.; Rooney, D. Techno-economic evaluation of biogas production from food waste via anaerobic digestion. *Sci. Rep.* **2020**, *10*, 15719. [CrossRef] [PubMed]
7. Gebrezgabher, S.A.; Meuwissen, M.P.; Prins, B.A.; Lansink, A.G.O. Economic analysis of anaerobic digestion—A case of green power biogas plant in The Netherlands. *NJAS Wagening. J. Life Sci.* **2010**, *57*, 109–115. [CrossRef]

8. National Renewable Energy Laboratory. *Feasibility Study of Anaerobic Digestion of Food Waste in St. Benard, Louisina*; Produced under Direction of the US Environmental Protection Agency (EPA), NREL/TP-7A30-57082; National Renewable Energy Laboratory: Golden, CO, USA, 2013.

9. Obileke, K.; Mamphweli, S.; Meyer, E.; Makaka, G.; Nwokolo, N. Development of mathematical model and validation for methane production using cow dung as substrate in the underground biogas digester. *Processes* **2021**, *9*, 643. [CrossRef]

10. Kabyanga, M.; Balana, B.B.; Mugisha, J.; Walekhwa, P.N.; Smith, J.; Glenk, K. Economic potential of flexible balloon biogas digester among smallholder farmers: A case study from Uganda. *Renew. Energy* **2018**, *120*, 392–400. [CrossRef]

11. Mukumba, P.; Makaka, G.; Mamphweli, S.; Misi, S. A possible design and justification for a biogas plant at Nyazura Adventist High School, Rusape, Zimbabwe. *J. Energy S. Afr.* **2013**, *24*, 12–20. [CrossRef]

12. Al-Maghalseh, M. Techno-economic assessment of biogas energy from animal wastes in central areas of Palestine; Bethleham perpective. *Int. J. Energy Appl. Technol.* **2018**, *5*, 119–126. [CrossRef]

13. Patinvoh, R.; Arulappan, J.; Johansson, F.; Taherzadeh, M. Biogas digesters from plastics and brick to textile bioreactor—A review. *J. Energy Environ. Sustain.* **2017**, *4*, 31–35.

14. Zhang, L.; Tsui, T.; Loh, K.; Dai, Y.; Tong, Y. Effects of plastics on reactor performance and microbial communities during acidogenic fermentation of food waste for production of volatile fatty acids. *Bioresour. Technol.* **2021**, *337*, 125481. [CrossRef] [PubMed]

15. Obileke, K.; Mamphweli, S.; Meyer, E.; Makaka, G.; Nwokolo, N. Design, and fabrication of a plastic biogas digester for the production of biogas from cow dung. *J. Eng.* **2020**, *2020*, 1848714. [CrossRef]

16. Nijaguna, B.T. *Biogas Technology*; New Age International Publishers: Kemalpasa, Turkey, 2002.

17. Obileke, K.; Mamphweli, S.; Makaka, G.; Nwokolo, N. Slurry utilization and impact of mixing ratio in biogas production. *Chem. Eng. Technol.* **2017**, *40*, 1742–1749. [CrossRef]

18. Wresta, A.; Andriani, D.; Saepudin, A.; Sudibyo, H. Economic analysis of cow manure biogas as energy source for electricity power generation in small ranch. *Energy Procedia* **2015**, *68*, 122–131. [CrossRef]

19. Egevet. Hayvancilik San ve Ticaret Limited Sirketi, Kaba Yem Kalitesi–Türkiye'de Inek Performansini Arttiran En Önemli Faktör. 2008. Available online: http://www.egevet.com.tr/kaba_yem_kalitesi_turkiyede_inek_performansini_artiran_en_onemli_faktor.htm (accessed on 22 April 2022).

20. Behzadi, A.; Houshfar, E.; Gholamain, E.; Ashjae, M.; Habibollahzade, A. Multi-criteria optimization and comparative performance analysis of a power plant fed by municipal solid waste using a gasifier or digester. *Energy Convers. Manag.* **2018**, *171*, 863–874. [CrossRef]

21. Choudhury, A.; Shelford, T.; Felton, G.; Lansing, S. Evaluation of hydrogen sulphide scrubbing systems for anaerobic digesters on two US dairy farms. *Energies* **2019**, *12*, 4605. [CrossRef]

Article

Optimum Biodiesel Production Using Ductile Cast Iron as a Heterogeneous Catalyst

Nada Amr El-Khashab, Marwa Mohamed Naeem and Mai Hassan Roushdy *

Chemical Engineering Department, Faculty of Engineering, The British University in Egypt, El-Sherouk City 11837, Egypt; nada173490@bue.edu.eg (N.A.E.-K.); marwa.mohamed@bue.edu.eg (M.M.N.)
* Correspondence: mai.hassan@bue.edu.eg

Abstract: Biofuels production become a target for many researchers nowadays. Biodiesel is one the most important biofuels that are produced from biomass using economics and modern techniques. The ductile cast iron solid waste dust is one of the wastes produced by the cast iron industry which has a bad effect on the environment. This paper investigates the possibility of reusing ductile cast iron solid waste as a biodiesel heterogeneous catalyst used in its production from sunflower waste cooking oil. Four reaction parameters were chosen to determine their effect on the reaction responses. The reaction parameters are M:O ratio, reaction time and temperature, and catalyst loading. The reaction responses are the biodiesel and glycerol conversions. The upper and lower limits are selected for each reaction parameter such as (50–70 °C) reaction temperature, (5–20) methanol to oil molar ratio, (1–5%) catalyst loading, and (1–4 h) reaction time. Optimization was done with economic and environmental targets which include lowering the biodiesel production cost, increasing the volume of biodiesel produced, and decreasing the amount of resulting glycerol. The optimum reactions are 20:1 M:O molar ratio, 65 °C reaction temperature, 5 wt% catalyst loading, 2 h reaction time, and a stirring rate of 750 rpm. The biodiesel conversion resulting at this optimum reaction conditions is 91.7 percent with agreed with all biodiesel standards. The catalyst usability test was done it was found the catalyst can be used up to 4 times after that a fresh catalyst is required to be used.

Keywords: biodiesel; ductile cast iron; solid waste; MgO; response surface methodology; optimization

Citation: El-Khashab, N.A.; Naeem, M.M.; Roushdy, M.H. Optimum Biodiesel Production Using Ductile Cast Iron as a Heterogeneous Catalyst. *ChemEngineering* **2022**, 6, 40. https://doi.org/10.3390/chemengineering6030040

Academic Editors: Venko N. Beschkov, Konstantin Petrov and Dmitry Y. Murzin

Received: 18 April 2022
Accepted: 23 May 2022
Published: 27 May 2022

Publisher's Note: MDPI stays neutral with regard to jurisdictional claims in published maps and institutional affiliations.

1. Introduction

Economic and social evolution relies extremely on energy. The global energy crisis and rising awareness of the vitality of environment preservation, are the originators behind the development and exploration of renewable energy to act as a replacement for non-renewable energy which is facing rapid depletion. Currently, noteworthy attention has been directed towards biofuels as a renewable energy resource. Biodiesel, which mostly comprises fatty acid methyl esters, possesses beneficial qualities such as low sulfur content, low toxicity, and low carbon dioxide and carbon monoxide emissions along with being biodegradable and renewable [1,2].

The production process of biodiesel could be converted to being mostly green through the employment of wastes. As a demonstration, wastes could be utilized in two different forms in the process which are as follows; at first, the utilization of waste cooking oil which refers to second-hand vegetable oil as a feedstock. It was proposed that waste cooking oil would act as an efficient, cost-effective, available feedstock along with offering an environmental advantage along with an economical one owing to the 60–70% reduction in feedstock costs which represents 70–95% of total production cost [3–6].

Several studies were conducted to confirm the viability of such a theory. A study was conducted by Sahar et al. where a waste cooking oil was utilized in a trans-esterification reaction in the presence KOH as an alkali catalyst to produce biodiesel. The yield of Fatty acid methyl ester attained was about 94% in the presence of a 1% catalyst at a temperature

50 °C. The characterization results confirmed the possibility of employing waste cooking oils in biodiesel production based on ASTM standards [7]. Moreover, another study was performed by da Silva et al. which affirmed and used the waste cooking oil as a raw material to produce high-quality biodiesel and to provide a feasibility condition to use the residual glycerol [8].

At present, the production of biodiesel is undergone in the presence of homogeneous catalysts such as potassium and sodium hydroxide due to their availability and feasibility. However, the total process cost had suffered a significant increase due to the major limitations it possesses which can only be minimized through the utilization of a heterogeneous catalyst. Heterogeneous catalysts are known to be non-corrosive, ecological, with superior selectivity, and activity, separated with ease from liquid products and minimal problems in the disposal. Furthermore, there are numerous types of heterogeneous catalysts such as acids, bases, and enzymes [9–11].

Many researchers focused their efforts on the employment of industrial and municipal wastes as heterogeneous catalysts for biodiesel. The employment of waste-derived catalyst which could be industrial, or municipal is designated as being highly advantageous as it introduces the conversion of wastes that are readily available and requires disposal to a significant asset for biodiesel production and thus, achieving solid waste management and economic efficiency due to its low cost along with being environmentally friendly. These materials are readily available and constitute some active metal oxides such as CaO and MgO making them an appealing option [12].

There were various research and studies concerning the employment of waste-derived catalysts. At first, the electric arc furnace dust solid waste which is significantly hazardous was analyzed in a study constructed by Khodary et.al. This study examined and confirmed the economical production of biodiesel from sunflower oil in the presence of the aforementioned solid waste as a heterogeneous catalyst with bearing mind that this catalyst is composed mainly of oxides specifically ZnO, CaO, Fe_2O_3 and SiO_2, and the optimum biodiesel yield was 96% at conditions of 20:1 methanol: oil molar ratio, 1 h as a reaction time, 57 °C as a reaction temperature and 5% catalyst loading [1]. Another solid waste was employed as a heterogeneous catalyst for biodiesel production which was waste iron filling. Ajala et al. analyzed the utilization of solid waste in the production process of biodiesel from waste cooking oil. The waste iron filling was utilized in synthesizing α-Fe_2O_3 through co-precipitation into acidic solid catalysts and achieved a yield of biodiesel 87, 90, and 92% respectively at conditions of 12:1 methanol: oil molar ratio, 3 h as a reaction time, 80 °C as a reaction temperature and 6% catalyst loading [13]. Furthermore, Rasouli & Esmaeili established a study that examined biodiesel production by transesterification of goat fat in the presence of a magnesium oxide (MgO) nano-catalyst at a temperature of 70 °C, a methanol/oil molar ratio of 12:1, a catalyst content of 1 wt. percent and a reaction period of 3 h, the maximum biodiesel yield of 93.12 percent was attained [14].

The ductile cast iron industry is a prosperous industry where the manufacturing had seen rapid growth due to ductility, elevated strength, and impact toughness in comparison to other steel grades, corrosion, and wear resistance because of graphite morphology modification which involves pure or an alloy of magnesium addition converting lamellar to a globular shape when crystallized. Initially, the core wire technique is defined as a graphite morphology modification method of simple mannerism where the core wire is injected into molten cast iron. The difficulty arises in magnesium's reaction with molten iron due to Mg's lower boiling point leading to spontaneous MgO fumes release and low yield in adsorption. The aversion to air pollution through a collection of dust formed by filtration leads to solid waste creation which requires handling to avert risks of land contamination and respiratory diseases [15].

This paper examines the utilization of ductile cast iron solid waste as a heterogeneous catalyst in a trans-esterification reaction to produce biodiesel using optimum, low energy, and economic process. This research examines biodiesel production using waste cooking oil, and ductile cast iron solid wastes which are considered dangerous materials to the environment so this research will have environmental benefit in addition to the economic benefit because using waste materials as a replacement for raw materials.

2. Methodology for Research

2.1. Raw Materials

The materials used in this research are described as follows:

(a)   Ductile cast iron supplied from Cairo Great Foundries, Cairo, Egypt.
(b)   Methanol 99% was supplied by Morgan Chemical company Ltd., Cairo, Egypt.
(c)   Sunflower waste cooking oil provided (SFWCO) by Egyptian restaurants and cafes which is characterized by chemical and physicochemical properties as shown in Figure 1 and Table 1. The method used for determining the Physicochemical properties of oil was mentioned in Roushdy [16].

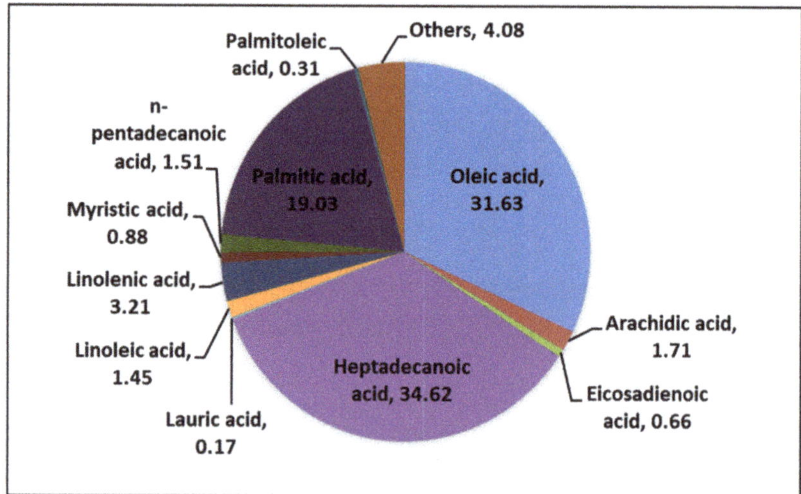

Figure 1. Chemical composition of oil.

Table 1. Physicochemical properties of oil.

Property	Molecular Weight	Acid Value (mg of KOH/g of Oil)	Density of 25 °C (kg/m^3)	Viscosity at 40 °C
Value	822.7268	1.3	887	35.8

2.2. Solid Waste Preparation

The solid molds were collected from the ductile cast iron factory from the dust accumulated around the furnace that was used to produce the ductile cast iron using the core wire technique.

2.3. Assessment of Solid Waste

The used characterization methods for the solid waste are described in the Table 2.

Table 2. Assessment method for the used solid waste.

Method/Technique	Importance	Description	Standard
X-ray fluorescence (XRF)	Evaluate the amounts of each oxide in the solid waste	The analysis was done at a humidity of 44% ± 1% and a temperature of 22 °C ± 1 °C	ASTM guidelines (C114-18) [17]
X-ray Diffraction (XRD)	Identify the phases contained in a substance	Using a PANalytical computer-certified program (X' Pert High Score Software 2006—Licensed modules: PW3209) with the aid of the International Center of Diffraction Database (ICDD) received with the X-ray diffraction equipment (X' Pert Pro PANalytical-Manufactured by Panalytical B.V Company, Almelo, Netherlands. The anode material was copper, and the scan was continuous. 30 mA and 40 KV were the default settings.	(ISO 9001/14001 KEMA–0.75160) [18,19]
Particle size distribution (PSD)	The particle size distribution determination	The particle size distribution is determined using a set of standard screens with a standard opening.	• ASTM D 422/2007 for the method [20] • ASTM E 11/2009 for the sieves [21]

2.4. Collection and Preparation of Waste Sunflower Cooking Oil

Sunflower waste cooking oil (SFWCO) was a discarded item in many households. A centrifuge and filter were used to remove any suspended particulates, fried food particles, and other pollutants, and it was then dried at 105 °C for two hours to eliminate the water.

2.5. Experimental Work Done to Produce Biodiesel

The experimental step that was used for biodiesel production as shown in Figure 2 can be described as follow:

Figure 2. Experimental setup.

1. Round bottom flask was used as a batch biodiesel reactor
2. Magnetic stirrer on which the biodiesel reactor is put. This stirrer is used for providing a good reaction mixing.
3. Heater that is provided with the stirrer to provide the required reaction temperature for the transesterification reaction.
4. Thermometer is used to measure the reaction temperature
5. A reflux fitted with the batch reactor to prevent methanol escape by condensation.

The oil, methanol, and the solid catalyst were added to the batch biodiesel reactor taking into consideration the required catalyst percentage and methanol to oil ratio than the reaction temperature was adjusted, and the reaction timer started and adjust for a certain time. When the reaction ended the solid catalyst was removed by the filter media and then glycerol was separated from the resulted biodiesel using a separating funnel finally the excess methanol was removed using 80 °C, 30 min drying. The biodiesel conversion was calculated by the weight ratio between the resulted biodiesel and the used SFWCO.

2.6. Experimental Design

The surface methodological technique (RSM) was utilized to design the experimental work, and a detailed analysis of the process was generated using Design-Expert version 13 [20]. The process response is the conversion of biodiesel and glycerol while the reaction variables are shown in Table 3.

Table 3. Reaction parameters and their limits.

Process Parameter	Unit	Lower Limit	Upper Limit
Reaction time	h	1	4
Methanol to oil molar ratio	-	5	20
Catalyst loading	%	1	5
Reaction temperature	°C	50	70
Stirring rate	rpm	750	

Based on reaction parameters used by Rasouli & Esmaeili in their paper, the processing parameters and ranges were chosen [14]. Thirty experimental runs were generated by the design expert program [22] using the central composite design technique (CCD) as shown in Table 4. The conditions in experimental runs 25 to 30 represent the design center point. The optimization process was done based on the economic purpose to maximize biodiesel production while minimizing production cost. This target was reached by minimizing both reaction time and temperature, maximizing the biodiesel production rate, and minimizing the glycerol production rate.

Table 4. Design expert suggested experiments.

No.	Temperature, °C	Reaction Time, h	Catalyst Loading, %	Methanol/Oil Ratio
1	50	1	1	5
2	50	4	1	5
3	50	1	1	20
4	50	4	1	20
5	50	1	5	5
6	50	4	5	5
7	50	1	5	20
8	50	4	5	20
9	70	1	1	5
10	70	4	1	5
11	70	1	1	20
12	70	4	1	20
13	70	1	5	5
14	70	4	5	5
15	70	1	5	20
16	70	4	5	20
17	60	0.5	3	12.5
18	60	5.5	3	12.5
19	60	2.5	3	2.5
20	60	2.5	3	27.5
21	60	2.5	1	12.5
22	60	2.5	7	12.5
23	40	2.5	3	12.5
24	80	2.5	3	12.5
25–30	60	2.5	3	12.5

2.7. Optimum Biodiesel Sample Analysis

Two important tests must be done to make sure that the resulted product is biodiesel and complied with the standards required. The first test is gas chromatography (GC) which determines the amount of total fatty acid methyl ester (FAME), glycerol, and triglycerides in the biodiesel sample and compared it with the standards EN 14103 [23] and EN 14105 [24].

The second test is physicochemical determination and compares their results with the standards ASTM D6751 [25] and European Biodiesel Standard, EN 14214 [26].

2.8. Reusability Test of Biodiesel Catalyst

A reusable test was done using two methods under the resulted optimum conditions. Once the reaction ended the reaction product was filtered to remove the heterogeneous catalyst. The used method can be summarized in Table 5. The reaction conversion was calculated at the point of reuse to determine the catalyst efficiency and strength.

Table 5. Reusability Test.

Step	Details
Step 1: Washing Chemical Treatment Method for contaminations removal	Washed with methanol
Step 2: Drying	Dried at 80 °C for 30 min

3. Results and Discussion

3.1. Ductile Cast Iron Solid Waste Characterization

3.1.1. Chemical Analysis

The chemical analysis of the used ductile cast iron solid waste is shown in Table 6. Solid waste contains 88% MgO with negligible amounts of other oxides. Cast iron's high melting temperature causes more magnesium to be eliminated, resulting in a higher MgO percentage. It's due to the decomposition of the core wire structure in the foundry, which mostly releases MgO. The waste produced in an oxidation atmosphere and at high melting temperatures contains intermediate levels of zinc and iron oxides contaminated with carbonate elements, resulting in destruction and gas conversion as indicated by the L.O.I. %. This result indicates that the ductile cast iron solid waste is a promising biodiesel catalyst as MgO is good biodiesel based on previous researchers like Rasouli and Esmaeili [14].

Table 6. Chemical analysis of the ductile cast iron solid waste.

Oxide	Percentage, %
MgO	88
Fe_2O_3	2.28
ZnO	4.2
Na_2O	0.4
SiO_2	0.2
CaO	0.24
MnO	0.04
TiO_2	0.02
K_2O	0.01
P_2O_2	0.01
L.O.I	4.54

3.1.2. Mineralogical Analysis

The mineralogical analysis of the solid waste as shown in Figure 3 shows the major phase is Periclase which is the cubic form of magnesium oxide (MgO) [27]. Periclase is a relatively high-temperature mineral which confirmed what is mentioned in XRF analysis about Cast iron's high melting temperature which causes more magnesium to be eliminated, resulting in a higher MgO percentage. XRD analysis indicates that the MgO is found in its oxide shape not hydroxide or carbonated so no need for any heat or chemical treatment to be done on the catalyst before its usage so it will be used as it. There are also minor phases which are Zincite (ZnO) and Osbornite (TiN). The colors of peaks of periclase, zincite, and osbornite are blue, green, and grey respectively.

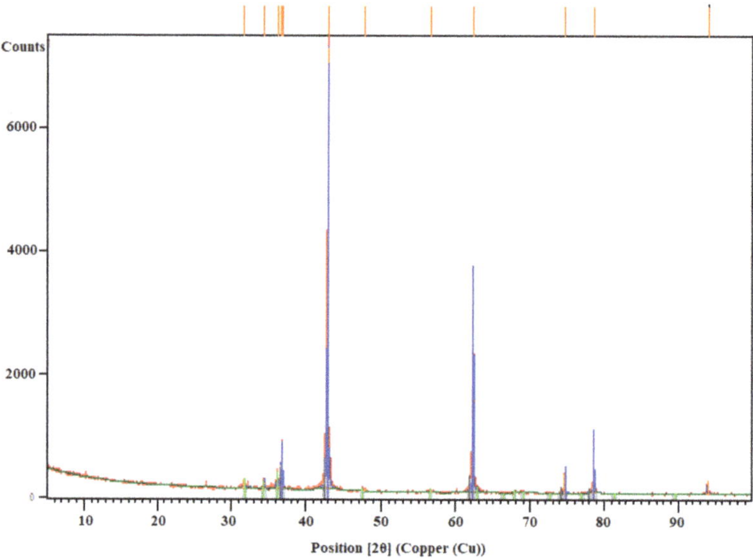

Figure 3. Mineralogical analysis of the ductile cast iron solid waste.

3.1.3. Screen Analysis

The cumulative screen analysis curve of the ductile cast iron solid waste is shown in Figure 4. The catalyst is extremely fine, and in the nano range as shown in this diagram. The average particle size was 0.098 µm. This indicates that this catalyst will be highly active as it has a high surface area and large number of active centers.

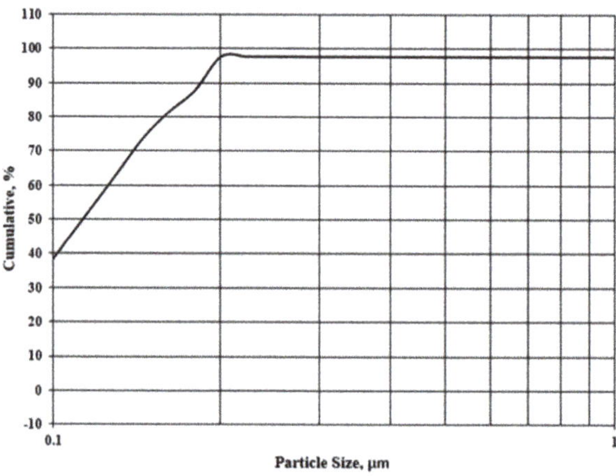

Figure 4. Cumulative screen analysis curve of the ductile cast iron solid waste.

3.2. Process Modelling Using Design Expert

The conversion of both biodiesel and glycerol was determined using the previously mentioned experimental run. The design expert with version 13 generated models which represent the relation between the process or reaction parameters and biodiesel and glycerol conversion as a process response. ANOVA method is used at a confidence level of 95% to determine if the resulted models are significant or suitable or not by determining p and F values. The optimum significant model for biodiesel conversion is the two factors' interactions model (2FI) while for the glycerol conversion the significant model is the quadratic one. Because several terms are not significant in the model as their p-values are bigger than 0.1 so the models are simplified to reduced ones. The two modules are shown in Equations (1) and (2). The result table, which summarizes the ANOVA analysis, is a Tables 6 and 7. As shown in Figures 4 and 5, the calculated and experimental results for both biodiesel and glycerol conversions exhibit reasonable agreement as confirmed by Figures 5 and 6, and the values of R in Tables 7 and 8. This agreement confirms the adequacy of the models.

$$X = 3.157\,A + 4.077\,B + 0.107\,C + 2.131\,D - 0.037\,BD - 88.022 \tag{1}$$

$$Y = 259.257 - 7.522\,A - 4.233\,B - 0.212\,C - 4.316\,D + 0.076\,AD + 0.04\,BD + 0.016\,D^2 \tag{2}$$

where X denotes biodiesel conversion and Y denotes glycerol conversion, with reaction time A, methanol to oil ratio B, catalyst loading C, and reaction temperature D affecting both. All reaction parameters have a positive impact on the biodiesel conversion while have a negative impact on glycerol conversion.

Table 7. Results of ANOVA analysis for biodiesel response.

Source	Sum of Squares	df	Mean Square	F-Value	p-Value
Model	11,635.89	5	2327.18	312.43	<0.0001
A-Reaction Time	487.88	1	487.88	65.50	<0.0001
B-Methanol/oil ratio	4312.90	1	4312.90	579.03	<0.0001
C-Catalyst loading	0.9689	1	0.9689	0.1301	0.7215
D-Temperature	6715.29	1	6715.29	901.56	<0.0001
BD	121.00	1	121.00	16.24	0.0005
Residual	178.76	24	7.45		
Lack of Fit	178.76	19	9.41		
Pure Error	0.0000	5	0.0000		
Cor Total	11,814.65	29	**Predicted R^2**	0.975	
R^2	0.985		**Adjusted R^2**	0.982	

Table 8. Results of ANOVA analysis for glycerol response.

Source	Sum of Squares	df	Mean Square	F-Value	p-Value
Model	11,518.89	7	1645.56	389.81	<0.0001
A-Reaction Time	436.59	1	436.59	103.42	<0.0001
B-Methanol/oil ratio	4095.97	1	4095.97	970.27	<0.0001
C-Catalyst loading	3.76	1	3.76	0.8896	0.3558
D-Temperature	6708.60	1	6708.60	1589.16	<0.0001
AD	20.57	1	20.57	4.87	0.0380
BD	144.36	1	144.36	34.20	<0.0001
D^2	76.26	1	76.26	18.06	0.0003
Residual	92.87	22	4.22		
Lack of Fit	92.87	17	5.46		
Pure Error	0.0000	5	0.0000		
Cor Total	11,611.76	29	**Predicted R^2**	0.984	
R^2	0.992		**Adjusted R^2**	0.9895	

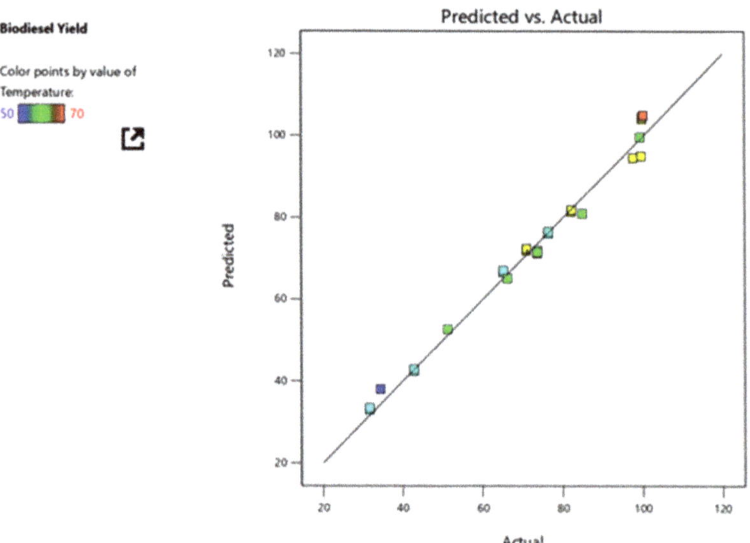

Figure 5. The link between the predicted and experimental biodiesel conversion.

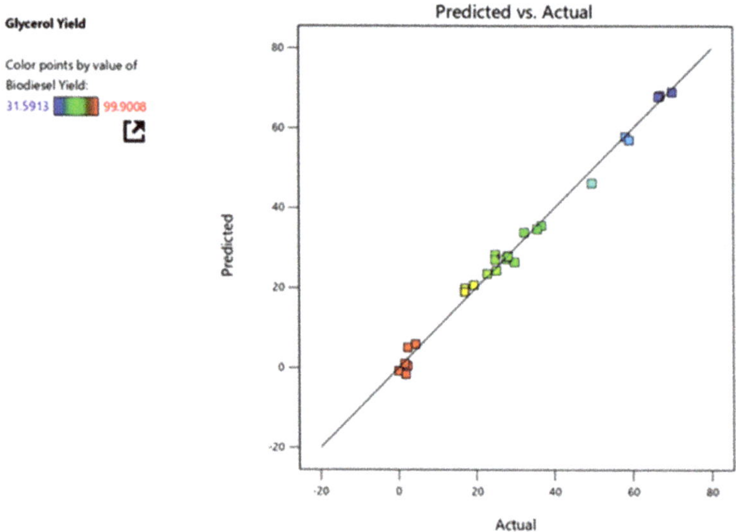

Figure 6. The link between the predicted and experimental glycerol conversion.

3.3. Variation of Biodiesel and Glycerol Conversions with Reaction Conditions

Figures 7 and 8 show the effect of each reaction parameter on both biodiesel and glycerol conversions. The two figures show that the reaction temperature and the M:O ratio have the greatest impact on both biodiesel and glycerol conversions than the reaction time while the amount of catalyst added to the reaction mixture has approximately no effect as it is not a significant factor as indicated by ANOVA analysis because its p-value is more than 0.05.

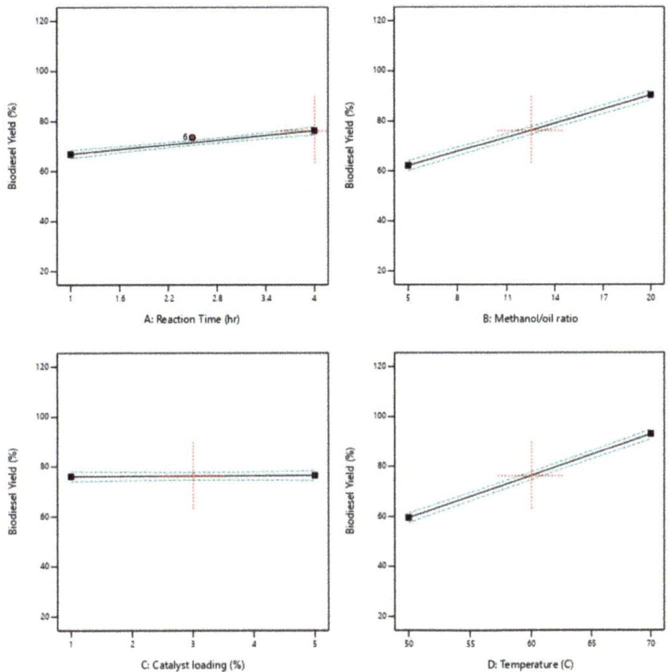

Figure 7. The impact of reaction parameters on biodiesel Yield.

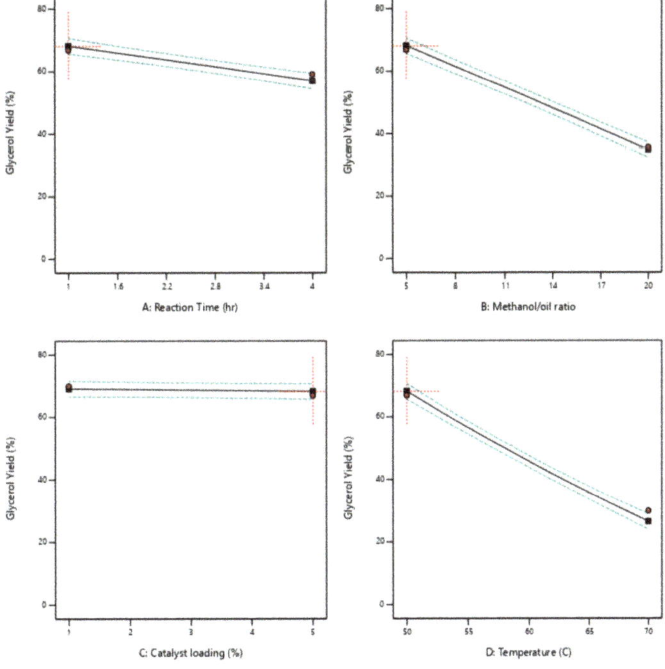

Figure 8. The impact of reaction parameters on glycerol conversion.

The methanol to oil ratio is an important reaction parameter in the transesterification reaction for biodiesel production. According to the balanced equation for the transesterification reaction, 3 moles of methanol are required to react with 1 mole of oil (Triglyceride, TG) to form 3 moles of biodiesel (fatty acid methyl ester, FAME) so the minimum methanol to oil ratio is 3. The used ratio must be more than 3 which means excess methanol which is important for reaction enhancement in the forward direction, so it was chosen to be in the range between 5 to 20. Increasing the methanol to oil ratio will enhance the reaction and thus increases the biodiesel conversion. The methanol to oil ratio has a negative effect on glycerol yield.

The reaction temperature is also an important reaction parameter. Increasing the reaction temperature will increase the reactant collision and decrease the viscosity of oil to increase the biodiesel conversion. The reaction temperature has a negative effect on glycerol yield.

Increasing the reaction time has positive effect on biodiesel yield as it gives more time for reactants to react. The reaction time has a negative effect on glycerol yield.

3.4. The Reaction Parameters Interactions with Both Biodiesel and Glycerol Conversion

Figure 9 shows the relationship between biodiesel conversion and the M:O ratio and the reaction temperature interaction (BD). Figure 10 shows the relationship between glycerol conversion and the reaction temperature and time interaction (AD). Figure 11 shows the relationship between glycerol conversion and the M:O ratio and the reaction temperature interaction (BD).

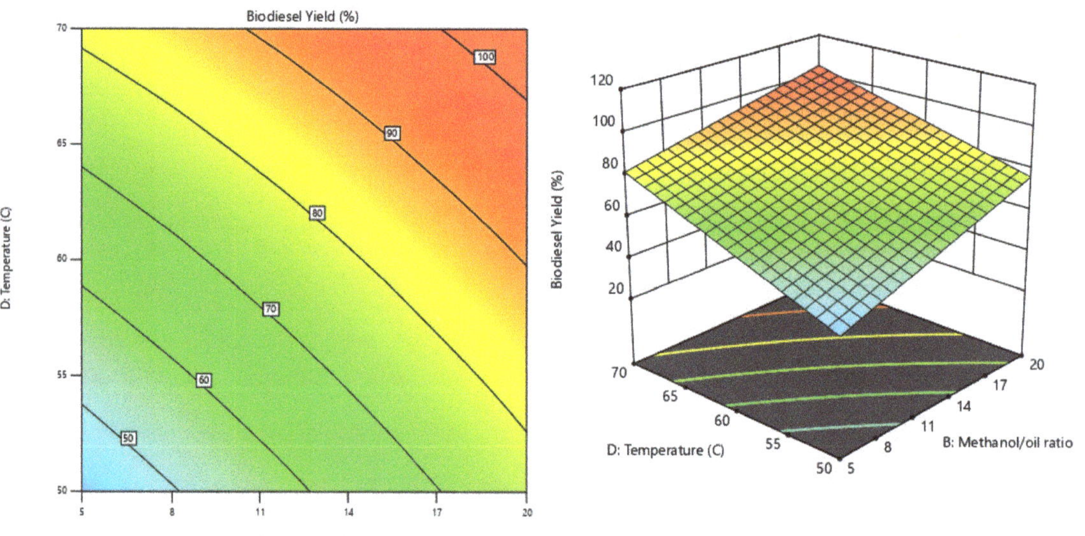

Figure 9. The link between biodiesel conversion, M:O ratio, and reaction temperature interactions as a contour and surface graph.

3.5. Process Optimization

Based on the goals shown in Table 9, the design expert program generated 10 suggested solutions with different desirability and then select the optimum solution with the highest desirability as shown in Table 9.

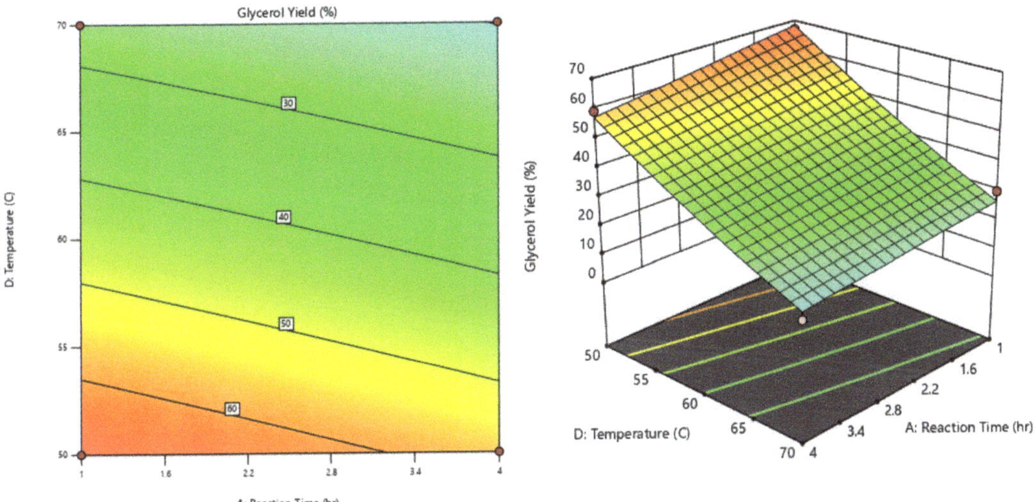

Figure 10. The link between glycerol conversion, reaction time, and reaction temperature interactions is given as a contour and surface graph.

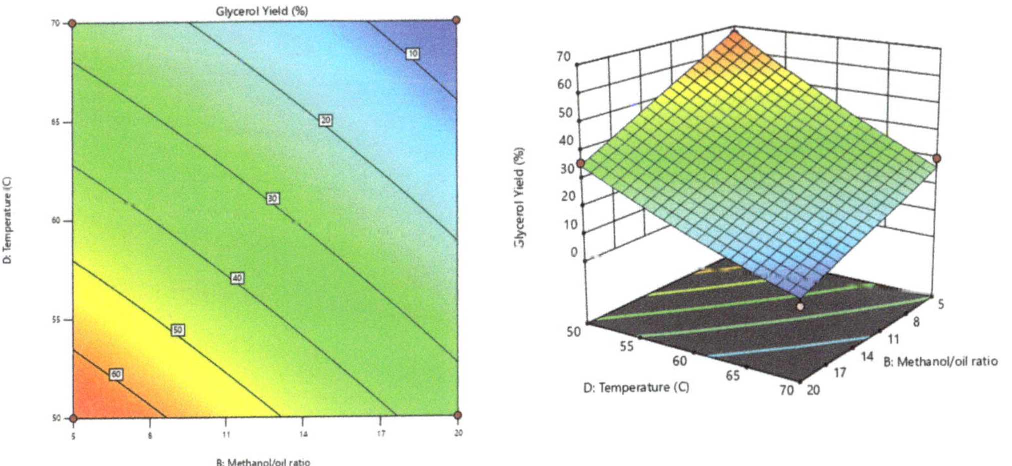

Figure 11. The link between glycerol conversion, M:O ratio, and the reaction temperature interactions is given as a contour and surface graph.

Table 9. Optimization Goals and Results.

Reaction Parameter/Response	Goal	Resulted Value
A: Reaction Time	minimize	2 h
B: Methanol/oil ratio	is in range	20
C: Catalyst loading	is in range	5%
D: Temperature	minimize	65 °C
Biodiesel Conversion	maximize	91.7%
Glycerol Conversion	minimize	8.3%

3.6. Analysis for Resulted Optimum Sample

The physicochemical properties of this optimum sample were determined and compared with its standards both EN14214 [23] and ASTM D 6751 [24] as shown in Table 10. All measured properties are agreed with the required standard.

Table 10. Biodiesel Physicochemical properties and standards.

Physicochemical Properties	Standard Method	Results	EN14214	ASTM D6751
Pour point (°C)	ASTM D-97 [28]	−22		
Cloud point (°C)	ASTM D-97 [29]	−10	<−4	
Flashpoint (°C)	ASTM D-93 [29]	155	>101	>130
Kinematic viscosity at 40 °C (cSt)	ASTM D-445 [30]	4.9	3.5–5.0	1.9–6.0
Density at 15 °C (g/cm^3)	ASTM D-4052 [31]	0.87	0.86–0.9	
Calorific value (MJ/kg)	ASTM D-5865 [2]	40.18	>32.9	

Table 11 shows the results of the Gas Chromatography (GC) experiments for the optimum sample which shows also the agreement with the standards of biodiesel which are EN 14103 [25] and EN 14105 [26].

Table 11. GC results and standards.

Composition		Specification Range	Results
Total FAME		more than 96.5%	98.1%
Glycerol	Total	less than 0.25	0.018%
	Free	less than 0.02	0.015%
Glycerides	Tri-	less than 0.02	0.0156%
	Di-	less than 0.02	0.0108%
	Mono	less than 0.08	0.02%

3.7. Catalyst Reusability

The reusability test of the catalyst showed that the catalyst up to 4 times after a fresh catalyst must be used as shown in Figure 12. The reasons for this phenomenon of catalyst reactivity change are the following:

1. Glycerol deposition on the catalyst active center.
2. MgO slaking into less active carbonate, bio-carbonate, and hydroxides compounds.
3. Catalyst loss during washing and filtration steps.

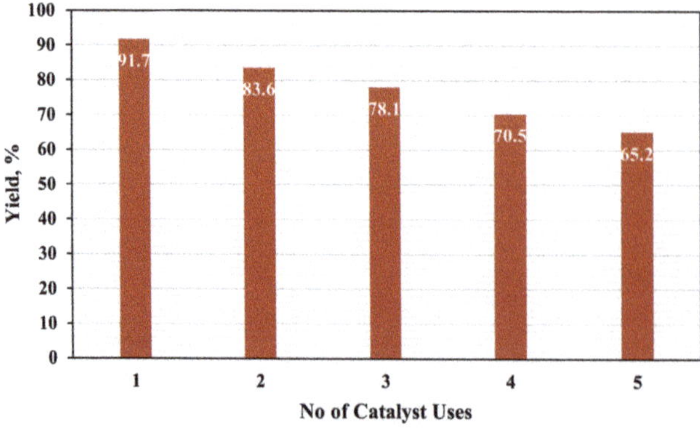

Figure 12. Ductile cast iron solid waste reusability test.

4. Current Research Comparison with the Previous Ones

This current study is the best compared with the previous studies as shown in Table 12. The advantages of this study can be summarized as follow:

1. The used catalyst is solid waste so need for catalyst preparation.
2. The used catalyst is heterogeneous, so it needs simple separation techniques.
3. The used oil is waste cooking oil.
4. The biodiesel conversion is high at minimum reaction conditions and energy and cost.
5. This study reused a dangerous solid waste and waste cooking oil to produce biodiesel so the process cost will be minimum and save the environment at the same time.

Table 12. Research comparison.

Study	Used Catalyst	Catalyst Preparation	Reaction Conditions				Biodiesel Conversion	Reference
			Methanol/Oil Ratio	Catalyst Loading	Reaction Temperature	Reaction Time		
1	MgO nano-catalyst	No need	12:1	1 wt.%	70 °C	3 h	93.12%	[14]
2	MgO catalysts	Chemical preparation	20:1	5 wt.%	70 °C	8 h	97–98%	[32]
3	MgO nanocatalyst	sol-gel method	10:1	2 wt.%	60 °C	2 h	80%	[33]
4	MgO Loaded with KOH	Chemical preparation	12:1	20 wt.%	70 °C	8 h	95.05%	[34]
5	magnesium oxide nanocatalyst	Chemical preparation	24:1	2 wt.%	65 °C	1 h	93.3%	[35]
6	Ductile cast iron solid waste (88%MgO)	No need	20:1	5 wt.%	65 °C	2 h	91.7%	(Present work)

5. Conclusions

This paper examined the utilization of ductile cast iron solid waste as a heterogeneous catalyst in a trans-esterification reaction to produce biodiesel using optimum, low energy, and economic process. This research examined biodiesel production using waste cooking oil, and ductile cast iron solid wastes which are considered dangerous materials to the environment, so this research has environmental benefit in addition to the economic benefit because using waste materials as a replacement for raw materials. Four reaction parameters were chosen to determine their effect on the reaction responses. The reaction parameters are M:O ratio, reaction time and temperature, and catalyst loading. The reaction responses are the biodiesel and glycerol conversions. The design expert program was used in the analysis, models generation, and optimization. It generated 25 different experimental runs and determine the impact of each reaction parameter using resulted models, 2D graphs, 3D plots, and contour figures. Optimization was done with economic and environmental targets. 100 possible optimum solutions which include lowering the cost of biodiesel production, increasing the volume of biodiesel produced, and decreasing the amount of resulting glycerol. The optimum reactions are 20:1 M:O molar ratio, 65 °C reaction temperature, 5 wt% catalyst loading, 2 h reaction time, and a stirring rate of 750 rpm. The biodiesel conversion resulting at this optimum reaction conditions is 91.7 percent with agreed with all biodiesel standards. The catalyst usability test was done it was found the catalyst can be used up to 4 times after that a fresh catalyst is required to be used.

Author Contributions: Conceptualization, M.M.N. and M.H.R.; Data curation, N.A.E.-K.; Formal analysis, M.H.R.; Investigation, N.A.E.-K. and M.H.R.; Methodology, M.M.N. and M.H.R.; Project administration, M.H.R.; Supervision, M.H.R.; Writing—original draft, M.H.R.; Writing—review & editing, M.H.R. All authors have read and agreed to the published version of the manuscript.

Funding: This research received no external funding.

Data Availability Statement: This article contains all the data generated or evaluated throughout the research.

Acknowledgments: M.H.R., the last author, wishes to express her heartfelt gratitude to Prof. Magdi F. Abadir of Cairo University.

Conflicts of Interest: The authors declare no conflict of interest.

References

1. Khodary, K.E.; Naeem, M.M.; Roushdy, M.H. Utilization of Electric Arc Furnace Dust as a Solid Catalyst in Biodiesel Production. Available online: https://link.springer.com/article/10.1007/s10098-021-02174-0 (accessed on 1 March 2022).
2. Mahapatra, D. A Review on Steam Coal Analysis -Calorific Value. *Am. Int. J. Res. Sci. Technol. Eng. Math.* **2016**, *1*, 143–152.
3. Linganiso, E.C.; Tlhaole, B.; Magagula, L.P.; Dziike, S.; Linganiso, L.Z.; Motaung, T.E.; Moloto, N.; Tetana, Z.N. Biodiesel Production from Waste Oils: A South African Outlook. *Sustainability* **2022**, *14*, 1983. [CrossRef]
4. Erchamo, Y.S.; Mamo, T.T.; Workneh, G.A.; Mekonnen, Y.S. Improved biodiesel production from waste cooking oil with mixed methanol–ethanol using enhanced eggshell-derived CaO nano-catalyst. *Sci. Rep.* **2021**, *11*, 6708. [CrossRef] [PubMed]
5. Hussien, M.; Abdul Hameed, H. Biodiesel production from used vegetable oil (sunflower cooking oil) using eggshell as bio catalyst. *Iraqi J. Chem. Pet. Eng.* **2019**, *20*, 21–25. [CrossRef]
6. Sarno, M.; Iuliano, M. Biodiesel production from waste cooking oil. *Green Processing Synth.* **2019**, *8*, 828–836. [CrossRef]
7. Sahar Sadaf, S.; Iqbal, J.; Ullah, I.; Bhatti, H.N.; Nouren, S.; Habib-ur-Rehman Nisar, J.; Iqbal, M. Biodiesel production from waste cooking oil: An efficient technique to convert waste into biodiesel. *Sustain. Cities Soc.* **2018**, *41*, 220–226. [CrossRef]
8. Da Silva, C.A.; Dos Santos, R.N.; Oliveira, G.G.; De Souza Ferreira, T.P.; De Souza, N.L.G.D.; Soares, A.S.; De Melo, J.F.; Colares, C.J.G.; De Souza, U.J.B.; De Araújo-Filho, R.N.; et al. Biodiesel and Bioplastic Production from Waste-Cooking-Oil Transesterification: An Environmentally Friendly Approach. *Energies* **2022**, *15*, 1073. [CrossRef]
9. Aghel, B.; Mohadesi, M.; Razmehgir, M.H.; Gouran, A. Biodiesel production from waste cooking oil in a micro-sized reactor in the presence of cow bone-based KOH catalyst. *Biomass Convers. Biorefinery* **2022**, *4*, 1–15. [CrossRef]
10. Topi, D. Transforming waste vegetable oils to biodiesel, establishing of a waste oil management system in Albania. *SN Appl. Sci.* **2020**, *2*, 513. [CrossRef]
11. Laskar, I.B.; Rajkumari, K.; Gupta, R.; Chatterjee, S.; Paul, B.; Rokhum, L. Waste snail shell derived heterogeneous catalyst for biodiesel production by the transesterification of soybean oil. *RSC Adv.* **2018**, *8*, 20131–20142. [CrossRef] [PubMed]
12. Al-Sakkari, E.G.; El-Sheltawy, S.T.; Abadir, M.F.; Attia, N.K.; El-Diwani, G. Investigation of cement kiln dust utilization for catalyzing biodiesel production via response surface methodology. *Int. J. Energy Res.* **2017**, *41*, 593–603. [CrossRef]
13. Ajala, E.O.; Ajala, M.A.; Ayinla, I.K.; Sonusi, A.D.; Fanodun, S.E. Nano-synthesis of solid acid catalysts from waste-iron-filling for biodiesel production using high free fatty acid waste cooking oil. *Sci. Rep.* **2020**, *10*, 13256. [CrossRef] [PubMed]
14. Rasouli, H.; Esmaeili, H. Characterization of MgO nanocatalyst to produce biodiesel from goat fat using transesterification process. *3 Biotech* **2019**, *9*, 429. [CrossRef] [PubMed]
15. Pourrahim, S.; Salem, A.; Salem, S.; Tavangar, R. Application of solid waste of ductile cast iron industry for treatment of wastewater contaminated by reactive blue dye via appropriate nano-porous magnesium oxide. *Environ. Pollut.* **2020**, *256*, 113454. [CrossRef] [PubMed]
16. Roushdy, M.H. Heterogeneous Biodiesel Catalyst from Steel Slag Resulting from an Electric Arc Furnace. *Processes* **2022**, *10*, 465. [CrossRef]
17. *ASTM C114-15*; Standard Test Methods for Chemical Analysis of Hydraulic Cement. ASTM International: West Conshohocken, PA, USA, 2018. Available online: https://www.astm.org/standards/c114 (accessed on 1 March 2022).
18. *ISO 9001*; Quality Management System-Requirements. International Organization for Standardization: Geneva, Switzerland, 2015.
19. International Standardization Organization (ISO). Environmental Management Systems—Requirements with Guidance for Use. *ISO Cent. Secr.* **2015**, *2015*, 1–46. Available online: www.iso.org (accessed on 1 April 2022).
20. *ASTM D422-63*; Particle Size Analysis. ASTM International: West Conshohocken, PA, USA, 2017; pp. 70–138.
21. *ASTM E11-20*; Standard Specification for Woven Wire Test Sieve Cloth and Test Sieves. ASTM International: West Conshohocken, PA, USA, 2017; pp. 1–11. Available online: https://www.astm.org/e0011-20.html (accessed on 1 March 2022). [CrossRef]
22. Montgomery, D.C. *Montgomery Design and Analysis of Experiments*, 8th ed.; Wiley: New York City, NY, USA, 2013; Volume 2009.
23. *EN 14103*; Determination of Total FAME and Linolenic Acid Methyl Ester in FAME with AC Biodiesel All in One Solution. European Standards: Milan, Italy, 2011; pp. 1–3.
24. *EN 14105*; Determination of Free and Total Glycerol and Mono-, Di, -Triglycerides in Fatty Acid Methyl Esters (FAME). European Standards: Milan, Italy, 2011; pp. 2–4.
25. *ASTM D6751-15c*; Standard Specification for Biodiesel Fuel Blend Stock (B100) for Middle Distillate Fuels. ASTM International: West Conshohocken, PA, USA, 2010; pp. 1–10.
26. *EN 14214:2013 V2+A1*; Liquid Petroleum Products-Fatty Acid Methyl Esters (FAME) for Use in Diesel Engines and Heating Applications—Requirements and Test Methods. European Standards: Milan, Italy, 2013; p. 14538.

27. Bowles, J.F.W. Oxides. In *Encyclopedia of Geology*; Alderton, D., Elias, S., Eds.; Academic Press: Cambridge, MA, USA, 2021; pp. 428–441. Available online: https://www.sciencedirect.com/science/article/pii/B9780081029084001855 (accessed on 12 April 2022).
28. *FDUS 2055: 2018*; Final Draft Uganda Standard. ASTM International: West Conshohocken, PA, USA, 2013.
29. *ASTM D93-20*; Standard Test Methods for Flash Point by Pensky-Martens Closed Cup Tester. ASTM International: West Conshohocken, PA, USA, 2004.
30. *ASTM D445-06*; Standard Test Method for Kinematic Viscosity of Transparent and Opaque Liquids (and Calculation of Dynamic Viscosity). ASTM International: West Conshohocken, PA, USA, 2008; pp. 1–10.
31. *D40052-15*; Standard Test Method for Density, Relative Density, and API Gravity of Liquids by Digital Density Meter. ASTM International: West Conshohocken, PA, USA, 2013; pp. 1–8. [CrossRef]
32. Margellou, A.; Koutsouki, A.; Petrakis, D.; Vaimakis, T.; Manos, G.; Kontominas, M.; Pomonis, P.J. Enhanced production of biodiesel over MgO catalysts synthesized in the presence of Poly-Vinyl-Alcohol (PVA). *Ind. Crops Prod.* **2018**, *114*, 146–153. [CrossRef]
33. Amirthavalli, V.; Warrier, A.R. Production of biodiesel from waste cooking oil using MgO nanocatalyst. *AIP Conf. Proc.* **2019**, *2115*, 30609. [CrossRef]
34. Ilgen, O.; Akin, A.N. Transesterification of Canola Oil to Biodiesel Using MgO Loaded with KOH as a Heterogeneous Catalyst. *Energy Fuels* **2009**, *23*, 1786–1789. [CrossRef]
35. Ashok, A.; Kennedy, L.J.; Vijaya, J.J.; Aruldoss, U. Optimization of biodiesel production from waste cooking oil by magnesium oxide nanocatalyst synthesized using coprecipitation method. *Clean Technol. Environ. Policy* **2018**, *20*, 1219–1231. [CrossRef]

Article

Synthesis of Biodiesel via Interesterification Reaction of *Calophyllum inophyllum* Seed Oil and Ethyl Acetate over Lipase Catalyst: Experimental and Surface Response Methodology Analysis

Ratna Dewi Kusumaningtyas [1,*], **Normaliza Normaliza** [1], **Elva Dianis Novia Anisa** [1], **Haniif Prasetiawan** [1], **Dhoni Hartanto** [1], **Harumi Veny** [2], **Fazlena Hamzah** [2] and **Miradatul Najwa Muhd Rodhi** [2]

[1]   Chemical Engineering Department, Faculty of Engineering, Universitas Negeri Semarang, Kampus Sekaran, Gunungpati, Semarang 50229, Indonesia
[2]   School of Chemical Engineering, College of Engineering, Universiti Teknologi Mara (UiTM), Shah Alam 40450, Selangor, Malaysia
*   Correspondence: ratnadewi.kusumaningtyas@mail.unnes.ac.id

Citation: Kusumaningtyas, R.D.; Normaliza, N.; Anisa, E.D.N.; Prasetiawan, H.; Hartanto, D.; Veny, H.; Hamzah, F.; Rodhi, M.N.M. Synthesis of Biodiesel via Interesterification Reaction of *Calophyllum inophyllum* Seed Oil and Ethyl Acetate over Lipase Catalyst: Experimental and Surface Response Methodology Analysis. *Energies* **2022**, *15*, 7737. https://doi.org/10.3390/en15207737

Academic Editor: Jung Rae Kim

Received: 15 September 2022
Accepted: 17 October 2022
Published: 19 October 2022

Publisher's Note: MDPI stays neutral with regard to jurisdictional claims in published maps and institutional affiliations.

Abstract: Biodiesel is increasingly being considered as an alternative to the fossil fuel as it is renewable, nontoxic, biodegradable, and feasible for mass production. Biodiesel can be produced from various types of vegetable oils. *Calophyllum inophyllum* seed oil (CSO) is among the prospective nonedible vegetable oils considered as a raw material for biodiesel synthesis. The most common process of the biodiesel manufacturing is the transesterification of vegetable oils which results in glycerol as a by-product. Thus, product purification is necessary. In this work, an alternative route to biodiesel synthesis through interesterification reaction of vegetable oil and ethyl acetate was conducted. By replacing alcohol with ethyl acetate, triacetin was produced as a side product rather than glycerol. Triacetin can be used as a fuel additive to increase the octane number of the fuel. Therefore, triacetin separation from biodiesel products is needless. The interesterification reaction is catalyzed by an alkaline catalyst or by a lipase enzyme. In this study, biodiesel synthesis was carried out using a lipase enzyme since it is a green and sustainable catalyst. The interesterification reaction of CSO with ethyl acetate in the presence of a lipase catalyst was conducted using the molar ratio of CSO and ethyl acetate of 1:3. The reaction time, lipase catalyst concentration, and reaction temperature were varied at 1, 2, 3, 4, 5 h, 10%, 15%, 20%, and 30 °C, 40 °C, 50 °C, 60 °C, respectively. The experimental results were also analyzed using response surface methodology (RSM) with the Box–Behnken design (BBD) model on Design Expert software. Data processing using RSM revealed that the highest conversion within the studied parameter range was 41.46%, obtained at a temperature reaction of 44.43 °C, a reaction time of 5 h, and a lipase catalyst concentration of 20%.

Keywords: biodiesel; *Calophyllum inophyllum* seed oil; interesterification; enzymatic; Box–Behnken design

1. Introduction

Biodiesel is an alternative fuel for diesel engines which is synthesized from vegetable oils or animal fats. Biodiesel is one of the modern bioenergies which has several advantages. It has non-toxic properties, low emission rates and no sulfur gas, just to name a few. Thus, biodiesel is a prospective alternative as an environmentally friendly diesel fuel [1]. Biodiesel is commonly derived from vegetable oil which has renewable characteristics. The main components of vegetable oils and animal fats are triglycerides, which can be converted into mono alkyl esters of long chain fatty acid or termed as fatty acid alkyl ester (biodiesel). Various types of vegetable oils can be applied as prospective raw materials for biodiesel preparation [2]. The widely used raw material for biodiesel synthesis in Indonesia is crude palm oil (CPO). CPO is abundantly available since it is produced on a large scale. However, the production of palm oil biodiesel competes with food needs. Therefore, other alternative vegetable oils are needed as raw materials for biodiesel. Nonedible oil feedstocks are favorable to ensure the sustainability biodiesel production [3,4].

There are several nonedible oils that have been widely investigated as biodiesel feedstocks, such as silk–cotton or *Ceiba pentandra* seed oil [5], jatropha oil [4,6], castor oil [7], rubber seed oil [8], karanja oil [9], mahua oil, neem oil [10], waste cooking oil [11], palm fatty acid distillate [12] and *Calophyllum inophyllum* seed oil [13]. One among the potential nonedible vegetable oils in Indonesia is *Calopyllum inophyllum* seed oil (CSO). It is also known as tamanu or nyamplung seed oil. Akram et al. [1] reported that *Calopyllum inophyllum* (tamanu) seed has an oil content of 65–75%, which is higher than other nonedible seed oil plants, such as jatropha seed oil (27–40%), rubber seed kernel oil (40–50%), and castor oil (45–65%). Rasyid et al. [14] reported that the fatty acids composition of CSO comprises of oleic acid, palmitic acid, linoleic acid, and stearic acid. According to Adenuga et al. [13], CSO biodiesel meets the Australian, ASTM and EN standards. Therefore, CSO is suitable to be used as a feedstock in the production of biodiesel. However, CSO contains high gummy substances, especially in the form of lipoid A (the hydratable phosphatides) and lipoid B (the non-hydratable phosphatides [15,16], waxes, and other impurities [17]. The existence of gum is unfavorable since it may lower the reaction rate which in consequence reduces the biodiesel yield. At an industrial level, this condition leads to an economic disadvantage. Besides, the presence of phosphorus compounds can generate the carbon particle deposition in diesel engine which causes plugging in the engine filter, line, and injectors, as well as reducing engine performance [17,18]. Thus, a degumming process is essential as the pretreatment step of biodiesel feedstocks with a high phosphorus content [19].

The most general method for making biodiesel is an alkaline-catalyzed transesterification reaction by reacting vegetable oils with short-chain alcohols such as methanol or ethanol [20,21]. However, the transesterification reaction has a limitation in terms of the potential for a saponification side reaction if the oil feedstock contains high free fatty acid (FFA) 2 mgKOH/g, as described by Kusumaningtyas et al. [4] and Sebastian et al. [22]. Besides, the transesterification reaction results in glycerol as a by-product, which needs a separation process to obtain high purity biodiesel. Biodiesel purification from glycerol by-products requires a series of separation steps which brings in high operation costs. To overcome these problems, it is necessary to apply the process of producing biodiesel which is glycerol-free through an interesterification reaction [23]. In the interesterification reaction, the use of methanol or ethanol is replaced with methyl acetate or ethyl acetate. Interesterification of triglyceride yields a triacetin co-product instead of glycerol [24]. Triacetin has an advantage as a fuel additive, a good anti-knocking, and can enhance the octane number. The addition of triacetin to the fuel also offers an environmental benefit since it reduces the exhaust smoke and NO emissions to some extent. It can be stated that the existence of the triacetin co-product in biodiesel contributes to engine performance improvement [25], as the addition of triacetin increases the oxygen content of the fuel [26]. Thus, it is unnecessary to separate triacetin from biodiesel products. A comparison between transesterification and interesterification reaction schemes are presented in Equations (1) and (2), respectively.

$$
\begin{array}{l}
H_2C-O-\overset{\overset{O}{\|}}{C}-R_1 \\
HC-O-\overset{\overset{O}{\|}}{C}-R_2 \;+\; 3CH_3CH_2COCH_3 \;\underset{\longleftarrow}{\overset{Catalyst}{\longrightarrow}}\; C_2H_5-O-\overset{\overset{O}{\|}}{C}-R_1 \;+\; H_2C-\overset{\overset{O}{\|}}{C}-O-CH_3 \\
H_2C-O-\overset{\overset{O}{\|}}{C}-R_3
\end{array}
$$

Triglyceride   Ethyl Acetate   Fatty Acid Ethyl Esters (FAEE)   Triacetin   (1)

$$\begin{array}{ccccccc}
CH_2\!-\!OOC\!-\!R_1 & & & & R_1\!-\!COO\!-\!R & & CH_2\!-\!OH \\
| & & & \text{Catalyst} & | & & | \\
CH\;\!-\!OOC\!-\!R_2 & + & 3ROH & \rightleftharpoons & R_2\!-\!COO\!-\!R & + & CH\;\!-\!OH \\
| & & & & | & & | \\
CH_2\!-\!OOC\!-\!R_3 & & & & R_3\!-\!COO\!-\!R & & CH_2\!-\!OH \\
\text{Triglyceride} & & \text{Alcohol} & & \text{Alkyl esters} & & \text{Glycerol}
\end{array} \qquad (2)$$

To date, the development of green energy through environmentally friendly processes has become a priority. One alternative towards a greener process is reducing the use of chemical catalysts and switching to enzymatic catalysts (biocatalysts). The enzyme-catalyzed biodiesel synthesis shows superiority compared to the chemically catalyzed one, mainly in terms of the lower energy requirement and natural conservation aspect. Lipase is the most common enzyme employed for biodiesel synthesis. There are two principal classes of lipase, free enzymes and immobilized enzymes. Free lipase enzymes are cheaper and simpler, thus it is feasible for large-scale applications [27]. Besides that, the use of lipase enzymes is very promising to overcome the disadvantages of alkaline catalysts related to the occurrence of the undesired saponification reaction when high FFA oil used as raw material. Lipase enzyme catalysts can be applied with high FFA feedstocks without any necessity of FFA removal as a pretreatment step. However, lipase biocatalysts are easily deactivated in an alcoholic environment since short-chain alcohols often cause irreversible loss of enzyme activity [28]. Therefore, the non-alcoholic interesterification route is preferable in order to maintain high biocatalyst activity and stability during the reaction. In the interesterification reaction, alcohol can be replaced by methyl acetate or ethyl acetate, which is harmless for the lipase catalyst [29].

In this work, biodiesel was synthesized through lipase-catalyzed interesterification of *Calopyllum inophyllum* seed oil (CSO) with ethyl acetate. The lipase enzyme used was the liquid-free lipase from Novozyme. The effects of the main parameters on the reaction conversion were studied experimentally. Analysis using the response surface methodology was also carried out to investigate the best reaction conversion obtained within the values range of the studied independent variables. Process optimization was also conducted using the response surface methodology, a collection of mathematical and statistical calculation techniques based on the compatibility of the empirical model with the experimental data obtained. Generally, RSM is useful for the modeling and analysis of problems that influences the response variables with the purpose of optimizing the response. RSM is also beneficial when determining the operating conditions to achieve the desired conversion [30]. There are several experimental design models used in RSM, such as the Box–Behnken Design (BBD), control composite design (CCD) and Doehlert Design. The BBD has been broadly applied for biodiesel production optimization since it is effective, requires the smallest number of trials than the other designs [31]. In this investigation, analysis using BBD in RSM was performed to determine the best reaction operation condition which resulted in the highest reaction conversion of the interesterification reaction of CSO in the presence of lipase enzyme catalyst within the ranges of the parameter values studied.

2. Materials and Methods

2.1. Materials

The materials used were: *Calophyllum inophyllum* seed oil (CSO) from UMKM Samtamanu Cilacap, Indonesia, ethyl acetate p.a (Merck), liquid lipase catalyst (Eversa Transform 2.0 from Novozyme), phosphoric acid p.a (Merck), KOH p.a (Merck), oxalic acid p.a (Merck), ethanol p.a (Merck), and aquadest from Indrasari chemicals store, Semarang, Indonesia. The catalyst used was liquid (free form) lipase Eversa Transform 2.0 (ET 2.0) from Novozyme, which is made from genetically modifed Aspergillus oryzae microorganism [32]. Commonly, lipase are active in broad range of temperature from 20–60 °C [33,34] and the ET 2.0 was reported to work satisfactorily up to the reaction temperature of 60 °C [35].

2.2. Method

2.2.1. Degumming

Crude CSO as much as 100 mL was put into a 250 mL flat bottom flask. The oil was then heated using hot plate equipped with a magnetic stirrer at a speed of 600 rpm until the temperature reached 70 °C. Phosphoric acid was subsequently added to the hot oil with the amount of 0.3% (*w/w*) of CSO and heated for 25 min. The oil was cooled at room temperature and put into a 500 mL separatory funnel. Distilled water at a temperature of 40 °C was thereafter added with a volume of 5% of the CSO volume. Subsequent to the degumming process, the mixture was settled in the separatory funnel for 24 h to attain the completion of the separation between the CSO and the residue. After the 24 h decantation process, the two layers were appeared. The top layer of brownish yellow color was the refined CSO, while the dark brown bottom layer was gum, impurities, and water which should be separated. Let the mixture in the separatory funnel for to attain the completion of the separation between the oil and the residue. The degummed CSO was then separated from the residue and was heated in the oven with a temperature of 105 °C to reduce the water content [30]. CSO was then analyzed to determine its physical properties (density, viscosity, and acid number).

2.2.2. Interesterification Reaction

After the degumming process, CSO was used as the raw material for biodiesel synthesis via interesterification reaction. CSO as much as 100 mL was introduced into a 250 mL flat bottom three neck flask and added with 10% (*w/w*) liquid lipase. The mixture was heated using a hot plate equipped with a magnetic stirrer with the stirring speed of 600 rpm until it reached the desired temperature. On the other hand, ethyl acetate was warmed up separately up to the similar temperature. When the reactants attained the specified temperature, ethyl acetate was subsequently poured into the reactor and mixed with the CSO and lipase. This incident was recorded as the reaction time of zero (t = 0 h). The ratio of CSO to ethyl acetate was fixed at 1:3. The stoichiometric ratio of the reactants was employed to investigate the optimum reaction conversion that can be obtained without any excess reactant (ethyl acetate) as it was conducted by Manurung et al. [36]. The reaction temperature was varied at 30 °C, 40 °C, 50 °C and 60 °C.The reaction temperature was varied at 30 °C, 40 °C, 50 °C and 60 °C. After the specified temperature was reached, the ethyl acetate with the similar temperature was added with the ratio of CSO to ethyl acetate of 1:3 (stoichiometric ratio). The reaction time was 1–5 h. Sample was taken periodically every 1 h of the reaction time. Prior to the analysis, sample was centrifuged with the centrifugation speed of 40,000 rpm for 15 min to separate between biodiesel product and the catalyst residue. The sample analysis was carried out using Gas Chromatography-Mass Spectroscopy (GC-MS). The independent variables for the experimental work is shown in Table 1.

Table 1. Independent Variables.

Variable	Values
Reaction Time	1, 2, 3, 4, and 5 h
Reaction Temperature	30 °C, 40 °C, 50 °C, and 60 °C
Catalyst Concentration	10%, 15%, and 20%

2.2.3. Gas Chromatography-Mass Spectroscopy (GC-MS)

Analysis of the fatty acid composition of the biodiesel was determined using a Gas Chromatography-Mass Spectrometer (GC-MS QP2010 SE) with a column flow rate of 1.20 mL/min, an oven column temperature of 65 °C, a pressure of 74.5 kPa, and an injecttion temperature of 250 °C with split injection mode. The internal standard used was heptadecanoic acid, methyl ester.

2.2.4. Response Surface Methodology Analysis

The GC-MS data was utilized to determine the reaction conversion. The results were subsequently analyzed using response surface methodology (RSM) with the Box–Behnken Design (BBD) model on Design Expert software to determine the optimum operation condition. The BBD is a vigorous and extensively applied model for biodiesel synthesis optimization and parameter analysis [31]. The levels of the tested parameters for the BBD experiment are presented in Table 2.

Table 2. Levels of Tested Parameters for Box–Behnken Design (BBD).

Independent Factor	Units	Symbol	Level	
			Low	High
Temperature	°C	A	30	50
Reaction Time	hours	B	1	5
Catalyst Concentration	%	C	10	20

3. Results and Discussion

3.1. Effects of the Degumming Process on the CSO Properties

Naturally, crude CSO contains phospholipids, sterols, free fatty acids, waxes, oil-soluble pigments and hydrocarbons. Degumming is the initial and essential refining process of crude vegetable oil to remove phospholipids and gums [37]. To date, there are various methods that can be employed to degum vegetable oil, explicitly water, acid, enzymatic, and membrane degumming techniques. Water degumming and acid degumming are the most used methods for vegetable degumming at an industrial scale [38]. Those methods are simple, easy, and low cost. However, water degumming is less effective compared to the acid method. The gum is comprised of hydratable (HP) and non-hydratable phosphatides (NHP). HP can be eliminated using water. In contrast, NHP cannot be eradicated via water degumming [15,16]. Thus, the acid degumming method is more advantageous in that it can remove both the HP and NHP. However, water degumming is less effective compared to acid degumming. Thus, the acid degumming method is more advantageous. The most used acids for the degumming process are phosphoric or citric acid, with a suggested concentration between 0.05–2% *w/w* oil [39]. In this work, 0.3% phosphoric acid was applied [23,40]. After the degumming process, the black color of the CSO turns reddish-yellow due to the loss of the phospholipids compound in crude CSO. Phospholipids can form dark colors in vegetable oil as a result of the autoxidation process of these compounds during storage. The condensation reaction between the amino groups in phospholipids with aldehydes will yield melanophosphatides compounds which give a dark color to the oil [41]. Thus, removal of phospholipids in CSO will reduce the occurrence of the abovementioned reaction and diminish the color intensity.

The degumming process can improve the properties of CSO. In this work, the density, viscosity, and acid number of crude and degummed CSO were tested experimentally. The effects of the degumming process on the physicochemical properties of CSO is indicated in Table 3. It was obvious that the density and viscosity of the CSO were slightly decreasing after the degumming process due to the removal of gum and other impurities. The effect of the degumming process on the CSO properties is indicated in Table 2. It was revealed that the density and viscosity of the CSO slightly decreased after the degumming process. It occurred with the removal of gum and other impurities. Besides, degumming also leads to a lessening of the acid number of the oil which is attributable to the decrease of the acid number and the existence of free fatty acids (FFA) in the oil during the degumming process. This phenomenon is in line with the data reported by Adekunle et al. [42]. The fatty acid composition of the CSO after undergoing the degumming process is presented in the Table 4, based on the interpretation of the chromatogram shown in Figure 1.

Table 3. Effects of Degumming Process on the Properties of CSO.

CSO Properties	Before Degumming	After Degumming
Density (kg/m^3)	941	937
Viscosity (mm^2/s)	63.42	59.73
Acid Number (mg KOH/g CSO)	64.62	48.24
Free Fatty Acid Content (%)	32.47	24.25

Table 4. Fatty Acid Composition of the CSO after Degumming Process.

Fatty Acid	Molecular Weight (g/mol)	Area (%)
Palmitic Acid	256.22	7.82
Linoleic Acid	280.45	16.82
Oleic Acid	282.52	26.62
Stearic Acid	284.47	8.86
Arachidic Acid	312.54	0.31

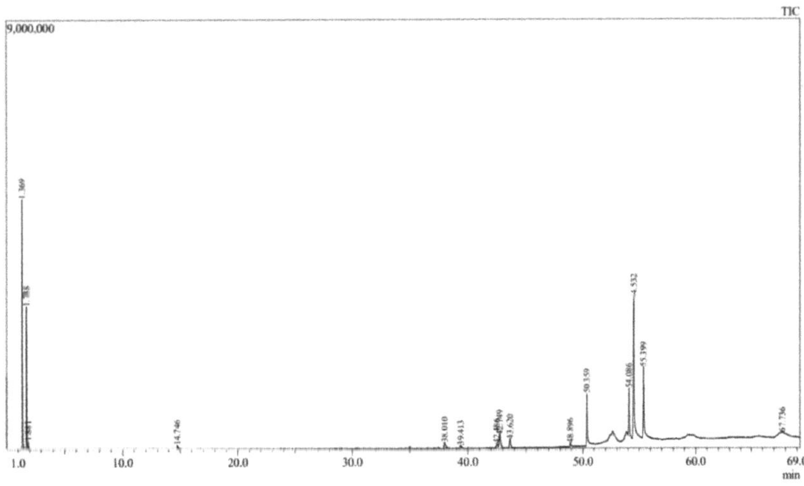

Figure 1. Chromatogram of the CSO after Degumming Process.

3.2. Effects of the Temperature on the Interesterification Reaction Conversion

After the degumming process, CSO directly underwent the interesterification process with ethyl acetate in the presence of the lipase catalyst. As depicted in Table 2, the crude and degummed CSO contained high FFA of 32.47% and 24.25%, respectively, which were over 2 mgKOH/g. However, the enzymatic reaction is generally insensitive to the existence of the FFA impurities in the oil feedstock. The possibility of the undesired saponification reaction can be neglected. Therefore, a specific pre-treatment for reducing the FFA content of CSO was considered unnecessary since the FFA presentation in the oil does not affect the enzymatic catalyzed reaction [22,43].

The influences of the reaction time and temperature on the reaction conversion were observed at a fixed lipase catalyst concentration of 10% (*w/w*). The result is exhibited in Figure 2. Basically, lipases are active in a temperature range of 20–60 °C [33,34] and the ET 2.0 lipase can work up to 60°C [35]. Moreover, the reaction time is often studied for up to 12 h, but the significant reaction rate is in the range of 0–5 h [44]. In this work, it was disclosed that the highest reaction conversion was 54.99%, obtained at a reaction temperature of 40 °C and a reaction time of 3 h. Commonly, the reaction conversion is enhanced with the increasing of the reaction temperature and reaction time, since at the higher temperature the viscosity of the liquid decreases, causing the higher solubility of reactants, mass diffusion, rate of reaction, and conversion [45]. However, in this work, the reaction conversion decreased when the reaction was performed at 50 °C for 5 h. This trend was in accordance with the results recorded by Gusniah et al. [46], where it was reported that the lipase enzyme catalyst activity in the transesterification reaction of waste cooking oil reached an optimum performance at 40 °C and it declined to some extent when the temperature rose to 50 °C, which happened when the lipase catalyst began to denature, causing damage of the active part of the enzyme at the higher temperature. Beyond the optimum temperature, the lipase catalyst becomes unstable and misplaces its tertiary structure, which causes the shortfall in its activity [45]. It thus leads to the decline of the reaction rate and conversion. Generally, the lipase enzyme reached the highest activity at the temperature of 40–45 °C as stated by Murtius et al. [47], Ayinla et al. [48], and Yazid [49].

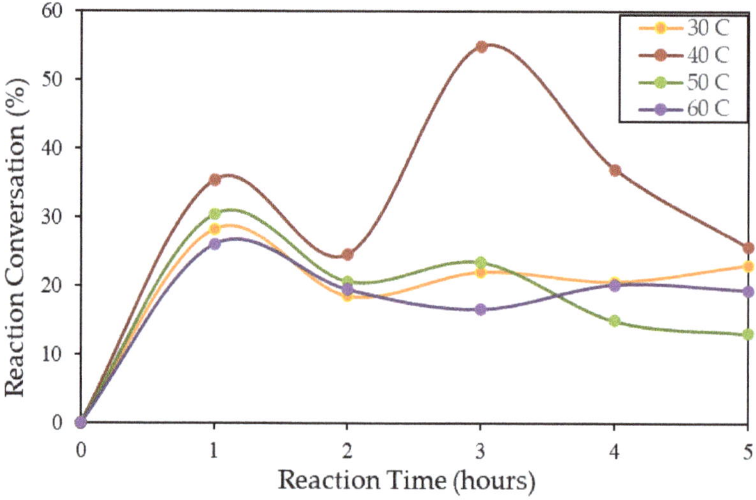

Figure 2. Effects of the reaction time and temperature on the reaction conversion at a fixed lipase catalyst concentration of 10% (*w/w*).

3.3. Effects of the Catalyst Concentration E on the Interesterification Reaction Conversion

Investigation on the influence of the lipase catalyst concentration was carried out at the molar ratio of oil to ethyl acetate of 1:3, a reaction temperature of 40 °C and a reaction time of 5 h. The lipase concentrations studied were 10%, 15%, and 20%. Figure 3 has shown that the optimum catalyst concentration was 15%, resulting in the reaction conversion of 28.445%. This result was in agreement with the results obtained by Gusniah et al. [46] which indicated that the highest lipase catalyst loading was 15%. However, the reaction conversion in this work was still low compared to those obtained in other research as the reaction was conducted with the low molar ratio of oil and ethyl acetate (1:3) and a short reaction time (5 h). By comparison, Sun et al. [44] achieved a 94.2% yield for the transesterification of semen Abutili seed oil in the presence of a similar lipase catalyst but at an excess molar ratio of oil and alcohol (1:7) and a longer reaction time (11 h).

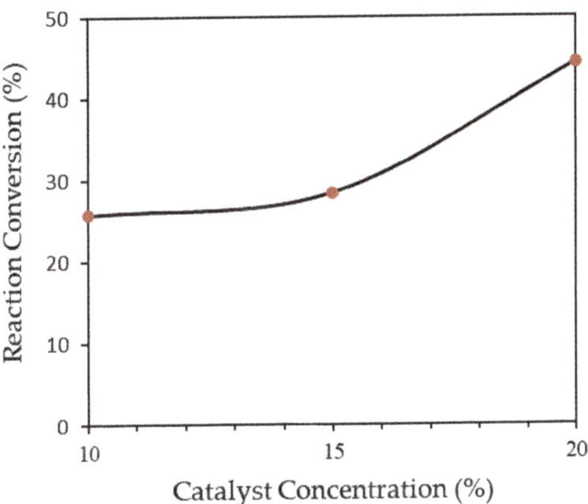

Figure 3. Effects of the lipase catalyst concentration at the reaction temperature of 40 °C and reaction time of 5 h.

Subhedar and Gogate [29] carried out the interesterification reaction of waste cooking oil and methyl acetate using immobilized lipase obtained from *Thermomyces lanuginosus* (Lipozyme TLIM) as a catalyst. The optimum reaction was demonstrated at the oil: methyl acetate molar ratio of 1:12, enzyme concentration of 6% (*w/v*), reaction temperature of 40 °C and reaction time of 24 h. Biodiesel yield was 90.1% under these operation conditions, which was higher than the results obtained in this work. A higher reaction conversion was obtained as the reaction was conducted on an excessive molar ratio and an extensively longer reaction time (24 h). Besides, this reaction employed an immobilized lipase, in which the lipase was attached on the carrier substrate, which led to higher product purity. Thus, it can be deduced that the lipase-catalyzed reaction in this work resulted in a lower conversion due to a shorter reaction time and low molar ratio of the reactants, which did not exceed the stoichiometric ratio. Besides that, utilization of the liquid lipase can reduce the apparent concentration of biodiesel yield since it can act as an impurity in the final biodiesel product.

3.4. Response Surface Methodology (RSM) Analysis

Response surface methodology (RSM) is the broadest employed method of experimental design for optimization. This technique applies mathematical and statistical methods for modeling and analyzing a process which involves numerous parameters. The purpose of the RSM is to optimize the response variables [29]. The factors that influence the process are entitled dependent variables. Meanwhile, the responses are named as dependent variables. Therefore, analysis using RSM is useful in designing operating conditions to achieve the targeted conversion. Many studies have shown that the most applicable tools in RSM for optimization of biodiesel production process are the Box–Behnken Design (BBD) and Central Composite Design (CCD) tools [50]. BBD is favored over CCD since it provides higher efficiency but more economical [51,52]. To evade the failure and extreme reaction variables value, BBD is limited to three levels. Hence, BBD typically has no factorial points. The BBD tool was also utilized by Sharma et al. [53] and Rokni et al. [54].

In this study, BBD was applied to obtain the best operating conditions and highest conversion for biodiesel synthesis via interesterification reaction within the ranges of the parameter values studied. The Box–Behnken Design (BBD) tool in RSM was used to evaluate the effect of the independent variables (reaction time, reaction temperature, and catalyst concentration). The experimental data of the reaction conversion compared with the prediction using BBD is exhibited in Table 5.

The parameters used to obtain the appropriate model to predict the response results must match the observed model to the experimental data. Analysis of variance (ANOVA) is accomplished using the most common types of polynomial models, namely linear, interactive (2FI), quadratic and cubic can be applied to predict the response variables of experimental data. Several parameters such as sequential *p*-value, lack of fit *p*-value, adjusted R¬2, predicted R2 and Adiq precision were used to conclude the most suitable type of model for the optimization of the conversion of biodiesel produced from CSO. In this study, optimization using RSM was initially performed using the quadratic model. The ANOVA quadratic model is presented in Table 6.

Table 5. Experimental data of the reaction conversion and the prediction using BBD.

Run	Temperature, °C	Reaction Time, h	Catalyst Concentration, % *w/w*	Reaction Conversion, % Experiment	Reaction Conversion, % Prediction	Error, %
1	40	3	15	12.92	18.15	40.4
2	30	3	10	22.08	19.877	9.97
3	40	5	20	48.92	48.03	1.8
4	30	3	20	27.75	27.33	1,5
5	40	5	10	25.77	26.66	3.4
6	40	3	15	18.27	18.15	0.6
7	30	1	15	23.41	24.72	5.6
8	50	5	15	14.90	13.59	8.8
9	40	1	20	13.55	12,66	6.5
10	30	5	15	22.10	23.41	5.9
11	40	3	15	23.25	18.15	21.9
12	40	1	10	35.43	36.32	2.5
13	50	1	15	26.82	25.51	4.8
14	50	3	20	12.01	14.22	18.3
15	50	3	10	23.53	23.96	1.7

Table 6. ANOVA of the Quadratic Model.

Source	Sum of Squares	DF	Mean Square	F-Value	*p*-Value	
Model	1007.96	9	112.00	2.13	0.2098	not significant
A-Suhu	40.83	1	40.83	0.7760	0.4187	
B-Waktu	19.45	1	19.45	0.3696	0.5698	
C-Katalis	2.61	1	2.61	0.0497	0.8325	
AB	28.14	1	28.14	0.5348	0.4974	
AC	73.90	1	73.90	1.40	0.2892	
BC	507.01	1	507.01	9.64	0.0267	
A^2	32.27	1	32.27	0.6134	0.4690	
B^2	161.73	1	161.73	3.07	0.1399	
C^2	139.90	1	139.90	2.66	0.1639	
Residual	263.06	5	52.61			
Lack of Fit	209.74	3	69.91	2.62	0.2881	not significant
Pure Error	53.32	2	26.66			
Cor Total	1271.02	14				

Based on the quadratic model, it was revealed that the *p*-value was 0.2098, which did not meet the requirement of the *p*-value < 0.05 as demonstrated in Table 6. Hence, the quadratic model was not significant for this case. To overcome this obstacle, a modified model was developed to obtain the significant model [55]. Modification of the model showed that the reduced cubic model was a significant model (Table 7).

Table 7. ANOVA of the reduced cubic model.

Source	Sum of Squares	DF	Mean Square	F-Value	*p*-Value	
Model	1197.58	10	119.76	6.52	0.0428	significant
A-Suhu	40.83	1	40.83	2.22	0.2102	
B-Waktu	165.26	1	165.26	9.00	0.0399	
C-Katalis	2.61	1	2.61	0.1423	0.7252	
AB	28.14	1	28.14	1.53	0.2834	
AC	73.90	1	73.90	4.03	0.1153	
BC	507.01	1	507.01	27.62	0.0063	
A^2	32.27	1	32.27	1.76	0.2555	
B^2	161.73	1	161.73	8.81	0.0412	
C^2	139.90	1	139.9	7.62	0.0508	
A^2B	189.62	1	189.62	10.33	0.0325	
Residual	73.44	4	18.36			
Lack of Fit	20.12	2	10.06	0.3773	0.7260	not significant
Pure Error	53.32	2	26.66			
Cor Total	1271.02	14				
Adeq Precision	9.6401					

It was found that the *p*-value of the modified cubic model was 0.0428, which met the requirement of *p*-value > 0.05. Furthermore, the value of lack-of-fit was examined. Lack-of-fit is a test that analyzes how satisfactory the full models suit with the data. Models with a significant lack-of-fit should not be applied for predictions. In this work, the lack-of-fit was not significant, which meant that the model could be employed for the prediction. The value of adeq precision is the measure of the ratio of the signal to the disturbance. The expected value of the ratio is > 4 [50]. This model provided the value of the Adeq precision of 9.6401, indicating that the model is proper. Therefore, the reduced cubic model can be used to describe the response on the reaction conversion. The empirical correlation of the reduced cubic model in the form of polynomial order for the reaction conversion in the CSO esterification is displayed in Equation (3).

$$\text{Conversion (\%)} = 18.15 - 2.26\,A + 6.43\,B - 0.5715\,C - 2.65\,AB - 4.30\,AC + 11.26\,BC - 2.96\,A2 + 6.62\,B2 + 6.16\,C2 - 9.74\,A2B \tag{3}$$

where, A is the reaction temperature (°C), B is the reaction time (min) and C is the Catalyst Concentration (%). Based on the reduced cubic model, analysis on the effects and the process variables interaction on the response variable were carried out using 3D RSM graphs. The results are presented in Figure 4.

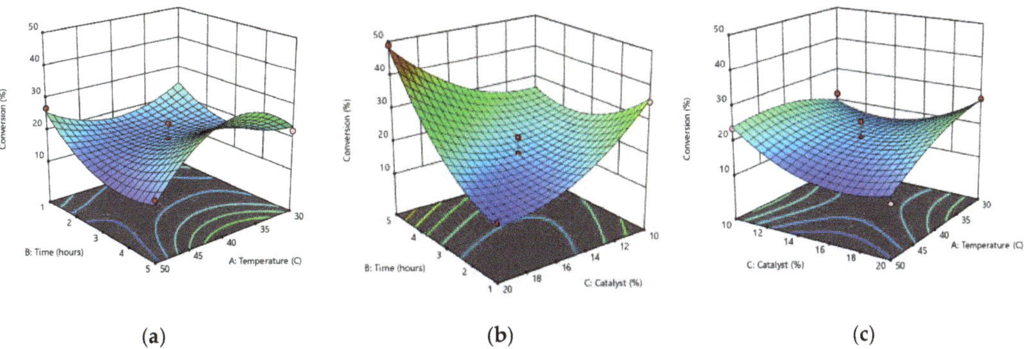

(a)   (b)   (c)

Figure 4. The 3D RSM graph representing the effects of (a) reaction time and temperature; (b) reaction time and catalyst concentration; (c) reaction temperature and catalyst concentration to the reaction conversion.

Figure 4a shows the interaction between temperature and reaction time which gives a significant change to the increase in conversion. It was found that there was an increase in conversion at the temperature of 40 °C and reaction time of 5 h. This phenomenon is in line with the investigation of Handayani et al. [55], which reported that the optimum conversion was obtained at the temperature of 45 °C and reaction time of 6 h. Figure 4b discloses that the longer the reaction time employed, the higher the conversion achieved, as the reaction time is directly proportional to the amount of product yielded. Li et al. [56] observed that the reaction conversion of *Pseudomonas cepacia* lipase was low at the reaction time of 3 h. (44%). On the contrary, it rose to 73% when the reaction time was extended to 12 h. Figure 4c bares that the lower the reaction temperature applied, the higher the conversion. This occurred because the lipase enzyme activity worked at the temperature of 30–45 °C. In the enzymatic reaction, the increase of the temperature will promote the reaction rate and accordingly the reaction conversion due to the decrease of the viscosity which enhance the solubility of oil and methanol. However, at a certain temperature, the reaction conversion declined, invoking the enzyme deactivation [57]. Meanwhile, the addition of the catalyst concentration increased the reaction conversion. This result was in line with the description of Gusniah et al. [46]. This indicated that increasing the amount of the enzyme employed will provide a higher amount of the available active sites for the reaction, leading to a higher reaction conversion.

The Derringer method is utilized to optimize the response which is characterized by the presence of a desirability function. Desirability shows how close the optimization results are to the optimum point. The desirability function (DF) is constructed on the transformation of all the acquired responses from different scales into a scale-free value. The values of DF range from 0 to 1. The value 0 means that the factors provide an undesirable response. On the other hand, the value 1 denotes the optimal performance of the evaluated factor [58]. Hence, the desirability value which is close to 1 is the expected value. However, the RSM analysis is not only to find the desirability value, which is equal to 1, but to determine the conditions that match expectations. The multiple correlation coefficient (R^2) was closed to 1. It denoted that the models have good predictive ability. Based on the RSM analysis results, the highest reaction conversion was 41.46%, attained at a temperature of 44.43 °C, a reaction time of 5 h, and a catalyst concentration of 20%, with a desirability value of 0.733 as shown in Figure 5. This finding can be stated as the local optimum within the certain parameter range studied, which is worthwhile as a basis for developing process intensification in order to obtain a higher conversion.

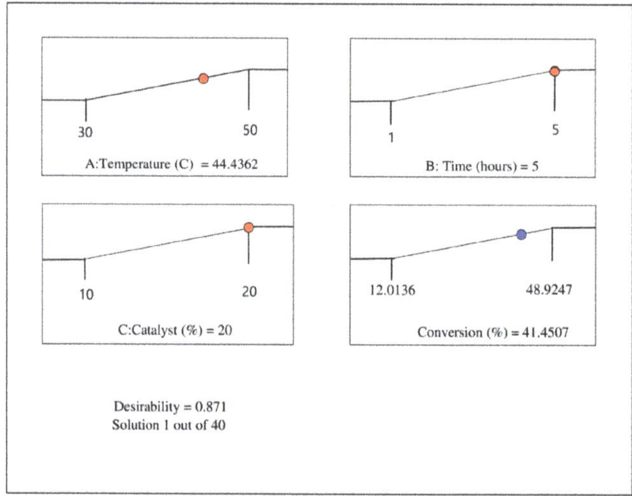

Figure 5. The highest reaction conversion based on the RSM analysis.

4. Conclusions

Based on the RSM analysis, it was revealed that the reaction time and catalyst concentration had significant effects on the conversion of the interesterification reaction of CSO with ethyl acetate in the presence of lipase catalysts for biodiesel synthesis. The results of the analysis showed that the highest reaction conversion was 41.46% at the reaction temperature, reaction time, and catalyst concentration of 44.40 °C, 5 h, and 20%, respectively, which can be stated as the local optimum. While the results are good, it is not economically feasible for industrial-scale biodiesel production and purposes in its current condition. Therefore, further process intensification is necessary to enhance the reaction conversion and biodiesel yield.

Author Contributions: Conceptualization, R.D.K. and H.V.; methodology, D.H. and F.H.; software, H.P.; validation, R.D.K., D.H. and H.P.; formal analysis, N.N. and E.D.N.A.; investigation, R.D.K., H.P., N.N. and E.D.N.A.; resources, R.D.K.; data curation, N.N. and E.D.N.A.; writing—original draft preparation, R.D.K.; writing—review and editing, R.D.K., H.P. and M.N.M.R.; visualization, N.N. and E.D.N.A.; supervision, R.D.K.; funding acquisition, R.D.K. All authors have read and agreed to the published version of the manuscript.

Funding: This research was funded by the Research and Community Service Institute (LPPM) of Universitas Negeri Semarang through International Collaboration Research Scheme (UNNES-UiTM) 2021 with the Contract Number of 10.26.4/UN37/PPK.3.1/2021.

Data Availability Statement: Not applicable.

Conflicts of Interest: The authors declare no conflict of interest.

References

1. Akram, F.; Haq, I.U.; Raja, S.I.; Mir, A.S.; Qureshi, S.S.; Aqeel, A.; Shah, F.I. Current Trends in Biodiesel Production Technologies and Future Progressions: A Possible Displacement of the Petro-Diesel. *J. Clean. Prod.* **2022**, *370*, 133479. [CrossRef]
2. Nayah, R.; Imran, M.; Ramzan, M.; Tariq, M.; Taj, M.B.; Akhtar, M.N.; Iqbal, H.M.N. Sustainable Biodiesel Production via Catalytic and Non-Catalytic Transesterification of Feedstock Materials—A Review. *Fuel* **2022**, *328*, 125254. [CrossRef]
3. Khan, I.W.; Naeem, A.; Farooq, M.; Ghazi, Z.A.; Saeed, T.; Perveen, F.; Malik, T. Biodiesel Production by Valorizing Waste Non-Edible Wild Olive Oil Using Heterogeneous Base Catalyst: Process Optimization and Cost Estimation. *Fuel* **2022**, *320*, 123828. [CrossRef]
4. Kusumaningtyas, R.D.; Ratrianti, N.; Purnamasari, I.; Budiman, A. Kinetics Study of Jatropha Oil Esterification with Ethanol in the Presence of Tin (II) Chloride Catalyst for Biodiesel Production. *AIP Conf. Proc.* **2017**, *1788*, 30086. [CrossRef]
5. Rahul Soosai, M.; Moorthy, I.M.G.; Varalakshmi, P.; Yonas, C.J. Integrated Global Optimization and Process Modelling for Biodiesel Production from Non-Edible Silk-Cotton Seed Oil by Microwave-Assisted Transesterification with Heterogeneous Calcium Oxide Catalyst. *J. Clean. Prod.* **2022**, *367*, 132946. [CrossRef]
6. Chang, A.; Pan, J.H.; Lai, N.C.; Tsai, M.C.; Mochizuki, T.; Toba, M.; Chen, S.Y.; Yang, C.M. Efficient Simultaneous Esterification/Transesterification of Non-Edible Jatropha Oil for Biodiesel Fuel Production by Template-Free Synthesized Nanoporous Titanosilicates. *Catal. Today* **2020**, *356*, 56–63. [CrossRef]
7. Gohar Khan, S.; Hassan, M.; Anwar, M.; Zeshan; Masood Khan, U.; Zhao, C. Mussel Shell Based CaO Nano-Catalyst Doped with Praseodymium to Enhance Biodiesel Production from Castor Oil. *Fuel* **2022**, *330*, 125480. [CrossRef]
8. Lüneburger, S.; Lazarin Gallina, A.; Cabreira Soares, L.; Moter Benvegnú, D. Biodiesel Production from Hevea Brasiliensis Seed Oil. *Fuel* **2022**, *324*, 124639. [CrossRef]
9. Amriya Tasneem, H.R.; Ravikumar, K.P.; Ramakrishna, H.V. Performance and Wear Debris Characteristics of Karanja Biodiesel and Biolubricant as a Substitute in a Compression Ignition Engine. *Fuel* **2022**, *319*, 123870. [CrossRef]
10. Sayyed, S.; Das, R.K.; Kulkarni, K. Experimental Investigation for Evaluating the Performance and Emission Characteristics of DICI Engine Fueled with Dual Biodiesel-Diesel Blends of Jatropha, Karanja, Mahua, and Neem. *Energy* **2022**, *238*, 121787. [CrossRef]
11. Kanwar Gaur, R.; Goyal, R. A Review: Effect on Performance and Emission Characteristics of Waste Cooking Oil Biodiesel- Diesel Blends on IC Engine. *Mater. Today Proc.* **2022**, *63*, 643–646. [CrossRef]
12. Buchori, L.; Widayat, W.; Hadiyanto, H.; Satriadi, H.; Chasanah, N.; Kurniawan, M.R. Modification of Magnetic Nanoparticle Lipase Catalyst with Impregnation of Activated Carbon Oxide (ACO) in Biodiesel Production from PFAD (Palm Fatty Acid Distillate). *Bioresour. Technol. Rep.* **2022**, *19*, 101137. [CrossRef]
13. Adenuga, A.A.; Oyekunle, J.A.O.; Idowu, O.O. Pathway to Reduce Free Fatty Acid Formation in Calophyllum Inophyllum Kernel Oil: A Renewable Feedstock for Biodiesel Production. *J. Clean. Prod.* **2021**, *316*, 128222. [CrossRef]

 *processes*

Article

Steel Slag Decorated with Calcium Oxide and Cerium Oxide as a Solid Base for Effective Transesterification of Palm Oil

Jichao Sun [1], Hewei Yu [1,*], Peisen Zhang [1], Gaoyu Qi [1], Xiuxiu Chen [2], Xiaohui Liang [2] and Hongyu Si [2,*]

[1] School of Energy and Power Engineering, Qilu University of Technology (Shandong Academy of Sciences), Jinan 250353, China
[2] Energy Research Institute, Qilu University of Technology (Shandong Academy of Sciences), Jinan 250013, China
* Correspondence: yhw@qlu.edu.cn (H.Y.); sihyusderi@163.com (H.S.)

Abstract: For further resource utilization of solid waste steel slag and the reduction in biodiesel production costs, this study used steel slag as a carrier to synthesize a $CaO\text{-}CeO_2$/slag solid base catalyst for the effective transesterification of palm oil into fatty acid methyl esters (FAMEs). The synthesis involved a two-step impregnation of steel slag with nitrate of calcium and cerium and thermal activation at 800 °C for 180 min. Then, the catalysts' textural, chemical, and CO_2 temperature-programmed desorption properties were characterized. The catalytic activity depended highly on the ratio of Ca-Ce to steel slag mass; the $CaO\text{-}CeO_2$/slag-0.8 catalyst showed outstanding performance. Characterization showed that the surface area and total basicity of the Ca-Ce/slag-0.8 catalyst were 3.66 m^2/g and 1.289 mmol/g, respectively. The reactivity results showed that FAMEs obtained using 7 wt.% catalyst, 9:1 of methanol-to-palm-oil molar ratio, 180 min reaction duration, and 70 °C reaction temperature was optimum (i.e., 95.3% yield). In addition, the $CaO\text{-}CeO_2$/slag-0.8 catalyst could be reused for at least three cycles, retaining 91.2% of FAMEs yield after n-hexane washing. Hence, the catalyst exhibits an excellent potential for cost-effective and environmentally friendly biodiesel production.

Keywords: steel slag; calcium oxide; cerium oxide; transesterification; biodiesel

Citation: Sun, J.; Yu, H.; Zhang, P.; Qi, G.; Chen, X.; Liang, X.; Si, H. Steel Slag Decorated with Calcium Oxide and Cerium Oxide as a Solid Base for Effective Transesterification of Palm Oil. *Processes* **2023**, *11*, 1810. https://doi.org/10.3390/pr11061810

Academic Editor: Aldo Muntoni

Received: 22 May 2023
Revised: 7 June 2023
Accepted: 12 June 2023
Published: 14 June 2023

1. Introduction

The energy shortage and pollutant emission crisis have prompted researchers to develop new energy sources to replace traditional fossil fuels. Biofuel refers to solid, liquid, or gas fuel made of or extracted from biomass, and is an essential direction in developing and using renewable energy. Biodiesel has become one of the clean energy substitutes for ordinary petrochemical diesel because of its reproducibility, high calorific value, and environmental friendliness. Biodiesel is chemically classified as fatty acid methyl esters (FAMEs) [1]. It is usually obtained by the transesterification of triglycerides (e.g., vegetable oil, animal fat, microalgae oil, and waste oil) and alcohol (typically methanol) in the presence of a catalyst (Figure 1) [2].

Catalysis is crucial to biodiesel technology. Homogeneous acid and base catalysts have high catalytic efficiencies, but they portend recycling and pH-neutral wastewater challenges. Solid acid catalysts can synergistically catalyze the esterification and transesterification of waste oil. However, they usually require a much higher reaction temperature or longer reaction time. Solid base catalysts are easy to separate, and the transesterification conditions are mild, conducive to biodiesel's continuous and large-scale sustainable production with in-depth research value.

Several types of solid bases (such as alkali earth metal oxides, alkali metal-supported catalysts, solid base catalysts supported by molecular sieves, hydrotalcite, anion-exchange resins, and so on) can be used in transesterification catalysis to prepare biodiesel [3]. Popular among them are the alkaline earth metal oxides (such as CaO, MgO, SrO, and BaO). They are solid base catalysts whose basic sites mainly come from oxygen with negative electric lattice and hydroxyl group generated by water adsorption on the surface. CaO is considered one of the most promising solid base catalysts for biodiesel production due to its high basicity, availability, and economy. Yaşar et al. [4] used CaO derived from calcinated waste eggshell to catalyze rapeseed oil and methanol under the optimized reaction conditions of 4% catalyst, 1 h reaction time at 60 °C, to achieve 95.12% biodiesel yield. However, in the reaction system with only a CaO catalyst, the leaching of Ca^{2+} species into the reaction media was a crucial problem affecting the stability of the catalyst [5]. Pandit et al. [6] confirmed that in the catalytic transesterification of microalgal oil and methanol by CaO catalyst, Ca^{2+} leaching would reduce the availability of the active site of reactants, reducing the conversion from 89.7% to 78.2% after three cycles.

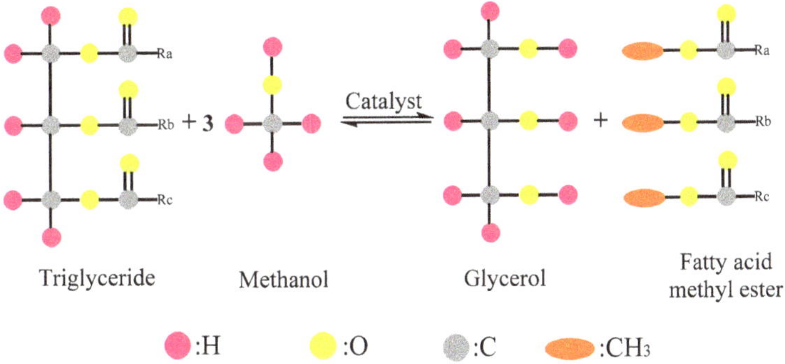

Figure 1. Schematic diagram of transesterification reaction.

Currently, CeO_2 is the most widely used rare earth metal oxide, capable of forming oxygen vacancies. It has excellent catalytic oxidation performance due to its unique crystal structure and high capacity for storing and releasing oxygen. Meanwhile, Ce exhibits acid-base dual properties, which can improve the catalyst's stability. Dehghani et al. [7] synthesized an active and stable catalyst of CaO-loaded Ce-MCM-41 to transform waste vegetable oil into biodiesel. A 96.8% conversion of triglycerides can be achieved at a methanol-to-oil molar ratio of 9 with 5 wt.% of catalyst loading at 60 °C and for 6 h. Elsewhere, Zhang et al. [8] prepared a novel $CeO_2@CaO$ catalyst via a hydrothermal method, achieving 98% biodiesel yield after 6 h. Repeated tests indicated that >80% of biodiesel yield could be obtained after nine reaction cycles, establishing the stability of $CeO_2@CaO$. These studies showed that catalyst stability could be improved after complexing CaO with CeO_2. Yet, the catalytic activity of this catalyst type should be further optimized to shorten the reaction time.

China is the leading steel producer and supplier nation, accounting for over half of the global steel production [9]. As a by-product of steel production, steel slag accounts for 10–20% of crude steel. China's annual output of steel slag is about 70 million tons; the cumulative inventory exceeds 1 billion tons, and the utilization rate of steel slag is only about 30% [10]. Unutilized steel slag occupies a large amount of industrial land and causes excellent pressure on the atmosphere, soil, and water. Therefore, seeking comprehensive utilization of steel slag conforms to the requirement of sustainable development. Additionally, it is a basis for developing a circular economy and building green homes.

The steel slag contains majorly CaO, SiO_2, Al_2O_3, MnO, MgO, Fe_2O_3, P_2O_5, and free calcium oxide (f-CaO) [11]. For resource utilization of steel slag, Liu et al. [12] studied the synthesis of hydrotalcite-type mixed oxide catalysts from waste steel slag for transesterification. Casiello et al. [13] verified the feasibility of steel slag as a catalyst for synthesizing fames from soybean oil. Elsewhere, Kang et al. [14] prepared a novel CeO_2-loaded porous alkali-activated steel slag-based photocatalyst used for photocatalytic water-splitting for hydrogen production. Lin et al. [15] used acid leaching–electrolyzation–calcination to treat steel slag in preparing catalytic H_2O_2 degradation of dye wastewater as a catalyst. The above studies fully demonstrated the effective use of steel slag in catalyst research.

The present study introduced steel slag as a carrier to combine the active components of CaO and CeO_2 to prepare a solid base catalyst. We investigated the effect of the mass ratio of $CaO-CeO_2$ to steel slag on the catalytic performance of $CaO-CeO_2$/slag catalysts. Various technologies, including N_2 adsorption–desorption, X-ray diffraction (XRD), X-ray photoelectron spectroscopy (XPS), field-emission scanning electron microscopy coupled with energy-dispersive spectroscopy (SEM-EDS), Fourier-transform infrared spectroscopy (FTIR), and CO_2-temperature-programmed desorption (TPD) clarified the physical and chemical properties of the catalysts. Reaction parameters (such as catalyst dosage, methanol-to-oil molar ratio, reaction temperature, and duration) were optimized for catalytic transesterification to achieve the highest biodiesel yield. Finally, the acid resistance and reusability of the $CaO-CeO_2$/slag catalyst were conducted to evaluate the catalyst's feasibility for biodiesel production.

2. Materials and Methods

2.1. Materials

The analytical grade reagents of cerium(III) nitrate hexahydrate ($Ce(NO_3)_3 \cdot 6H_2O$) (>99.0% purity), calcium(II) nitrate tetrahydrate ($Ca(NO_3)_2 \cdot 4H_2O$) (>98.5% purity), anhydrous methanol (>99.5% purity), anhydrous ethanol (>99.7% purity) were purchased from Sinopharm Chemical Reagent Co. Ltd. (Shanghai, China). The chromatographic grade reagents n-hexane (>99.9% purity) and methyl salicylate (>99.5% purity) were supplied by Aladdin Reagent Co., Ltd. (Shanghai, China).

Steel slag was collected from a steel mill at Yanshan City located in Hebei Province, China. The chemical composition of the original steel slag (Table 1) was analyzed by X-ray fluorescence (XRF). The palm oil was procured from Guangdong Province, China, which was native to Malaysia. Its compositions were quantitatively analyzed with myristic acid (C14:0, 1.2%), palmitic acid (C16:0, 39.51%), lauric acid (C17:0, 0.27%), stearic acid (C18:0, 7.8%), arachidic acid (C20:0, 0.33%), oleic acid (C18:1, 39.71%), and linoleic acid (C18:2, 11.13%). Its acid value (*AV*) and saponification value (*SV*), seen in Table 2, were determined through GB 5009.229-2016 and GB/T 5534-2008 criteria, respectively. Besides, the palm oil's average relative molecular weight (*M*, g/mol) can be estimated using Equation (1).

$$M = \frac{1000 \times 3 \times 56.1}{SV - AV} \tag{1}$$

Table 1. Chemical composition of the original steel slag/%.

CaO	SiO$_2$	Al$_2$O$_3$	MgO	SO$_3$	TiO$_2$	P$_2$O$_5$	Fe$_2$O$_3$	MnO	K$_2$O	Na$_2$O	Other
41.36	31.09	13.34	8.85	2.30	1.11	0.40	0.38	0.36	0.34	0.34	0.13

Table 2. Basic indices of the palm oil used in the current study.

Reagent	*AV* mgKOH/g	*SV* mgKOH/g	*M* g/mol
Palm oil	0.42	182.60	923.98

2.2. Catalyst Synthesis and Evaluation of Activity

The steel slag was ground into a fine powder using a pulverizer, and sieved using a 125 μm standard sieve. $Ca(NO_3)_2 \cdot 4H_2O$ and $Ce(NO_3)_3 \cdot 6H_2O$, which provided the active site components, and the steel slag as their carriers were successively dissolved in deionized water in a particular proportion and impregnated at room temperature for 120 min using a magnetic agitator. After soaking, it was dried in a water bath at 90 °C. Afterward, the dried catalyst precursors were placed in a muffle furnace and heated to 800 °C for 180 min at a

5 °C/min heating rate to obtain the Ca-Ce/slag-x catalyst. Here, x represents the mass ratio of CaO to steel slag, calculated from the mass of Ca(NO$_3$)$_2$·4H$_2$O. During the experiment, x was 0.4, 0.6, 0.8, and 1.0. Here, the mass ratio of CaO to CeO$_2$ was fixed at 1:1.

The synthesized Ca-Ce/slag-x catalyst catalyzed the transesterification of palm oil and methanol, and the catalytic activity was evaluated. The entire experimental flow chart is exhibited in Figure 2. A total of 20 g of palm oil was first poured into a 250 mL three-necked flask. Then, the Ca-Ce/slag-x catalyst and methanol (in proportion to palm oil) were transferred to the flask. The reaction flask was placed in the microwave reactor. A thermocouple was inserted to measure the reaction temperature, while magnetic stirring ensured the homogeneity of the reactants. After the transesterification, the Ca-Ce/slag-x catalyst and liquid products were separated via centrifugation. The liquid products were transferred into a separatory funnel to realize the layering of crude biodiesel and glycerin. Then, the crude biodiesel was dried at 105 °C to remove the unreacted methanol before gas chromatographic (GC) analysis.

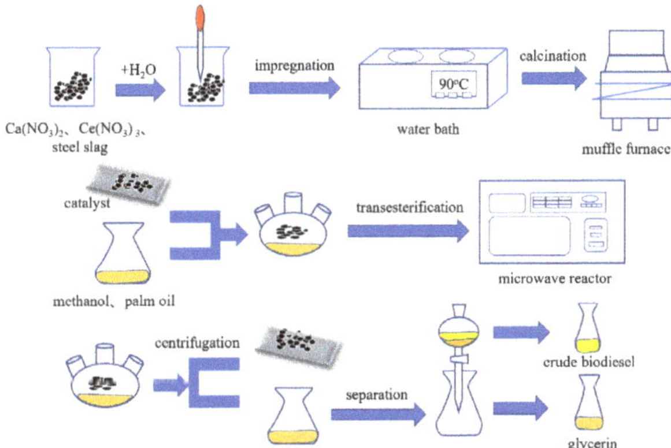

Figure 2. The experimental flow chart of catalyst synthesis and transesterification.

The FAMEs were assessed using an 8890 gas chromatograph (Agilent, California, USA) equipped with a flame ionization detector (FID) and an HP-5 (30 m × 320 μm × 0.25 μm) column and determined by internal standardization. n-hexane was the solvent used to prepare the GC sample, while methyl salicylate served as the internal standard. The injector and FID temperatures were 250 and 300 °C, respectively. The programmed temperature of the column oven was set as follows: start at 140 °C (keep for 4 min) and ramp at 10 °C/min to 260 °C (keep for 12 min). High-purity nitrogen was the carrier gas with a split ratio of 25:1. FAMEs' yield of the products was analyzed according to the reference method with methyl salicylate, and calculated using the following Equation (2) [16]:

$$\text{FAMEs' yield}(\%) = \frac{\Sigma f_{ester} A_{ester}}{A_{internal}} \times \frac{m_{internal}}{m_{sample}} \times 100\% \tag{2}$$

where A_{ester} is the peak area of FAME, $A_{internal}$ is the peak area of internal standard, $m_{internal}$ is the mass of methyl salicylate, m_{sample} is the mass of the sample used, and f_{ester} is the correction factor of the FAME.

2.3. Catalyst Characterization

The XPS analyses were carried out on a Thermo escalab 250Xi (Thermo Fisher, Waltham, MA, USA) with Al Kα radiation (hν = 1486.6 eV). The test results were corrected with C1s as 284.8 eV. Further, XRD spectra were obtained using a SmartLab apparatus (Rigaku, Tokyo, Japan) to investigate the crystalline phases in the 2θ range from 5° to 90° with a step size of 0.02° and at a scanning speed of 5°/min. The tube voltage and electricity current were 40 kV and 30 mA, respectively. Additionally, the average crystalline size of the catalyst was evaluated using the Debye–Scherrer equation, expressed as $D = K\gamma/\beta\cos\theta$, where K is the Scherrer constant, γ is the wavelength of the X-ray, β is the half-peak width of the diffraction peak of the measured sample, and θ is the Bragg angle. The catalyst's morphology was inspected by SUPRA™ 55 FE-SEM (Zeiss, Oberkochen, Germany) with an accelerating voltage of 5 kV. Additionally, the elemental distribution of the catalyst was synchronously detected by the EDS instrument. Here, the sample was sprayed with Au to enhance its electrical conductivity. In addition, the textural properties of the catalyst were determined by N_2 adsorption and desorption using ASAP 2460 surface area and a porosity analyzer (Micromeritics, Norcross, Georgia, USA) at -196 °C. Before the adsorption test, the sample was degassed at 100 °C for 3 h under a vacuum. The Brunauer–Emmer–Teller (BET) absorption curve and the Barrett–Joyner–Halenda (BJH) model determined the catalyst's surface area and pore size distribution. The FTIR spectroscopy evaluated the functional groups of the catalyst on an iS50/6700 spectrometer (Thermo Fisher Scientific, Waltham, MA, USA) with a wavenumber of 400 to 4000 cm^{-1}. The basic strength and basicity of the catalysts were quantitated by CO_2-TPD, using a TP-5080 (Xianquan, Tianjin, China) chemical adsorption instrument. Briefly, helium gas (pre-treated) first purged the sample at 300 °C (ramping rate of 10 °C/min) for 60 min before lowering it to 50 °C. Then, the carrier/reference gas was switched to CO_2 and adsorbed at a flow rate of 30 mL/min for 30 min before the carrier gas was switched back to helium at 50 °C for 60 min. After the baseline was stabilized, the temperature was raised to 900 °C at 10 °C/min. Ten signals were recorded every 1 s.

3. Results

3.1. Optimization of the Ca-Ce/Slag Catalyst Synthesis

We studied the effect of the Ca-Ce-to-slag mass ratio on the catalytic capability of the Ca-Ce/slag-*x* in the transesterification of palm oil and methanol under the following conditions: catalyst dosage = 7 wt.%, methanol-to-oil molar ratio = 15, reaction temperature = 65 °C, and reaction duration = 120 min. The results are illustrated in Figure 3. No FAMEs' yield was obtained when steel slag was used as catalyst. A 36.7% FAMEs' yield was only attained when the Ca-Ce-to-slag mass ratio was 0.4, probably due to insufficient active component content on the catalyst surface, leading to poor catalytic activity. When the mass ratio increased from 0.6 to 0.8, the FAMEs' yield gradually optimized (from 55.0%) to 85.2%, attributed to available basic sites for transesterification. However, the yield reduced significantly to 76.2% when the mass ratio was >0.8. This occurrence may be due to excess CaO and CeO_2 agglomeration and the tethering of active sites onto clusters. Thus, the Ca-Ce/slag-1.0 catalytic performance was severely hindered [17]. Borah et al. [18] also observed excess loading of Zn on CaO, which decreased the conversion because of the structural distortion of the parent catalyst. Thus, the 0.8 mass ratio of Ca-Ce-to-slag was selected as optimum for the catalyst synthesis.

3.2. Catalyst Characterization

3.2.1. XRD Analysis

The XRD patterns of slag and Ca-Ce/slag-*x* catalysts are shown in Figure 4. The slag shows a considerably broad peak at 2θ of 20–40°, ascribed to aromatic carbon sheets [19], indicating the presence of uncombined carbon. For the Ca-Ce/slag-*x* catalysts, the peaks at $2\theta = 32.18°, 37.34°, 53.84°, 64.12°, 67.34°$, and $79.62°$ are attributable to the crystal faces of (111), (200), (220), (311), (222), and (400) for CaO, whereas the peaks at $2\theta = 28.52°, 33.08°, 47.48°, 56.30°, 59.06°, 69.36°, 76.68°, 79.00°$, and $88.42°$ are assigned to the crystal faces of (111), (200), (220), (311), (222), (400), (331), (420), and (422) for CeO_2 [20]. We observed that the diffraction peak intensity of CaO and CeO_2 increased correspondingly with the Ca and Ce loading. Moreover, using the Debye–Scherrer equation, the average crystalline sizes of CaO and CeO_2 in the catalyst were calculated as 55.63, 63.41, 69.89, and 14.01, 14.98, 16.47 nm for the Ca-Ce/slag-0.4 catalyst, Ca-Ce/slag-0.6, and Ca-Ce/slag-1.0 catalysts, respectively. The small average crystalline size enhances the specific surface area and pore volume of the catalyst, thus guaranteeing the catalytic activity. For the Ca-Ce/slag-1.0 catalyst, numerous accumulations, and agglomerations of CaO and CeO$_2$ resulted in a large average crystalline size, explaining the excessive addition of active Ca and Ce components that lowered the catalyst's activity (Figure 3).

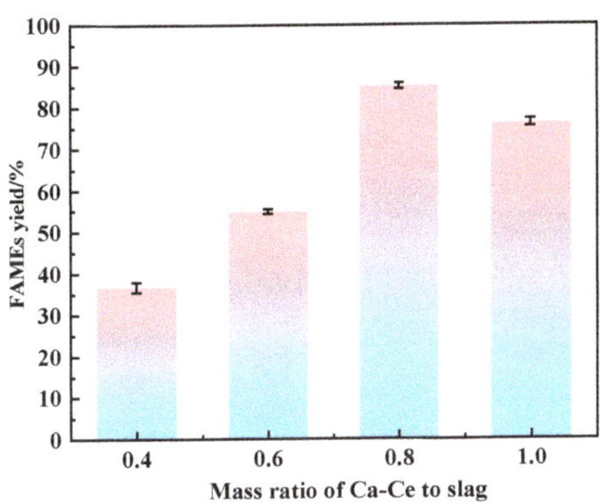

Figure 3. Effect of Ca-Ce-to-slag mass ratio.

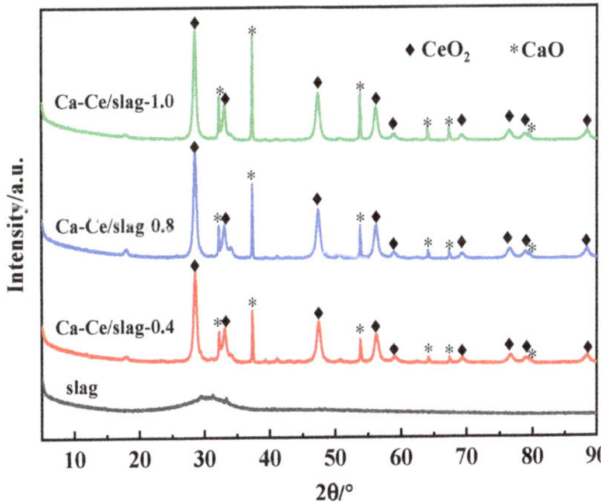

Figure 4. XRD patterns of the slag and Ca-Ce/slag-*x* catalysts.

3.2.2. N$_2$ Adsorption–Desorption Analysis

The microstructural properties of the slag and Ca-Ce/slag-*x* catalysts are listed in Table 3. The BET surface area of the slag, Ca-Ce/slag-0.4, and Ca-Ce/slag-0.8 catalysts are 0.27, 1.29, and 3.66 m^2/g, respectively. These values indicate that CaO and CeO$_2$ determine the BET surface area of the catalysts within a specific numerical range. Moreover, the pore volume for the slag, Ca-Ce/slag-0.4, and Ca-Ce/slag-0.8 catalysts are 0.00049, 0.022, and 0.043 cm^3/g, respectively. In general, the microstructural properties of Ca-Ce/slag-*x* are superior to those of the steel slag carrier. However, those of Ca-Ce/slag-0.8 are better than Ca-Ce/slag-0.4 catalyst, especially the average pore diameter (39.27 nm for the former; 22.19 nm for the latter). The molecular size of typical triglycerides is 5 nm [21]; hence, larger average pore sizes are more conducive to their transportation, providing better accessibility to the catalysts' active sites and favoring transesterification. The N$_2$ adsorption–desorption curve and pore size distribution plot of the Ca-Ce/slag-0.8 catalyst are depicted in Figure 5.

The catalyst exhibits a type V isotherm, having H3-type hysteresis loops, confirming the slit formed by the accumulation of flake particles [22]. The pore size distribution of the Ca-Ce/slag-0.8 extends from 10 to 100 nm, showing the predominance of mesopores (2–50 nm) and macrospores (>50 nm). As previously established, these pore size distributions could be crucial to the catalytic biodiesel production process, conducive to mass transfer and diffusion [23].

Table 3. Microstructural properties of the slag and Ca-Ce/slag-*x* catalysts.

Samples	BET Surface Area (m^2/g)	Pore Volume (cm^3/g)	Average Pore Diameter (nm)
slag	0.27	0.00049	-
Ca-Ce/slag-0.4	1.29	0.022	22.19
Ca-Ce/slag-0.8	3.66	0.043	39.27

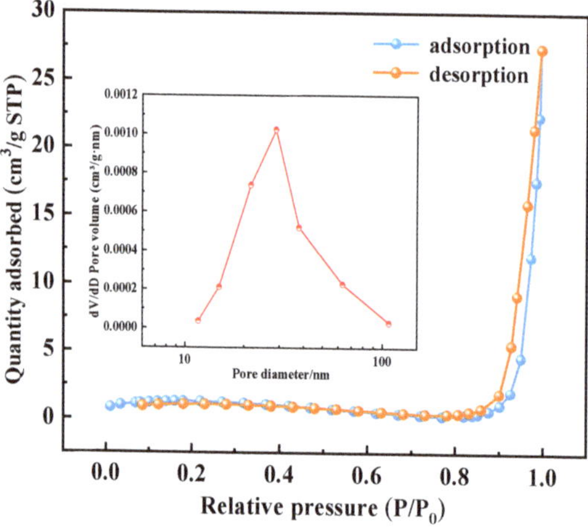

Figure 5. N$_2$ adsorption–desorption curve of the Ca-Ce/slag-0.8 catalyst.

3.2.3. SEM-EDS Analysis

The SEM characterized the morphologies of the Ca-Ce/slag-0.8 catalyst. As illustrated in Figure 6a, the catalyst appears as uneven and irregular strips or blocks. These strip or block structures pile on each other, with a few accumulated slit pores in between. Figure 6b depicts the images of the catalyst's local surface magnified 30,000×. We observed that the catalyst's surface contains some fine particles with pores. These pores facilitate efficient contact between the reactants and the active sites of the Ca-Ce/slag-0.8 catalyst.

Additionally, the EDS spectrum (Figure 6c) reflects the elemental composition and proportion of the Ca-Ce/slag-0.8 catalyst. Due to the nature of metal oxide on the steel slag carrier of the catalyst, oxygen exhibited the highest atomic percentage in the catalyst. The weight ratio of Ca and Ce was 34.82% and 34.47%, respectively, consistent with the parameter setting in the catalyst preparation, thereby confirming the homogeneity of the catalyst. Meanwhile, we identified minute amounts of Mg, Al, Si, and Cl in the catalyst, arising from the steel slag carrier.

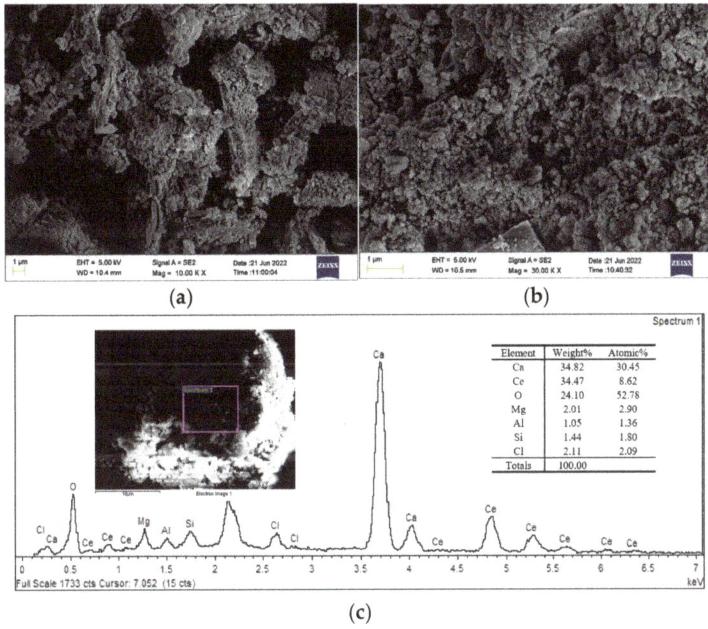

Figure 6. The SEM-EDS images of the Ca-Ce/slag-0.8 catalyst. (**a**) SEM image with a 10,000× magnification, (**b**) SEM image with a 30,000× magnification, (**c**) EDS element spectrum.

3.2.4. FTIR Analysis

The change in the chemical functional groups between slag and Ca-Ce/slag-x catalysts was characterized by FTIR (Figure 7). For steel slag carrier, the bands at 1435 and 875 cm^{-1} were assigned to the in-plane bending and out-of-plane bending of the carbonate groups, respectively [24]. For Ca-Ce/slag-x catalysts, a new band at 3641 cm^{-1} belonging to the stretching of the -OH groups can be observed in the FTIR spectra due to the water-absorbing characteristics of the CaO and CeO$_2$ components. Significantly, the –OH groups of the Ca-Ce/slag-0.8 catalyst showed a sharper band, indicating more water-absorbing potential. In addition, the bands at 800–400 cm^{-1} belonged to Ca–O and Ce–O vibration bonds.

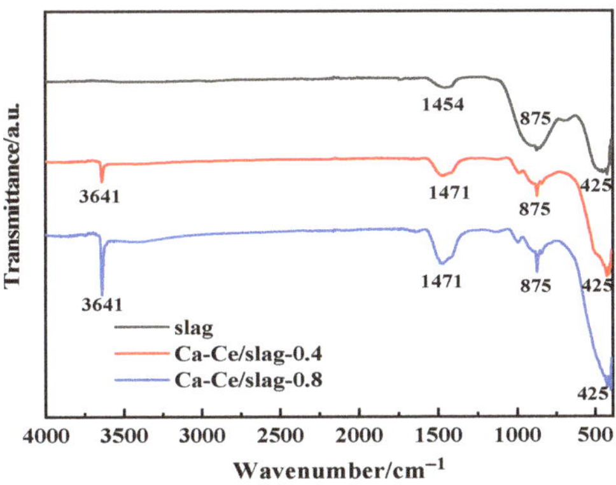

Figure 7. FTIR spectra of the slag and Ca-Ce/slag-x catalysts.

3.2.5. XPS Analysis

The XPS was deduced from a chemical state of the metal species of the Ca-Ce/slag-0.8 catalyst (Figure 8). Figure 8a shows the XPS survey with photoelectron peaks of Ce 3d, O 1s, Ca 2p, C 1s, and Si 2p at the binding energies of 930.1, 545.1, 360.1, 298.1, and 110.1 eV, respectively, reflecting the primary elements of the catalyst. Further, high-resolution XPS spectra of O 1s (Figure 8b) can be deconvoluted into two peaks. One peak centered at 531.4 eV corresponded to the oxygen species of surface hydroxyl or carbonate groups; the other peak centered at 528.7 eV was ascribed to the O^{2-} in the lattice of metal oxides [25]. Moreover, the high-resolution XPS spectra in Figure 8c confirmed the presence of Ca, evidenced by the bands assigned to Ca 2p 1/2 at 350.5 eV and Ca 2p 3/2 at 347.0 eV. Compared to the CaO peak (Ca 2p 1/2 at 349.9 eV and Ca 2p3/2 at 346.6 eV) reported in the literature, the binding energies of Ca 2p in the Ca-Ce/slag-0.8 catalyst shifted towards a higher values, suggesting the interaction between Ca and other elements (such as Ce), enhancing the stability of active components [26]. The Ce 3d spectrum of the catalyst is shown in Figure 8d. The peaks at 916.2, 907.1, 900.4, 897.8, 888.7, and 881.9 eV are typically assigned to the Ce^{4+} species. In contrast, other peaks at 902.6 and 883.9 eV were attributed to the Ce^{3+} species [27]. More Ce^{3+} species could generate more oxygen vacancies because of the unsaturated chemical bonds and charge imbalance [28], further increasing the active sites and, eventually, the activity on the catalyst. Hence, the concentration ratio of $Ce^{3+}/(Ce^{3+} + Ce^{4+})$ of the Ca-Ce/slag-0.8 catalyst was estimated as 27.9%, confirming its catalytic performance.

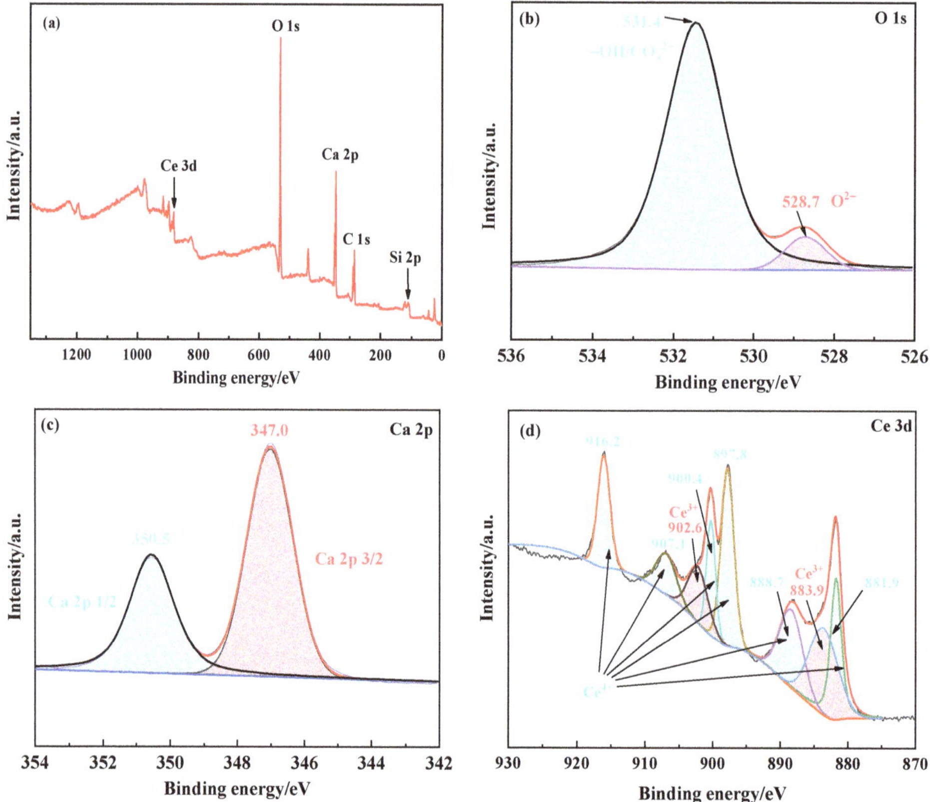

Figure 8. XPS spectra of the Ca-Ce/slag-0.8 catalyst. (**a**) XPS survey; (**b**) O 1s; (**c**) Ca 2p; (**d**) Ce 3d.

3.2.6. CO_2-TPD Analysis

The CO_2-TPD measurement studied the surface basicity of the slag and Ca-Ce/slag-*x* catalysts and the results are exhibited in Figure 9 and Table 4. Typically, surface basicity is a crucial index in determining the catalytic activity of solid base catalysts. The TPD profile of the steel slag exhibited three desorption peaks, with a basicity of 0.009, 0.110, and 0.069 mmol/g during the temperature intervals of 50–530, 530–745, and 745–900 °C, respectively. The total basicity of the slag added up to 0.188 mmol/g. It was challenging to support catalytic transesterification based on this low value. The Ca-Ce/slag-0.4 and Ca-Ce/slag-0.8 catalysts exhibited no desorption peak at 50–340 °C, demonstrating the negligible weak basic sites. For the Ca-Ce/slag-0.4 catalyst, two distinct desorption peaks were revealed at temperature intervals of 350–540 and 540–750 °C, correlating to the moderate and strong base sites, with respective basicities of 0.235 and 0.193 mmol/g. For the Ca-Ce/slag-0.8 catalyst, the corresponding temperature intervals of the moderate and strong base sites were shown at 340–560 and 560–800 °C, with corresponding basicities of 0.758 and 0.509 mmol/g. With the increased additive of active components Ca and Ce, the temperature range of desorption peaks moved to the high-temperature region, and the basicity increased correspondingly. The total basicity of the Ca-Ce/slag-0.4 and Ca-Ce/slag-0.8 catalysts was 0.445 and 1.289 mmol/g, respectively. In summary, the adequate basicity of the Ca-Ce/slag-0.8 catalyst can guarantee a high catalytic performance in transesterification. This finding compares with the HNTS-La /Ca catalyst, whose total basicity is 0.809 mmol/g [29].

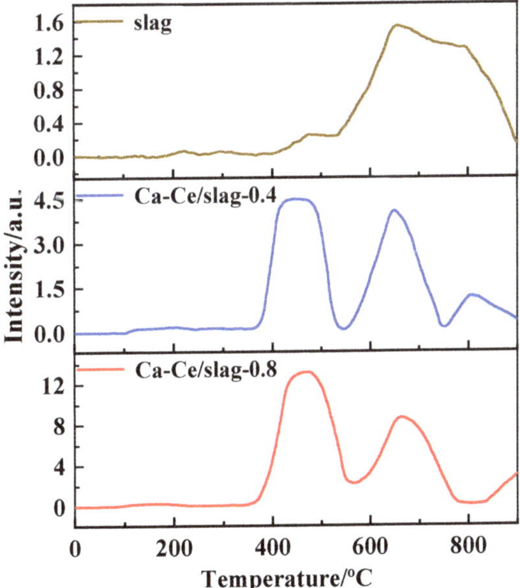

Figure 9. CO_2-TPD profiles of the slag and Ca-Ce/slag-*x* catalysts.

Table 4. CO_2-TPD data for the slag and Ca-Ce/slag-*x* catalysts.

Samples	Temperature Interval/°C	Peak Temperature/°C	Basicity /(mmol/g)	Total Basicity/(mmol/g)
Slag	50–530	476	0.009	
	530–745	656	0.110	0.188
	745–900	745	0.069	
Ca-Ce/slag-0.4	50–350	206	0.017	
	350–540	446	0.235	0.445
	540–750	649	0.193	
Ca-Ce/slag-0.8	50–340	168	0.022	
	340–560	469	0.758	1.289
	560–800	664	0.509	

3.3. Catalytic Activity

3.3.1. Transesterification Parameters Optimization

Using the Ca-Ce/slag-0.8 catalyst, the transesterification of palm oil with methanol to obtain the corresponding FAMEs' yields was investigated by studying the reaction parameters: catalyst dosage, methanol-to-palm-oil molar ratio, reaction temperature, and reaction duration. The optimization results are shown in Figure 10.

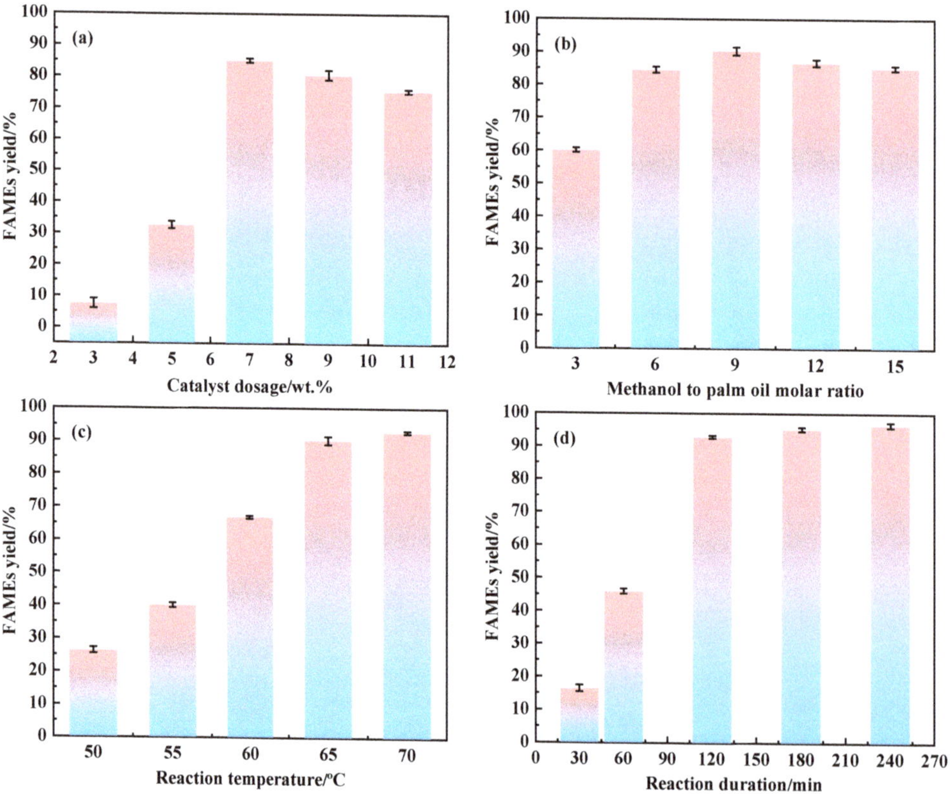

Figure 10. Effects of transesterification parameters on FAMEs yield: (**a**) Catalyst dosage; (**b**) Methanol-to-palm-oil molar ratio; (**c**) Reaction temperature; (**d**) Reaction duration.

The optimization study of the catalyst dosage was performed using 3, 5, 7, 9, and 11 wt.% dosages for 120 min at 65 °C with a methanol-to-palm-oil molar ratio of 15. As depicted in Figure 10a, the catalyst dosage dramatically affects the FAMEs yield. When the catalyst dosage was 3 wt.%, the concentration in the reaction system was sparse, making the reaction incomplete, achieving a mere 7.5% yield. The number of O^{2-} anion sites is proportional to the catalyst dosage. Higher dosages can provide more sites for adsorbing H^+ from methanol to form active centers [30], thus improving the contact of the reactants with the active centers, ultimately improving the transesterification efficiency. So, the FAMEs' yield was optimized (from 32.7%) to 85.2% as the catalyst dosage ranged from 5 to 7 wt.%. However, the yield started to decrease with further catalyst dosing. Figure 10a depicts that the FAMEs' yield depreciated to 75.5% when the catalytic dosage was further increased to 11 wt.%. This result could be attributed to excessive catalyst dosage causing saponification, affecting the main reaction's transesterification, thereby lowering the reaction yield [31]. In summary, a 7 wt.% catalyst dosage was selected as optimum.

The methanol-to-palm-oil molar ratio optimization was carried out under 3, 6, 9, 12, and 15 ratios for 120 min at 65 °C with 7 wt.% catalyst dosage. The methanol-to-oil molar ratio is vital in transesterification, where the stoichiometric ratio of transesterification between palm oil and methanol is at least 1:3 [16]. Higher positive FAMEs' yield can be obtained by increasing methanol dosage due to the reversible nature of transesterification. As shown in Figure 10b, when the molar ratio of methanol to palm oil is on a 3-to-9 scale, the increase in the molar ratio was beneficial in improving the FAMEs yield, where the FAMEs yield increased from 60.1% to 90.3%. Increasing the methanol content is conducive to reducing the viscosity of the reaction system, thereby reducing the mass transfer resistance between reactants, and further improving the conversion efficiency of reactants. However, a depressing trend of FAMEs yield was observed when the molar ratio exceeded 9. Excessive methanol lowers the contact between the catalyst and palm oil, leading to a reduced yield.

Furthermore, the reaction temperature was optimized under 50, 55, 60, 65, and 70 °C at 120 min duration using a 7 wt.% catalyst dosage and a methanol-to-palm-oil molar ratio of 9. When the temperature was held at 50 °C (Figure 10c), the reaction between methanol and palm oil in the presence of the catalyst resulted in the lowest FAMEs' yield because the temperature was lower than that required for conversion reaction into methyl ester molecules [32]. In the transesterification of macromolecular triglycerides, sufficient temperature is needed to provide enough energy to overcome the threshold. On the other hand, increasing the temperature can reduce the reactant viscosity and improve the miscibility, thus decreasing the mass-transfer resistance [33]. The FAMEs' yield increased substantially from 26.3% to 90.3% by raising the reaction temperature from 50 to 65 °C. Further increasing the reaction temperature to 70 °C improved the FAMEs yield to 92.9%. At 70 °C, the molecular movement rate of the reaction system was fast enough to enhance the effective collision chance between the reactants and the catalyst, thus obtaining a higher FAMEs' yield.

To optimize the reaction duration within 30–240 min, we set other conditions at their optimized values, i.e., 7 wt.% catalyst dosage, nine methanol-to-palm-oil molar ratio, and 70 °C. The transesterification rate is slow, especially the solid catalysts. Palm oil and methanol are immiscible, and the whole reaction system is a catalyst–methanol–palm oil three-phase state, cumbersome to homogenize. At the same time, the FAMEs' yield of transesterification was low in the initial reaction stage due to the palm oil and methanol needing to be mixed and diffused for a long duration to adsorb to the active center on the catalyst. As evidenced in Figure 10d, a 16.4% FAMEs' yield was attained after the initial 30 min of transesterification. It increased significantly to 92.9% when the reaction reached the 120 min mark. When the reaction duration was extended to 180 min, the FAMEs' yield slowly enhanced to 95.3%. After that, a puny increase of 1.3% FAMEs' yield was achieved upon increasing the duration to 240 min. We opine that the transesterification reached equilibrium when the reaction duration exceeded 180 min.

Table 5 presents the catalytic performance in the transesterification compared with some recent reports on solid base catalysts. These studies considered the production conversion (or yield) and transesterification operating parameters. A >90% conversion or biodiesel yield was obtained from various types of solid base catalysts under different transesterification conditions. Based on the scientific literature, our results revealed that the Ca-Ce/slag-0.8 catalyst has satisfactory catalytic performance and relatively moderate transesterification conditions compared to other heterogeneous base catalysts. Moreover, the Ca-Ce/slag-0.8 catalyst optimizes the steel slag waste, saving preparation costs, and making it highly applicable industrially.

Table 5. Comparison of performance efficiencies of various base catalysts.

Catalyst	Oil Feedstock	Transesterification Parameters				Con. or Yield [4]/%	Ref.
		C. Dosage [1] /wt.%	Molar Ratio [2]	Temp [3] /°C	Time /min		
CaO/CeO_2	Waste seed oil	4	9	70	90	90.14	[34]
$ZnO/BiFeO_3$	Canola oil	4	15	65	360	95.43	[35]
$K_2CO_3/\gamma\text{-}Al_2O_3$	Sunflower oil	5	12	80	240	99.3	[36]
Ca-Mg-Al	Sunflower oil	2.5	15	60	360	95	[30]
$NaOH/Chitosan\text{-}Fe_3O_4$	Waste cooking oil	0.5	6	25	270	92	[37]
$biochar/CaO\text{-}K_2CO_3$	Waste edible oil	4	18	65	200	98.83	[23]
MgO/CaO nanorods	Castor oil	6	15	70	70	96.2	[38]
CuO/ZnO	Waste cooking oil	5	9	65	120	93.5	[39]
Acai seed ash	Soybean oil	12	18	100	60	98.5	[40]
Ca-Ce/slag-0.8	Palm oil	7	9	70	180	95.3	This study

[1] C. dosage—catalyst dosage; [2] Molar ratio—alcohol-to-oil molar ratio; [3] Temp—temperature; [4] Con. or yield—conversion or yield.

3.3.2. Effect of FFAs Content

To verify the acid resistance effect of the Ca-Ce/slag-0.8 catalyst, we added 2, 5, and 8 wt.% of oleic acid to the palm oil under optimal conditions: 7 wt.% catalyst dosage, 9 methanol-to-oil molar ratio, and reaction at 70 °C for 180 min. With 2 and 5 wt.% oleic acid added to the palm oil, the FAMEs' yield decreased from 95.3% to 84.8% and 78.9%, respectively (Figure 11). However, a 63.1% FAMEs' yield was still achieved even when up to 8 wt.% oleic acid was added. We found that the Ca-Ce/slag-0.8 catalyst has specific acid resistance in transesterification, but the oil's free fatty acids' content should not be higher than 5 wt.%.

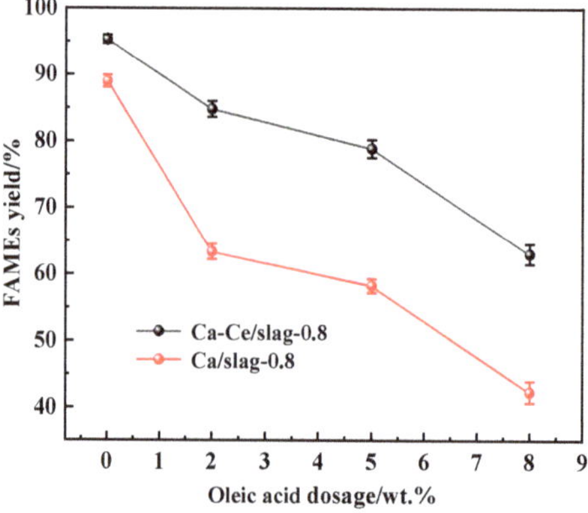

Figure 11. Effect of oleic acid dosage on the FAMEs yield.

The Ca/slag-0.8 catalyst was synthesized under the same preparation conditions for comparison, and its acid resistance was studied (Figure 11). With the addition of oleic acid, the FAMEs' yield of Ca/slag-0.8 catalyst decreased faster than that of the Ca-Ce/slag-0.8 catalyst. Only 42.4% of the FAMEs' yield could be obtained when 8 wt.% of oleic acid was added to the transesterification reaction. This finding indicates that the acid resistance of Ca-Ce/slag-0.8 was more enhanced than that of Ca/slag-0.8.

3.4. Catalyst Reusability

The reusability of solid catalysts is an attractive property relative to homogeneous catalysts. Transesterification was conducted at optimized conditions for three recycling tests. During each cycle, the Ca-Ce/slag-0.8 catalyst was reused by processing in two different ways: (1) Separated from the transesterification mixture by filtration and dried at 105 °C for 5 h; (2) Separated from the transesterification mixture by centrifugation and washed thrice with n-hexane to remove polar and non-polar adsorbed compounds before drying at 105 °C for 5 h. As depicted in Figure 12, for the direct drying method, the Ca-Ce/slag-0.8 catalyst exhibited good stability for the first two cycles, achieving 95.3% and 90.1% FAMEs' yields. After that, the yield depreciated with the recycling number. Only 82.1% of the FAMEs' yield was achieved in the third cycle. In contrast, the FAMEs' yield slowly decreased from 95.3% to 93.8%, and 91.2% after the three cycles, with the one washed with n-hexane. This observation indicated that the active site on the catalyst was covered by organic matter, which hindered the catalytic activity of the Ca-Ce/slag-0.8. Lee et al. [41] reported that the palm oil conversion was reduced to 70% after the third cycles catalyzed by waste obtuse horn shell-derived CaO catalyst. In summary, the Ca-Ce/slag-0.8 catalyst synthesized in the present study exhibited an appreciable FAMEs' yield.

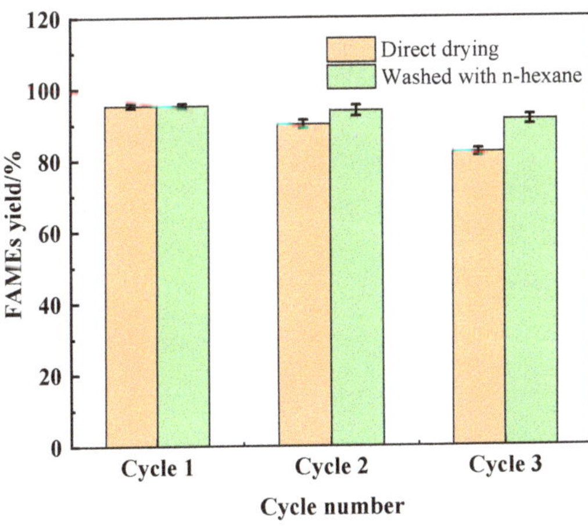

Figure 12. Effect of recycling of the Ca-Ce/slag-0.8 catalyst on the FAMEs' yield.

4. Discussion

This study synthesized a newly cost-effective base catalyst of Ca-Ce/slag to generate biodiesel from palm oil. The effect of the mass ratio of Ca-Ce to steel slag on the transesterification activity was investigated. Moreover, diversified characterization techniques were employed to analyze the catalyst properties (such as crystalline phases, microscopic pore structure, morphology, functional groups, and surface basicity). The results showed that the mass ratio of Ca-Ce to steel slag at 0.8 (i.e., Ca-Ce/slag-0.8 catalyst) exhibited excellent physical and chemical properties. According to the range of transesterification conditions selected in this study, the optimal transesterification conditions are as follows: methanol: oil molar ratio of 9, catalyst dosage of 7 wt.%, and reaction at 70 °C for 180 min, yielding 95.3% of FAMEs. Moreover, the Ca-Ce/slag-0.8 catalyst exhibited stable catalytic performance after three cycles with >90% FAMEs' yield. Consequently, this study provides a new view on the resource utilization of steel slag and an excellent prospect for developing low-cost solid biodiesel catalysts.

Author Contributions: Conceptualization, J.S. and H.Y.; methodology, G.Q.; software, G.Q. and P.Z.; validation, P.Z., J.S. and X.C.; formal analysis, X.L.; investigation, H.Y.; resources, J.S.; data curation, H.Y.; writing—original draft preparation, H.S.; writing—review and editing, J.S.; visualization, H.S.; supervision, X.C.; project administration, H.Y. and H.S; funding acquisition. All authors have read and agreed to the published version of the manuscript.

Funding: This research was funded by Open Project of Shandong Key Laboratory of Biomass Gasification Technology [BG-KFX-08], the National Natural Science Foundation of China [52206262], the Science, Education and Industry Training New Project of Qilu University of Technology [2022PX091], the Natural Science Foundation of Shandong Province, China [ZR2020QE210], "20 New Universities" Funding Project of Jinan [2021GXRC053] and Undergraduate Training Program for Innovation and Entrepreneurship Program of Shandong Province.

Data Availability Statement: The data that support the findings of this study are available within the article.

Conflicts of Interest: The authors declare no conflict of interest.

References

1. Woranuch, W.; Ngaosuwan, K.; Kiatkittipong, W.; Wongsawaeng, D.; Appamana, W.; Powell, J.; Rokhum, S.L.; Assabumrungrat, S. Fine-tuned fabrication parameters of CaO catalyst pellets for transesterification of palm oil to biodiesel. *Fuel* **2022**, *323*, 124356. [CrossRef]
2. Li, H.; Wang, Y.; Ma, X.; Guo, M.; Li, Y.; Li, G.; Cui, P.; Zhou, S.; Yu, M. Synthesis of CaO/ZrO$_2$ based catalyst by using UiO–66(Zr) and calcium acetate for biodiesel production. *Renew. Energy* **2022**, *185*, 970–977. [CrossRef]
3. Zul, N.A.; Ganesan, S.; Hamidon, T.S.; Oh, W.D.; Hussin, M.H. A review on the utilization of calcium oxide as a base catalyst in biodiesel production. *J. Environ. Chem. Eng.* **2021**, *9*, 105741. [CrossRef]
4. Yaşar, F. Biodiesel production via waste eggshell as a low-cost heterogeneous catalyst: Its effects on some critical fuel properties and comparison with CaO. *Fuel* **2019**, *255*, 115821–115828. [CrossRef]
5. Teo, S.H.; Rashid, U.; Taufiq-Yap, Y.H. Heterogeneous catalysis of transesterification of jatropha curcas oil over calcium-cerium bimetallic oxide catalyst. *RSC. Adv.* **2014**, *4*, 48836–48847. [CrossRef]
6. Pandit, P.R.; Fulekar, M.H. Biodiesel production from microalgal biomass using CaO catalyst synthesized from natural waste material. *Renew. Energy* **2019**, *136*, 837–845. [CrossRef]
7. Dehghani, S.; Haghighi, M.; Vardast, N. Structural/texture evolution of CaO/MCM nanocatalyst by doping various amounts of cerium for active and stable catalyst: Biodiesel production from waste vegetable cooking oil. *Int. J. Energy. Res.* **2019**, *43*, 3779–3793. [CrossRef]
8. Zhang, N.; Xue, H.; Hu, R. The activity and stability of CeO$_2$@CaO catalysts for the production of biodiesel. *RSC. Adv.* **2018**, *8*, 32922. [CrossRef]
9. Shu, K.; Sasaki, K. Occurrence of steel converter slag and its high value-added conversion for environmental restoration in China: A review. *J. Clean. Prod.* **2022**, *373*, 133876. [CrossRef]
10. Gan, Y.; Li, C.; Ke, W.; Deng, Q.; Yu, T. Study on pavement performance of steel slag asphalt mixture based on surface treatment. *Case. Stud. Constr. Mat.* **2022**, *16*, e1131. [CrossRef]
11. Song, B.; Wang, Z.; Li, J.; Ma, Y. Preparation and electrocatalytic properties of kaolin/steel slag particle electrodes. *Catal. Commun.* **2021**, *148*, 106177. [CrossRef]
12. Liu, G.; Yang, J.; Xu, X. Synthesis of hydrotalcite-type mixed oxide catalysts from waste steel slag for transesterification of glycerol and dimethyl carbonate. *Sci. Rep-UK* **2020**, *10*, 10273. [CrossRef]
13. Casiello, M.; Losito, O.; Aloia, A.; Caputo, D.; Fusco, C.; Attrotto, R.; Monopoli, A.; Nacci, A.; D'Accolti, L. Steel Slag as New Catalyst for the Synthesis of Fames from Soybean Oil. *Catalyst* **2021**, *11*, 619. [CrossRef]
14. Kang, L.; Zhang, Y.J.; Zhang, L.; Zhang, K. Preparation, characterization and photocatalytic activity of novel CeO$_2$ loaded porous alkali-activated steel slag-based binding material. *Int. J. Hydrogen. Energy.* **2017**, *42*, 17341–17349. [CrossRef]
15. Lin, S.; Zhang, T.; Cao, X.; Liu, X. Recovery of converter steel slag to prepare catalytic H$_2$O$_2$ degradation of dye wastewater as a catalyst. *J. Mater. Sci-Mater. El.* **2021**, *32*, 24889–24901. [CrossRef]
16. Gao, L.; Teng, G.; Xiao, G.; Wei, R. Biodiesel from palm oil via loading KF/Ca–Al hydrotalcite catalyst. *Biomass Bioenergy* **2010**, *34*, 1283–1288. [CrossRef]
17. Kesserwan, F.; Ahmad, M.N.; Khalil, M.; El-Rassy, H. Hybrid CaO/Al$_2$O$_3$ aerogel as heterogeneous catalyst for biodiesel production. *Chem. Eng. J.* **2020**, *385*, 123834. [CrossRef]
18. Borah, M.J.; Das, A.; Das, V.; Bhuyan, N.; Deka, D. Transesterification of waste cooking oil for biodiesel production catalyzed by Zn substituted waste egg shell derived CaO nanocatalyst. *Fuel* **2019**, *242*, 345–354. [CrossRef]
19. Ning, Y.; Niu, S. Preparation and catalytic performance in esterification of a bamboo-based heterogeneous acid catalyst with microwave assistance. *Energy Convers. Manag.* **2017**, *153*, 446–454. [CrossRef]

20. Qu, T.; Niu, S.; Zhang, X.; Han, K.; Lu, C. Preparation of calcium modified Zn-Ce/Al_2O_3 heterogeneous catalyst for biodiesel production through transesterification of palm oil with methanol optimized by response surface methodology. *Fuel* **2021**, *284*, 118986. [CrossRef]

21. Yadav, M.; Sharma, Y.C. Process optimization and catalyst poisoning study of biodiesel production from kusum oil using potassium aluminum oxide as efficient and reusable heterogeneous catalyst. *J. Clean. Prod.* **2018**, *199*, 593–602. [CrossRef]

22. Zhang, Y.; Niu, S.; Kan, H.; Li, Y.; Lu, C. Synthesis of the SrO–CaO–Al_2O_3 trimetallic oxide catalyst for transesterification to produce biodiesel. Renew. *Energy* **2021**, *168*, 981–990.

23. Foroutan, R.; Mohammadi, R.; Razeghi, J.; Ramavandi, B. Biodiesel production from edible oils using algal biochar/CaO/K_2CO_3 as a heterogeneous and recyclable catalyst. *Renew. Energy* **2021**, *168*, 1207–1216. [CrossRef]

24. Murguía-Ortiz, D.; Cordova, I.; Manriquez, M.E.; Ortiz-Islas, E.; Cabrera-Sierra, R.; Contreras, J.L.; Alcantar-Vazquez, B.; Trejo-Rubio, M.; Vazquez-Rodriguez, J.T.; Castro, L.V. Na-CaO/MgO dolomites used as heterogeneous catalysts in canola oil transesterification for biodiesel production. *Mater. Lett.* **2021**, *291*, 129587. [CrossRef]

25. Niu, S.; Zhang, X.; Ning, Y.; Zhang, Y.; Qu, T.; Hu, X.; Gong, Z.; Lu, C. Dolomite incorporated with cerium to enhance the stability in catalyzing transesterification for biodiesel production. *Renew. Energy* **2020**, *154*, 107–116. [CrossRef]

26. Wang, J.; Yang, L.; Luo, W.; Yang, G.; Miao, C.; Fu, J.; Xing, S.; Fan, P.; Lv, P.; Wang, Z. Sustainable biodiesel production via transesterification by using recyclable $Ca_2MgSi_2O_7$ catalyst. *Fuel* **2017**, *196*, 306–313. [CrossRef]

27. Ma, S.; Tan, H.; Li, Y.; Wang, P.; Zhu, Y. Excellent low-temperature NH_3-SCR NO removal performance and enhanced H_2O resistance by Ce addition over the $Cu_{0.02}Fe_{0.2}CeyTi_{1-y}O_x$ (y = 0.1, 0.2, 0.3) catalysts. *Chemosphere* **2019**, *243*, 125309. [CrossRef]

28. Ma, Y.; Li, W.; Wang, H.; Chen, J.; Wen, J.; Xu, S.; Tian, X.; Gao, L.; Hou, Z.; Zhang, Q. Enhanced performance of iron-cerium NOx reduction catalysts by sulfuric acid treatment: The synergistic effect of surface acidity and redox capacity. *Appl. Catal. A-Gen.* **2021**, *621*, 118200. [CrossRef]

29. Lin, T.; Zhao, S.; Niu, S.; Lyu, Z.; Hu, X. Halloysite nanotube functionalized with La-Ca bimetallic oxides as novel transesterification catalyst for biodiesel production with molecular simulation. *Energy Convers. Manag.* **2020**, *220*, 113138. [CrossRef]

30. Dahdah, E.; Estephane, J.; Haydar, R.; Youssef, Y.; Khoury, B.E.; Gennequin, C.; Aboukais, A.; Abi-Aad, E.; Aouad, S. Biodiesel production from refined sunflower oil over Ca–Mg–Al catalysts: Effect of the composition and the thermal treatment. *Renew. Energy* **2020**, *146*, 1242–1248. [CrossRef]

31. Li, H.; Niu, S.; Lu, C.; Liu, M.; Huo, M. Transesterification catalyzed by industrial waste—Lime mud doped with potassium fluoride and the kinetic calculation. *Energy Convers. Manag.* **2014**, *86*, 1110–1117. [CrossRef]

32. Al-Muhtaseb, A.H.; Osman, A.I.; Jamil, F.; Mehta, N.; Al-Haj, L.; Coulon, F.; Al-Maawali, S.; Al Nabhani, A.; Kyaw, H.H.; Myint, M.T.Z.; et al. Integrating life cycle assessment and characterisation techniques: A case study of biodiesel production utilising waste Prunus Armeniaca seeds (PAS) and a novel catalyst. *J. Environ. Manag.* **2022**, *304*, 114319. [CrossRef]

33. Farooq, M.; Ramli, A.; Naeem, A. Biodiesel production from low FFA waste cooking oil using heterogeneous catalyst derived from chicken bones. *Renew. Energy* **2015**, *76*, 362–368. [CrossRef]

34. Al-Muhtaseb, A.H.; Osman, A.I.; Kumar, P.S.M.; Jamil, F.; Al-Haj, L.; Al Nabhani, A.; Kyaw, H.H.; Myint, M.T.Z.; Mehta, N.; Rooney, D.W. Circular economy approach of enhanced bifunctional catalytic system of CaO/CeO_2 for biodiesel production from waste loquat seed oil with life cycle assessment study. *Energy Convers. Manag.* **2021**, *236*, 114040. [CrossRef]

35. Salimi, Z.; Hosseini, S.A. Study and optimization of conditions of biodiesel production from edible oils using ZnO/$BiFeO_3$ nano magnetic catalyst. *Fuel* **2019**, *239*, 1204–1212. [CrossRef]

36. Junior, E.S.; Perez, V.H.; Reyero, I.; Serrano-Lotina, A.; Justo, O.R. Biodiesel production from heterogeneous catalysts based K_2CO_3 supported on extruded γ-Al_2O_3. *Fuel* **2019**, *241*, 311–318. [CrossRef]

37. Helmi, M.; Hemmati, A. Synthesis of magnetically solid base catalyst of NaOH/Chitosan-Fe_3O_4 for biodiesel production from waste cooking oil: Optimization, kinetics and thermodynamic studies. *Energy Convers. Manag.* **2021**, *248*, 114807. [CrossRef]

38. Abukhadra, M.R.; Mohamed, A.S.; El-Sherbeeny, A.M.; Soliman, A.T.A.; Elgawad, A.E.E.A. Sonication induced transesterification of castor oil into biodiesel in the presence of MgO/CaO nanorods as a novel basic catalyst: Characterization and optimization. *Chem. Eng. Process.* **2020**, *154*, 108024. [CrossRef]

39. Guo, M.; Jiang, W.; Ding, J.; Lu, J. Highly active and recyclable CuO/ZnO as photocatalyst for transesterification of waste cooking oil to biodiesel and the kinetics. *Fuel* **2022**, *315*, 123254. [CrossRef]

40. Mares, E.K.L.; Gonçalves, M.A.; Lourenço, E.K.; Luz, P.T.S.; Filho, G.N.R.; Zamian, J.R.; Conceição, L.R.V. Acai seed ash as a novel basic heterogeneous catalyst for biodiesel synthesis: Optimization of the biodiesel production process. *Fuel* **2021**, *299*, 120887. [CrossRef]

41. Lee, S.L.; Wong, Y.C.; Tan, Y.P.; Yew, S.Y. Transesterification of palm oil to biodiesel by using waste obtuse horn shell-derived CaO catalyst. *Energy Convers. Manag.* **2015**, *93*, 282–288. [CrossRef]

Article

Improving Biomethanol Synthesis via the Addition of Extra Hydrogen to Biohydrogen Using a Reverse Water–Gas Shift Reaction Compared with Direct Methanol Synthesis

Kuntima Krekkeitsakul [1], Rujira Jitrwung [2], Weerawat Patthaveekongka [3] and Teerasak Hudakorn [1,*]

[1] Department of Mechanical Engineering, Faculty of Engineering and Industrial Technology, Silpakorn University, Nakorn Pathom 73000, Thailand; kuntima22@gmail.com

[2] Expert Center of Innovative Clean Energy and Environment, Research and Development Group for Sustainable Development, Thailand Institute of Scientific and Technological Research (TISTR), Khlong Luang, Pathum Thani 12120, Thailand; rujira_j@tistr.or.th

[3] Department of Chemical Engineering, Faculty of Engineering and Industrial Technology, Silpakorn University, Nakorn Pathom 73000, Thailand; patthaveekongka_w@su.ac.th

* Correspondence: teehudakorn@gmail.com; Tel.: +66-61-190-3999; Fax: +66-3427-0401

Abstract: Conventionally, methanol is derived from a petroleum base and natural gas, but biomethanol is obtained from biobased sources, which can provide a good alternative for commercial methanol synthesis. The fermentation of molasses to produce biomethanol via the production of biohydrogen (H_2 and CO_2) was studied. Molasses concentrations of 20, 30, or 40 g/L with the addition of 0, 0.01, or 0.1 g/L of trace elements (TEs) ($NiCl_2$ and $FeSO_4 \cdot 7H_2O$) were investigated, and the proper conditions were a 30 g/L molasses solution combined with 0.01 g/L of TEs. H_2/CO_2 ratios of 50/50% (v/v), 60/40% (v/v), and 70/30% (v/v) with a constant feed rate of 60 g/h for CO_2 conversion via methanol synthesis (MS) and the reverse water–gas shift (RWGS) reaction were studied. MS at temperatures of 170, 200, and 230 °C with a $Cu/ZnO/Al_2O_3$ catalyst and pressure of 40 barg was studied. Increasing the H_2/CO_2 ratio increased the maximum methanol product rate, and the maximum H_2/CO_2 ratio of 70/30% (v/v) resulted in methanol production rates of 13.15, 17.81, and 14.15 g/h, respectively. The optimum temperature and methanol purity were 200 °C and 62.9% (wt). The RWGS was studied at temperatures ranging from 150 to 550 °C at atm pressure with the same catalyst and feed. Increasing the temperature supported CO generation, which remained unchanged at 21 to 23% at 500 to 550 °C. For direct methanol synthesis (DMS), there was an initial methanol synthesis (MS) reaction followed by a second methanol synthesis (MS) reaction, and for indirect methanol synthesis (IMS), there was a reverse water–gas shift (RWGS) reaction followed by methanol synthesis (MS). For pathway 1, DMS (1st MS + 2nd MS), and pathway 2, IMS (1st RWGS + 2nd MS), the same optimal H_2/CO_2 ratio at 60/40% (v/v) or 1.49/1 (mole ratio) was determined, and methanol production rates of 1.04 (0.033) and 1.0111 (0.032) g/min (mol/min), methanol purities of 75.91% (wt) and 97.98% (wt), and CO_2 consumptions of 27.32% and 57.25%, respectively, were achieved.

Keywords: biohydrogen; biomethanol; catalytic conversion; molasses fermentation; reverse water–gas shift

Citation: Krekkeitsakul, K.; Jitrwung, R.; Patthaveekongka, W.; Hudakorn, T. Improving Biomethanol Synthesis via the Addition of Extra Hydrogen to Biohydrogen Using a Reverse Water–Gas Shift Reaction Compared with Direct Methanol Synthesis. *Processes* **2023**, *11*, 2425. https://doi.org/10.3390/pr11082425

Academic Editor: Davide Papurello

Received: 12 June 2023
Revised: 8 August 2023
Accepted: 9 August 2023
Published: 11 August 2023

1. Introduction

Biofuels have begun to replace fossil fuels due to their potential to reduce greenhouse gas emissions by replacing fossil fuels. Biofuels thereby have the potential to both decrease CO_2 emissions and increase energy security [1]. The first example is biodiesel (derived from the transesterification/esterification of vegetable oil), which is mixed with diesel fuel. The second example is biogas (derived from the decomposition of organic/agricultural waste), which generates methane (CH_4), which has been used as a replacement for natural gas in the generation of heat and electricity and in vehicles. Bioethanol (produced via the fermentation of sugar and starch feedstocks) has been used for over three decades because it can be blended with gasoline. It is more prevalent than biomethanol due to the compatibility of bioethanol/gasoline blends with current internal combustion engines. However, biomethanol is methanol which is obtained from bio-sources such as biogas and biohydrogen, and it can also be produced from CO_2. Therefore, it may be more desirable in the future and replace bioethanol due to the CO_2 utilization potential [2]; higher specific energy yield, which can be explained by the fact that oxygenated fuels have a better combustion efficiency and more enrichment of oxygen than ethanol, owing to methanol [3]; and high volumetric energy density [4]. Accordingly, biomethanol is easy to store and transport and can be readily used as a raw material for synthesizing a variety of useful organic compounds of industrial importance. Methanol is a valued chemical, and it can be used in various sectors as a solvent and in biodiesel, biofuels, and additives [5]. Regarding its direct usage, methanol is used as a solvent and raw material by reacting it with vegetable oil to produce biodiesel. Regarding its indirect usage, methanol is transformed into other chemicals such as formaldehyde, methyl tertiary butyl ether (MTBE)/TAME (a blending component of gasoline), dimethyl ether (DME), MTO (methanol–olefin), and MTP (methanol–paraffin). Biogas and biohydrogen derived from molasses play important roles as raw gases for biomethanol production. Biogas normally contains methane (CH_4), carbon dioxide (CO_2), and varied amounts of hydrogen sulfide (H_2S) depending on the variety of raw materials. Therefore, when using biogas as a raw material for biomethanol, it is necessary to remove H_2S via a H_2S separation process [6], whereas biohydrogen comprises only hydrogen (H_2) and CO_2. For this reason, when using biohydrogen as a raw material for biomethanol production, a H_2S separation unit is not needed (saving costs), and the method is attractive due to not only reducing the CO_2 problem by using biohydrogen but also increasing the use of molasses to produce biohydrogen. Furthermore, this will provide another option for the sugar industry to extract value from the by-product of molasses by making biohydrogen, representing an alternative source of raw materials for biomethanol production. The idustrial-scale production of methanol via thermo-chemical processes has been achieved using petroleum sources, but there is awareness that this releases CO_2 into the environment. On the other hand, the implementation of the biological conversion process at the laboratory scale is still being investigated and lacks sufficient information. The biochemical production of methanol requires the microbes and metabolic pathways and enzymes that are involved to be studied to properly understand the bioconversion process and determine the essential parameters for scaling up the laboratory process to full-size production units. Integrating these biological and thermo-chemical processes could provide an opportunity to make the production of biomethanol feasible.

Agricultural/organic wastes, such as manure, feed, fruit, cassava, bagasse, and molasses, are abundant resources that can be turned into biofuels. Molasses is a plentiful resource obtained from sugar production and is used to produce fermented ethanol, which has been promoted for blending with gasoline [7]. Electric vehicles are becoming a substitute for vehicles that require the use of fossil fuels for power, especially gasoline vehicles, which normally use ethanol-blended gasoline. As a result, increased electric vehicle usage will lead to a reduction in ethanol production. Ethanol production will be disrupted and decreased due to decreasing demand, which means that the demand for molasses as a raw material will be reduced, affecting the income of sugarcane farmers. To compensate for this situation, transforming molasses into hydrogen via the fermentation process is desirable as it only generates H_2, i.e., biohydrogen, and CO_2 [8,9]. Hydrogen can be used in hydrogen fuel cells, for the hydrogen treatment of biodiesel/vegetable oil to produce bio-oleochemicals, and to transform carbon dioxide into an important precursor to petrochemicals, such as methane or methanol [10,11].

Conventionally, methanol production requires a mixture of synthesis gases (H_2, CO, and CO_2) that is obtained from the steam reforming of natural gas (CH_4 and CO_2). In the case of using a biohydrogen source (H_2 and CO_2) for producing biomethanol, direct methanol synthesis (DMS) can be performed, which is the reaction of CO_2 and H_2 using an appropriate catalyst, specific temperature, and pressure, resulting in methanol and water as the products, following Equation (1). However, CO_2 and H_2 can be reacted through a reverse water–gas shift (RWGS) reaction, obtaining CO and H_2O, following Equation (3). An advantage of RWGS is that it can be applied for transforming CO_2 and H_2 into CO and H_2O as products, meaning that the H_2O liquid can be easily separated, and then CO will react with the excess H_2, following Equation (2), which is called indirect methanol synthesis (IMS), resulting in methanol as a product; this methanol is purer than that obtained from the reaction shown in Equation (1). However, there is a side reaction called a water–gas shift (WGS), shown in Equation (4), which is the reverse reaction of RWGS [3,12–14]. Biohydrogen sources (H_2 and CO_2) can be used as raw materials for producing methanol.

Methanol synthesis (MS)

$$CO_2 + 3H_2 \rightleftharpoons CH_3OH + H_2O \quad \Delta H = -49.43 \ kJ/mol \tag{1}$$

$$CO + 2H_2 \rightleftharpoons CH_3OH \quad \Delta H = -90.55 \ kJ/mol \tag{2}$$

Reverse water–gas shift (RWGS)

$$CO_2 + H_2 \rightleftharpoons CO + H_2O \quad \Delta H = 41.12 \ kJ/mol \tag{3}$$

Water–gas shift (WGS)

$$CO + H_2O \rightleftharpoons CO_2 + H_2 \quad \Delta H = -41.12 \ kJ/mol \tag{4}$$

The purpose of this article is to guide the optimization of biohydrogen in terms of the molasses concentration and the concentration of trace elements. The transformation of biohydrogen into biomethanol was studied. The effect of the H_2/CO_2 ratio on CO_2 hydrogenation and the RWGS was investigated. Two pathways for transforming biohydrogen and CO_2 were investigated, and both pathways involved a two-step reaction: pathway 1 included two-step methanol synthesis involving direct CO_2 hydrogenation (Equation (1)) comprising a first methanol synthesis (1st MS) and second methanol synthesis (2nd MS), with both steps connected in a series fixed-bed reactor operated using the same catalyst and controlled temperature, as shown in the Materials and Methods section; pathway 2 was an innovative route comprising a reverse water–gas shift (RWGS) as shown in Equation (3) and then methanol synthesis (MS) as shown in Equation (2). The same catalyst was used for each step, but each step had a different temperature and pressure, as described in the Materials and Methods section. The methanol production rate, methanol purity, and CO_2 consumption were obtained for both pathways.

2. Materials and Methods

2.1. Biohydrogen Experiment

2.1.1. Microorganism, Culture Conditions, and Raw Materials

- The bacterial strain *Enterobacter aerogenes* (E.A.) (TISTR 1540) was obtained from the Biodiversity Research Center of the Thailand Institute of Scientific and Technological Research (TISTR), maintained at 0 °C with 15% purified glycerol, and used for H_2 production from molasses.
- Molasses obtained from the Konkaen sugar company was dissolved in deionized water at concentrations of 20, 30, and 40 g/L.
- The following nutrient medium was prepared: beef extract (1.0 g/L), yeast extract (2.0 g/L), peptone (5.0 g/L), and NaCl (5.0 g/L). This was then placed in deionized water and sterilized in an autoclave for 15 min at 121 °C. After that, the bacteria were incubated in the nutrient medium for 18 h.
- The synthetic medium contained the following components: $(NH_4)2SO_4$ (4.0 g/L), KH_2PO_4 (5.5 g/L), tryptone (5 g/L), yeast extract (5 g/L), $(NH_4)2SO_4$ (1g/L), $MgSO_4 \cdot 7H_2O$

(0.25g/L), $CaCl_2 \cdot 2H_2O$ (0.020g/L), and $Na_2MoO_4.2H_2O$ (0.12g/L). The trace elements were $NiCl_2$ (0.01 and 0.1 g/L) and $FeSO_4 \cdot 7H_2O$ (0.01 and 0.1 g/L).

- N_2 gas and gas mixtures of CO_2 and H_2 were supplied by the Thai Special Gas Company.

2.1.2. Apparatus and Operation

- A 10 L bioreactor with a 5 L working volume was used in the biohydrogen experiment and obtained from Marubishi (model: MDFT-10 L), as shown in Figure 1.

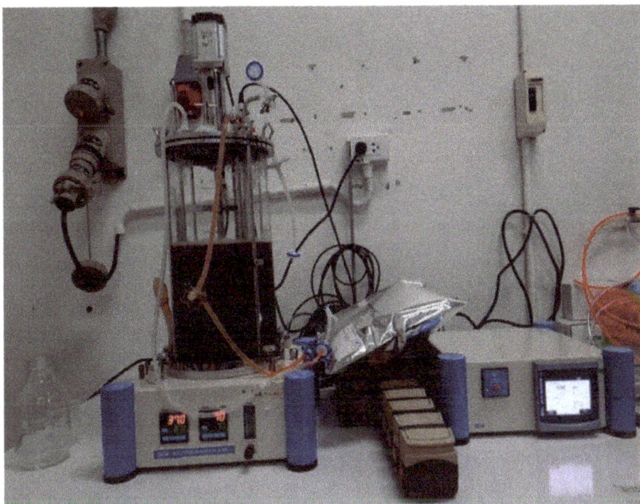

Figure 1. Ten-liter bioreactor for biohydrogen production.

- The fermentation conditions were a 37 °C temperature, agitation by rotation at 70 rpm, and a pH of 6.0–7.0.
- Molasses solutions (20, 30, and 40 g/L) were prepared and dissolved in solutions of a medium with a volume of 5L containing different quantities of trace elements (0, 0.01, and 0.1 g); the pHs were adjusted to 7.0 and then the solutions were sterilized (for 15 min at 121 °C) in an autoclave.
- After removal from the autoclave, the solutions were cooled and then 10% of each seed culture was inoculated into the bioreactor.
- To prepare the limited aerobic conditions, nitrogen (N_2) was used to purge the bioreactor until the oxygen (O_2) was detected to be below 1%, the temperature was set at 37 °C, and the contents were agitated at 70 rpm. The gas product was collected using an aluminum gas bag and a sample was taken every 6 h. The fermentation time was 72 h, continually, per batch.

2.2. Biomethanol Experiment

2.2.1. Raw Materials

- The commercial catalyst $Cu/ZnO/Al_2O_3$ was supplied by Xi'an Sunward Aeromat Co., Ltd., Xi'an, China. An amount of 5 kg of the catalyst was placed in reactor 1 (RX1) and in reactor 2 (RX2). The activation conditions for the catalyst were feeding 15% H_2 mixed with N_2 at a temperature of 230 °C and pressure of 3 barg for 18 h.
- The raw gas composition (H_2 and CO_2) obtained using the biohydrogen reactor was measured using an Agilent gas chromatograph (model: 7890 B).
- The gas compositions in the feeds and products of the biomethanol synthesis and reverse water–gas shift experiments were measured using a gas analyzer (MRU model,

Vario luxx), as shown in Figure 2, which measured the gas compositions (H_2, CO, CO_2, O_2, N_2, and CH_4) in % by volume, and the sum of the gases was 100% by volume.

Figure 2. Gas analyzer (MRU model, Vario luxx).

2.2.2. Apparatus and Operation

- Reactor 1 (RX1) and reactor 2 (RX2) were identical fixed-bed reactors used for the biomethanol synthesis experiments. The fixed-bed reactors had an inside diameter of 16 cm and a length of 30 cm, were made from 304 stainless steel, and were filled with 5 kg of $Cu/ZnO/Al_2O_3$. Each heater was used to supply heat to each reactor with a program controlling the temperature from 30 to 600 °C.
- Mass flow controllers were used to control all the gases feeding the reactors: mass flow of biohydrogen (MFBH), mass flow of hydrogen (MFH), mass flow of carbon dioxide (MFC), mass flow of nitrogen (MFN), and mass flow of syngas (MFSG).
- Other equipment: cool separator 1 (CS1) was used to separate the liquid product from the gas product from RX1. Cool separator 2 (CS2) was used to separate the liquid product from the gas product from RX2. A low-pressure tank (LPT) was used to collect the gas product from CS1. A compressor was used to build up and compress the low-pressure gas to obtain high-pressure gas and then it was stored in a high-pressure tank (HPT). Cool water from a chiller was supplied to CS1 and CS2 for liquid–gas separation.

The two-step biomethanol synthesis (TSBS) process was set up with two identical fixed-bed reactors (RX1 and RX2) connected in series and linked with all the apparatus, as shown in Figure 3. It was used for the biomethanol synthesis experiments described in Sections 3.2–3.4 (note that Section 3.1 describes the biohydrogen experiment).

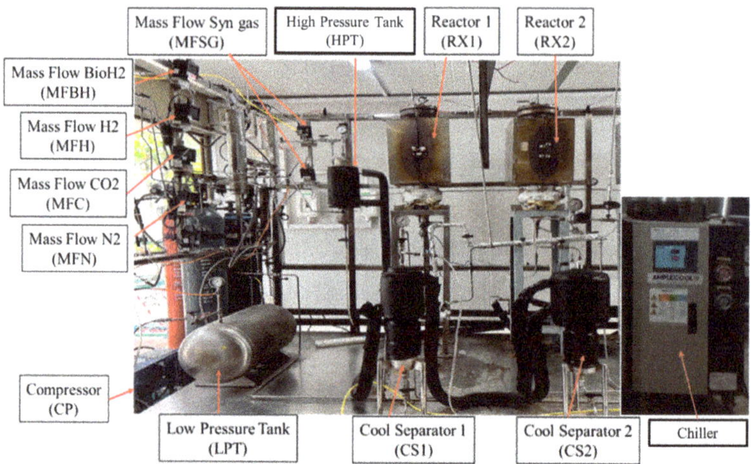

Figure 3. Two-step methanol synthesis process.

Section 3.2: In this experiment, direct CO_2 hydrogenation using RX2 only was studied.

Section 3.3: In this experiment, the transformation of CO_2 into CO via the RWGS reaction using RX1 only was studied.

Section 3.4: In this experiment, the transformation of biohydrogen into biomethanol via pathway 1—direct methanol synthesis (DMS) + direct methanol synthesis (DMS) (1st MS + 2nd MS)—and pathway 2—indirect methanol synthesis (IMS) comprised of RWGS + direct methanol synthesis (DMS) (1st RWGS + 2nd MS)—was studied.

Pathway 1: The process diagram for DMS+DMS, methanol synthesis reaction step 1 (1st MS), and methanol synthesis reaction step 2 (2nd MS) is shown in Figure 4. The experiment was started by feeding a gas comprising biohydrogen ($H_2 + CO_2$) using MFBH and the gas composition was adjusted with CO_2 or H_2 using MFC or MFH; then, all the gases were blended in a gas mixer until the following volume ratios of H_2/CO_2 were obtained: 50/50% (v/v), 60/40% (v/v), and 70/30% (v/v). The gas was transferred by MFSG to reactor 1 (RX1), in which the temperature was controlled at 200 °C and the pressure was controlled at 40 barg. After the reaction, hot fluid flowed out of the first reactor and cooled down in CS1 and then the liquid was removed by opening the bottom valve connected to CS1. Gas flowed out from the top of CS1 and was fed continuously into reactor 2 (RX2), in which the temperature was controlled at 200 °C and the pressure was controlled at 40 barg for methanol synthesis. After the reaction in RX2, the fluid was cooled in CS2. The liquid was removed from the bottom of CS2 and the gas was extracted. The compositions of the gases were measured using a gas analyzer, MRU, which was connected in three positions, as shown in Figure 4.

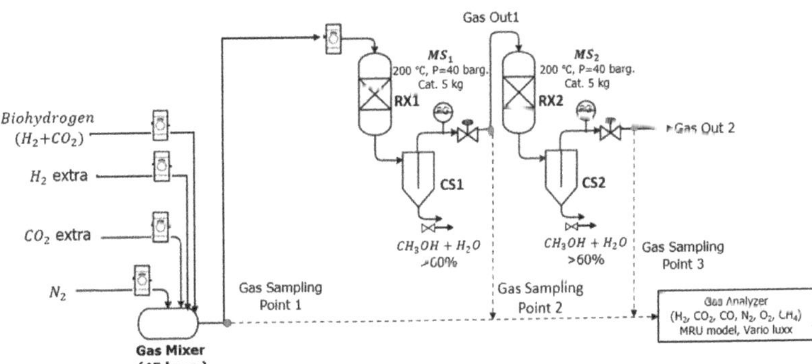

Figure 4. Process flow diagram of the H_2/CO_2 experiment via pathway 1: MS1 and MS2 in fixed-bed reactor.

Pathway 2: The process diagram of IMS + DMS, the reverse water–gas shift reaction (RWGS), and methanol synthesis reaction (MS) were carried out as shown in Figure 5. The experiment was started using a set-up pressure of 3 barg for all the feeding gases, which comprised biohydrogen ($H_2 + CO_2$), which was fed by MFBH, and the gas composition was adjusted using CO_2 or H_2 by MFC or MFH; then, all the gases were blended in a gas mixer until the following volume ratios of H_2/CO_2 were obtained: 50/50% (v/v), 60/40% (v/v), and 70/30% (v/v). The gas was sent by MFSG to reactor 1 (RX1), in which the temperature was controlled at 500 °C and the pressure was controlled at atmospheric pressure. After the reaction, hot fluid flowed out of the first reactor and cooled down in CS1 and then the liquid was removed by opening the bottom valve connected to CS1. Gas flowed out from the top of CS1 and was collected in LPT, and the minimum and maximum pressures were controlled at 0.1 and 0.5 barg, respectively. The compressor (CP) started to compress the gas into HPT when the gas pressure in the LPT reached the maximum pressure and stopped when the minimum pressure was reached. The gas was collected in the HPT until the pressure reached over 45 barg and was prepared to be fed continuously with MFSG into reactor 2 (RX2), the pressure in which was controlled at 40 barg by a gas back-pressure regulator. After the reaction in RX2, the fluid was cooled in CS2. The liquid was removed from the bottom of CS2 and the gas was drawn out. The compositions of the gases were measured using a gas analyzer, MRU, which was connected in three positions, as shown in Figure 5.

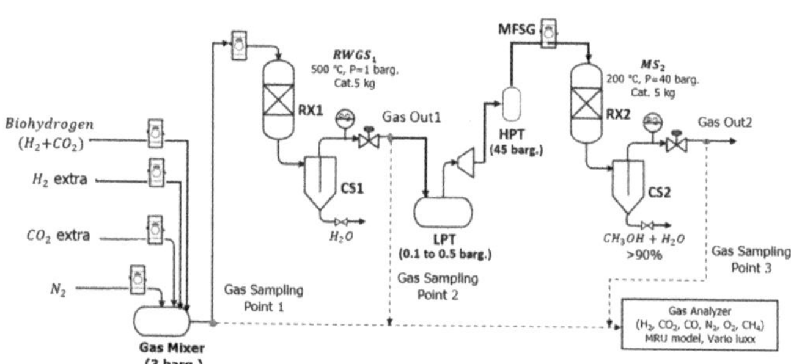

Pathway 2 : Biohydrogen combined with extra H_2
then RWGS1 and following with MS2

Figure 5. Process flow diagram of the H_2/CO_2 experiment via pathway 2: RWGS1 and MS2 in fixed-bed reactor.

3. Results

3.1. Optimization of Biohydrogen with Varying Molasses Concentrations versus Trace Element (TE) Concentrations with Enterobacter Aerogenes

The production of biohydrogen was studied by investigating two parameters: the molasses concentration (MC) and the trace element (TE) concentration. The experiments were performed by comparing molasses concentrations of 20, 30, and 40 g/L corresponding to low, medium, and high levels, respectively, and then combining them with the two metal sources of $NiCl_2$ and $FeSO_4 \cdot 7H_2O$ (TEs) at varying concentrations of 0, 0.01, 0.05, and 0.1 g/L, as shown in Table 1.

Table 1. Biohydrogen experimental conditions.

No.	Conditions	Total Gas (L)			H_2/CO_2		
		1st Batch	2nd Batch	Ave.	1st Batch	2nd Batch	Ave.
1.	20 g/L Mol. [2], 5 g/L Try [3], TE [1] not added	10.50	9.50	10.00	0.37	0.33	0.35
2.	20 g/L Mol., 5 g/L Try, 0.01 g/L TE	10.00	11.00	10.50	1.01	1.05	1.03
3.	20 g/L Mol., 5 g/L Try, 0.05 g/L TE	9.87	10.59	10.23	0.36	0.32	0.34
4.	20 g/L Mol., 5 g/L Try, 0.10 g/L TE	9.20	9.82	9.51	0.29	0.25	0.27
5.	30 g/L Mol., 5 g/L Try, TE not added	14.01	14.67	14.34	0.43	0.41	0.42
6.	30 g/L Mol., 5 g/L Try, 0.01 g/L TE	31.00	31.72	31.36	0.99	0.95	0.97
7.	30 g/L Mol., 5 g/L Try, 0.05 g/L TE	20.48	21.32	20.90	0.35	0.41	0.38
8.	30 g/L Mol., 5 g/L Try, 0.10 g/L TE	11.50	9.38	10.44	0.32	0.3	0.31
9.	40 g/L Mol., 5 g/L Try, TE not added	11.60	13.70	12.65	0.53	0.49	0.51
10.	40 g/L Mol., 5 g/L Try, 0.01 g/L TE	26.70	27.14	26.92	0.51	0.49	0.50
11.	40 g/L Mol., 5 g/L Try, 0.05 g/L TE	17.12	16.68	16.90	0.50	0.5	0.50
12.	40 g/L Mol., 5 g/L Try, 0.10 g/L TE	6.30	7.46	6.88	0.52	0.48	0.50

[1] TE = trace element, [2] Mol. = molasses, [3] Try = tryptone.

The results show that the molasses concentration affected biohydrogen generation. It was found that using 30 g/L of molasses resulted in the best biohydrogen generation, better than molasses concentrations of 20 and 40 g/L, for every concentration of TEs added. Increasing the TE concentration in the system from 0, 0.01, 0.05, and 0.1 g/L demonstrated that 0.01 g/L of TEs enhanced both the maximum biohydrogen and the H_2/CO_2 ratio. As shown in Table 1, the optimal combination was a 30 g/L molasses concentration and 0.01 g/L TE concentration, which produced 31.36 L of biohydrogen and a 0.97 H_2/CO_2 mole ratio. Implementing a TE concentration above this did not increase biohydrogen production at any molasses concentration because TE concentrations above this inhibited hydrogenase activities, resulting in both lower hydrogen production and H_2/CO_2 ratios. In addition, the results show that adding an appropriate amount of TEs not only enhanced biohydrogen development but also increased the ratio of H_2/CO_2 up to 1. In Alshiyab et al.'s study, they reported that metal ions, such as Fe^{2+}/Fe^{3+} and Ni^{2+}, facilitated an increase in hydrogen production during dark fermentation [15].

3.2. Effect of H_2/CO_2 Ratio on Direct CO_2 Hydrogenation upon Adding Extra H_2 in Methanol Synthesis

This biohydrogen experiment obtained a gas composition with a maximum H_2/CO_2 ratio of approximately 1 (50/50% (v/v) H_2/CO_2). Following Equation (1), direct methanol synthesis normally requires a mole ratio for H_2/CO_2 of approximately 3. Extra hydrogen was added to the system. After that, the effects of three different H_2/CO_2 ratios, 50/50, 60/40, and 70/30% (v/v), were studied. Direct CO_2 hydrogenation was carried out at temperatures of 170, 200, and 230 °C with the TISTR optimum conditions (the pressure was fixed at 40 barg with under 5 kg of $CuZnO/Al_2O_3$ at a total gas feed flow rate of 1 g/min (60 g/h)). As can be seen in Figure 6, the results showed that increasing the H_2/CO_2 ratio resulted in an increase in the methanol production rate at every temperature applied: at 170 °C, the methanol production rates were 9.91, 12.01, and 13.15 g/h; at 200 °C, the production rates were 16.24, 17.00, and 17.81 g/h; and at 230 °C, the production rates were 10.91, 13.01, and 14.15 g/h. The optimum temperature was 200 °C, resulting in the maximum methanol amount. However, there is no evidence that a difference in the purity of the methanol product, which ranged from 61 to 63%, was related to CO_2 hydrogenation, as was also reported by Sarp S. in 2021 [16]. The methanol percentage obtained via the experiment showed that direct CO_2 hydrogenation followed Equation (3). The methanol percentage was over 50% because, during CO_2 hydrogenation, a side reaction involving the transformation of CO_2 into CO via RWGS appeared in parallel, as shown in Equation (4), which is related to the findings discussed in [17]. The gas mixture containing an amount of CO in the reaction resulted in some CO hydrogenation, which had a positive effect on the methanol yield of over 50%.

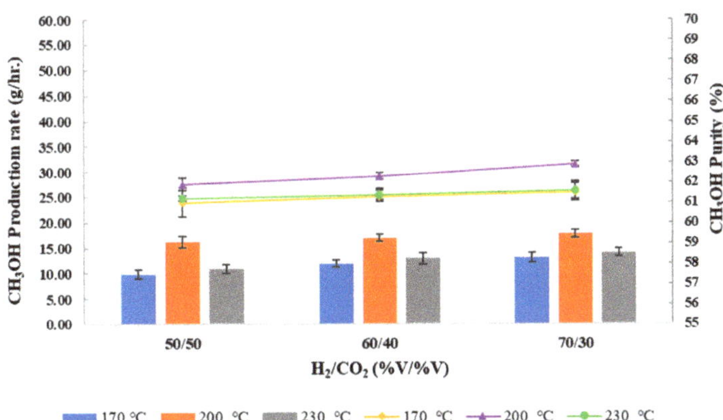

Figure 6. Direct methanol synthesis via CO_2 hydrogenation with varying H_2/CO_2 ratios at a feed flow rate of 1 g/min. The bars represent methanol production (g/h), and the curves represent methanol purity in % by weight.

3.3. Effect of H_2/CO_2 Ratio on CO Generation in RWGS with Varying Temperatures

Direct methanol synthesis via CO_2 hydrogenation gave a low methanol production rate, and the product was contaminated with a large amount of water, as shown in Figure 6. Therefore, indirect methanol synthesis via the transformation of CO_2 into CO followed by CO hydrogenation may be a method for increasing methanol purity. The desired method was to apply the RWGS reaction (Equation (3)) to transform CO_2 into CO before feeding it into the methanol synthesis process following Equation (2) for CO hydrogenation. This was expected to produce a higher methanol production rate than direct methanol synthesis. Therefore, the optimum ratio of H_2/CO_2 would be the best for CO conversion via the RWGS. As in the previous section, the biohydrogen yielded a H_2/CO_2 ratio of approximately 1. Therefore, the relationship between the H_2/CO_2 feed ratio and the mixed gas product composed of H_2, CO_2, and CO generated via the RWGS was studied. The results in the previous section demonstrated that adding H_2 to the biohydrogen increased the normal H_2/CO_2 % v/v or (mole ratio) of 50/50 (1/1) to 60/40 (1.49/1) and 70/30 (2.33/1). Three H_2/CO_2 ratios for the RWGS with $Cu/ZnO/Al_2O_3$ at temperatures ranging from 150 to 550 °C were studied. The results in Figure 7 show that the change in the % by vol. of H_2 and CO_2 (raw material gases) for all ratios decreased with an increase in the temperature from 150 to 500 °C and then CO_2 and H_2 slowly declined when a temperature between 500 and 550 °C was applied. The generation of CO increased from 150 to 500 °C and then slightly stabilized from 500 to 550 °C. This phenomenon follows Le Châtelier's principle as the reaction was endothermic; it was thermodynamically favored at higher temperatures until the CO and CO_2 contents were balanced in equilibrium in the RWGS and WGS following the reverse reactions in Equations (3) and (4). In addition, increasing the H_2/CO_2 ratio increased CO_2 conversion and favored the RWGS reaction, as discussed by Chinchen [18]. The results showed that the RWGS equilibrium in Equation (3) dominated over the WGS in Equation (4), showing a % by vol. (mol) CO generation of approximately 21.20, 21.71, and 23.31 at 550 °C. Under all conditions, CO generation increased according to the increasing temperature, but the final amounts of CO conversion differed in terms of minimum numbers. This result can be explained by the moles of CO and CO_2 being balanced in the RWGS and WGS conditions relating to Equations (3) and (4) [19–22]. In addition, the $Cu/ZnO/Al_2O_3$ catalyst had a maximum temperature of approximately 550 °C, which followed the specification of the catalyst company. Therefore, the optimum temperature for CO_2/H_2 in the RWGS was 500 °C, which was used in the RWGS reaction for safety reasons and generated the maximum CO and minimum CO_2 contents.

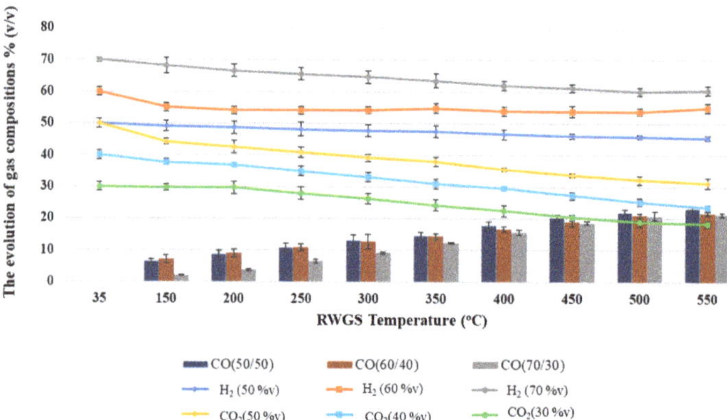

Figure 7. Changes in gas compositions with varying H_2/CO_2 ratios, and CO generation when applying different RWGS temperatures at a feed flow rate of 5 g/min.

3.4. The Comparison of H_2/CO_2 Ratios for Methanol Synthesis via Two Pathways: (1) MS and MS and (2) RWGS and MS

Pathway 1 (DMS + DMS) was conducted as follows: two-step direct methanol synthesis was set up with two methanol synthesis reactors with a feed flow rate of 5 g/min, temperature of 200 °C, and pressure of 40 barg, as shown in Tables 2–4. The H_2/CO_2 feed in unit % v/v (mol ratios) of 50/50 (1.00), 60/40 (1.50), and 70/30 (2.33) added to the first MS reactor resulted in methanol production rates of 0.43, 0.49, and 0.49 g/min. The feed contained only CO_2 and H_2 and then the reactions were followed by direct CO_2 hydrogenation via Equation (1), which produced a mixture of methanol and water around 61.43% (wt), 62.03% (wt), and 63.64% (wt), respectively. The off gas that came out of the first reactor contained mixed $H_2/CO_2/CO$ % $v/v/v$ (mol/min) of 44.17/45.45/10.38 (0.085/0.083/0.018), 53.52/36.80/9.67 (0.120/0.078/0.020), and 64.93/24.71/10.36 (0.195/0.070/0.028), respectively. There was clear evidence that an increase in the H_2/CO_2 feed increased the % (wt) methanol purity and methanol production rate but generated almost no difference in CO generation, which was obtained from the side reaction following Equation (3). The mixed gas that was released from the first MS reactor turned into the feed in the second MS reactor in series. The results showed that the series of continual H_2/CO_2 feed ratios of 50/50, 60/40, and 70/30 (%v) generated a percentage of CO in the mixed gas and resulted in higher mole ratios for H_2/CO (4.70, 6.11, and 6.92) than for H_2/CO_2 (1.03, 1.54, and 2.78), which supported the methanol synthesis via CO hydrogenation following Equation (2). As a result, the methanol concentration in the second step showed a higher yield compared with that in the first step, in which the methanol purity was 93.29% (wt), 94.16% (wt), and 80.39% (wt) and the methanol production rates were 0.50, 0.55, and 0.60 g/min. The mixed $H_2/CO_2/CO$ % $v/v/v$ (mol/min) values derived from the second reactor were 35.59/62.43/1.98 (0.050/0.082/0.003), 49.02/49.35/1.63 (0.080/0.077/0.003), and 63.94/31.64/4.41 (0.140/0.065/0.009), respectively. There was evidence that the mixed feed gas contained CO, which affected the selectivity of CO hydrogenation (Equation (2)) or CO_2 hydrogenation (Equation (1)). The first effect was that a lower percentage of CO_2 in the feed gas increased both the methanol production rates and the methanol purity (%), and the second effect was that a higher H_2/CO ratio in the feed gas composition resulted in the reaction of Equation (2) competing with that of Equation (1). However, whilst the H_2/CO_2 ratio was close to 3, the H_2/CO ratio was over 2 following stoichiometry, and the methanol synthesis favored the reaction in Equation (1) over that in Equation (2), as shown in the 70/30% (v/v) H_2/CO_2 feed case in the second reactor.

Table 2. Pathway 1: combining 1st MS and 2nd MS at a constant feed flow rate of 5 g/min and H_2/CO_2 ratio of 50/50 (% v/v).

MS1_MS2	IN1	Out1	Out2	IN1	Out1	Out2	Total
	%v (mol/min)	%v (mol/min)	%v (mol/min)	g/min	g/min	g/min	g/min (mol/min)
H_2	50 (0.11)	44.17 (0.085)	35.59 (0.050)	0.23	0.17	0.1	0
CO_2	50 (0.11)	45.45 (0.083)	62.43 (0.082)	4.77	3.63	3.59	0
CO	0	10.38 (0.018)	1.98 (0.003)	0	0.51	0.07	0
Total gas	100 (0.22)	100 (0.186)	100 (0.134)	5	4.3	3.77	0
CH_3OH	0	61.7 (0.013)	93.29 (0.016)	0	0.43	0.5	0.93 (0.029)
H_2O	0	38.3 (0.015)	6.71 (0.002)	0	0.27	0.04	0.31 (0.017)
Total liquid	0	100 (0.028)	100 (0.018)	0	0.7	0.54	1.24 (0.046)
CO_2 conversion (g/min)							1.18
CO_2 consumption (%)							24.71
CH_3OH concentration (%)					61.43	92.59	78.69
H_2/CO_2 (mol/mol)	1	1.03					
H_2/CO (mol/mol)		4.7					

Table 3. Pathway 1: combining 1st MS and 2nd MS at a constant feed flow rate of 5 g/min and H_2/CO_2 ratio of 60/40 (% v/v).

MS1_MS2	IN1	Out1	Out2	IN1	Out1	Out2	Total
	%v (mol/min)	%v (mol/min)	%v (mol/min)	g/min	g/min	g/min	g/min (mol/min)
H_2	60 (0.170)	53.52 (0.120)	49.02 (0.080)	0.34	0.24	0.16	0
CO_2	40 (0.106)	36.8 (0.078)	49.35 (0.077)	4.66	3.42	3.39	0
CO	0	9.67 (0.020)	1.63 (0.003)	0	0.55	0.07	0
Total gas	100 (0.276)	100 (0.217)	100 (0.160)	5	4.21	3.63	0
CH_3OH	0	62.2 (0.015)	94.16 (0.017)	0	0.49	0.55	1.04 (0.033)
H_2O	0	37.8 (0.017)	5.84 (0.002)	0	0.3	0.03	0.33 (0.018)
Total liquid	0	100 (0.032)	100 (0.019)	0	0.79	0.58	1.37 (0.051)
CO_2 conversion (g/min)							1.27
CO_2 consumption (%)							27.32
CH_3OH concentration (%)					62.03	94.83	75.91
H_2/CO_2 (mol/mol)	1.5	1.54					
H_2/CO (mol/mol)		6.11					

Table 4. Pathway 1: combining 1st MS and 2nd MS at a constant feed flow rate of 5 g/min and H_2/CO_2 ratio of 70/30 (% v/v).

MS1_MS2	IN1	Out1	Out2	IN1	Out1	Out2	Total
	%v (mol/min)	%v (mol/min)	%v (mol/min)	g/min	g/min	g/min	g/min (mol/min)
H_2	70 (0.250)	64.93 (0.195)	63.94 (0.140)	0.5	0.39	0.28	0
CO_2	30 (0.102)	24.71 (0.070)	31.64 (0.065)	4.5	3.06	2.88	0
CO	0	10.36 (0.028)	4.41 (0.009)	0	0.78	0.25	0
Total gas	100 (0.352)	100 (0.292)	100 (0.214)	5	4.23	3.48	0
CH_3OH	0	63.3 (0.015)	80.39 (0.019)	0	0.49	0.6	1.09 (0.034)
H_2O	0	36.7 (0.016)	19.61 (0.008)	0	0.28	0.15	0.43 (0.024)
Total liquid	0	100 (0.031)	100 (0.027)	0	0.77	0.75	1.52 (0.058)
CO_2 conversion (g/min)							1.62
CO_2 consumption (%)							36.00
CH_3OH concentration (%)					63.64	80.00	71.71
H_2/CO_2 (mol/mol)	2.33	2.78					
H_2/CO (mol/mol)		6.92					

Pathway 2 (IMS + DMS) was conducted as follows: the RWGS followed by methanol synthesis was set up. The first reaction with the RWGS reactor was operated at a feed flow rate of 5 g/min, a temperature of 500 °C, and atm pressure with H_2/CO_2 feed ratios of 50/50 (1.00), 60/40 (1.50), and 70/30 (2.33). Then, it was connected with the second methanol synthesis reactor in series at 200 °C and 40 barg, and the results are shown in Tables 5–7. The result was clear: feeding a higher ratio of H_2/CO_2 into the RWGS resulted in higher conversion following Equation (3), which resulted in a methanol concentration of 0%; water yields of 0.82, 1.79, and 2.54 g/min; and mixed $H_2/CO_2/CO$ gas containing % ($v/v/v$) or (mol/min) 37.08/39.26/23.66 (0.065/0.067/0.039), 51.23/26.73/22.05 (0.095/0.046/0.036), and 62.68/17.70/19.62 (0.115/0.030/0.032), respectively. The RWGS converted CO_2 into CO following Equation (3), equating with the WGS following Equation (4), which resulted in the balancing of the ratio of CO_2 to CO. The equilibrium of RWGS and WGS resulted in the % CO (mol) generation being slightly reduced to 23.66 (0.039), 22.05 (0.036), and 19.62 (0.032), whereas it boosted the H_2/CO ratios to 1.73, 2.56, and 3.53, respectively, which supported methanol synthesis via CO hydrogenation following Equation (2). The mixed gas that was released from the first RWGS reactor was collected in the LPT tank and the pressure was increased using the CP and then fed into the second MS reactor in series. The methanol concentration in the second step showed higher purity percentages of 95.81% (wt), 97.98% (wt), and 92.93% (wt) and methanol yields of 0.69, 1.01, and

1.05 g/min. The mixed $H_2/CO_2/CO$ gas derived from the second reactor contained % $v/v/v$ or (mol/min) 18.24/64.12/17.64 (0.020/0.067/0.018), 32.01/61.15/6.84 (0.025/0.045/0.005), and 57.54/40.10/2.36 (0.045/0.029/0.002), respectively. The results are the same as those for pathway 1, wherein the mixed feed gas containing CO then influenced the selectivity of CO hydrogenation (Equation (2)) or CO_2 hydrogenation (Equation (1)). The main effect was that a lower percentage of CO_2 in the feed gas increased both the methanol yield and the methanol purity %, and a higher H_2/CO in the feed gas composition promoted the dominance of Equation (2) over Equation (1). However, the H_2/CO_2 mole ratio was close to or over 3, and the reaction preferred CO_2 hydrogenation following Equation (1), resulting in a reduction in the methanol purity, as shown by the H_2/CO_2 ratio for the 70/30 feed case in the second reactor.

Table 5. Pathway 2: combining 1st RWGS + 2nd MS at a constant feed flow rate of 5 g/min and H_2/CO_2 ratio of 50/50 (% v/v).

RWGS1_MS2	IN1	Out1	Out2	IN1	Out1	Out2	Total
	%v (mol/min)	%v (mol/min)	%v (mol/min)	g/min	g/min	g/min	g/min (mol/min)
H_2	50 (0.11)	37.08 (0.065)	18.24 (0.020)	0.23	0.13	0.04	0
CO_2	50 (0.11)	39.26 (0.067)	64.12 (0.067)	4.77	2.96	2.93	0
CO	0	23.66 (0.039)	17.64 (0.018)	0	1.09	0.49	0
Total gas	100 (0.22)	100 (0.171)	100 (0.104)	5	4.18	3.46	0
CH_3OH	0	0	95.81 (0.022)	0	0	0.69	0.69 (0.022)
H_2O	0	100 (0.046)	4.19 (0.002)	0	0.82	0.03	0.85 (0.047)
Total liquid	0	100 (0.046)	100 (0.024)	0	0.82	0.72	1.54 (0.069)
CO_2 conversion (g/min)							1.84
CO_2 consumption (%)							38.61
CH_3OH concentration (%)						0	95.81	95.81
H_2/CO_2 (mol/mol)	1	1					
H_2/CO (mol/mol)		1.73					

Table 6. Pathway 2: combining 1st RWGS + 2nd MS at a constant feed flow rate of 5 g/min and H_2/CO_2 ratio of 60/40 (% v/v).

RWGS1_MS2	IN1	Out1	Out2	IN1	Out1	Out2	Total
	%v (mol/min)	%v (mol/min)	%v (mol/min)	g/min	g/min	g/min	g/min (mol/min)
H_2	60 (0.170)	51.23 (0.095)	32.01 (0.025)	0.34	0.19	0.05	0
CO_2	40 (0.106)	26.73 (0.046)	61.15 (0.045)	4.66	2.01	1.99	0
CO	0	22.05 (0.036)	6.84 (0.005)	0	1.01	0.14	0
Total gas	100 (0.276)	100 (0.177)	100 (0.075)	5	3.21	2.18	0
CH_3OH	0	0	97.98 (0.032)	0	0	1.01	1.01 (0.032)
H_2O	0	100 (0.099)	2.02 (0.001)	0	1.79	0.02	1.81 (0.001)
Total liquid	0	100 (0.099)	100 (0.033)	0	1.79	1.03	1.54 (0.069)
CO_2 conversion (g/min)							2.67
CO_2 consumption (%)							57.25
CH_3OH concentration (%)						0	97.98	97.98
H_2/CO_2 (mol/mol)	1.49	2.03					
H_2/CO (mol/mol)		2.56					

Table 7. Pathway 2: combining 1st RWGS + 2nd MS at a constant feed flow rate of 5 g/min and H_2/CO_2 ratio of 70/30 (% v/v).

RWGS1_MS2	IN1	Out1	Out2	IN1	Out1	Out2	Total
	%v (mol/min)	%v (mol/min)	%v (mol/min)	g/min	g/min	g/min	g/min (mol/min)
H_2	70 (0.250)	62.68 (0.115)	57.54 (0.045)	0.5	0.23	0.09	0
CO_2	30 (0.102)	17.7 (0.030)	40.1 (0.029)	4.5	1.33	1.26	0
CO	0	19.62 (0.032)	2.36 (0.002)	0	0.9	0.05	0
Total gas	100 (0.352)	100 (0.177)	100 (0.075)	5	2.46	1.33	0
CH_3OH	0	0	92.93 (0.033)	0	0	1.05	1.05 (0.033)
H_2O	0	100 (0.141)	7.07 (0.004)	0	2.54	0.08	2.62 (0.146)
Total liquid	0	100 (0.141)	100 (0.037)	0	2.54	1.13	3.67 (0.178)
CO_2 conversion (g/min)							3.24
CO_2 consumption (%)							72.02
CH_3OH concentration (%)						0	92.93	92.93
H_2/CO_2 (mol/mol)	2.33	3.74					
H_2/CO (mol/mol)		3.53					

4. Discussion

A comparison of the two-step methanol synthesis between pathway 1, DMS (MS + MS), and pathway 2, IMS (RWGS + MS), showed that pathway 2 resulted in lower methanol production rates (g/h) of 0.69, 1.01, and 1.05 than the 0.93, 1.04, and 1.09 obtained via pathway 1, but it resulted in higher average methanol purities % (wt) of 95.81, 97.98, and 92.93 than the purities obtained via pathway 1 (78.69, 79.10, and 72.71). It is apparent that the IMS pathway (adding RWGS before MS) for biohydrogen (H_2 and CO_2) could improve the purity of the methanol more effectively than the DMS pathway because CO_2 was transformed into CO by the RWGS reaction following Equation (3) and limited by the WGS reaction, as in Equation (4). The evolution of CO_2 into CO depended on the conditions (CO/CO_2 mole ratio, temperature, and pressure). The CO_2 was transformed by reacting it with H_2, and CO was generated, obtaining the optimum H_2/CO ratio of around 2 as the stoichiometry followed Equation (2). These were the optimum conditions for generating high-purity biomethanol (97.98%), as obtained in the experiment with H_2/CO_2 at 60/40% (v/v). Otherwise, the mole ratio of H_2 to CO over 2 was partly caused by the higher ratio of H_2/CO_2 in the raw gas feed, at 70/30% (v/v), or 2.33/1 (mol/mol); in this case, there was a higher ratio, over 3, for both H_2/CO_2 (3.74) and H_2/CO (3.53). As a result, CO_2 hydrogenation (Equation (1)) competed with CO hydrogenation (Equation (2)), and water was generated and mixed with methanol, resulting in the methanol purity being reduced to 92.93%. Therefore, the IMS would be expected to produce a higher methanol yield than DMS.

Although the IMS was beneficial in terms of both methanol purity and CO_2 consumption, the RWGS step in IMS was an endothermic reaction that required a higher temperature (500 °C) than that required for the MS step (200 °C); hence, it consumed more energy in methanol production. Last but not least, the overall energy consumption should be studied in greater depth for overall processes, including methanol synthesis and methanol refinery.

5. Conclusions

The focus of this work was on the added value of turning molasses into biohydrogen over biogas for biomethanol synthesis. It was clear that CO_2 utilization achieved by adding a small amount of H_2 for methanol synthesis would create an opportunity to use molasses to produce biomethanol. Our results reveal that RWGS performed before DMS supported CO hydrogenation, and the transformation of biohydrogen into biomethanol (1) provided higher methanol concentrations, approaching the concentration of commercial methanol (99.9%), (2) necessitated a small refinery, and (3) increased CO_2 consumption, which creates an opportunity to apply this knowledge for the management of CO_2, called CO_2 utilization.

Taken together, this novel frontier of the implementation of the RWGS process will better support commercial methanol synthesis.

Author Contributions: Conceptualization, T.H.; Methodology, R.J.; Formal analysis, K.K. and W.P.; Investigation, K.K.; Resources, R.J.; Writing—original draft, K.K.; Writing—review & editing, R.J., W.P. and T.H.; Project administration, T.H. All authors have read and agreed to the published version of the manuscript.

Funding: This research was supported by Silpakorn University Research, Innovation and Creative Fund, and the Department of Mechanical Engineering, Faculty of Engineering and Industrial Technology, Silpakorn University, Thailand.

Data Availability Statement: Not applicable.

Acknowledgments: This research was supported by Silpakorn University Research, Innovation and Creative Fund, and the Department of Mechanical Engineering, Faculty of Engineering and Industrial Technology, Silpakorn University, Thailand. The author also gives special thanks to the Thailand Institute of Scientific and Technological Research (TISTR) for all the support provided.

Conflicts of Interest: The authors declare no conflict of interest.

Nomenclature

%	Percent
% (v/v)	Percent by volume
% (wt).	Percent by weight
°C	Celsius
1st MS	First methanol synthesis
2nd MS	Second methanol synthesis
Ave.	Average
barg	Bar (guage pressures)
cm	Centimeter
CS1	Cool separator 1
CS2	Cool separator 2
DME	Dimethyl ether
DMS	Direct methanol synthesis
E.A.	*Enterobacter aerogenes*
g	Gram
g/L	Grams per liter
H_2	Hydrogen
H_2O	Water
HPT	High pressure tank
h	Hour
IMS	Indirect methanol synthesis
kg	Kilogram
kJ/mol	Kilojoules per mole
LPT	Low-pressure tank
MC	Molasses concentration
MFBH	Mass flow of biohydrogen
MFC	Mass flow of carbon dioxide
MFH	Mass flow of hydrogen
MFN	Mass flow of nitrogen
MFSG	Mass flow of syngas
min	Minute
Mol.	Molasses
MS	Methanol synthesis
MTBE	Methyl tertiary butyl ether
MTO	Methanol to olefin

MTP	Methanol to paraffin
RWGS	Reverse water–gas shift
RX1	Reactor 1
RX2	Reactor 2
TAME	Tert-amyl methyl ether
TE	Trace elements
Try.	Tryptone
TSBS	Two-step biomethanol synthesis
WGS	Water–gas shift

References

1. Kazamia, E.; Smith, A.G. Assessing the environmental sustainability of biofuels. *Trends Plant Sci.* **2014**, *19*, 615–618.
2. Hasegawa, F.; Yokoyama, S.; Imou, K. Methanol or ethanol produced from woody biomass: Which is more advantageous? *Bioresour. Technol.* **2010**, *101*, S109–S111. [CrossRef] [PubMed]
3. Demirbas, A. Biomethanol Production from Organic Waste Materials. *Energy Sources Part A Recovery Util. Environ. Eff.* **2008**, *30*, 565–572. [CrossRef]
4. Verhelst, S.; Turner, J.W.G.; Sileghem, L.; Vancoillie, J. Methanol as a fuel for internal combustion engines. *Prog. Energy Combust. Sci.* **2019**, *70*, 43–88. [CrossRef]
5. Dalena, F.; Senatore, A.; Marino, A.; Gordano, A.; Basile, M.; Basile, A. Methanol Production and Applications: An Overview. *Methanol* **2018**, *1980*, 3–28.
6. Rashed Al Mamun, M.; Torii, S. Removal of Hydrogen Sulfide (H_2S) from Biogas Using Zero-Valent Iron. *J. Clean Energy Technol.* **2015**, *3*, 428–432. [CrossRef]
7. Goldemberg, J.; Coelho, S.T.; Guardabassi, P. The sustainability of ethanol production from sugarcane. *Energy Policy* **2008**, *36*, 2086–2097. [CrossRef]
8. Ozgur, E.; Mars, A.E.; Peksel, B.; Louwerse, A.; Yucel, M.; Gunduz, U.; Claassen, P.A.M.; Eroglu, I. Biohydrogen production from beet molasses by sequential dark and photofermentation. *Int. J. Hydrog. Energy* **2010**, *35*, 511–517. [CrossRef]
9. Rohani, R.; Chung, Y.T.; Mohamad, I.N. Purification of Biohydrogen Produced from Palm Oil Mill Effluent Fermentation for Fuel Cell Application. *Korean Chem. Eng. Res. J.* **2019**, *57*, 469–474.
10. Ghimire, A.; Frunzo, L.; Pirozzi, F.; Trably, E.; Escudie, R.; Lens, P.N.L.; Esposito, G. A review on dark fermentative biohydrogen production from organic biomass: Process parameters and use of by-products. *Appl. Energy* **2015**, *144*, 73–95.
11. Bakonyi, P.; Nemestothy, N.; Belafi-Bako, K. Biohydrogen purification by membranes: An overview on the operational conditions affecting the performance of non-porous, polymeric and ionic liquid based gas separation membranes. *Int. J. Hydrogen Energy* **2013**, *38*, 9673–9687. [CrossRef]
12. Gaikwa, R.; Reymond, H.; Phongprueksathat, N.; Rohr, P.R.V.; Urakawa, A. From CO or CO_2?: Space-resolved insights into high-pressure CO_2 hydrogenation to methanol over $Cu/ZnO/Al_2O_3$. *Catal. Sci. Technol* **2020**, *10*, 2763–2768. [CrossRef]
13. Leonzio, G.; Zondervan, E.; Foscolo, P.U. Methanol production by CO_2 hydrogenation: Analysis and simulation of reactor performance. *Int. J. Hydrogen Energy* **2019**, *44*, 7915–7933. [CrossRef]
14. Ren, M.; Zhang, Y.; Wang, X.; Qiu, H. Catalytic Hydrogenation of CO_2 to Methanol: A Review. *Catalysts* **2022**, *12*, 403.
15. Alshiyab, H.; Kalil, M.S.; Hamid, A.A.; Yusoff, W.M.W. Trace Metal Effect on Hydrogen Production Using *C. acetobutylicum*. *J. Biol. Sci.* **2008**, *8*, 1–9. [CrossRef]
16. Sarp, S.; Hernandez, S.G.; Chen, C.; Sheehan, S.W. Alcohol Production from Carbon Dioxide: Methanol as a Fuel and Chemical Feedstock. *Joule* **2021**, *5*, 59–76.
17. Slotbooma, Y.; Bosa, M.J.; Piepera, J.; Vrieswijka, V.; Likozarb, B.; Kerstena, S.R.A.; Brilman, D.W.F. Critical assessment of steady-state kinetic models for the synthesis of methanol over an industrial $Cu/ZnO/Al_2O_3$ catalyst. *Chem. Eng. J.* **2020**, *389*, 124181. [CrossRef]
18. Chinchen, G.C.; Denny, P.J.; Parker, D.G.; Spencer, M.S.; Nhan, D.A. Mechanism of methanol synthesis from $CO_2/CO/H_2$ mixtures over copper/zinc oxide/alumina catalyst: Use of 14C-labelled reactants. *Appl. Catal.* **1987**, *30*, 333–338. [CrossRef]
19. Takagawa, M.; Ohsugi, M. Study on Reaction Rates for Methanol Synthesis from Carbon Monoxide, Carbon Dioxide, and Hydrogen. *J. Catal.* **1987**, *107*, 161–172. [CrossRef]
20. Rafiee, A. Modelling and optimization of methanol synthesis from hydrogen and CO_2. *J. Environ. Chem. Eng.* **2020**, *8*, 104314. [CrossRef]
21. Mcneil, M.A.; Schack, C.J.; Rinker, R.G. Methanol Synthesis from Hydrogen, Carbon Monoxide and Carbon Dioxide over a $CuO/ZnO/Al_2O_3$ Catalyst. *Appl. Catal.* **1989**, *50*, 265–285. [CrossRef]
22. Klier, K.; Chatikavanij, V.; Herman, R.G.; Simmons, G.W. Catalytic Synthesis of Methanol from CO/H, IV. The Effects of Carbon Dioxide. *J. Catal.* **1982**, *74*, 343–360. [CrossRef]

chemengineering

MDPI

Article

Numerical Study of Dry Reforming of Methane in Packed and Fluidized Beds: Effects of Key Operating Parameters

Fahad Al-Otaibi [1,2], Hongliang Xiao [1,3], Abdallah S. Berrouk [1,2,*] and Kyriaki Polychronopoulou [1,2]

[1] Department of Mechanical Engineering, Khalifa University of Science and Technology, Abu Dhabi P.O. Box 127788, United Arab Emirates; kyriaki.polychrono@ku.ac.ae (K.P.)
[2] Center for Catalysis and Separation (CeCas), Khalifa University of Science and Technology, Abu Dhabi P.O. Box 127788, United Arab Emirates
[3] College of Mechanical and Transportation Engineering, China University of Petroleum, Beijing 102249, China
* Correspondence: abdallah.berrouk@ku.ac.ae

Abstract: Replacing the conventionally used steam reforming of methane (SRM) with a process that has a smaller carbon footprint, such as dry reforming of methane (DRM), has been found to greatly improve the industry's utilization of greenhouse gases (GHGs). In this study, we numerically modeled a DRM process in lab-scale packed and fluidized beds using the Eulerian–Lagrangian approach. The simulation results agree well with the available experimental data. Based on these validated models, we investigated the effects of temperature, inlet composition, and contact spatial time on DRM in packed beds. The impacts of the side effects on the DRM process were also examined, particularly the role the methane decomposition reaction plays in coke formation at high temperatures. It was found that the coking amount reached thermodynamic equilibrium after 900 K. Additionally, the conversion rate in the fluidized bed was found to be slightly greater than that in the packed bed under the initial fluidization regime, and less coking was observed in the fluidized bed. The simulation results show that the adopted CFD approach was reliable for modeling complex flow and reaction phenomena at different scales and regimes.

Keywords: dry reforming of methane; packed bed; fluidized bed; computational fluid dynamics; Eulerian–Lagrangian approach

Citation: Al-Otaibi, F.; Xiao, H.; Berrouk, A.S.; Polychronopoulou, K. Numerical Study of Dry Reforming of Methane in Packed and Fluidized Beds: Effects of Key Operating Parameters. *ChemEngineering* **2023**, 7, 57. https://doi.org/10.3390/chemengineering7030057

Academic Editors: Alírio E. Rodrigues and Aibing Yu

Received: 28 October 2022
Revised: 25 April 2023
Accepted: 8 May 2023
Published: 20 June 2023

1. Introduction

Carbon dioxide (CO_2) reformation (also known as dry reformation) is a method of producing synthesis gas, a mixture of hydrogen (H_2) and carbon monoxide (CO), from the reaction of carbon dioxide with hydrocarbons such as methane (CH_4) [1,2]. Synthesis gas is conventionally produced via a steam reforming reaction. Alarming concerns have grown over the last several years on the global warming caused by the influence of greenhouse gases (GHGs) and there has been elevated interest in replacing the reactant steam with CO_2 [3].

The most common commercial method and least expensive process for industrial hydrogen production is steam reforming of methane (SRM) [4]. Over 50% of the produced hydrogen comes from the SRM process. Despite its cost competitiveness and operational maturity, SRM still suffers from the drawback of producing high quantities of greenhouse gases (GHGs). As a result, many researchers and scientists have been investigating alternatives to SRM that can adequately address the challenging GHG emissions problem while maintaining the process's functionality in the petroleum refining and petrochemical industries.

In order to meet the current and future demand for syngas and hydrogen, it is necessary to produce it while maintaining minimal GHGs and cost competitiveness in OPEX and CAPEX simultaneously. Samantha Hillard [5] conducted an economic analysis of green hydrogen and found that green hydrogen technology is the most promising option for the far future, especially for electrolysis of the water splitting process, which requires enormous energy input, and the cost trend analysis showed that it had much higher costs compared to methane reforming. Therefore, the technology is still immature and needs intensive research and development.

Consequently, due to its volume and efficiency, natural gas reforming continues to be the primary source of hydrogen production. SRM, a mature technology, has endured as the primary driver of hydrogen production because of its capacity and minimized CAPEX and OPEX costs [6]. However, despite the wide industrial experience of SRM, it still has the highest reforming emissions compared to other natural-gas-reforming technologies, with a capacity of 7.05 kg CO_2/kg H_2, in addition to energy efficiency issues due to the need for water evaporation as feed steam for the process [7]. Thus, dry reforming of methane (DRM) has gained the attention of scientists as a possible replacement for SRM because of the attractiveness of CO_2 utilization, energy efficiency, and the minimization of GHG- and CO-rich syngas production for Fischer–Tropsch synthesis [8].

Despite DRM's positive environmental, economic, and energy aspects, it still suffers from high catalyst coking and rapid deactivation at prominent pressures using the industrial nickel-oxide-based catalyst [9]. Accordingly, extensive research has been carried out on experimentally developing DRM catalysts over the last few years [10–18]. The advancement of computational fluid dynamics (CFD) allows for the visualization of micro-reaction data and significantly lowers the cost of calculations [19]. The accessibility to powerful software and processors nowadays has reshaped the research and development of the field of gas–solid multiphase flow in various bed settings, making possible CFD simulations of DRM in packed- and fluidized-bed reactors.

Parker et al. emphasized the importance of selecting the proper numerical method in CFD to precisely capture a particle's interaction with other particles and the vessel's internal walls [20]. The two main approaches in CFD are the Eulerian–Eulerian and Eulerian–Lagrangian methods, and many existing CFD models and software can be used to estimate gas–solid multiphase flows. Each approach has its advantages and limitations.

In the Eulerian–Eulerian method, continuous phases are assumed for the whole domain, such that the solid particles are considered as a discretized phase, and both solid and gas phases are modeled as Navier–Stokes. This treatment takes much less computational time due to the solids' behavior acting as a pseudo fluid, and the sum of the solid and fluid volume fractions is unity [21,22]. However, not accounting for particle physical properties is a major disadvantage and cannot be overlooked in modeling gas–solid multiphase flow [23]. The Eulerian–Lagrangian method treats the fluid phase similarly to the Eulerian–Eulerian approach, and the differentiation is depicted in their treatment of the solid phase.

The computational particle fluid dynamics–multiphase particle in cell method (CPFD–MP-PIC) operates within the Eulerian–Lagrangian method and takes advantage of continuum and discrete models by mapping the particle properties from the Lagrangian coordinate to the Eulerian grid, where continuum terms are assessed and particle properties are mapped to the individual particles [24]. Thus, MP-PIC is practical for dense large-scale beds with less computational time. Barracuda is CPFD software that groups particles into parcels, making particle tracking slightly implicit. It has proven to be attractive in modeling circulating fluidized beds (CFBs) in recent years, due to its ability to account for an industrial-scale number of particles and design complexity, which effectively predicts the fluid structure [25–28]. Chen et al. simulated CFB risers using CPFD for Geldart B and compared them with Eulerian–Eulerian simulations. It was found that Barracuda correctly captured the riser's axial and radial fluid structure and gave a better prediction than the Eulerian–Eulerian method [29]. Berrouk et al. adopted the MP-PIC approach in modeling a real industrial FCC riser and reached good agreement with actual plant data in addition to performing various parametric studies on the riser design and operating conditions using Barracuda CPFD software [30]. Further studies have demonstrated the effectiveness of this method in describing the interaction and flow characteristics between gas and solid phases [31–34].

Packed-bed reactors are the most widely used in the refining and petrochemical industries, which hold merit advantages of a simpler design and lower capital cost. However, the thriving demand for valuable refining products and profitability was the main driving force for developing fluidized-bed reactors for catalytic cracking over packed-bed cracking. This determination resulted in the establishment of the FCC process [35]. Enos JL [36] described it as the largest research effort ever undertaken after the atomic bomb. The following advantages have been depicted in utilizing fluidized beds over packed beds:

- The catalyst moving in liquified form has enhanced the continuous operation process, which gives flexibility in temperature control and smoothly responds to changes in operation;
- Improvements in gas and solid mixing have led to isothermal operations free of cold and hot spots;
- More contact between the catalyst and reactant gases increases the conversion, mass, and heat rates;
- Lower pressure drop;
- Suitable for large-scale operation.

Due to the successful operation of beds, many other processes have adopted them. Some of them have reached commercial operation, and in this study, a bed-hosting chemical process was simulated to investigate and tap into its promising potential at a foundational level.

In this study, the Eulerian–Lagrangian approach was used to simulate the reaction of DRM in a packed bed and a fluidized bed. We first investigated how the temperature, inlet composition, and contact spatial time affected the DRM process in a packed bed. The coke formation was linked to the side reaction under various operating conditions, which helped include the coke amount and reactant conversion in the comparison of the reforming effects of the packed and fluidized beds.

2. Numerical Simulation

2.1. CFD Model and Simulation Setup

This simulation employed Barracuda Ver. 21.0 software [37], which is based on the MP-PIC method and operates within the Eulerian–Lagrangian framework. In addition, solid phase stress was employed to substitute the particle collision in order to further reduce the computational resources. MP-PIC simulations are carried out with the energy minimization sub-grid drag model to capture the mesoscale structures in a DRM reactor. Details of the energy minimization sub-grid drag model can be found in the literature [24,26,38,39]. The governing equations for the gas and the solid phases are summarized in Table 1.

2.1.1. Gas-Species Transport Equations

Each gas species is described by a transport equation. The entire fluid phase characteristics are solved from the individual gas-species mass fractions, $Y_{g,i}$. By convention, mass is exchanged between the gas species, and this is accounted for via the chemical source terms $\delta \dot{m}_{i,\text{chem}}$ in the discrete gas-species transport equation:

$$\frac{\partial(\theta_g \rho_g Y_{g,i})}{\partial t} + \nabla \cdot \left(\theta_g \rho_g Y_{g,i} \vec{u}_g\right) = \nabla \cdot \left(\theta_g D \rho_g \nabla Y_{g,i}\right) + \delta \dot{m}_{i,\text{chem}} \tag{1}$$

where θ_g is the gas volume fraction, $\vec{u}_g$ is the gas velocity, ρ_g is the gas density, and D is the turbulent mass diffusivity, which can be derived from the relation between the gas viscosity μ_g and the Schmidt number Sc:

$$Sc = \frac{\mu_g}{\rho_g D} \tag{2}$$

In this study, Sc was set to 0.9, a default value that is consistent with values published in the literature [25].

Table 1. Governing equations of the MP-PIC approach.

(1)   Continuity equation of gas phase:

$$\frac{\partial \theta_g \rho_g}{\partial t} + \nabla \cdot \left(\theta_g \rho_g \vec{u}_g \right) = 0 \tag{3}$$

(2)   Momentum equation of gas phase:

$$\frac{\partial \left(\theta_g \rho_g \vec{u}_g \right)}{\partial t} + \nabla \cdot \left(\theta_g \rho_g \vec{u}_g \vec{u}_g \right) = -\theta_g \nabla p - \vec{F} + \theta_g \mu_g \nabla^2 \vec{u}_g + \theta_g \rho_g \vec{g} \tag{4}$$

(3)   Interphase momentum exchange rate per volume:

$$\vec{F} = \iiint f V_p \rho_p \left[D \left(\vec{u}_g - \vec{u}_p \right) - \frac{1}{\rho_p} \nabla p \right] dV_p d\rho_p d\vec{u}_p \tag{5}$$

(4)   Liouville equation for finding the particle distribution function f at each time:

$$\frac{\partial f}{\partial t} + \nabla_{\vec{u}_p} \left(f \vec{A} \right) + \nabla \left(f \vec{u}_p \right) = 0 \tag{6}$$

(5)   Particle acceleration equation:

$$\vec{A} = D \left(\vec{u}_g - \vec{u}_p \right) - \frac{1}{\rho_p} \nabla p + \vec{g} - \frac{1}{\theta_p \rho_p} \nabla \tau_p \tag{7}$$

(6)   Particle normal stress model:

$$\tau_p = \frac{10 P_s \theta_p^\beta}{\max \left[\left(\theta_{cp} - \theta_p \right), \varepsilon \left(1 - \theta_p \right) \right]} \tag{8}$$

(7)   Particle volume fraction description:

$$\theta_p = \iiint f V_p dV_p d\rho_p d\vec{u}_p \tag{9}$$

2.1.2. Equations of Energy Conservation

Snider [39] presented the energy conservation equation of the gas phase:

$$\frac{\partial (\theta_g \rho_g h_g)}{\partial t} + \nabla \cdot \left(\theta_g \rho_g h_g \vec{u}_g \right) = \theta_g \left(\frac{\partial p}{\partial t} + \vec{u}_g \cdot \nabla p \right) + \Phi - \nabla \cdot \left(\theta_g q \right) + \dot{Q} + S_h + \dot{q}_D \tag{10}$$

where h_g is the gas enthalpy, Φ is the viscous dissipation, and $\dot{Q}$ is the energy source per volume. S_h is the conservative energy transfer from the solid phase to the fluid phase. The gas heat flux

$$q = -\lambda_g \nabla T_g \tag{11}$$

where λ_g is the gas thermal conductivity, which is the sum of the molecular and eddy conductivities from Reynolds stress mixing. The eddy conductivity λ_{t_t} is calculated based on the turbulent Prandtl number:

$$Pr_t = \frac{C_p}{\lambda_t} \tag{12}$$

Pr_t is set to 0.9, denoting the default value.

The enthalpy diffusion term $\dot{q}_D$ is

$$\dot{q}_D = \sum_{i=1}^{N_s} \nabla \cdot \left(h_{g,i} \theta_g \rho_g D \nabla Y_{g,i} \right) \tag{13}$$

where N_s is the sum of all gas species and $h_{g,i}$ is the enthalpy of gas-phase species i.

Furthermore, the pressure, enthalpy, temperature, density, and mass fractions of the gas phase are closely related via the state equation. In the MP-PIC method, the partial pressure of a gas species is obtained using the ideal gas state equation:

$$P_i = \frac{\rho_g Y_{g,i} R T_g}{Mw_i} \tag{14}$$

where R is the universal gas constant, T_g is the gas mixture temperature, and Mw_i is the molecular weight of gas species i. The relation between the gas thermodynamic pressure of the total average flow and the pressures of the gas species is

$$P = \sum_{i=1}^{N_s} P_i \tag{15}$$

The relation between the mixture enthalpy and the gas species enthalpies is

$$h_g = \sum_{i=1}^{N_s} Y_{g,i} h_{g,i} \tag{16}$$

The term C_p in Equation (17) is the mixture specific heat at constant pressure. It is defined as

$$C_p = \sum_{i=1}^{N_s} Y_{g,i} C_{p,i} \tag{17}$$

where $C_{p,i}$ denotes the specific heat of gas species i. The gas species enthalpies $h_{g,i}$ is a function of the gas temperature T_g:

$$h_{g,i} = \int_{T_{ref}}^{T_g} C_{p,i} dT + \Delta h_{g,i}, \tag{18}$$

where $\Delta h_{g,i}$ is the heat formation of gas species i at the reference temperature T_{ref}.

During the chemical reactions, no heat is released within a particle, which is an assumption required for the particle energy conservation equation. In addition, the heat liberated at the surfaces of the particles during chemical reactions is directly tied to the gas-phase energy and does not affect the surface energy balance. As a result, the lumped particle heat equation is

$$C_v \frac{dT_p}{dt} = \frac{1}{m_p} \frac{\lambda_g Nu_{g,p}}{2r_p} A_p (T_g - T_p) \tag{19}$$

where T_p is the solid catalyst temperature, T_g is the gas mixture temperature, C_v is the specific heat of the particle, $Nu_{g,p}$ is the Nusselt number for heat exchange between the gas and the solid particle, and λ_g is the gas thermal conductivity.

The energy exchanged between the solid and fluid phases is expressed as

$$S_h = \iiint f \left\{ m_p \left[D_p \left(\vec{u}_p - \vec{u}_g \right)^2 - C_v \frac{dT_p}{dt} \right] - \frac{dm_p}{dt} \left[h_p + \frac{1}{2} \left(\vec{u}_p - \vec{u}_g \right)^2 \right] \right\} dm_p d\vec{u}_p dT_p \tag{20}$$

where h_p is the solid enthalpy:

$$h_p = \left(2.0 + 1.2 Re^{0.5} Pr^{0.33} \right) \times \left(\frac{\lambda_g}{d_p} \right) \tag{21}$$

The packed bed's schematic diagram is shown in Figure 1a, and it comes from the experimental apparatus of Benguerba et al. [40]. Atmospheric pressure mixed gas with a molar ratio of $CH_4:CO_2:N_2 = 1:1:8$ controlled the device's inlet. This reactor had a diameter and length of 0.008 m and 0.22 m, respectively. Figure 1b displays the schematic diagram of the fluidized bed, which was taken from Durán et al.'s experimental setup [41]. Its diameter and height were 0.03 m and 0.3 m, respectively. Table 2 lists the packed- and fluidized-bed geometries, the main physical characteristics of the gas and solids phases, and the operating conditions.

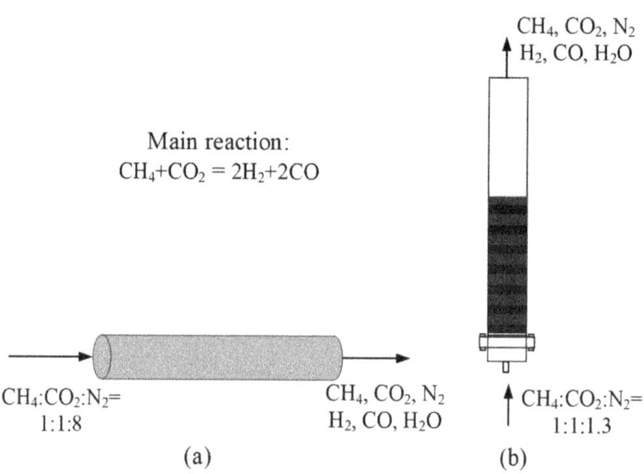

Figure 1. The schematic diagram of the reactor.

Table 2. Gas and solid properties, operating conditions, and geometries of the simulated reactors. (a) Packed bed [40]; (b) fluidized bed [41].

(a)		(b)	
Parameters	Value	Parameters	Value
Operating pressure, P (Pa)	101,325	Operating pressure, P (Pa)	101,325
Operating temperature, T (K)	723–923	Operating temperature, T (K)	773
Gas density, ρ_g (kg/m^3)	1.2	Gas density, ρ_g (kg/m^3)	0.456
Gas viscosity, μ (Pa·s)	1.8×10^{-5}	Gas viscosity, μ (Pa·s)	3.62×10^{-5}
Particle density, ρ_p (kg/m^3)	1500	Particle density, ρ_p (kg/m^3)	1250
Average particle diameter, $\overline{d_p}$ (mm)	0.32	Particle diameter, d_p (μm)	106–180
Particle volume fraction at close pack	0.6	Particle volume fraction at close pack	0.2
CH$_4$/CO$_2$ feed ratio	1:1	CH$_4$/CO$_2$ feed ratio	1:1
CH$_4$/N$_2$ feed ratio	1:8	CH$_4$/N$_2$ feed ratio	1:1.3
Diameter, D (m)	0.008	Diameter, D (m)	0.03
Length, L (m)	0.22	Length, L (m)	0.03
Inlet flow rate, mL/min	52.2	Column height, H (m)	0.3
Drag model	WenYu-Ergun [42]	Static bed height, H$_s$ (m)	0.15
-		Superficial gas U$_g$ velocity, (m/s)	0.0064–0.15
		Drag model	EMMS [43]

2.2. Reaction Kinetics

In modeling and simulating the dry reforming of methane, there are thousands of associated elementary chemical reaction steps with the heterogeneous catalyst. The involvement of the excess species and reaction intermediates in the detailed surface kinetic mechanism leads to an enormous network; this is called the microkinetic approach [44–47]. Microkinetic models give a precise report of the chemical system at the atomic level for varieties of operating conditions and concentrations since it avoids RDS assumptions. However, this large-scale network is impractical to imbed in CFD due to its expensive computational cost and inactive reaction path calculations. Another classification of kinetic modeling that avoids detailed information about all side reactions is the macrokinetic approach. Macrokinetic models are characterized in the form of a globalized power law that treats the chemical system at the molecular level as a black box, offering less computational time and power. The black box system risks overprediction, although it has been used in the scale-up designs of reactors for several years due to its straightforward implementation in CFD. The most widely used macrokinetic model in the industry has been the Langmuir–Hinshelwood–Hougen–Watson (LHHW) model. Built on continuum models that include simplifications in the assumption of a single rate-limiting step, the LHHW model is a more detailed assembly model than the black box system and can be derived from micro-kinetic models [48–50].

The following are the primary DMR reactions:

Main reaction:

$$CH_4 + CO_2 \leftrightarrow 2H_2 + 2CO \tag{22}$$

Reverse of water–gas shift (RWGS):

$$CO_2 + H_2 \leftrightarrow H_2O + CO \tag{23}$$

Methane decomposition:

$$CH_4 \leftrightarrow 2H_2 + C_{(s)} \tag{24}$$

Boudouard reaction:

$$C_{(s)} + CO_2 \leftrightarrow 2CO \tag{25}$$

Carbon gasification:

$$C_{(s)} + H_2O \leftrightarrow CO + H_2 \tag{26}$$

The following are the equations used to calculate the conversions of CH_4 and CO_2 and the yields of H_2, CO, and H_2O in DRM:

$$X_{CH_4} = \frac{[F_{CH_4}]_{in} - [F_{CH_4}]_{out}}{[F_{CH_4}]_{in}} \times 100\% \tag{27}$$

$$X_{CO_2} = \frac{[F_{CO_2}]_{in} - [F_{CO_2}]_{out}}{[F_{CO_2}]_{in}} \times 100\% \tag{28}$$

$$Y_{H_2} = \frac{[F_{H_2}]_{out}}{2[F_{CH_4}]_{in}} \times 100\% \tag{29}$$

$$Y_{CO} = \frac{[F_{CO}]_{out}}{[F_{CH_4}]_{in} + [F_{CO_2}]_{in}} \times 100\% \tag{30}$$

$$Y_{H_2O} = \frac{[F_{H_2O}]_{out}}{[F_{CH_4}]_{in}} \times 100\% \tag{31}$$

where $[F_{CH4}]_{in}$ and $[F_{CO2}]_{in}$ are the inlet molar flow rates of CH_4 and CO_2, respectively, and $[F_{CO}]_{out}$, $[F_{H2}]_{out}$, and $[F_{H2O}]_{out}$ are the outlet molar flow rates of CO, H_2, and H_2O, respectively. The yields of steam and coke were neglected as they were very negligible.

Although the SRM process has a high methane conversion rate and a low carbon deposition, it cannot help reduce global carbon dioxide emissions. DRM has attracted many researchers due to its vast capability to effectively valorize both natural gas and CO_2 without the need for water vaporization to produce the required syngas with a valuable ratio of H_2/CO that can be exploited in various refining processes. However, early investigations have shown that DRM's potential to replace SRM is challenged by some operational issues, such as excessive catalyst deactivation and difficult conversion to the desired product amount and ratio. The latter issues are particularly true for packed-bed DRM. To tackle these issues, the development of the DRM process in fluidized beds appears to be one of the solutions since fluidization is known for its high mass and heat transfers and better control over catalyst and gas flows. Therefore, this study researched each of them separately in order to study the difference in DRM between the packed bed and the fluidized bed.

3. Results and Discussion

3.1. Model Validation of Packed Bed

Figure 2 shows the conversions and yields of DRM over time under different temperatures. It should be noted that the conversion rates of CH_4 and CO_2 were close to one because only a few gases can be detected at the outlet at the start of the reaction. The conversions of CH_4/CO_2 and the yields of H_2/CO reached stable operations after 10 min, indicating the total reaction arrived at thermodynamic equilibrium. However, the author did not provide details of the simulation time. Hence, the data selected for this study were for after 15 min of operations for modal verification.

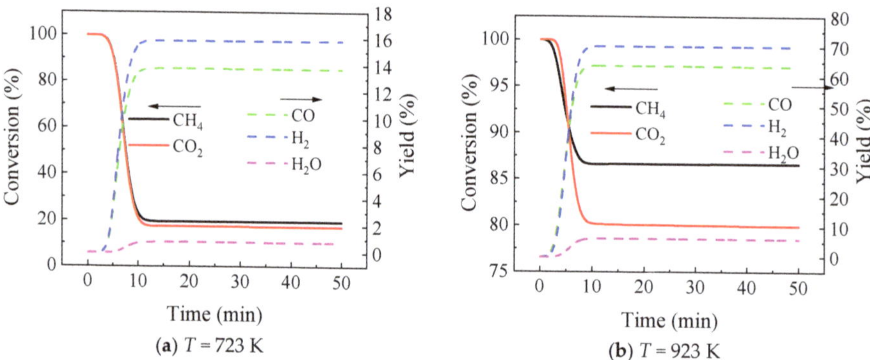

Figure 2. Rate of the conversion and yield over time at the reactor outlet. (**a**) $T = 723$ K, (**b**) $T = 923$ K.

3.2. Effect of Temperature

The conversion, yield, and ratio of H_2/CO are more susceptible to the reactant temperature. It is necessary to compare the simulation and experiment settings used in this study, particularly concerning how the reaction temperature affected the DRM process.

Figure 3 shows a comparison of the experimental and simulated CH_4 and CO_2 conversion at various temperatures. In the DRM process at equilibrium, the temperature affected the reactant conversion favorably within a given temperature range. Additionally, the obtained simulation conversions were close to Benguerba's numerical data and experimental results [40]. Although the CH_4 conversions in this study were slightly lower than the experimental data, the CO_2 conversions were very close to the experiment. Figure 3a shows that the methane conversion rate was lower than the experimental value regardless of our simulation value or Benguerba et al.'s simulation value. The kinetic equation explains why the simulated methane conversion was less than the observed measurement. The lower stoichiometric coefficient of the primary reaction of DRM was the most likely cause. Both employed DRM kinetic models were derived from various sources in the literature. The simulated materials, however, cannot be wholly equal to the catalysts described in the literature, and slight differences in the reaction coefficient could be a result of different catalysts.

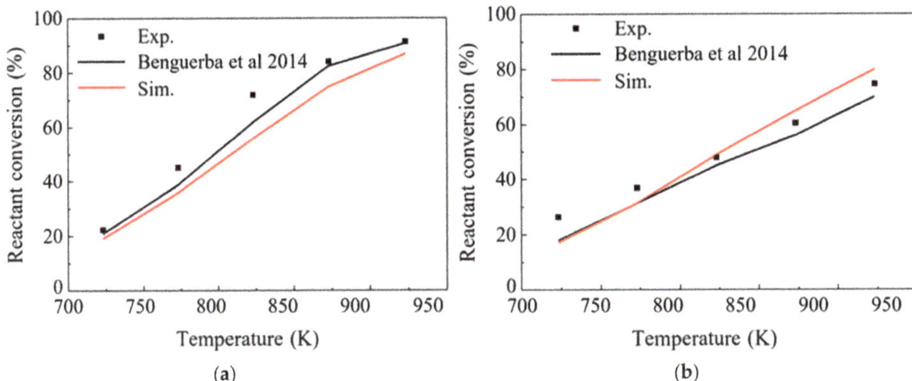

Figure 3. Conversion of experiment and simulation for different temperatures [40]. (a) CH_4; (b) CO_2.

As shown in Figure 4a, we immediately observed a change in the simulated yield with the change in temperature since the simulated reactant conversion rate in the example above is consistent with the experimental trend. The increased yields of CO and H_2 with the temperature are consistent with the trend shown in Figure 3, because the conversions of the reactants increased as the temperature rose. The molar ratio of H_2/CO is presented in Figure 4b. The obtained values were slightly greater than unity. This demonstrates that, in addition to the main reaction (reaction 1), the methane decomposition (reaction 3) contributed to hydrogen production. Methane decomposition is an endothermic reaction that becomes stronger as the temperature rises. Thermal methane cracking occurs as the temperature rises above 400 °C for most catalysts [18,51–53].

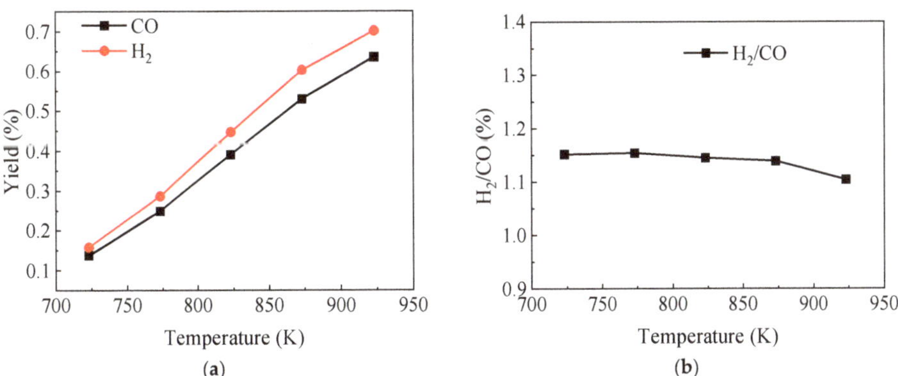

Figure 4. Effect of the reaction temperature on products. (a) Yield; (b) ratio of H_2/CO.

Numerous studies have demonstrated that carbon deposition on the catalyst is a necessary part of the DRM process and the primary cause of catalyst deactivation [18,40,54]. In the DRM process, the methane decomposition (3), Boudouard reaction (4), and carbon gasification (5) are the primary contributors to coke formation. Of these reactions, methane decomposition (3) produces substantially more coke than the other reactions [18,55,56], especially at high reaction temperatures. Because methane decomposition tends to have a more violent reaction in high-temperature environments, coke formation is facilitated by higher reaction temperatures. This correspondingly explains why, in most cases, the ratio of H_2/CO is slightly greater than one.

The trend of carbon deposition at various temperatures is depicted in Figure 5, showing that the amount of carbon deposition increased with an increasing reaction temperature. As the temperature reached 850 °C, the carbon deposition barely shifted. The Boudouard reaction (4) and carbon gasification (5) consumed some coke even though the temperature rise accelerated methane cracking. This indicates that the momentum, mass, and energy exchange between particles and the surrounding environment achieved a relatively stable equilibrium. Therefore, once the system reached a stable state, it was in a thermodynamic equilibrium condition.

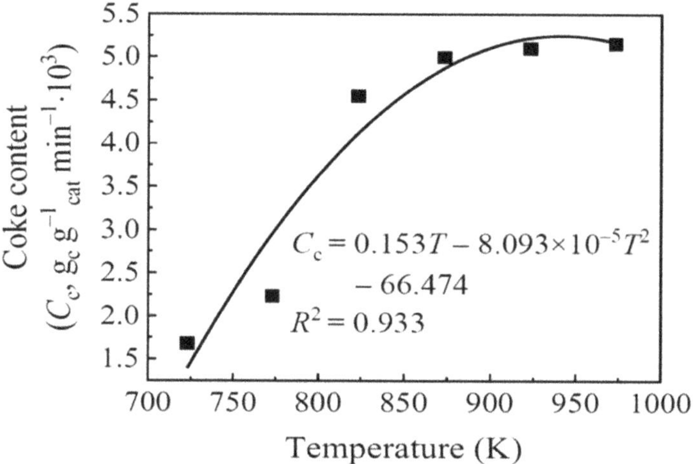

Figure 5. Effect of temperature on coke deposition, with the coke content represented by the coking amount per minute per gram of catalyst.

3.3. Effect of Components

One of the most critical parameters in the DRM process is the composition of the gas inlet. The leading gases at the reactor inlet are methane and carbon dioxide, with nitrogen serving as an inert gas. Nitrogen oxides were not produced because the reaction temperature was generally less than 1500 K. As shown in Figure 5, the carbon deposition was the lowest at 923 K, so the following simulation conditions were all at this temperature.

As shown in Figure 6, the first set of gas composition data was CH_4:CO_2:N_2= 1:1:1, and it was used as a baseline to study the effect of each component on the reaction by changing a single gas. The conversion of reactants was essentially unaffected by altering the nitrogen content at the fixed molar ratio of CH_4/CO_2. Nitrogen is typically exclusively employed as a reaction's dilution gas or shielding gas.

Figure 6 also depicts the effect of the ratio of CH_4/CO_2 on the reactant conversion. The ratio of CH_4/CO_2 was 0.50:0.67:1.0, and the corresponding CH_4 and CO_2 conversions were 88.1%:84.2%:78.6% and 66.1%:70.3%:75.4%, respectively. As a result, the inlet components had significant control over the pace at which the reactants converted. The ratio of CH_4/CO_2 may be increased to achieve a higher carbon dioxide conversion rate; it may be decreased to achieve a higher methane conversion rate. The CH_4 and CO_2 had a near conversion rate when the ratio was around unity.

Products, in general, tend to be more focused on the DRM process. Figure 7 shows the effects of the ratio of CH_4/CO_2 on the yields of H_2 and CO. The highest results for H_2 and CO occurred when the component was CH_4:CO_2:N_2= 1:1:2, and the corresponding yields were 51.8% and 47.1%. The following component was CH_4:CO_2:N_2= 1:1:1.5, with matching yields of 48.7% and 44.7%. Thus, both yields were not significantly different when the molar ratio CH_4/CO_2 was at unity. The molar rates of CH_4 and CO_2 at the inlet should be kept constant or close to each other from a balance standpoint to have a higher yield.

Although a lower ratio of CH_4/CO_2 resulted in a higher CH_4 conversion rate, it did not effectively increase the amount of product; therefore, side reactions, i.e., reactions (2)–(5), should be considered.

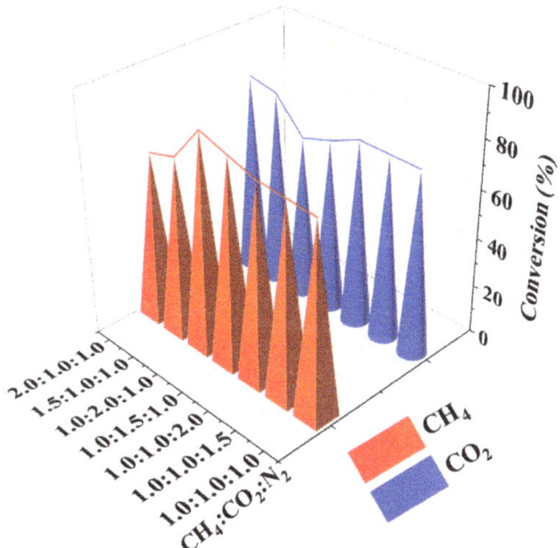

Figure 6. Conversion of CH_4 and CO_2 at $T = 923$ K under the different inlet components.

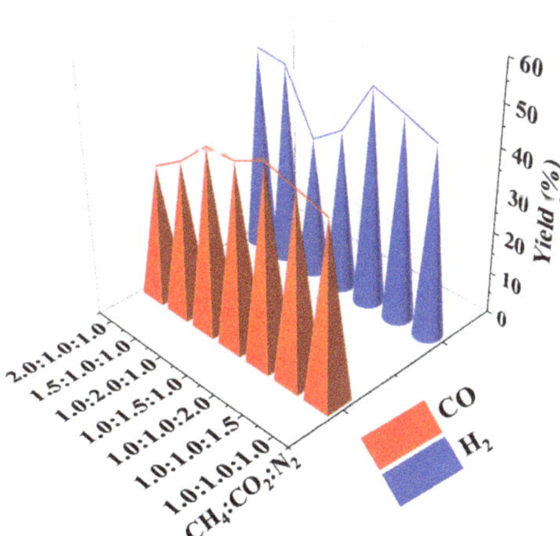

Figure 7. Yields of H_2 and CO at $T = 923$ K under the different inlet components.

Figure 8 shows the ratio of H_2/CO under the different inlet components; obviously, this value can be controlled by a variety of elements. The proportions of H_2/CO were 0.76, 1.31, and 1.53, while the ratios of CH_4/CO_2 were 0.5, 1.0, and 2.0. Therefore, if DRM had a high H_2/CO ratio, it would have to run on more CH_4. Similarly, it would be essential to function with excess CO_2 if DRM had a low H_2/CO ratio, and the product ratio would be close to unity at $CH_4:CO_2 = 1:1$. As a result, the desired balance of H_2/CO can be obtained by adjusting the feed percentage of the reactant gases.

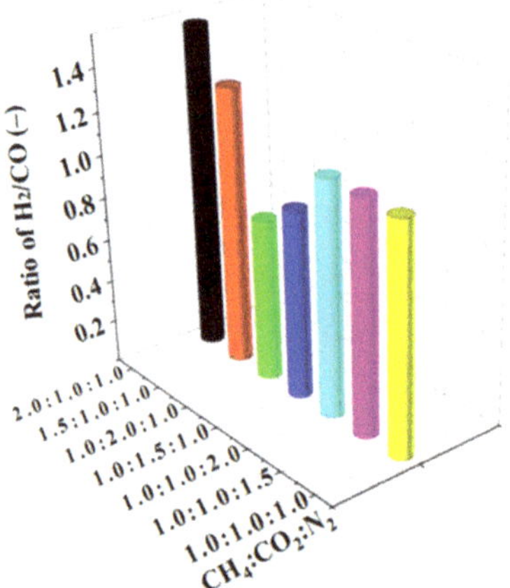

Figure 8. Ratio of H_2 over CO at $T = 923$ K under the different inlet components.

3.4. Coke Formation

The catalyst channel will be blocked after coking on the catalyst surface, resulting in catalyst deactivation. As a result, coke formation is the primary cause of catalyst deactivation [50,57]. Coke is primarily produced by methane cracking in a high-temperature environment, though the Boudouard reaction and carbon gasification will consume some coke as well. The contact spatial time is crucial to the DRM process in a packed bed. In this study, different contact spatial times were obtained by changing the inlet flow rate of CH_4 but keeping the solid mass constant.

Figure 9 depicts the coke content as a function of the contact spatial time. The coke content was significantly affected by the contact spatial time; when the coke increased with the contact spatial time, it reached a maximum and then decreased. Many researchers have identified this phenomenon [18,58]. This complex phenomenon was related to methane decomposition, indicating a maximum coking range between the methane flow and catalyst loading. The contact spatial time may need to be appropriately increased if the coke formation is severe.

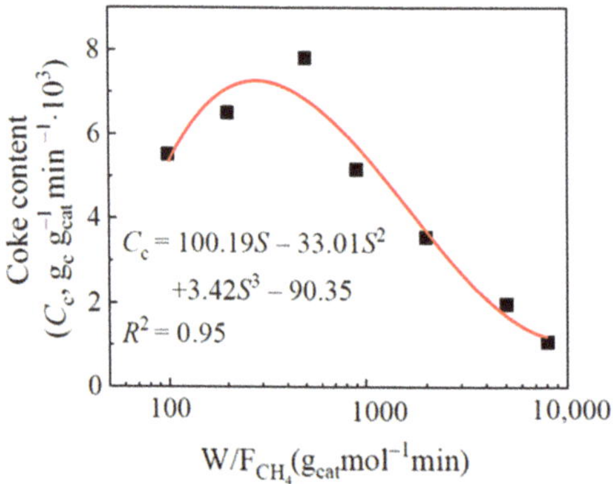

Figure 9. Effects of contact spatial time on coke content ($T = 923$ K, CH_4:CO_2:N_2 = 1:1:1.5). Here, the coke content represents the coking amount per minute per gram of catalyst. W represents the catalyst mass.

3.5. Comparison of Conversion in Packed and Fluidized Beds

As discussed above, the issue of carbon deposition is particularly true for packed-bed DRM. To tackle these issues, the development of a DRM process in fluidized beds appears to be one of the solutions since fluidization is known for its high mass and heat transfers and better control over catalyst and gas flows [49,59–63]. In addition to the catalytic performance, mechanical stability is crucial for use in the fluidized bed [48]. Comparing the differences in the DRM process between the packed bed and fluidized bed under proximate operating conditions was therefore another purpose of this study.

Using Abrahamsen and Geldart's correlation [64], the minimum fluidization velocity was estimated as follows:

$$U_{mf} = \frac{0.0009(\rho_P - \rho_g)^{0.934} g^{0.934} d_P^{1.8}}{\mu^{0.87} \rho_g^{0.066}} = 0.0055 \text{ m/s}$$

Durán et al.'s fluidized-bed experimental setup served as the basis for the simulation. Figure 1b displays the experimental schematic diagram, and Table 2b lists the simulation parameters.

Table 3 contrasts the two reactors' typical results. Because the kinetic equation employed in this study was developed from other catalyst materials, it may explain why the simulation results for the methane conversion are slightly bigger than the experimental values. Notably, the focus was to compare the reforming effects of packed and fluidized beds. It can be noticed that, under the initial fluidization regime, the conversion rate in the fluidized bed was higher than in the packed bed with less coking. In order to provide practical direction for the design, scale-up, and industrial application of the fluidized-bed DRM process, the next task is to traverse more fluidized beds and examine the conversion rates, yields, and coking phenomena under various flow regimes.

Table 3. Durán et al. [41] verified the fluidized bed's performance against a packed bed with similar operating conditions.

Bed	P (bar)	U_g (m/s)	T (K)	CH$_4$:CO$_2$:N$_2$ (Mole)	X_{CH4}, Exp.	X_{CH4}, Sim.	Coke Content (g$_c$/g$_{cat}$/min/10^3)
Packed bed	1.0	0.005	773.0	1:1:1.3	-	26.6%	2.30
Bubbling fluidized bed	1.0	0.006	773.0	1:1:1.3	21.0%	29.7%	1.92

4. Conclusions

In this study, the Eulerian–Lagrangian approach was used to simulate the reaction of DRM in packed- and fluidized-bed settings. The simulation results agree well with the published experimental data. The validated models were used to study the impact of the temperature, inlet composition, and contact spatial time on the performance of the DRM process. The key element that directly influenced the conversion and yield was found to be the temperature, as the methane cracking worsened as it rose. The impact of the side effects on the DRM process was analyzed. It was demonstrated that the excess of carbon dioxide greatly influenced the methane conversion rate. The DRM yields were not significantly affected when the molar ratio of CH_4/CO_2 was close to unity for different operating conditions. The desired ratio of H_2/CO could be obtained by adjusting the feed percentage of the reactant gases. Furthermore, coke formation could properly be reduced by changing the contact spatial time. Finally, it was found that the conversion rate in a fluidized-bed setting was marginally higher than that in the packed-bed setting, but less coking was formed in the case of the fluidized bed.

Author Contributions: Conceptualization, F.A.-O. and A.S.B.; methodology, F.A.-O. and H.X.; software, F.A.-O. and H.X.; validation, F.A.-O. and H.X.; formal analysis, F.A.-O., H.X. and A.S.B.; investigation, F.A.-O. and H.X.; resources, A.S.B. and K.P.; data curation, F.A.-O.; writing—original draft preparation, F.A.-O.; writing—review and editing, A.S.B.; visualization, F.A.-O. and H.X.; supervision, A.S.B.; project administration, A.S.B. and K.P.; funding acquisition, A.S.B. and K.P. All authors have read and agreed to the published version of the manuscript.

Funding: This research was funded by Khalifa University grant number [RC2-2018-024] And The APC was funded by [RC2-2018-024].

Data Availability Statement: The data presented in this study are available on request from the corresponding author. The data are not publicly available due to restrictions imposed by RC2 on sharing data generated by projects sponsored by the Center.

Acknowledgments: The authors acknowledge the financial support from the Center for Catalysis and Separation, Khalifa University of Science and Technology, through grant No. RC2-2018-024. The second author acknowledges the financial support (travel fund) of China Scholarships Council (No. 202106440082).

Conflicts of Interest: The authors declare no conflict of interest.

References

1. Fan, M.-S.; Abdullah, A.Z.; Bhatia, S. Catalytic Technology for Carbon Dioxide Reforming of Methane to Synthesis Gas. *ChemCatChem* **2009**, *1*, 192–208. [CrossRef]
2. Wang, S.; Lu, G.Q.; Millar, G.J. Carbon Dioxide Reforming of Methane To Produce Synthesis Gas over Metal-Supported Catalysts: State of the Art. *Energy Fuels* **1996**, *10*, 896–904. [CrossRef]
3. Sazali, N. Emerging technologies by hydrogen: A review. *Int. J. Hydrogen Energy* **2020**, *45*, 18753–18771. [CrossRef]
4. Noh, Y.S.; Lee, K.-Y.; Moon, D.J. Hydrogen production by steam reforming of methane over nickel based structured catalysts supported on calcium aluminate modified SiC. *Int. J. Hydrogen Energy* **2019**, *44*, 21010–21019. [CrossRef]
5. Hilliard, S. Water Splitting Photoelectrocatalysis: The Conception and Construction of a Photoelectrocatalytic Water Splitting Cell. Doctoral Dissertation, Université Pierre et Marie Curie-Paris VI, Paris, France, 2016.
6. Turner, J.A. Sustainable hydrogen production. *Science* **2004**, *305*, 972–974. [CrossRef]
7. Bahzad, H.; Shah, N.; Dowell, N.M.; Boot-Handford, M.; Soltani, S.M.; Ho, M.; Fennell, P.S. Development and techno-economic analyses of a novel hydrogen production process via chemical looping. *Int. J. Hydrogen Energy* **2019**, *44*, 21251–21263. [CrossRef]
8. Gangadharan, P.; Kanchi, K.C.; Lou, H.H. Evaluation of the economic and environmental impact of combining dry reforming with steam reforming of methane. *Chem. Eng. Res. Des.* **2012**, *90*, 1956–1968. [CrossRef]
9. Şener, A.N.; Günay, M.E.; Leba, A.; Yıldırım, R. Statistical review of dry reforming of methane literature using decision tree and artificial neural network analysis. *Catal. Today* **2018**, *299*, 289–302. [CrossRef]
10. Al-Fatesh, A.S.; Naeem, M.A.; Fakeeha, A.H.; Abasaeed, A.E. Role of La_2O_3 as Promoter and Support in Ni/γ-Al_2O_3 Catalysts for Dry Reforming of Methane. *Chin. J. Chem. Eng.* **2014**, *22*, 28–37. [CrossRef]
11. Aramouni, N.A.K.; Touma, J.G.; Tarboush, B.A.; Zeaiter, J.; Ahmad, M.N. Catalyst design for dry reforming of methane: Analysis review. *Renew. Sustain. Energy Rev.* **2018**, *82*, 2570–2585. [CrossRef]
12. Arora, S.; Prasad, R. An overview on dry reforming of methane: Strategies to reduce carbonaceous deactivation of catalysts. *RSC Adv.* **2016**, *6*, 108668–108688. [CrossRef]
13. Bang, S.; Hong, E.; Baek, S.W.; Shin, C.-H. Effect of acidity on Ni catalysts supported on P-modified Al_2O_3 for dry reforming of methane. *Catal. Today* **2018**, *303*, 100–105. [CrossRef]
14. Chen, X.; Sheng, Z.; Murata, S.; Zen, S.; Kim, H.-H.; Nozaki, T. CH_4 dry reforming in fluidized-bed plasma reactor enabling enhanced plasma-catalyst coupling. *J. CO2 Util.* **2021**, *54*, 101771. [CrossRef]
15. Fujitsuka, H.; Kobayashi, T.; Tago, T. Development of Silicalite-1-encapsulated Ni nanoparticle catalyst from amorphous silica-coated Ni for dry reforming of methane: Achieving coke formation suppression and high thermal stability. *J. CO2 Util.* **2021**, *53*, 101707. [CrossRef]
16. Pakhare, D.; Spivey, J. A review of dry (CO_2) reforming of methane over noble metal catalysts. *Chem. Soc. Rev.* **2014**, *43*, 7813–7837. [CrossRef]
17. Yusuf, M.; Farooqi, A.S.; Alam, M.A.; Keong, L.K.; Hellgardt, K.; Abdullah, B. Response surface optimization of syngas production from greenhouse gases via DRM over high performance Ni–W catalyst. *Int. J. Hydrogen Energy* **2021**, *47*, 31058–31071. [CrossRef]
18. Zambrano, D.; Soler, J.; Herguido, J.; Menéndez, M. Kinetic Study of Dry Reforming of Methane Over Ni–Ce/Al_2O_3 Catalyst with Deactivation. *Top. Catal.* **2019**, *62*, 456–466. [CrossRef]
19. Wang, J. Continuum theory for dense gas-solid flow: A state-of-the-art review. *Chem. Eng. Sci.* **2020**, *215*, 115428. [CrossRef]
20. Parker, J.; LaMarche, K.; Chen, W.; Williams, K.; Stamato, H.; Thibault, S. CFD simulations for prediction of scaling effects in pharmaceutical fluidized bed processors at three scales. *Powder Technol.* **2013**, *235*, 115–120. [CrossRef]

21. Zhang, Y.; Lan, X.; Gao, J. Modeling of gas-solid flow in a CFB riser based on computational particle fluid dynamics. *Pet. Sci.* **2012**, *9*, 535–543. [CrossRef]
22. Abbasi, A.; Islam, M.A.; Ege, P.E.; de Lasa, H.I. CPFD flow pattern simulation in downer reactors. *AIChE J.* **2013**, *59*, 1635–1647. [CrossRef]
23. Chen, X.; Wang, J. A comparison of two-fluid model, dense discrete particle model and CFD-DEM method for modeling impinging gas–solid flows. *Powder Technol.* **2014**, *254*, 94–102. [CrossRef]
24. Andrews, M.J.; O'Rourke, P.J. The multiphase particle-in-cell (MP-PIC) method for dense particulate flows. *Int. J. Multiph. Flow* **1996**, *22*, 379–402. [CrossRef]
25. Zhao, P.; O'Rourke, P.J.; Snider, D. Three-dimensional simulation of liquid injection, film formation and transport, in fluidized beds. *Particuology* **2009**, *7*, 337–346. [CrossRef]
26. Snider, D.; Banerjee, S. Heterogeneous gas chemistry in the CPFD Eulerian–Lagrangian numerical scheme (ozone decomposition). *Powder Technol.* **2010**, *199*, 100–106. [CrossRef]
27. Karimipour, S.; Pugsley, T. Application of the particle in cell approach for the simulation of bubbling fluidized beds of Geldart A particles. *Powder Technol.* **2012**, *220*, 63–69. [CrossRef]
28. Abbasi, A.; Ege, P.E.; de Lasa, H.I. CPFD simulation of a fast fluidized bed steam coal gasifier feeding section. *Chem. Eng. J.* **2011**, *174*, 341–350. [CrossRef]
29. Chen, C.; Werther, J.; Heinrich, S.; Qi, H.-Y.; Hartge, E.-U. CPFD simulation of circulating fluidized bed risers. *Powder Technol.* **2013**, *235*, 238–247. [CrossRef]
30. Berrouk, A.S.; Pornsilph, C.; Bale, S.S.; Du, Y.; Nandakumar, K. Simulation of a Large-Scale FCC Riser Using a Combination of MP-PIC and Four-Lump Oil-Cracking Kinetic Models. *Energy Fuels* **2017**, *31*, 4758–4770. [CrossRef]
31. Oloruntoba, A.; Zhang, Y.; Xiao, H. Study on effect of gas distributor in fluidized bed reactors by hydrodynamics-reaction-coupled simulations. *Chem. Eng. Res. Des.* **2022**, *177*, 431–447. [CrossRef]
32. Oloruntoba, A.; Zhang, Y.; Xiao, H. Hydrodynamics-reaction-coupled simulations in a low-scale batch FCC regenerator: Comparison between an annular and a free-bubbling fluidized beds. *Powder Technol.* **2022**, *407*, 117608. [CrossRef]
33. Xiao, H.; Zhang, Y.; Wang, J. Virtual error quantification of cross-correlation algorithm for solids velocity measurement in different gas fluidization regimes. *Chem. Eng. Sci.* **2021**, *246*, 117013. [CrossRef]
34. Xiao, H.; Zhang, Y.; Wang, J. Correlating measurement qualities of cross-correlation based solids velocimetry with solids convection-mixing competing mechanism in different gas fluidization regimes. *Chem. Eng. Sci.* **2022**, *253*, 117602. [CrossRef]
35. Yates, J.; Lettieri, P. *Fluidized-Bed Reactors: Processes and Operating Conditions*; Springer: Cham, Switzerland, 2016; Volume 26.
36. Enos, J.L. *Petroleum, Progress and Profits: A History of Process Innovation*; MIT Press: Cambridge, MA, USA, 1962.
37. CPFD Software. Available online: https://cpfd-software.com/ (accessed on 12 January 2023).
38. Snider, D.M. Three fundamental granular flow experiments and CPFD predictions. *Powder Technol.* **2007**, *176*, 36–46. [CrossRef]
39. Snider, D.M. An Incompressible Three-Dimensional Multiphase Particle-in-Cell Model for Dense Particle Flows. *J. Comput. Phys.* **2001**, *170*, 523–549. [CrossRef]
40. Benguerba, Y.; Dehimi, L.; Virginie, M.; Dumas, C.; Ernst, B. Modelling of methane dry reforming over Ni/Al$_2$O$_3$ catalyst in a fixed-bed catalytic reactor. *React. Kinet. Mech. Catal.* **2014**, *114*, 109–119. [CrossRef]
41. Durán, P.; Sanz-Martínez, A.; Soler, J.; Menéndez, M.; Herguido, J. Pure hydrogen from biogas: Intensified methane dry reforming in a two-zone fluidized bed reactor using permselective membranes. *Chem. Eng. J.* **2019**, *370*, 772–781. [CrossRef]
42. Wen, C.-Y. Mechanics of Fluidization. In *Chemical Engineering Progress Symposium Series*; American Institute of Chemical Engineers: New York, NY, USA, 1966.
43. Yang, N.; Wang, W.; Ge, W.; Li, J. CFD simulation of concurrent-up gas–solid flow in circulating fluidized beds with structure-dependent drag coefficient. *Chem. Eng. J.* **2003**, *96*, 71–80. [CrossRef]
44. Wehinger, G.D.; Eppinger, T.; Kraume, M. Detailed numerical simulations of catalytic fixed-bed reactors: Heterogeneous dry reforming of methane. *Chem. Eng. Sci.* **2015**, *122*, 197–209. [CrossRef]
45. Wehinger, G.D.; Eppinger, T.; Kraume, M. Evaluating Catalytic Fixed-Bed Reactors for Dry Reforming of Methane with Detailed CFD. *Chem. Ing. Tech.* **2015**, *87*, 734–745. [CrossRef]
46. Wehinger, G.D.; Kraume, M.; Berg, V.; Korup, O.; Mette, K.; Schlögl, R.; Behrens, M.; Horn, R. Investigating dry reforming of methane with spatial reactor profiles and particle-resolved CFD simulations. *AIChE J.* **2016**, *62*, 4436–4452. [CrossRef]
47. Karst, F.; Maestri, M.; Freund, H.; Sundmacher, K. Reduction of microkinetic reaction models for reactor optimization exemplified for hydrogen production from methane. *Chem. Eng. J.* **2015**, *281*, 981–994. [CrossRef]
48. Olsbye, U.; Wurzel, T.; Mleczko, L. Kinetic and Reaction Engineering Studies of Dry Reforming of Methane over a Ni/La/Al$_2$O$_3$ Catalyst. *Ind. Eng. Chem. Res.* **1997**, *36*, 5180–5188. [CrossRef]
49. Wurzel, T.; Malcus, S.; Mleczko, L. Reaction engineering investigations of CO$_2$ reforming in a fluidized-bed reactor. *Chem. Eng. Sci.* **2000**, *55*, 3955–3966. [CrossRef]
50. Yin, L.; Wang, S.; Lu, H.; Ding, J.; Mostofi, R.; Hao, Z. Simulation of effect of catalyst particle cluster on dry methane reforming in circulating fluidized beds. *Chem. Eng. J.* **2007**, *131*, 123–134. [CrossRef]
51. Abbas, S.Z.; Dupont, V.; Mahmud, T. Kinetics study and modelling of steam methane reforming process over a NiO/Al$_2$O$_3$ catalyst in an adiabatic packed bed reactor. *Int. J. Hydrogen Energy* **2017**, *42*, 2889–2903. [CrossRef]

52. de la Cruz-Flores, V.G.; Martinez-Hernandez, A.; Gracia-Pinilla, M.A. Deactivation of Ni-SiO$_2$ catalysts that are synthetized via a modified direct synthesis method during the dry reforming of methane. *Appl. Catal. A Gen.* **2020**, *594*, 117455. [CrossRef]
53. Mark, M.F.; Maier, W.F.; Mark, F. Reaction kinetics of the CO$_2$ reforming of methane. *Chem. Eng. Technol.* **1997**, *20*, 361–370. [CrossRef]
54. Benguerba, Y.; Dehimi, L.; Virginie, M.; Dumas, C.; Ernst, B. Numerical investigation of the optimal operative conditions for the dry reforming reaction in a fixed-bed reactor: Role of the carbon deposition and gasification reactions. *React. Kinet. Mech. Catal.* **2015**, *115*, 483–497. [CrossRef]
55. Chen, S.; Zaffran, J.; Yang, B. Dry reforming of methane over the cobalt catalyst: Theoretical insights into the reaction kinetics and mechanism for catalyst deactivation. *Appl. Catal. B Environ.* **2020**, *270*, 118859. [CrossRef]
56. Nikoo, M.K.; Amin, N.A.S. Thermodynamic analysis of carbon dioxide reforming of methane in view of solid carbon formation. *Fuel Process. Technol.* **2011**, *92*, 678–691. [CrossRef]
57. Niu, J.; Liland, S.E.; Yang, J.; Rout, K.R.; Ran, J.; Chen, D. Effect of oxide additives on the hydrotalcite derived Ni catalysts for CO$_2$ reforming of methane. *Chem. Eng. J.* **2019**, *377*, 119763. [CrossRef]
58. Kumar, M.; Srivastava, V.C. Simulation of a Fluidized-Bed Reactor for Dimethyl Ether Synthesis. *Chem. Eng. Technol.* **2010**, *33*, 1967–1978. [CrossRef]
59. Prasad, P.; Elnashaie, S.S.E.H. Novel Circulating Fluidized-Bed Membrane Reformer for the Efficient Production of Ultraclean Fuels from Hydrocarbons. *Ind. Eng. Chem. Res.* **2002**, *41*, 6518–6527. [CrossRef]
60. Jing, Q.; Lou, H.; Mo, L.; Zheng, X. Comparative study between fluidized bed and fixed bed reactors in methane reforming with CO$_2$ and O$_2$ to produce syngas. *Energy Convers. Manag.* **2006**, *47*, 459–469. [CrossRef]
61. Thiemsakul, D.; Kamsuwan, C.; Piemjaiswang, R.; Piumsomboon, P.; Chalermsinsuwan, B. Computational fluid dynamics simulation of internally circulating fluidized bed reactor for dry reforming of methane. *Energy Rep.* **2022**, *8*, 817–824. [CrossRef]
62. Xiao, H.; Zhang, Y.; Hua, Y.; Oloruntoba, A. Quantitative comparison of measurement quality of cross-correlation based particle velocity instruments in different gas fluidization regimes. *Adv. Powder Technol.* **2021**, *32*, 3915–3926. [CrossRef]
63. Aguedal, L.; Semmar, D.; Berrouk, A.S.; Azzi, A.; Oualli, H. 3D vortex structure investigation using large eddy simulation of flow around a rotary oscillating circular cylinder. *Eur. J. Mech.-B Fluids* **2018**, *71*, 113–125. [CrossRef]
64. Abrahamsen, A.R.; Geldart, D. Behavior of gas-fluidized beds of fine powder part I: Homogeneous expansion. *Powder Technol.* **1980**, *26*, 35–46. [CrossRef]

chemengineering

MDPI

Article

Thermogravimetric Kinetic Analysis of Non-Recyclable Waste CO_2 Gasification with Catalysts Using Coats–Redfern Method

Ahmad Mohamed S. H. Al-Moftah [1,2,*], Richard Marsh [1] and Julian Steer [1]

[1] Cardiff School of Engineering, Cardiff University, Queen's Buildings, The Parade, Cardiff CF24 3AA, UK; marshr@cardiff.ac.uk (R.M.); steerj1@cardiff.ac.uk (J.S.)
[2] Qatar National Research Fund, Qatar Foundation, Doha P.O. Box 5825, Qatar
*   Correspondence: al-moftaha@cardiff.ac.uk

Abstract: In the present study, the effect of dolomite and olivine as catalysts on the carbon dioxide (CO_2) gasification of a candidate renewable solid recovered fuel, known as Subcoal™ was determined. Thermogravimetric analysis (TGA) was used to produce the TGA curves and derivative thermogravimetry (DTG) for the gasification reaction at different loadings of the catalyst (5, 10, 15 wt.%). The XRD results showed that the crystallinity proportion in Subcoal™ powder and ash was 42% and 38%, respectively. The Arrhenius constants of the gasification reaction were estimated using the model-fitting Coats–Redfern (CR) method. The results showed that the mass loss reaction time and thermal degradation decreased with the increase in catalyst content. The degradation reaction for complete conversion mainly consists of three sequences: dehydration, devolatilisation, and char/ash formation. The complete amount of thermal degradation of the Subcoal™ sample obtained with dolomite was lower than with olivine. In terms of kinetic analysis, 19 mechanism models of heterogeneous solid-state reaction were compared by the CR method to identify the most applicable model to the case in consideration. Among all models, G14 provided excellent linearity for dolomite and G15 for olivine at 15 wt.% of catalyst. Both catalysts reduced the activation energy (E_a) as the concentration increased. However, dolomite displayed higher CO_2 gasification efficiency of catalysis and reduction in E_a. At 15 wt.% loading, the E_a was 41.1 and 77.5 kJ/mol for dolomite and olivine, respectively. Calcination of the mineral catalyst is substantial in improving the activity through enlarging the active surface area and number of pores. In light of the study findings, dolomite is a suitable mineral catalyst for the industrial-scale of non-recyclable waste such as Subcoal™ gasification.

Keywords: non-recyclable waste; Subcoal™; non-isothermal; olivine; dolomite; TGA; CO_2 gasification; Qatar national vision 2030

Citation: Al-Moftah, A.M.S.H.; Marsh, R.; Steer, J. Thermogravimetric Kinetic Analysis of Non-Recyclable Waste CO_2 Gasification with Catalysts Using Coats–Redfern Method. *ChemEngineering* **2022**, *6*, 22. https://doi.org/10.3390/chemengineering6020022

Academic Editor: Anker Degn Jensen

Received: 26 January 2022
Accepted: 3 March 2022
Published: 4 March 2022

Publisher's Note: MDPI stays neutral with regard to jurisdictional claims in published maps and institutional affiliations.

1. Introduction

Climate change and global warming from the emission of greenhouse gases (GHG) are currently defined as major threats to humanity. A series of international measures and actions have been taken to mitigate the GHG effect. Recently, in Glasgow, the 26th UN Climate Change Conference of the Parties (COP26) was conducted. The participants emphasised the commitment to deep GHG emissions cuts and support the 1.5 °C global warming limit goal [1]. One of the most effective ways to cut emissions is using the thermal conversion of sustainable biomass into energy, where this biomass is assumed to utilise renewable carbon within its lifecycle.

Municipal solid waste (MSW) is one of the most inexpensive and available alternative fuels. The landfilling of non-recycled MSW in some countries such as Qatar is a challenging disposal method due to the lack of land and emissions from landfills [2]. Alternatively to landfill, the non-recycled MSW in Qatar can potentially be gasified to produce valuable synthesis gas (syngas) consisting of CO, CO_2, CH_4, and H_2 [3].

The study herein was driven by the Qatar National Vision 2030 for better MSW management. Adopting biomass gasification in Qatar will minimise the air pollutants emitted from the combustion of fossil fuels and landfills. Currently, the Domestic Solid Waste Management Centre (DSWMC) in Doha generates 30 MW of electricity from MSW as a part of the Waste-to-Energy (WtE) program [4].

Subcoal™ is a promising fossil fuel substitute consisting of a non-recyclable mixture of plastics and paper waste and is produced by the N + P Group [5]. It comes with unique properties such as high energy content, low sulphur content, and good hydrophobicity. Due to these properties, this material has been used as a solid fuel in different industries such as steel and cement production. The kinetics of Subcoal™ gasification have received limited attention in the reported literature [6]. It can be prepared from the MSW in Qatar as a feedstock for a gasification plant.

Biomass fluidised bed gasification takes place widely over bed materials, which are a mixture of sand and catalyst particles. However, the production of tar during gasification may lead to serious problems in the system such as condensation and clogging, therefore, a tar removal method is usually required [7]. In the case of municipal solid waste, the amount of material involved is likely to be very large, so the reactor residence time and length scales are the major design criteria of technical interest [2]. Therefore, a catalyst is necessary to enhance the reaction for the production of syngas. Another role of the catalyst is the cracking of tar including thermal and hydrocracking [5]. The hydrocarbons are adsorbed in their dissociative form after hydrogen is removed catalytically [6].

Synthetic and natural catalysts have been used in biomass gasification. Natural mineral catalysts are found in the Earth's crust such as dolomite and olivine [8–10]. They are commonly used because they are abundantly available and are inexpensive compared to refined and precious metals [8]. The formation of dolomite in the Khor Al-Adaid sabkha in Qatar has been reported in the work of DiLoreto et al. [11].

Olivine is a rock-forming mineral that is naturally available with a general chemical formula of magnesium iron silicate $(Mg_x, Fe_{1-x})_2 SiO_4$. The catalytic activity of olivine for tar reduction can be related to iron oxide (Fe_2O_3), magnesite (MgO), and nickel (Ni) portions. Iron is effective when it is found on the surface of the catalyst. Oxidation and/or calcination of olivine helps to transfer iron to the surface. Olivine is mainly deactivated by the formation of coke, which covers the active sites and reduces the surface area of a catalyst [12]. The catalytic influence of olivine on biomass gasification has been examined in several studies [13–16].

Dolomite is a widespread inexpensive mineral that forms in rocks over a significant underground area. It is a common tar conversion catalyst that is composed of calcium magnesium carbonate $CaMg(CO_3)_2$ [17,18]. The composition of dolomite may vary depending on the geographical location [19], however, the major compounds found in dolomite are calcium oxide (CaO), magnesium oxide (MgO), and CO_2 [20]. Dolomite may also contain traces of silica (SiO_2), aluminium oxide (Al_2O_3), and Fe_2O_3 [21].

This paper aimed to investigate the effects of dolomite and olivine concentrations on the Subcoal™ CO_2 gasification using TGA. The XRD data of Subcoal™ fuel and ash will be provided as part of the catalytic effect investigation. The thermal behaviour and kinetic parameters in the presence of the catalyst will be examined using the Coats–Redfern (CR) method. In addition, this study is regarded as foundational knowledge for future Subcoal™ gasification with catalyst research. However, this paper aimed to investigate the effects of mineral-based catalysts, namely, dolomite and olivine on the Subcoal™ CO_2 gasification using TGA. Subcoal™ as a fuel is a novel and new material type that is segregated from municipal solid waste (MSW), which can be used effectively in gasification technology to generate power in countries suffering from MSW landfilling and fossil fuel emissions such as in Qatar. In our previous studies, we presented the behaviour and kinetics of uncatalysed Subcoal™ gasification using a model free method. Some mineral catalysts such as dolomite and olivine are abundant in the Middle East region. It would be useful to achieve an understanding of the catalysis performance of these minerals in the gasification process. The influence of inexpensive mineral catalysts on the gasification of Subcoal has received insufficient or no attention. This is the first study that also included reaction kinetics investigations, which contribute to the better design of a Subcoal™ gasifier in the presence of CO_2 and a mineral catalyst. CO_2 as a gasifying agent offers unique features over air or steam to ensure better gasification performance and as a way of CO_2 utilisation. Finally, the present study is regarded as foundational knowledge for future Subcoal™ gasification with catalyst research.

2. Methodology

2.1. Materials

Pulverised Subcoal™ pellets are a commercially available waste-derived material made of non-recyclable paper and plastic waste [22]. A sample of this material was provided by N + P Group B.V in the Netherlands. To study the effect of catalyst loadings on Subcoal™ by TGA, pulverised Subcoal™ pellets were milled to less than 3 mm using a knife mill grinding machine (Fritish GmbH, Idar-Obersten, Germany). This milled product is known as a pulverised alternative fuel (PAF). The mineral content in the Subcoal™ ash (as received) was determined using X-ray fluorescence (XRF), as shown in Table 1. The minerals and elements in ash may potentially provide catalytic effects on the gasification reaction [23].

Table 1. XRF analysis of the major minerals in Subcoal™ ash.

Elements	Al_2O_3	CaO	Fe_2O_3	K_2O	SiO_2
wt.%	18.3	38	2.7	0.1	22.7

The XRD analysis of Subcoal™ powder <3 mm and ash were carried using X-ray diffraction (Diffraktometer D5000, Siemens, Munich, Germany) with a 2θ range of 20° to 90° to identify the crystallinity and composition. The sample was mounted on the diffractometer sample holder in a flat layout. A typical sample holder is a 2 mm thick plate with a 20 mm square hole in the centre. The proportion of the crystalline components in the fuel (crystallinity) was determined according to standard test method ASTM D5758 and results evaluated using the following formula [24]:

$$\text{Crystallinity \%} = \frac{A_c}{A_c + A_a} \times 100 \tag{1}$$

where A_c is the area under the peaks representing the total crystalline region, and A_a is the area under the peaks representing the total amorphous region [24]. In addition, the sample was placed onto the holder in the machine sample stage and compressed using a glass slide to obtain a flat and even surface. The holder was transferred into the XRD goniometer and the door was closed. The chiller unit and XRD unit were then switched on, respectively. The default power was set at 200 W (20 mA and 10 kV), which was increased to 1400 W (35 mA and 40 kV). The shutter door was kept open to allow X-rays to reach the sample. Then, the scanned data were inserted in the computer-based software (Panalytical X'Pert HighScore Plus) that organises the scanning process. Finally, the scan program was sent to the XRD machine and the scan commenced. The results were transferred to OriginPro® software and the crystallinity of the sample was calculated.

The XRD analysis of the Subcoal™ powder produced a large peak of amorphous compounds at 2θ of 23° composed of barium (Ba), calcium (Ca), and silica (Si), as shown in Figure 1. The first amorphous peak compounds consisted mainly of Ba, Ca, and Si. The next peak confirmed a crystalline compound at 27° with an intensity of 161 a.u. while the second amorphous peak appeared at 30°. Previous research in this area, specifically in relation to the analysis of polymers with XRD, by Marsh [25] found that amorphous peaks were dominant, as shown by their defined peaks. However, the amount of crystalline matters in plastic and paper waste exceeded the randomly distributed amorphous ones.

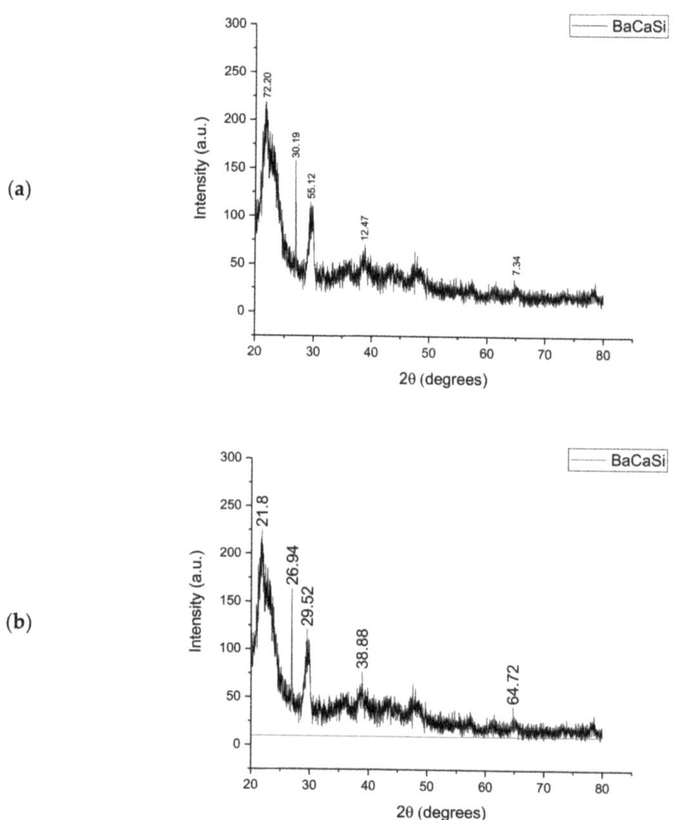

Figure 1. XRD analysis of the Subcoal™ powder: (**a**) peak height, and (**b**) peak area. The red line has been added to calculate the area under the curve.

Small crystalline peaks appeared at 2θ of 40° and 65° with the intensity of 75.5 and 26.5 au., respectively, which consisted of trace elements. It was found that the crystallinity proportion in Subcoal™ powder based on the total area of all peaks was 42%. XRD is also a substantial tool for the semi-quantitative evaluation of crystalline phases and mineral components in ash [26]. The content of amorphous materials in the biomass sample or ash determines the decomposition temperature. The higher the glass (amorphous) content, the lower the gasification temperature [27]. The non-crystallinity is attributed to the presence of amorphous aluminosilicate [28].

Figure 2 shows the XRD analysis of Subcoal™ ash in terms of peak height (a) and area (b) that was used to identify the crystallinity of the sample. The area of a peak is calculated by multiplying the FWHM (full width at half maximum) times the height. The area under the peak provides information on the percentage of crystallinity. The height of the main peaks indicates the intensity of each polymorphic phase. The largest crystalline peak of calcium chloride (CaCl) was obtained at 38.5° with an intensity of 245 au. Semi-crystalline peaks were also detected between 30° and 35°. The remaining signals included small crystalline peaks and experimental noise. The crystallinity percentage in ash based on the total area of all peaks was found to be 37.67%. In comparison with the Subcoal™ sample, the reduction in the crystallinity can be attributed to the decomposition of crystalline cellulose fibres [29]. The crystalline minerals and metal oxides in the ash may behave as a catalyst of thermal decomposition reactions [30].

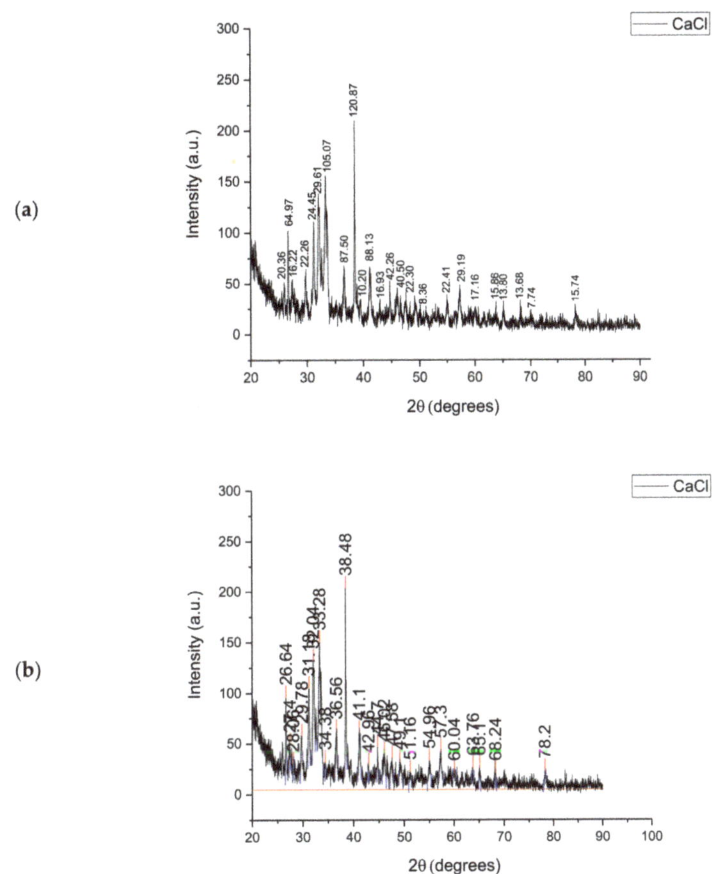

Figure 2. XRD analysis of the Subcoal™ ash: (**a**) peak height, and (**b**) peak area. The red line has been added to calculate the area under the curve.

Dolomite and olivine were supplied by the Tarmac Company (Solihull, UK) from Port Talbot in the form of powder with a particle size of <1 mm. Catalyst particle size was reduced and sieved to 106–212 µm according to BS 1377-9:1990 using a ring mill machine (Labtech Essa 100100, Australia). The chemical composition of olivine and dolomite is listed in Table 2 as supplied by the manufacturer. As can be seen, the dispersion of the active species (CaO and MgO) on the dolomite and olivine particles were around 48.72% and 49.26%, respectively.

2.2. Experimental Procedure

The experiments of CO_2 gasification of Subcoal™ were carried out at Cardiff University using TGA (Mettler Toledo AG TGA/SDTA 851e, USA). CO_2 converts char (carbon) to CO according to the slightly endothermic Boudouard reaction, which is favourable at high temperature [31]. However, using air as a gasifying agent leads to syngas dilution with nitrogen, which reduces the heating value of the fuel gas. In contrast to steam, CO_2 requires no heat for water evaporation [32]. The gasification study in TGA provides data on the thermal decomposability of Subcoal™ in relatively less expensive and complex experiments. Under non-isothermal conditions, TGA and DTG analysis were conducted for a temperature range of 25 to 900 °C following the BS EN ISO 11358:1997 standard. A constant heating rate of 20 °C/min was adopted to test the influence of different catalyst loadings (0%, 5%, 10%, and 15% on a weight basis). However, low loadings of the catalyst (e.g., 1 wt.%) were considered in this study due to insignificant changes in the gasification performance that is due to poor mixing with Subcoal and the gasification agent. A 10 ± 5 mg of a Subcoal™ PAF sample was placed into a cylindrical alumina crucible. The crucible and sample were loaded onto the TGA carousel using tweezers. CO_2 was used as a gasifying agent with a flow rate of 100 mL/min [26]. The weight loss was recorded against the time and temperature to produce TGA and DTG curves. Once the experiment was completed, the temperature, time, and mass change data were exported to a spreadsheet.

Table 2. Chemical composition of dolomite and olivine (wt.%).

Specie	Dolomite	Olivine
Fe_2O_3	2.23	6.9
SiO_2	2.4	41.7
CaO	30.57	0.06
MgO	18.15	49.2
Al_2O_3	0.71	0.45
Na_2O	0.05	-
K_2O	0.158	-
Zn	0.002	-
C	14.64	-
S	0.005	-
NiO	-	0.31
Cr_2O_3	-	0.3

2.3. Kinetic Model Description

A reaction kinetics study provides information on the reaction pathway and whether a reaction is reversible or irreversible as well as the intermediate steps [33]. The kinetics of a thermal reaction can be understood by performing experimental thermodynamic measurements under different operating conditions. Based on the overall reaction order, the rate of reaction depends on the reaction temperature. The rate of reaction (r) is expressed as follows:

$$r \propto [B]^n \tag{2}$$

where $[B]$ is a reactant concentration and n the overall reaction order. The rate of the reaction is proportional to the reaction constant (k):

$$r = k[B]^n \tag{3}$$

$$\frac{d[B]}{dt} = k[B]^n \tag{4}$$

The rate of the reaction in Equation (4) can be given as a function of conversion degree (x):

$$-[B]_o^{1-n} \frac{dx}{dt} = k[1-x]^n \tag{5}$$

A chemical reaction between molecules/particles takes place based on the theory of collision [34]. The E_a of the reaction determines the response of the reaction rate to the temperature and time. The functional form of the mathematical relationship between the reaction constant and absolute temperature ($\overline{T}$) was proposed by Arrhenius in 1889 [33]:

$$k = A\, e^{\frac{-E_a}{RT}} \tag{6}$$

Both pre-exponential constant A and E_a are independent of temperature. By combining Equations (5) and (6), the following equation is obtained:

$$-[B]_o^{1-n} \frac{dx}{dt} = \left[A\, e^{\frac{-E_a}{RT}} \right] [1-x]^n \tag{7}$$

The sample mass loss was used to calculate the conversion degree according to the following formula:

$$x = \frac{m_0 - m_t}{m_0 - m_f} \tag{8}$$

where m_0, m_t, and m_f are the mass of the sample at the start of the experiment, at the time (t), and at the end, respectively. The TGA was set to work at a constant heating rate (β) which can be introduced to Equation (9) for the non-isothermal conditions:

$$\beta = \frac{dT}{dt} \tag{9}$$

Then, the equation becomes:

$$- [B]_o^{1-n} \frac{dx}{dT} = \frac{A}{\beta} e^{\frac{-E_a}{RT}} [1 - x]^n \tag{10}$$

Integration of Equation (10) from an initial to a final value of temperature (T_i and T_f) and conversion (x_i and x_f) allows the E_a in the equation to be estimated as described below:

$$- [B]_o^{1-n} \int_{x_i}^{x_f} \frac{dX}{[1 - x]^n} = \frac{A}{\beta} \int_{T_i}^{T_f} e^{\frac{-E_a}{RT}} dT \tag{11}$$

In the present work, the CR method was used to evaluate the Arrhenius parameters [35]. The CR is an integral model-fitting method with good accuracy of estimation. According to Liu et al. [36], the error by the Coats–Redfern method does not exceed $\pm 2\%$ for a conversion degree between 20% and 80%. However, knowledge of the reaction mechanisms (overall reaction order) is required. The following approach was developed by modifying the conventional form to be effective for single or multiple heating rates [37]:

$$\ln\left(\frac{\ln(1 - x)}{T^2}\right) = \ln\left[\frac{R\,A}{\beta\,E_a}\left(1 - \frac{2\,R\,\overline{T}}{E_a}\right)\right] - \frac{E_a}{R\,T} \tag{12}$$

where $\overline{T}$ is the arithmetic mean temperature of an experiment. Equation (12) was generalised to be compatible with different reaction mechanisms. It was also assumed that the term $\frac{2\,R\,\overline{T}}{E_a} \ll 1$ to simplify the equation, hence the general form of the method becomes:

$$\ln\left(\frac{G(x)}{T^2}\right) = \ln\left[\frac{R\,A}{\beta\,E_a}\right] - \frac{E_a}{R\,T} \tag{13}$$

The E_a from Equation (13) above can be achieved by plotting term $\ln\left(\frac{G(x)}{T^2}\right)$ against $(1/T)$. The thermal decomposition mechanism function $G(x)$ can be approximated from 19 mechanism models for solid–gas reactions, as listed in Table 3. The accuracy of this method largely depends on the selection of the mechanism function and correlation coefficient (R^2). However, if the E_a is known, the reaction kinetics can be determined. The y-intercept of the E_a in the curve gives the value of the term $\ln\left[\frac{R\,A}{\beta\,E_a}\right]$, which can be rearranged to obtain the A value, as follows:

$$A = \frac{\beta\,E_a}{R} e^{y-intercept} \tag{14}$$

Table 3. Typical mechanisms for the solid–gas reaction,(LN: 5235900730034) [38].

Symbol	Reaction Mechanism	$f(x)$	$G(x)$
G1	One-dimensional diffusion, 1D	$1/2x$	x^2
G2	Two-dimensional diffusion (Valensi)	$[-\ln(1-x)]^{-1}$	$x + (1-x)\ln(1-x)$
G3	Three-dimensional diffusion (Jander)	$1.5(1-x)^{2/3}[1-(1-x)^{1/3}]^{-1}$	$[1-(1-x)^{1/3}]^2$
G4	Three-dimensional diffusion (G–B)	$1.5[1-(1-x)^{1/3}]^{-1}$	$1 - 2x/3 - (1-x)^{2/3}$
G5	Three-dimensional diffusion (A–J)	$1.5(1+x)^{2/3}[(1+x)^{1/3}-1]^{-1}$	$[(1+x)^{1/3}-1]^2$
G6	Nucleation and growth ($n = 2/3$)	$1.5(1-x)[-\ln(1-x)]^{1/3}$	$[-\ln(1-x)]^{2/3}$
G7	Nucleation and growth ($n = 1/2$)	$2(1-x)[-\ln(1-x)]^{1/2}$	$[-\ln(1-x)]^{1/2}$
G8	Nucleation and growth ($n = 1/3$)	$3(1-x)[-\ln(1-x)]^{2/3}$	$[-\ln(1-x)]^{1/3}$
G9	Nucleation and growth ($n = 1/4$)	$4(1-x)[-\ln(1-x)]^{1/3}$	$[-\ln(1-x)]^{1/4}$
G10	Autocatalytic reaction	$x(1-x)$	$\ln[x/(1-x)]$
G11	Mampel power law ($n = 1/2$)	$2x^{1/2}$	$x^{1/2}$
G12	Mampel power law ($n = 1/3$)	$3x^{2/3}$	$x^{1/3}$
G13	Mampel power law ($n = 1/4$)	$4x^{3/4}$	$x^{1/4}$
G14	Chemical reaction ($n = 3$)	$(1-x)^3$	$[(1-x)^{-2}-1]/2$
G15	Chemical reaction ($n = 2$)	$(1-x)^2$	$(1-x)^{-1}-1$
G16	Chemical reaction ($n = 1$)	$1-x$	$-\ln(1-x)$
G17	Chemical reaction ($n = 0$)	1	x
G18	Contraction sphere	$3(1-x)^{2/3}$	$1-(1-x)^{1/3}$
G19	Contraction cylinder	$2(1-x)^{1/2}$	$1-(1-x)^{1/2}$

Note: A–J: Anti–Jander; G–B: Ginstling–Brounshtein.

3. Results

3.1. Effects of Dolomite

The thermal decomposition in CO_2 gasification of Subcoal™ PAF was carried out in the presence of different loadings of dolomite catalysts, as shown in Figure 3a,b.

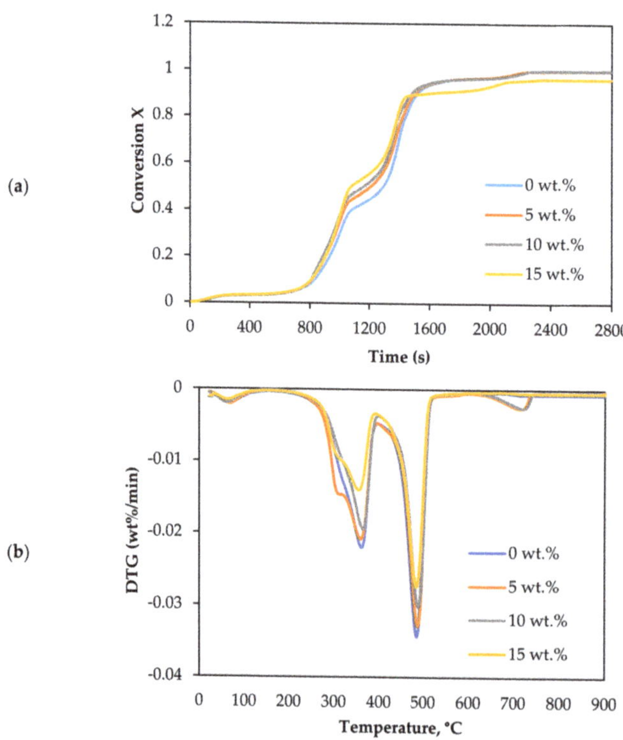

Figure 3. Effect of dolomite loading on the CO_2 gasification: (**a**) conversion degree, and (**b**) DTG curves.

The mass loss of the sample indicates that the rate of decomposition with 15 wt.% dolomite increased more steeply than without the catalyst, as shown in Figure 3a. Additionally, the conversion changed to lower temperature with the addition of the dolomite. The complete conversion reaction decreased from 792.7 to 748.4 °C as the dolomite concentration increased from 0 to 15 wt.%. The sample mainly decomposed in two sub-stages between 200 and 520 °C. Dolomite showed good performance in promoting the reaction rate of CO_2 gasification. It enhances the cracking of tar and increases the production of syngas [39]. However, at the concentration and conditions, dolomite catalysed the CO_2 gasification reaction slightly better than olivine [13].

Table 4 lists the T_m and thermal degradation DTG values. The degradation rate in the first and fourth stages was insignificant compared to the second and third stages (decomposition step). The T_m value reduced with the increase in the dolomite loading. T_m decreased from 484.8 to 477.2 °C as the dolomite loading increased from 0 to 15 wt.%. The DTG at the lowest T_m and 15 wt.% dolomite was −0.0270 wt.%/min.

Table 4. DTG and values of T_m of CO_2 gasification of Subcoal™ PAF with different loadings of the catalyst.

wt.% of Catalyst	Dolomite		Olivine	
	T_m (°C)	DTG (wt.%/min)	T_m (°C)	DTG (wt.%/min)
0	489.3	−0.0322	489.3	−0.0322
5	482.3	−0.0273	488.4	−0.0315
10	479.8	−0.0272	485.5	−0.0292
15	477.2	−0.027	483.8	−0.0285

3.2. Effects of Olivine

CO_2 gasification was conducted in the presence of different loadings of olivine. The decomposition and DTG reaction were plotted as a function of temperature in Figure 4a,b.

The olivine in CO_2 gasification exhibited a similar performance as that in dolomite. The olivine also promoted the decomposition of biomass. It can be seen from Figure 4a; the reaction time decreased as the catalyst loading increased. The complete conversion reaction decreased from 792.7 to 767.7 °C as the olivine concentration increased from 0 to 15 wt.%. The conversion curve also showed two regimes of decomposition: the first was between 0.2 and 0.45, and then from 0.45 to 0.8.

In terms of the DTG analysis, the curves drifted to a lower temperature as the loading of olivine increased. The DTG parameters are listed in Table 4. The value of T_m reduced from 489.3 to 483.8 °C as the catalyst loading increased from 0 to 15 wt.%. Regarding the degradation rate, the peaks decreased from −0.0322 to −0.0285 wt.%/min. The difference between the highest and lowest values of T_m was 5.5 °C at catalyst loadings of 15 wt.% olivine and 0 wt.%.

3.3. Kinetic Analysis

The CR method was used to determine the E_a and A of gasification for different mechanism models ($G(x)$) using Equations (13) and (14). Figure 5 displays the approximated curves of the CR method for ($G(x)$). The linear relationship between Ln ($G(x)/T^2$) and $1/T$ implies a single mechanism reaction. However, most lines showed that the reaction has a multi model mechanism. Furthermore, as the catalyst ratio increased, the reaction lines for all models decreased due to a shorter reaction time, as shown in Figure 5. The comparison of the E_a obtained from each model and test is shown in Figure 6. The effect of dolomite loading on the kinetic parameters was evaluated as listed in Table 5. The table also includes the goodness of fit coefficient to the regression model (R^2) value based on E_a value in order to demonstrate the appropriateness of the model fitting.

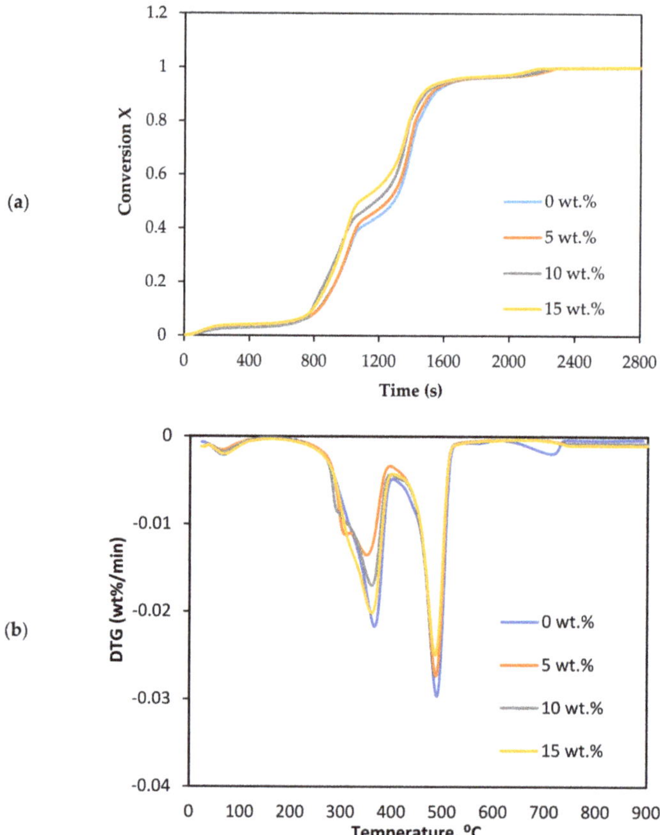

Figure 4. Effect of olivine loadings on the CO_2 gasification: (**a**) conversion degree, and (**b**) DTG curves.

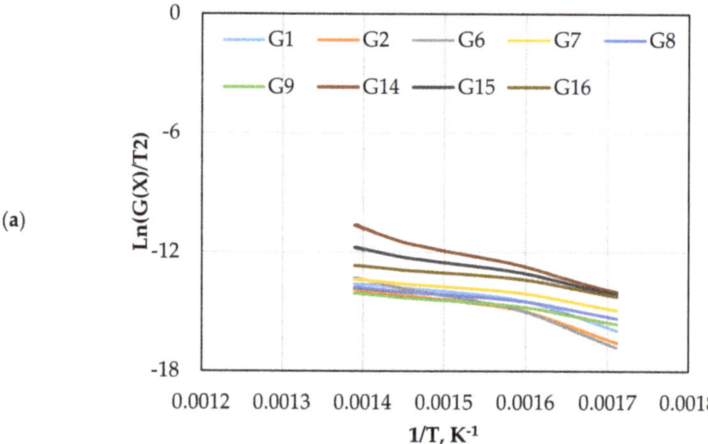

Figure 5. *Cont.*

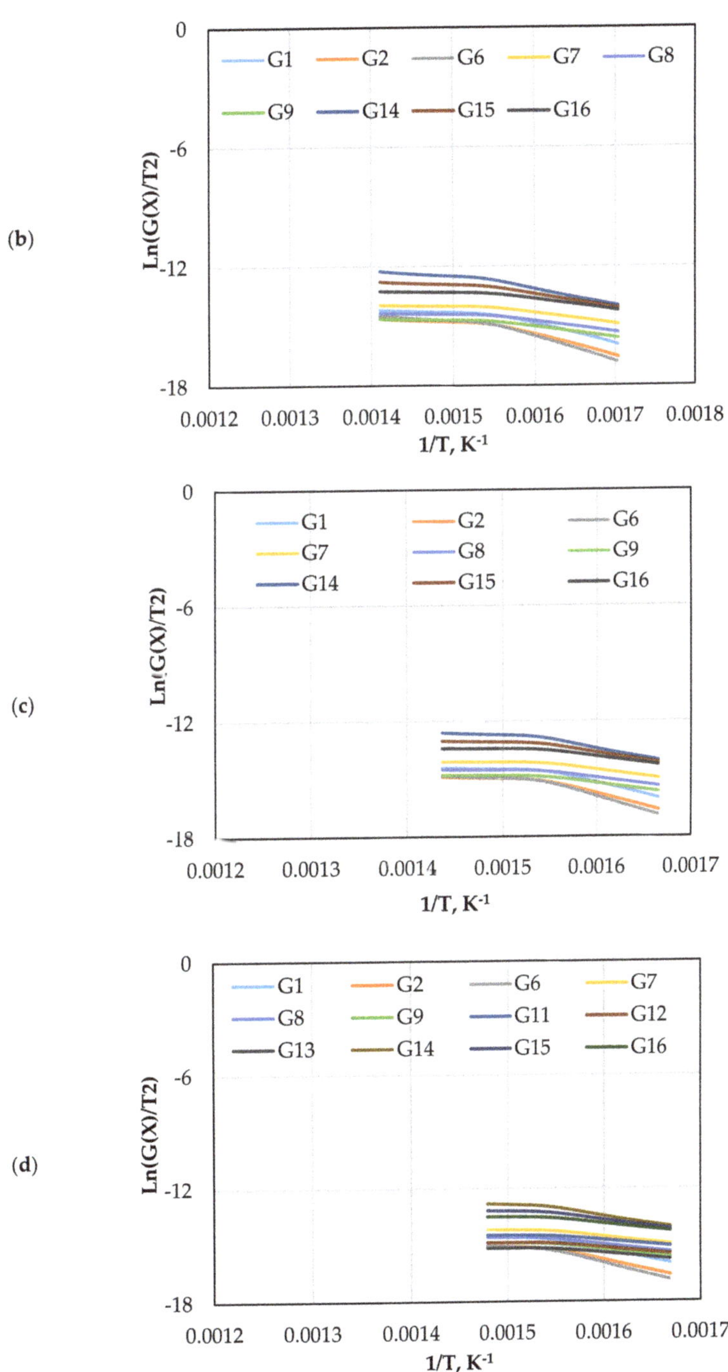

Figure 5. The CR method plots for CO_2 gasification at different dolomite loadings: (**a**) 0 wt.%, (**b**) 5 wt.%, (**c**) 10 wt.%, and (**d**) 15 wt.%.

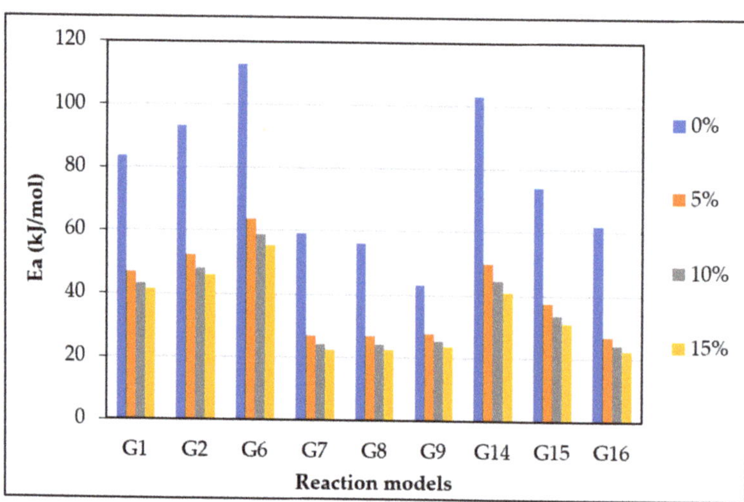

Figure 6. E_a from the most fitting reaction models of CO_2 gasification at different dolomite loadings.

Table 5. Kinetic parameters obtained by CR for the CO_2 gasification of Subcoal™ PAF at different dolomite loadings.

	0 wt.%			5 wt.%			10 wt.%			15 wt.%		
Model	E_a	A	R^2	E_a	A	R^2	E_a	A	R^2	E_a	A	R^2
G1	83.6	1.3×10^4	0.896	46.9	290	0.883	43.4	110	0.862	41.7	72	0.899
G2	93.3	5.7×10^4	0.918	52.2	520	0.891	48.2	180	0.829	46.1	110	0.858
G6	112.9	6.9×10^6	0.955	63.9	5.5×10^3	0.911	58.6	1.4×10^3	0.852	55.5	660	0.874
G7	59.2	290	0.943	26.9	6.4	0.883	24.2	2.98	0.798	22.6	1.98	0.888
G8	56.2	190	0.943	26.9	4.3	0.882	24.2	2.0	0.798	22.6	1.32	0.898
G9	43.3	140	0.943	27.9	3.2	0.881	25.2	1.5	0.798	23.6	0.99	0.928
G11	-	-	-	-	-	-	-	-	-	15.4	0.28	0.911
G12	-	-	-	-	-	-	-	-	-	13.5	0.16	0.871
G13	-	-	-	-	-	-	-	-	-	13.5	0.12	0.861
G14	103.1	2.9×10^7	0.992	50.1	3.4×10^3	0.941	44.8	890	0.890	41.1	270	0.982
G15	74.2	8.1×10^4	0.980	37.6	180	0.923	33.7	64	0.855	31.2	35	0.949
G16	62.2	570	0.943	26.9	13	0.881	24.2	5.9	0.798	22.6	3.96	0.928

The calculated E_a and A values with the highest R^2 values were 41.1 kJ/mol and 370 min^{-1}, which was found in G14 at 15 wt.% of dolomite loading, as shown in Table 5. However, the other models also showed high R^2 with different values of E_a due to the complexity of the Subcoal™ PAF gasification reactions, and the inhomogeneity of the material. The variation in the E_a was obtained due to the reaction mechanism models, as reported by Aboulkas and El Harfi [40].

Figure 7 displays the reaction kinetic graphs of gasification at different loadings of olivine. Figure 8 and Table 6 compare the values of E_a at various contents of catalyst, which was estimated using different mechanism models. Model G15 at 15 wt.% exhibited excellent linearity with the highest R^2 value of 0.992, which makes it a good candidate for E_a estimation and reaction mechanism.

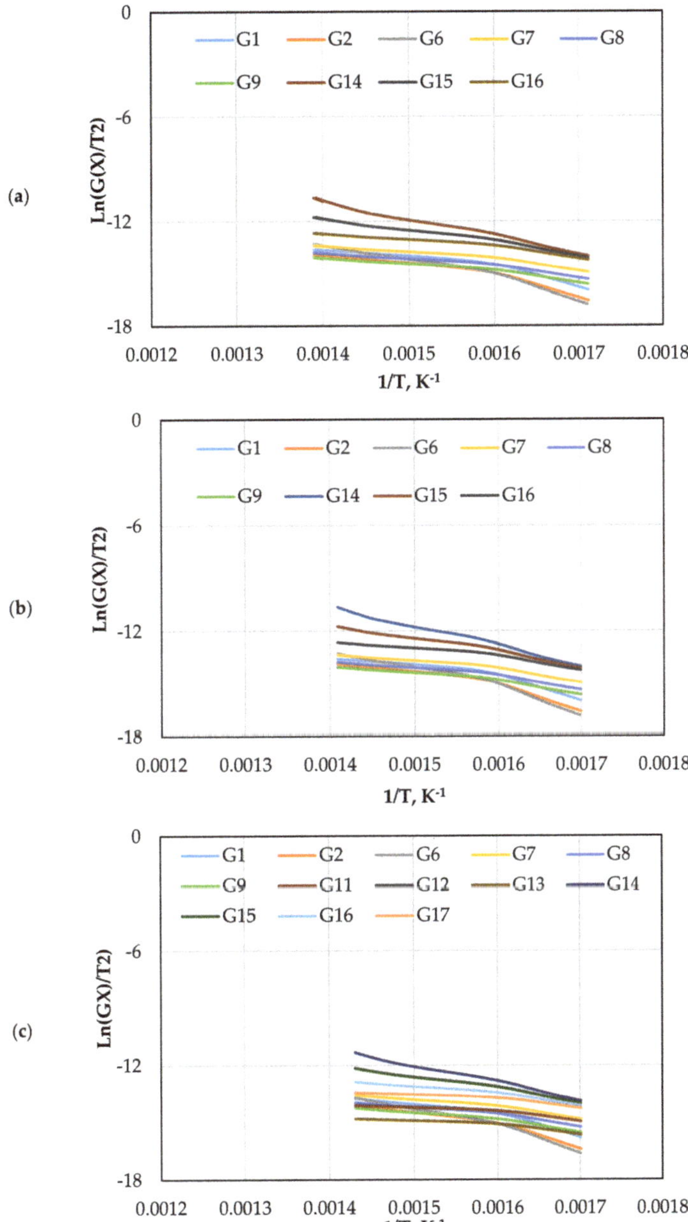

Figure 7. *Cont.*

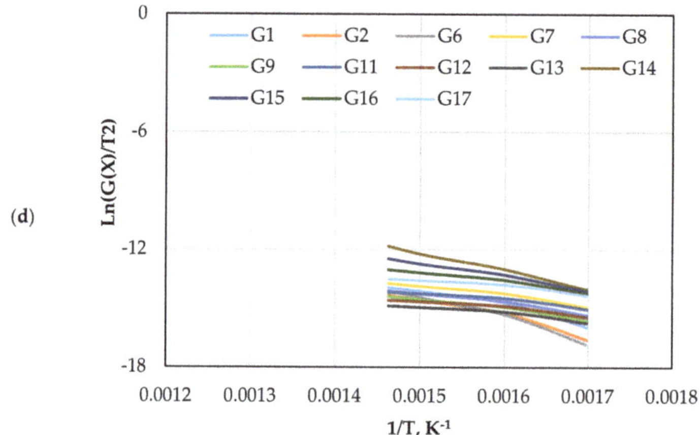

Figure 7. The CR method plots for CO_2 gasification at different olivine loadings: (**a**) 0 wt.%, (**b**) 5 wt.%, (**c**) 10 wt.%, and (**d**) 15 wt.%.

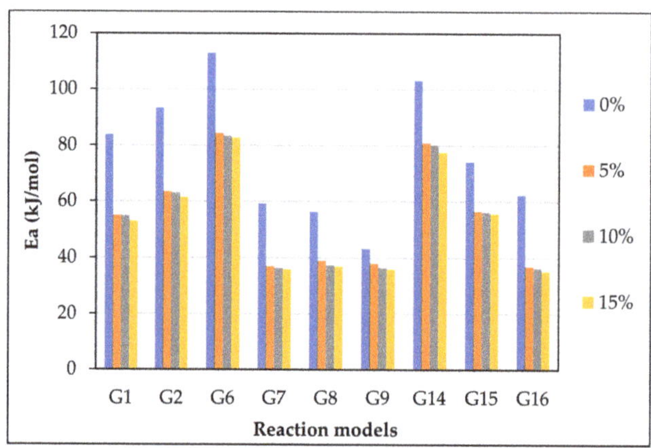

Figure 8. E_a from the most fitting reaction models of CO_2 gasification at different olivine loadings.

Table 6. Kinetic parameters obtained by CR for CO_2 gasification of Subcoal™ PAF at different olivine loadings.

	0 wt.%			5 wt.%			10 wt.%			15 wt.%		
Model	E_a	A	R^2	E_a	A	R^2	E_a	A	R^2	E_a	A	R^2
G1	83.9	1.3×10^4	0.896	54.9	2.1×10^3	0.904	54.8	2.0×10^3	0.926	52.9	2.7×10^3	0.949
G2	93.3	5.7×10^4	0.918	63.4	7.3×10^3	0.927	62.9	6.4×10^3	0.945	61.4	9.3×10^3	0.965
G6	112.9	6.9×10^6	0.955	84.3	5.4×10^5	0.963	83.2	4.1×10^5	0.975	82.6	6.8×10^5	0.986
G7	59.2	290	0.943	36.9	72	0.950	36.3	62	0.965	35.9	82	0.981
G8	56.2	190	0.943	38.9	48	0.950	37.3	41	0.965	36.9	54	0.981
G9	43.3	140	0.943	37.9	36	0.950	36.4	31	0.965	35.9	41	0.981
G11	-	-	-	-	-	-	23.0	1.8	0.886	22.0	2.2	0.921
G12	-	-	-	-	-	-	23.9	1.2	0.886	21.0	1.5	0.921
G13	-	-	-	-	-	-	21.9	0.91	0.886	21.0	1.1	0.921
G14	103.1	2.9×10^7	0.992	80.9	2.6×10^6	0.993	80.1	1.5×10^6	0.990	77.5	2.7×10^6	0.977
G15	74.2	8.1×10^4	0.980	56.4	1.3×10^4	0.984	56.1	9.4×10^3	0.989	55.6	1.4×10^4	0.992
G16	62.2	570	0.943	36.9	140	0.950	36.3	120	0.965	35.2	160	0.981
G17	-	-	-	-	-	-	23.0	3.7	0.886	22.0	4.4	0.921

4. Discussion

The gasification of Subcoal™ PAF in the TGA occurs according to consecutive steps, namely dehydration, devolatilisation, and char decomposition. In the first step, moisture was removed from a sample at a temperature between 150 to 250 °C. However, in this step, the sample mass insignificantly changed as a function of time. The second step consisted of two stages in which the volatile organic compounds decomposed into condensable and non-condensable gases. This process causes a sudden reduction in a sample mass in a temperature range from 250 to 650 °C [3]. In the final step, a small change in the conversion of biomass was observed due to the decomposition of char into non-condensable gases. The final step occurred from 600 °C to 750 °C. However, as the temperature was below 600 °C, the calcination of dolomite yielded active species such as CaO and MgO, as shown in the reactions (Equations (15) and (16)).

The CO_2 gasification in the presence of dolomite showed lower Tm peaks than olivine. The findings indicate that 15 wt.% of dolomite accelerates the reactions by lowering the Tm of Subcoal™ PAF decomposition. The abnormality of DTG peaks in the second stage can be attributed to the presence of O_2 and interactions of the catalyst [41]. However, the DTG analysis with olivine is quite organised and sharpened compared to the dolomite case [42].

Olivine and dolomite catalysts have a significant and measurable impact on the conversion rate when compared with the heating rate impact [22]. The presence of a catalyst reduced the induction time of the thermal decomposition. It also shifted the conversion curve to a lower temperature as the catalyst concentration increased, indicating a shorter decomposition reaction time. TGA plots of conversion degree were divided into two main degradation regions: 0.2 to 0.45, and then from 0.45 to 0.8. This can be explained by the successive reactions of decomposition.

The accurate estimation of the kinetic parameters of the heterogeneous and non-isothermal reaction is a substantial step in a design of a biomass gasifier. In this work, the model-fitting CR method was used to estimate the E_a and A of the CO_2 gasification process in the presence of olivine and dolomite as catalysts. The method models (G1–G19) were tested to obtain the most fitted graph. In the presence of dolomite or olivine, the E_a was significantly reduced with the increase in the catalyst loading [43]. However, dolomite had a better kinetic performance in comparison with olivine. The degradation process is a chemical reaction-controlled in the first step (reaction extent less than 0.2). In the second and third steps of degradation, when the reaction extent is between 0.2 and 0.8, the process is controlled by three-dimensional diffusion and chemical reaction. Finally, at a reaction extent greater than 0.8, the decomposition is solely controlled by the three-dimensional diffusion mechanism [44]. This reveals that the reaction mechanism of Subcoal™ PAF gasification in the fixed bed is complex and occurs in a multistage thermal decomposition. The higher the complexity of the dehydration mechanism, the greater the variation of E_a [45]. In the DTG analysis, the degradation rate in wt.%/min was plotted against the temperature. Four peaks appeared in the graph showing the amount of degradation that occurred at a specific temperature. Thermal gasification reaction with the catalyst was evaluated according to DTG and Tm values. The results showed two small peaks corresponding to degradation in the first and fourth stages, respectively, as well as two large peaks of the stages in the devolatilisation stage [46]. The Tm peaks overlapped, and the irregularity of the peak shape indicated the occurrence of simultaneous gasification reactions. Additionally, the presence of oxygen in the reaction mixture led to several oxidation reactions that impact the evaluation of DTG. Dolomite has a better degradation performance at constant catalyst concentration than olivine. For model G14, the E_a reduced by 62 kJ/mol as the dolomite concentration increased from 0 to 15 wt.%, while only a 25.6 kJ/mol reduction was obtained by olivine. The lower the E_a, the lower the gasification temperature. For both catalysts, the E_a decreased as the loading of the catalyst increased. The higher the loading of catalysts, the lower the Tm peak obtained [47]. Dolomite causes a higher reduction in Tm value and better DTG outcomes compared with olivine. The fitting of curves showed that models G14 and G15 provided relatively high correlation coefficient R^2 for both catalysts. The mechanism of the catalytic gasification accordingly varied between second- to third-order chemical reactions. Dolomite enhances tar cracking by promoting the water–gas reaction [48]. The tar elimination in the presence of dolomite can attain 100%, as demonstrated by Simell et al. [17]. Syngas products and an active carbon deposit are formed when a benzene ring is incorporated on the active sites of CaO. The active carbon deposit undergoes a reaction with steam or ends with coke formation on CaO. MgO breaks the H_2 bonding in water, forming OH, which is absorbed on the active sites of MgO. The OH group then combines with the remaining active carbon to form formate. A spill over of OH takes place, leading to decomposition of formate to syngas and the removal of coke [49,50]. Kim et al. [51] carried out a kinetic study on the CO_2 gasification of Chinese low-rank lignite coal and found that the dolomite minimises the E_a and reaction time of biomass gasification.

Dolomite calcination includes the formation of MgO and CaO by thermal cracking [20]. The following chemical reactions describe the calcination of dolomite over two ranges of temperatures [52]:

$$MgCa(CO_3)_2 \leftrightarrow Ca(CO_3)MgO + CO_2 \qquad (T > 600\ ^\circ C) \qquad (15)$$

$$Ca(CO_3)MgO \leftrightarrow CaO{\cdot}MgO + CO_2 \qquad (T > 900\ ^\circ C) \qquad (16)$$

As can be seen in the reactions, the equilibria of calcination are sensitive to CO_2 gas release. It is preferred that the partial pressure of CO_2 stays under the equilibrium pressure to avoid catalyst deactivation [53]. Besides the process pressure, temperature plays a crucial role in controlling the calcination and carbonation reactions [54]. Other factors that influence the calcination process include heating rate, the quantity of the catalyst, and particle size [55]. The purpose of dolomite calcination is to increase the size of the catalyst pores, which is then responsible for mass transfer enhancement. The time and temperature of calcination increase the surface area of the catalyst, hence the catalysis activity increases [56].

It has been revealed that olivine increases the yield of syngas and tar conversion and reduces the production of CH_4 and CO_2 [12,57]. The gasification in the CO_2 atmosphere causes the oxidation of olivine in the combustion zone, producing binary iron oxide hematite (Fe_2O_3), SiO_2, and iron-depleted olivine. The formed Fe_2O_3 is subsequently reduced by the organic compounds in biomass to FeO, CO, and H_2. In gasification with the synthesis of olivine $(Mg0.5\ Fe0.5)_2\ SiO_4$, about 10% of iron content in olivine is oxidised at 300 s according to thermogravimetric analysis [58]. Moreover, olivine is capable of accommodating excess oxygen from a gasifying agent, as shown in the reaction below:

$$(Mg,\ Fe)_2\ SiO_4 + \frac{\delta}{2}\,O_2\ \rightarrow\ (Mg,\ Fe)_2\ SiO_{4+\delta} \qquad (17)$$

The gained oxygen in olivine is used to partially oxidise methane in the gasification zone when the partial pressure of O_2 is low [58]:

$$(Mg, Fe)_2\ SiO_{4+\delta} + \delta\,CH_4\ \rightarrow\ (Mg,\ Fe)_2\ SiO_4 + \delta(CO + 2H_2) \qquad (18)$$

In the CO_2 biomass gasification, olivine undergoes a cycle of oxidation and reduction to produce the active iron and then to regenerate the catalyst. The oxidation of iron in olivine takes place during the calcination at a temperature of 400–1400 °C in two steps. The purpose of olivine oxidation is to produce the Fe_2O_3, wch promotes the tar removal reactions [59].

The crystallinity and mineral content in the Subcoal™ sample and ash may contribute to the catalysis process of CO_2 gasification. The XRF analysis showed that the Subcoal™ ash contained 38 and 18.3 wt.% CaO and Al_2O_3, respectively, which is considered as an effective catalyst of char decomposition reactions [30,60]. The crystallinity of ash and biomass minerals also plays an important role in catalytic activity. Furthermore, there is a positive correlation between the crystallinity of biomass and decomposition temperature and activation energy [61].

The gasification of MSW, specifically Subcoal™, vitally contributes to the Qatar National Vision 2030 through reducing the MSW landfill and generating renewable power. The availability of naturally occurring minerals in Qatar improves the potential of using Subcoal™ gasification as an alternative to natural gas power plants. Mineral catalysts such as dolomite enhance the gasification efficiency and producer gas quality.

5. Conclusions

- Mineral catalysts can play a vital role in reducing tar formation, operating cost, gasifier size, and improving the producer gas quality and quantity. Herein, the influence of different loadings of olivine and dolomite on the CO_2 gasification of Subcoal™ PAF was examined using TGA.
- The XRD results showed that the crystallinity proportion in Subcoal™ powder and ash was 42% and 37.67%, respectively. The primary crystalline compound in the ash was calcium chloride, which may pose a catalytic effect on the gasification reactions.
- The reaction kinetics and mechanism were evaluated using the CR method at the tested loadings of olivine and dolomite. The findings showed that the decomposition time and temperature decreased as the loading of olivine or dolomite increased from 0 to 15 wt.%.
- For the gasification without a catalyst, G14 was the most appropriate model with the highest R^2 value.
- Dolomite exhibited a better performance in terms of reaction time and mean reaction temperature. Regarding the kinetic evaluation, the probable mechanism model for the gasification reaction at 15 wt.% was G14, which represents the third-order chemical reaction for dolomite. Second-order chemical reaction (G15) showed the highest linearity at 15 wt.% of olivine. The E_a dropped as the loading of the catalyst increased over the ranges tested herein. At 15 wt.% loading, the E_a was 41.1 and 77.5 kJ/mol for dolomite and olivine, respectively.
- Naturally-occurring dolomite is an excellent candidate for promoting large-scale Subcoal™ gasification in Qatar. Gasification of non-recyclable paper and plastic catalysed by dolomite is a sustainable alternative option to mitigate the consequences of global warming and waste landfill in accordance with Qatar National Vision 2030.
- CO_2 as a gasifying agent offers unique features over air or steam to ensure better gasification performance and syngas composition. This work addressed an important aspect of the Subcoal™ gasification catalysis and kinetic parameters. However, further research efforts are required to achieve a full understanding of Subcoal™ gasification.

Author Contributions: Conceptualization, A.M.S.H.A.-M., R.M. and J.S.; Methodology, A.M.S.H.A.-M.; Investigation, A.M.S.H.A.-M.; Resources, A.M.S.H.A.-M.; Writing—original draft preparation, A.M.S.H.A.-M.; Writing—review and editing, R.M. and J.S.; Supervision, R.M. and J.S. All authors have read and agreed to the published version of the manuscript.

Funding: This work was financially supported by the Qatar National Research Fund, Graduate Sponsorship Research Award (GSRA), Qatar Science and Technology Park, tech 1 building, level 1. Member of the Qatar Foundation for Education, Science and Community Development for doctoral project number GSRA5-1-0404-18034. Doha, Qatar.

Institutional Review Board Statement: Not applicable.

Informed Consent Statement: Not applicable.

Data Availability Statement: Not applicable.

Acknowledgments: The authors thank Cardiff University for their technical support and the Qatar National Research Fund.

Conflicts of Interest: The authors declare that there are no conflict of interest.

Nomenclature

m_t	Instantaneous mass of sample, (mg)	t	Time, (s)
E_a	Activation energy, (kJ/mol)	$G(x)$	Reaction mechanism model
A	Pre-exponential constant, (min^{-1})	R	Gas constant, (kJ/mol. K)
m_o	Sample mass at the beginning, (mg)	x	Conversion
m_f	Final mass loss of sample, (mg)	k	Rate constant
T	Temperature, (°C)	T_m	Mean temperature, (°C)
R^2	Correlation coefficient	$\overline{T}$	Absolute temperature, (K)

References

1. Mountford, H.; Waskow, D.; Gonzalez, L.; Gajjar, C.; Cogswell, N.; Holt, M.; Fransen, T.; Bergen, M.; Gerholdt, R. COP26: Key Outcomes from the UN Climate Talks in Glasgow. Available online: https://www.wri.org/insights/cop26-key-outcomes-un-climate-talks-glasgow (accessed on 21 December 2021).
2. Magazzino, C.; Mele, M.; Schneider, N. The relationship between municipal solid waste and greenhouse gas emissions: Evidence from Switzerland. *Waste Manag.* **2020**, *113*, 508–520. [CrossRef] [PubMed]
3. Al-Moftah, A.M.S.H.; Marsh, R.; Steer, J. Life Cycle Assessment of Solid Recovered Fuel Gasification in the State of Qatar. *ChemEngineering* **2021**, *5*, 81. [CrossRef]
4. Khan, R. WTE Prospects in the Middle East. Available online: https://www.bioenergyconsult.com/tag/qatar/ (accessed on 21 December 2021).
5. N + P Group. Subcoal® – N + P Recycling. Available online: https://www.np-recycling.nl/en/alternative-fuels/subcoal.html (accessed on 27 September 2021).
6. Syguła, E.; Świechowski, K.; Hejna, M.; Kunaszyk, I.; Białowiec, A. Municipal Solid Waste Thermal Analysis—Pyrolysis Kinetics and Decomposition Reactions. *Energies* **2021**, *14*, 4510. [CrossRef]
7. Bridgwater, A.V.; Beenackers, A.A.C.M.; Sipila, K.; Zhenhong, Y.; Chuangzhi, W.; Li, S. *An Assessment of the Possibilities for Transfer of European Biomass Gasification Technology to China*; Office for Official Publications of the European Communities: Luxembourg, 1999.
8. Islam, M.W. A review of dolomite catalyst for biomass gasification tar removal. *Fuel* **2020**, *267*, 117095. [CrossRef]
9. Tian, Y.; Zhou, X.; Lin, S.; Ji, X.; Bai, J.; Xu, M. Syngas production from air-steam gasification of biomass with natural catalysts. *Sci. Total Environ.* **2018**, *645*, 518–523. [CrossRef]
10. Sirinwaranon, P.; Atong, D.; Sricharoenchaikul, V. Gasification of torrefied cassava rhizome with Ni/MCM-41 catalyst derived from illite waste. *Energy Rep.* **2020**, *6*, 537–547. [CrossRef]
11. DiLoreto, Z.A.; Bontognali, T.R.; Al Disi, Z.A.; Al-Kuwari, H.A.S.; Williford, K.H.; Strohmenger, C.J.; Sadooni, F.; Palermo, C.; Rivers, J.M.; McKenzie, J.A.; et al. Microbial community composition and dolomite formation in the hypersaline microbial mats of the Khor Al-Adaid sab-khas, Qatar. *Extremophiles* **2019**, *23*, 201–218. [CrossRef]
12. Rapagnà, S.; Jand, N.; Kiennemann, A.; Foscolo, P.U. Steam-gasification of biomass in a fluid-ised-bed of olivine particles. *Biomass Bioenergy* **2000**, *19*, 187–197. [CrossRef]
13. Corella, J.; Toledo, J.M.; Padilla, R. Olivine or dolomite as in-bed additive in biomass gasification with air in a fluidized bed: Which is better? *Energy Fuels* **2004**, *18*, 713–720. [CrossRef]
14. Nitsch, X.; Commandre, J.M.; Clavel, P.; Martin, E.; Valette, J.; Volle, G. Conversion of phenol-based tars over olivine and sand in a biomass gasification atmosphere. *Energy Fuels* **2013**, *27*, 5459–5465. [CrossRef]
15. Miccio, F.; Piriou, B.; Ruoppolo, G.; Chirone, R. Biomass gasification in a catalytic fluidized reactor with beds of different materials. *Chem. Eng. J.* **2009**, *154*, 369–374. [CrossRef]
16. Göransson, K.; Söderlind, U.; Engstrand, P.; Zhang, W. An experimental study on catalytic bed materials in a biomass dual fluidised bed gasifier. *Renew. Energy* **2015**, *81*, 251–261. [CrossRef]
17. Simell, P.; Kurkela, E.; Ståhlberg, P.; Hepola, J. Catalytic hot gas cleaning of gasification gas. *Catal. Today* **1996**, *27*, 55–62. [CrossRef]
18. Basu, P. *Biomass Gasification, Pyrolysis and Torrefaction: Practical Design and Theory*; Academic Press: Cambridge, MA, USA, 2018.
19. Dayton, D. Review of the Literature on Catalytic Biomass tar Destruction: Milestone Completion Report. 2002. Available online: https://www.sciencedirect.com/book/9780123964885/biomass-gasification-pyrolysis-and-torrefaction (accessed on 29 September 2021).
20. Richardson, Y.; Blin, J.; Julbe, A. A short overview on purification and conditioning of syngas produced by biomass gasification: Catalytic strategies, process intensification and new concepts. *Prog. Energy Combust. Sci.* **2012**, *38*, 765–781. [CrossRef]
21. Sutton, D.; Kelleher, B.; Ross, J.R.H. Review of literature on catalysts for biomass gasification. *Fuel Process. Technol.* **2001**, *73*, 155–173. [CrossRef]
22. Al-Moftah, A.M.S.H.; Marsh, R.; Steer, J. Thermal Decomposition Kinetic Study of Non-Recyclable Paper and Plastic Waste by Thermogravimetric Analysis. *ChemEngineering* **2021**, *5*, 54. [CrossRef]
23. González-Vázquez, M.D.P.; García, R.; Gil, M.V.; Pevida, C.; Rubiera, F. Unconventional biomass fuels for steam gasification: Kinetic analysis and effect of ash composition on reactivity. *Energy* **2018**, *155*, 426–437. [CrossRef]

24. Rahaman, M.H.A.; Khandaker, M.U.; Khan, Z.R.; Kufian, M.Z.; Noor, I.S.M.; Arof, A.K. Effect of gamma irradiation on poly (vinyledene difluoride)–lithium bis (oxalato) borate electrolyte. *Phys. Chem. Chem. Phys.* **2014**, *16*, 11527–11537. [CrossRef]

25. Marsh, R. *Plastic Film Recycling from Waste Sources*; Cardiff University: Cardiff, UK, 2005.

26. Febrero, L.; Granada, E.; Patiño, D.; Eguía, P.; Regueiro, A. A comparative study of fouling and bottom ash from woody biomass combustion in a fixed-bed small-scale boiler and evaluation of the analytical techniques used. *Sustainability* **2015**, *7*, 5819–5837. [CrossRef]

27. Ward, C.R.; French, D. Determination of glass content and estimation of glass composition in fly ash using quantitative X-ray diffractometry. *Fuel* **2006**, *85*, 2268–2277. [CrossRef]

28. Shoppert, A.; Valeev, D.; Loginova, I.; Chaikin, L. Complete Extraction of Amorphous Alumi-nosilicate from Coal Fly Ash by Alkali Leaching under Atmospheric Pressure. *Metals* **2020**, *10*, 1684. [CrossRef]

29. Kim, S.H.; Lee, C.M.; Kafle, K. Characterization of crystalline cellulose in biomass: Basic principles, applications, and limitations of XRD, NMR, IR, Raman, and SFG. *Korean J. Chem. Eng.* **2013**, *30*, 2127–2141. [CrossRef]

30. Wang, X.; Yao, K.; Huang, X.; Chen, X.; Yu, G.; Liu, H.; Wang, F.; Fan, M. Effect of CaO and biomass ash on catalytic hydrogasification behavior of coal char. *Fuel* **2019**, *249*, 103–111. [CrossRef]

31. Karasmanaki, E.; Ioannou, K.; Katsaounis, K.; Tsantopoulos, G. The attitude of the local com-munity towards investments in lignite before transitioning to the post-lignite era: The case of Western Macedonia, Greece. *Resour. Policy* **2020**, *68*, 101781. [CrossRef]

32. Xiang, Y.; Cai, L.; Guan, Y.; Liu, W.; Cheng, Z.; Liu, Z. Study on the effect of gasification agents on the integrated system of biomass gasification combined cycle and oxy-fuel combustion. *Energy* **2020**, *206*, 118131. [CrossRef]

33. Nakamura, S. Fundamentals of Chemical Reaction Kinetics. In *Solar to Chemical Energy Conversion*; Springer: Berlin/Heidelberg, Germany, 2016; pp. 57–65.

34. Christov, S.G. *Collision Theory and Statistical Theory of Chemical Reactions*; Springer Science and Business Media LLC: Berlin/Heidelberg, Germany, 1980.

35. Coats, A.W.; Redfern, J.P. Kinetic Parameters from Thermogravimetric Data. *Nature* **1964**, *201*, 68–69. [CrossRef]

36. Liu, N.; Chen, H.; Shu, L.; Statheropoulos, M. Error evaluation of integral methods by consideration on the approximation of temperature integral. *J. Therm. Anal. Calorim.* **2005**, *81*, 99–105. [CrossRef]

37. Vyazovkin, S.; Burnham, A.K.; Criado, J.M.; Pérez-Maqueda, L.A.; Popescu, C.; Sbirrazzuoli, N. Kinetics Committee recommendations for performing kinetic computations on thermal analysis data. *Thermochim. Acta* **2011**, *520*, 1–19. [CrossRef]

38. Gai, C.; Dong, Y.; Lv, Z.; Zhang, Z.; Liang, J.; Liu, Y. Pyrolysis behavior and kinetic study of phenol as tar model compound in micro fluidized bed reactor. *Int. J. Hydrog. Energy* **2015**, *40*, 7956–7964. [CrossRef]

39. Cho, M.H.; Mun, T.Y.; Kim, J.S. Air gasification of mixed plastic wastes using calcined dolomite and activated carbon in a two-stage gasifier to reduce tar. *Energy* **2013**, *53*, 299–305. [CrossRef]

40. Aboulkas, A.; El Harfi, K. Study of the Kinetics and Mechanisms of Thermal Decomposition of Moroccan Tarfaya Oil Shale and Its Kerogen. *Oil Shale* **2008**, *25*, 426–443. [CrossRef]

41. Zan, R.; Wang, W.; Xu, R.; Schenk, J.; Zheng, H.; Wang, H. Gasification characteristics and kinetics of unburned pulverized coal in blast furnaces. *Energies* **2019**, *12*, 4324. [CrossRef]

42. Lancee, R.J. Characterization and Reactivity of Olivine and Model Catalysts for Biomass Gasification. 2014. Available online: https://pure.tue.nl/ws/files/3946832/781405.pdf (accessed on 1 October 2021).

43. Marcilla, A.; Beltran, M.; Conesa, J.A. Catalyst addition in polyethylene pyrolysis: Thermogravimetric study. *J. Anal. Appl. Pyrolysis* **2001**, *58*, 117–126. [CrossRef]

44. Vasconcelos, G.D.C.; Mazur, R.L.; Ribeiro, B.; Botelho, E.C.; Costa, M.L. Evaluation of decomposition kinetics of poly (ether-ether-ketone) by thermogravimetric analysis. *Mater. Res.* **2014**, *17*, 227–235. [CrossRef]

45. Wong, F.F.; Lin, C.M.; Chen, K.L.; Shen, Y.H.; Huang, J.J. Improvement of the thermal latency for epoxy-phenolic resins by novel amphiphatic imidazole catalysts. *Macromol. Res.* **2010**, *18*, 324–330. [CrossRef]

46. Nelson, J.B. *Determination of Kinetic Parameters of Six Ablation Polymers by Thermogravimetric Analysis*; National Aeronautics and Space Administration: Washington, DC, USA, 1967.

47. Balasundram, V.; Ibrahim, N.; Samsudin, M.D.H.; Kasmani, R.M.; Hamid, M.K.A.; Isha, R.; Hasbullah, H. Thermogravimetric studies on the catalytic pyrolysis of rice husk. *Chem. Eng. Trans.* **2017**, *56*, 427–432.

48. Yu, Q.-Z.; Brage, C.; Nordgreen, T.; Sjöström, K. Effects of Chinese dolomites on tar cracking in gasification of birch. *Fuel* **2009**, *88*, 1922–1926. [CrossRef]

49. Rigo, V.A.; Metin, C.O.; Nguyen, Q.P.; Miranda, C.R. Hydrocarbon Adsorption on Carbonate Mineral Surfaces: A First-Principles Study with van der Waals Interactions. *J. Phys. Chem. C* **2012**, *116*, 24538–24548. [CrossRef]

50. LaValley, J. Infrared spectrometric studies of the surface basicity of metal oxides and zeolites using adsorbed probe molecules. *Catal. Today* **1996**, *27*, 377–401. [CrossRef]

51. Kim, S.K.; Park, J.Y.; Lee, D.K.; Hwang, S.C.; Lee, S.H.; Rhee, Y.W. Kinetic Study on Low-Rank Coal Char: Characterization and Catalytic CO_2 Gasification. *J. Energy Eng.* **2016**, *142*, 04015032. [CrossRef]

52. Woolcock, P.J.; Brown, R.C. A review of cleaning technologies for biomass-derived syngas. *Biomass Bioenergy* **2013**, *52*, 54–84. [CrossRef]

53. Torres, W.; Pansare, S.S.; Goodwin, J.G., Jr. Hot Gas Removal of Tars, Ammonia, and Hydrogen Sulfide from Biomass Gasification Gas. *Catal. Rev.* **2007**, *49*, 407–456. [CrossRef]

54. Zamboni, I.; Courson, C.; Kiennemann, A. Synthesis of Fe/CaO active sorbent for CO_2 absorption and tars removal in biomass gasification. *Catal. Today* **2011**, *176*, 197–201. [CrossRef]

55. Zhou, C.; Rosén, C.; Engvall, K. Fragmentation of dolomite bed material at elevated tempera-ture in the presence of H_2O & CO_2: Implications for fluidized bed gasification. *Fuel* **2020**, *260*, 116340.

56. Duffy, A.; Walker, G.; Allen, S. Investigations on the adsorption of acidic gases using activated dolomite. *Chem. Eng. J.* **2006**, *117*, 239–244. [CrossRef]

57. Erkiaga, A.; Lopez, G.; Amutio, M.; Bilbao, J.; Olazar, M. Steam gasification of biomass in a conical spouted bed reactor with olivine and γ-alumina as primary catalysts. *Fuel Process. Technol.* **2013**, *116*, 292–299. [CrossRef]

58. Pecho, K.; Sturzenegger, M. Elucidation of the Function of Olivine in Biomass Gasification. 2005. Available online: https://www.osti.gov/etdeweb/servlets/purl/20671631 (accessed on 11 December 2021).

59. Christodoulou, C.; Grimekis, D.; Panopoulos, K.; Pachatouridou, E.; Iliopoulou, E.; Kakaras, E. Comparing calcined and un-treated olivine as bed materials for tar reduction in fluidized bed gasification. *Fuel Process. Technol.* **2014**, *124*, 275–285. [CrossRef]

60. Song, Y.; Hu, J.; Liu, J.; Evrendilek, F.; Buyukada, M. Catalytic effects of CaO, Al_2O_3, Fe_2O_3, and red mud on Pteris vittata combustion: Emission, kinetic and ash conversion patterns. *J. Clean. Prod.* **2020**, *252*, 119646. [CrossRef]

61. Chen, H.; Liu, Z.; Chen, X.; Chen, Y.; Dong, Z.; Wang, X.; Yang, H. Comparative pyrolysis be-haviors of stalk, wood and shell biomass: Correlation of cellulose crystallinity and reaction kinetics. *Bioresour. Technol.* **2020**, *310*, 123498. [CrossRef]

chemengineering

MDPI

Article

Sensitivity Analysis and Cost Estimation of a CO_2 Capture Plant in Aspen HYSYS

Shirvan Shirdel, Stian Valand, Fatemeh Fazli, Bernhard Winther-Sørensen, Solomon Aforkoghene Aromada, Sumudu Karunarathne and Lars Erik Øi *

Department of Process, Energy and Environmental Technology, University of South-Eastern Norway, Kjølnes Ring 56, 3918 Porsgrunn, Norway; 238778@student.usn.no (S.S.); 231227@student.usn.no (S.V.); 238746@student.usn.no (F.F.); 227226@student.usn.no (B.W.-S.); solomon.a.aromada@usn.no (S.A.A.); sumudu.karunarathne@usn.no (S.K.)
* Correspondence: lars.oi@usn.no; Tel.: +47-35575141

Abstract: A standard CO_2 capture process is implemented in Aspen HYSYS, simulated, and evaluated based on available data from Fortum's waste burning facility at Klemetsrud in Norway. Since amine-based CO_2 removal has high costs, the main aim is cost-optimizing. A simplified carbon-capture unit with a 20-m absorber packing height, 90% CO_2 removal efficiency, and a minimum approach temperature for the lean/rich amine heat exchanger (ΔT_{min}) of 10 °C was considered the base case simulation model. A sensitivity analysis was performed to optimize these parameters. For the base case study, CO_2 captured cost was calculated as 37.5 EUR/t. When the sensitivity analysis changes the size, the Power Law method adjusts the equipment cost. A comparison of the Enhanced Detailed Factor (EDF) and the Power Law approach was performed for all simulations to evaluate the uncertainties in the findings from the Power Law method. The optimums calculated for ΔT_{min} and CO_2 capture rate were 15 °C and 87% for both methods, with CO_2 removal costs of 37 EUR/t CO_2 and 36.7 EUR/t CO_2, respectively. With 19 m of packing height to absorber, the minimum CO_2 capture cost was calculated as 37.3 EUR/t and 37.1 EUR/t for the EDF and Power Law methods, respectively. Since there was a difference between the Power Law method and the EDF method, a size factor exponent derivation was performed. The derivation resulted in the following exponents: for the lean heat exchanger 0.74, for the lean/rich heat exchanger 1.03, for the condenser 0.68, for the reboiler 0.92, for the pump 0.88, and for the fan 0.23.

Keywords: CO_2 capture; Aspen HYSYS; simulation; optimization; cost estimation

Citation: Shirdel, S.; Valand, S.; Fazli, F.; Winther-Sørensen, B.; Aromada, S.A.; Karunarathne, S.; Øi, L.E. Sensitivity Analysis and Cost Estimation of a CO_2 Capture Plant in Aspen HYSYS. *ChemEngineering* **2022**, 6, 28. https://doi.org/10.3390/chemengineering6020028

Academic Editor: Venko N. Beschkov

Received: 20 February 2022
Accepted: 6 April 2022
Published: 11 April 2022

1. Introduction

A significant amount of human-made CO_2 emissions has its origin in combustion processes such as the burning of coal, oil and gas [1]. CO_2 capture utilization and storage (CCUS) has been suggested to cope with such emissions as a possible solution. The term CCUS describes a process that involves the capture of CO_2 from extensive facilities that are burning fossil- or biomass fuel. The captured CO_2 can be used at the facility or compressed and stored in permanent storage locations. Such locations for storage can, for instance, be depleted oil and gas reservoirs. Approximately 230 Mt of CO_2 is utilized globally each year, largely for fertilizer production (roughly 125 Mt/year) and improved oil recovery (around 70–80 Mt/year). Food and beverage manufacturing, cooling, water treatment, and greenhouses are some of the other commercial uses of CO_2. Today, 20% of the global CO_2 emissions are emitted from heavy industries, and the world's carbon-capture facilities can capture more than 40 Mt of CO_2 [2]. This amount of captured CO_2 is expected to double because the interest in CO_2 capture is growing. More facilities for CO_2 capture are being built, and the technology is recognized as a significant contributor to reducing CO_2 emissions [2].

The drawback of CO_2 capture is that it is assimilated with enormous capital costs (CAPEX) and an energy-demanding process that results in prohibitive operational expenses (OPEX) [3]. Therefore, further work on reducing costs is of great importance.

1.1. Literature Review

Post-combustion CO_2 capture is the most industry-established capture method found in fossil fuel power plants and, to some extent, in the cement, steel, and iron industries [4]. Regarding a study by Patel et al. [5], CO_2 capture is the most expensive part of the overall carbon-capture and storage (CCS) operation, accounting for about 70% of the total cost [5]. A lot of work has been done on the simulation and cost optimization of CO_2 capture. The work in this study is a continuation of previous projects at the University of South-Eastern Norway on CO_2 capture simulation and cost estimation using Aspen HYSYS. Therefore, it is interesting to investigate which variables affect the plant cost.

According to Kallevik [6], the cost of a plant is mostly influenced by five factors. The first is the exhaust gas flow that goes into the absorption column. This affects the dimensions of the process equipment in the gas path, which is a large part of the total equipment. The second is the CO_2 content in the flue gas. A high concentration lowers the energy consumption due to a higher driving force. Third, an increase in CO_2 removal rate increases energy consumption. Fourth, the flow rate of the solvent determines the size of the equipment and the utility need. Fifth is the energy requirement of hot utility and electricity. A hot utility is required as the desorption is endothermic, and it needs heat to reverse the CO_2 absorption. The thermal energy need will increase with a high solvent flow rate. The electricity demand is mainly due to the flue gas transport through the process and will grow with the pressure requirement and volume flow.

Lars Erik Øi [7] used Aspen HYSYS to simulate a basic combined cycle gas power plant and a monoethanolamine (MEA)-based CO_2 removal process. The CO_2 removal (%) and energy consumption in the CO_2 removal plant were calculated as a function of amine circulation rate, absorption column height, absorption temperature, and steam temperature. The focus has been on MEA-based absorption and desorption calculation techniques for CO_2 collection of atmospheric exhaust gas of a gas-based power plant in Aspen HYSYS. Total CO_2 removal quality and heat consumption have been calculated as a function of circulation rate, absorber temperature, and other variables. One of the project's objectives was to determine the process' cost optimum parameters [7].

Rubin et al. [8] advocated using a uniform set of items to include in a cost estimate, a standard nomenclature to characterize each cost element, and a consistent technique of aggregating intermediate cost elements to arrive at a project's total capital and operational costs. To increase the clarity and uniformity of cost estimates for greenhouse gas mitigation methods, they propose a standard costing approach as well as standards for CCS cost reporting [8]. Van Der Spek et al. [9] studied the recent advances in CCS engineering and economic analysis in 2019. They evaluated equipment design and size, cost indices and location factors, process and project contingency costs, CO_2 transportation and storage costs, and uncertainty analysis and validation.

Xiaobo Luo, in his PhD thesis [10], provided modelling, simulation, and optimization research on the best design and operation of an MEA-based post-combustion carbon-capture (PCC) process and integrated system with a natural gas combined cycle (NGCC) power plant, with the goal of lowering the cost of PCC commercial deployment for NGCC power plants. He created a cost model, using vendor-provided key equipment costs from a benchmark report that included a comprehensive technical design [10].

In another study, Øi et al. [11] simulated several absorption and desorption configurations for 85 per cent amine-based CO_2 removal from a natural gas-fired power plant. Simulated processes include a standard procedure, split-stream, vapor recompression, and various combinations thereof. Equipment dimensioning, cost estimates, and process optimization have all been carried out via the simulations. A basic vapor recompression case was determined to be the most cost-effective arrangement of the analyzed cases [11].

Aromada and Øi indicate in their research [3] that both vapor recompression and vapor recompression paired with split-stream operations can minimize energy usage. Among the combinations studied, the vapor recompression technique was shown to be the most energy efficient. In addition, for the CO_2 capture parameters, which are based on a natural gas-based power plant project in Mongstad, Norway, energy optimization and economic analysis were undertaken [3].

Regarding the other methods of CO_2 removal, Roussanaly et al. [12] present a new systematic method for designing and optimizing CO_2 capture membrane systems that integrates both technological and economic principles. This approach generates graphical separation issue solutions in order to construct a cost-optimal membrane system that meets CO_2 capture ratio and purity standards. The technique is demonstrated through the design of a post-combustion CO_2 capture membrane system placed on an Advanced Super Critical (ASC) power station and compared to a MEA capture unit. Finally, a comparison is made between the cost model used and models found in the literature to show that the competitiveness of the membrane system described in this research is attributable to superior design rather than an underestimating of the membrane capture cost [12].

CO_2 capture science and technology based on adsorbents are discussed and reviewed in terms of chemistry and methodology by Patel et al. [5]. Six criteria are anticipated to be satisfied in a successful sorbent design: cost, capacity, selectivity, stability, recyclability, and fast kinetics [5].

To cost-optimize the plant, this study wishes to conduct a sensitivity analysis on the effect of the overall cost when process parameters are changed. The typical choices of process parameters are according to Øi [7]: the gas temperature into the absorber, the pressure into the absorber, the minimum temperature difference in the lean/rich heat exchanger, the reboiler temperature, the condenser temperature or the reflux ratio, the solvent circulation rate, the pressure in the desorber, and efficiency of the CO_2 removal rate of the process. The process parameters investigated are the efficiency of the absorber, the minimum temperature in the lean/rich heat exchanger, and the absorber packing height.

For efficiency, similar studies were conducted by Øi et al., [13], Kallevik [6], and Ali [14]. They all ended up with an optimum cleaning efficiency equal to 85%. Ali found an optimum efficiency of 87%.

For the minimum temperature (ΔT_{min}) in the lean/rich heat exchanger, Aromada et al. [15] calculated 15 °C as the optimum. However, Kallevik [6] and Øi et al. [13] found ΔT_{min} = 13 °C as the optimum. It is worth noticing that in Øi et al. [13], depending on the parameters, the optimal temperature difference in the primary heat exchanger was determined to be 10–15 °C, which also agrees with the result of Aromada et al. [15].

For the absorber packing height, Kallevik [6] performed a sensitivity analysis with 90% cleaning efficiency, with a flue gas inlet 3.7 mole% CO_2 and an optimum absorber packing height of 19 m. Additionally, Øi et al. [13] studied a 90% cleaning efficiency with a CO_2-content of 17.8 mole%. They ended up with an optimum packing height of 12 m.

1.2. Scope of the Study

This paper will investigate an amine absorption process based on available emission data from Fortum's FEED report [16]. There are no simulated and cost-estimated scenarios on CO_2 capture from waste incineration facilities reported. Fortum's waste-burning facility at Klemetsrud in Norway has started a carbon-capture pilot project based on amine absorption of CO_2. According to Fortum's FEED report from 2019, the carbon-capture process targets removal efficiency of 95% of the CO_2 emissions [16]. A base case is established in Aspen HYSYS based on Fortum's data, and then a dimensioning and cost estimation is performed. A sensitivity analysis executes cost optimization to minimize and reduce costs. A series of case studies conducts the technique to investigate the changes in price affected by the different variables. In this study, the variables manipulated are the absorber packing height, the CO_2 capture efficiency, and the lean/rich heat exchanger minimum temperature difference.

Further, the Power Law method is utilized to adjust the equipment cost when the sensitivity analysis changes the size. However, this way of calculating the equipment cost can be considered a "shortcut" as it is a time-saving process. To investigate the uncertainties in the results from the Power Law method, a comparison between the Enhanced Detailed Factor (EDF) using Aspen In-Plant Cost estimator and the Power Law method has been accomplished.

Lastly, to improve the usage of the Power Law method, a derivation of the individual size factor exponent for all the equipment—except the absorber and desorber—was executed. The derivation is a technique to find a unique exponent factor for each piece of equipment.

In this study, data from Fortum's waste burning facility at Klemetsrud in Norway has been used. There are no simulated and cost-estimated scenarios on CO_2 capture from waste incineration facilities reported in the open literature. Additionally, for all cases in the sensitivity analysis, a comparison between the Enhanced Detailed Factor (EDF) using Aspen In-Plant Cost estimator and the Power Law method has been accomplished to investigate the uncertainties in the results from the Power Law approach. In most literature, just the Power Law method has been utilized to adjust the equipment cost when the sensitivity analysis changes the size. Further, to improve the usage of the Power Law method, a derivation of the individual size factor exponent for all the equipment—except the absorber and desorber—was executed.

2. Materials and Methods

In this study, Aspen HYSYS version 12 was used to create a standard amine-based CO_2 capture process, and the simulated results were the foundation for equipment dimensioning and cost estimation with the same calculation approach as in literature [15,17]. The fluid package of the acid gas property package, including vapor and liquid equilibrium models for electrolytes, was used in all simulations. This package replaced the Amine property package, which has been commonly used in literature. Constant Murphree efficiencies were specified in the absorber and desorber.

2.1. Process Description and Available Parameters

Waste coming from households is solely responsible for 5% of global greenhouse emissions. A waste burning facility can convert waste to energy during an incineration process. One of the vital roles of a waste burning facility is to treat and burn trash that is not recyclable. CO_2 removal is expected to significantly reduce emissions in a waste burning facility. The waste treatment produces emissions, and carbon capture is a possible solution [16].

To establish the base case for our simulations, the values for summer H_2O in Fortum's CO_2 capture plant [16], which gives an average H_2O content of 16.3%, and then will be made saturated at 40 °C, have been used in the simulations. Since the flue gas pressure ranges from 0.95–1.05 bar(a), an average inlet pressure of 101 kPa is considered in the cases. The amine used in Fortum's CO_2 capture plant is Cansolv DC- 103 mixed with 50% water. The Cansolv solution is made by Shell, and its specifications are not available to the public [16]. Instead, MEA has been used as the amine in the solvent for this project.

The conventional amine-based CO_2 absorption–desorption mechanism is depicted in Figure 1. A primary absorber and desorber (stripper) with a reboiler and condenser, lean/rich amine heat exchanger, pumps, and a cooler make up the system. An amine solvent absorbs CO_2 from the exhaust gas (e.g., monoethanolamine-MEA) in the absorption column. After being heated in the lean/rich amine heat exchanger, the CO_2-rich amine solution from the absorption column is injected into the stripper for regeneration. The regenerated (lean) amine is returned to the absorption column to be reused. It initially passes through the lean/rich amine heat exchanger to heat up the rich stream before being cooled in the amine cooler [3].

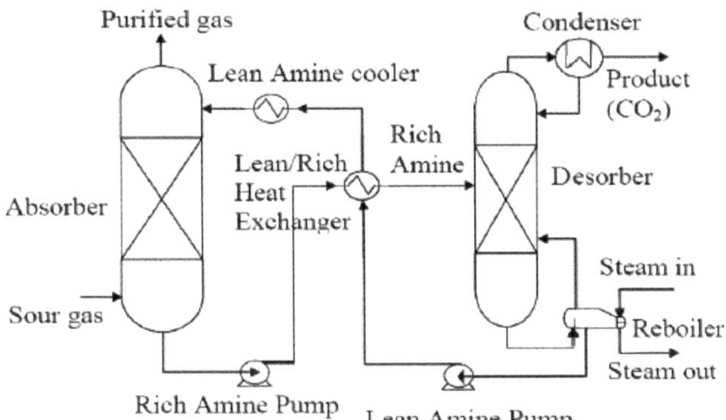

Figure 1. Process flow diagram of a standard amine-based CO_2 capture process [3].

2.2. Specifications and Simulation

The specifications in Table 1 correspond to a 90 per cent CO_2 removal efficiency and a minimum approach temperature of 10 °C in the lean/rich heat exchanger, which is considered the base case configuration.

Table 1. Aspen HYSYS model parameters and specifications for the base case alternative.

Items	Specifications (Unit)	Value
Inlet Flue Gas	Temperature (°C)	60
	Pressure (kPa)	101
	Molar flow rate (kmol/h)	17,110
	O_2 content (mole%)	9
	CO_2 content (mole%)	7.5
	H_2O content (mole%)	6.7
	N_2 content (mole%)	76.8
Flue gas to absorber	Temperature (°C)	40
	Pressure (kPa)	111
Lean MEA	Temperature (°C)	45
	Pressure (kPa)	101
	Molar flow rate (kmol/h)	42,110
	MEA content (W%)	29.48
	CO_2 content (W%)	5.58
Absorber	Number of stages	20
	Murphree efficiency (%)	15
	Rich amine pump pressure (kPa)	200
	Rich amine temp. out of Lean/Rich HEx (°C)	102.9
Desorber	Number of stages in stripper	8
	Murphree efficiency (%)	50
	Reflux ratio in the desorber	0.3
	Reboiler temperature (°C)	120
	Pressure (kPa)	200

The calculation method used is similar to previous works [3,13,15]. The absorption and desorption columns have been modelled as equilibrium stages with stage efficiencies. The absorber is modelled with 20 packing steps, but the desorber has just eight packing stages. For both columns, equilibrium stages of 1 m height are considered. For the absorption column, Murphree efficiencies of 15% were used. For all stages of the desorption column, a constant Murphree efficiency of 50% was given. The Modified HYSIM Inside-Out method was chosen in the columns since it aids in convergence. The pump and flue gas fan were specified to have an adiabatic efficiency of 75% [3,13,15].

An alternative to the equilibrium-based simulation approach is a column calculation based on the number and height of transfer units (NTU/HTU) or a rate-based approach [17]. A limitation of the alternative calculations is unknown parameters such as mass transfer and heat transfer numbers. The Murphree efficiency values in the equilibrium-based simulations can be based both on estimation from physical properties and on empirical values [7,17]. Another advantage with equilibrium-based simulations compared to rate-based simulations available in, e.g., Aspen Plus, is that they converge more easily [17].

Figure 2 shows the Aspen HYSYS simulation process flow diagram (PFD). Aspen HYSYS is an equation-based simulation program, meaning it can compute in-streams from out-streams. This study, on the other hand, uses a sequential calculating technique. To solve the flowsheet in Aspen HYSYS, one needs to include recycle blocks. This block compares the block's in-stream to the block's out-stream in the previous iteration [13].

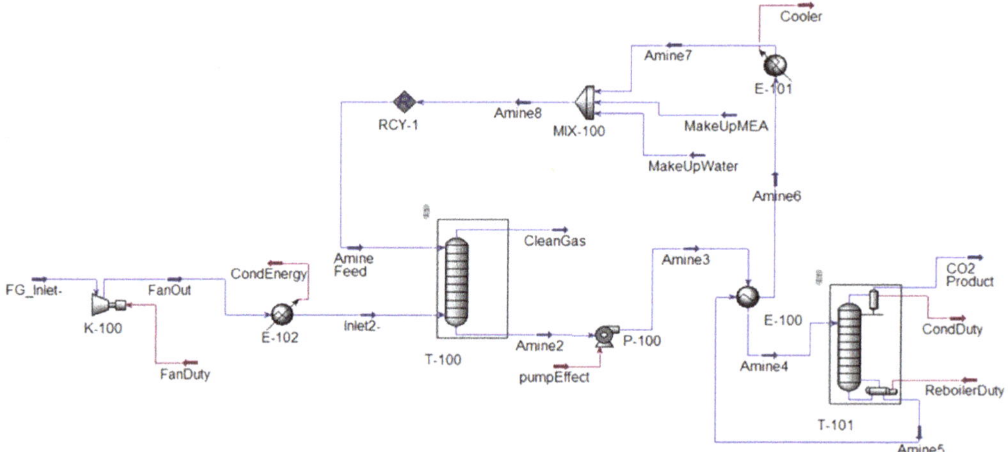

Figure 2. Aspen HYSYS flow sheet for the base case simulation.

A fan and a cooler in the process flow have been considered to adjust the inlet stream temperature and pressure into the absorber column. The fan will increase the pressure from 101 kPa to 111 kPa, which is assumed as sufficient and used in previous work such as by Øi [7]. The temperature of the flue gas will increase due to the pressure increase. This temperature increase is handled with a cooler before the absorber to obtain 40 °C at the absorber inlet. The cooler before the absorber will not be a part of further discussion as it is assumed that it will not significantly impact the cost estimates.

3. Dimensioning and Cost Estimation

A simplified dimensioning of the process equipment has been done. The calculations are based on the process flowsheet from Aspen HYSYS with equations of states, and energy and material balances.

3.1. Scope of Analysis

Except for the flue gas cooler, the cost analysis is confined to the equipment specified in the flowsheet in Figure 2. Pretreatment and especially the removal of SO_x and NO_x will probably be necessary. A possible process is given in Iliuta and Larachi [18]. There is no pre-treatment, such as inlet gas purification or cooling, and no post-treatment, such as compression, transport, or storage of CO_2 in this study. The cost estimate is only for the cost of the specified equipment installed. It excludes expenses such as land acquisition, preparation, service buildings, and ownership costs.

3.2. Dimensioning of Equipment

For the absorber and desorber cost estimation, the volume of packing and shell is used as the dimensioning factors. The most expensive component of a column is the packing, and in this study, structured packing is chosen because of its high efficiency, high capacity, and low-pressure drop [7]. A constant stage (Murphree) efficiency equivalent to 1 m of packing was assumed to calculate the packing height. Table 1 specifies Murphree efficiencies of 0.15 and 0.5 for the absorber and desorber, respectively. The absorption and desorption columns' total heights are estimated to be 40 and 20 m, respectively. The packing, liquid distributors, water wash, demister, gas inflow and outflow, and sump are all factored into the absorber height calculation. The condenser inlet, packing, liquid distributor, gas inlet, and sump are all factored into the desorber height calculation [13,19]. The absorption column diameter was determined using a gas velocity of 2.5 m/s, whereas the desorption column was computed using 1 m/s [19].

Based on the duties and temperature conditions extracted from the simulations and an ideal countercurrent flow assumption, the heat transfer areas of the heat exchangers are estimated. The heat transfer coefficients are assumed to be 500 $W/(m^2 \cdot K)$ for the lean/rich heat exchanger, 800 $W/(m^2 \cdot K)$ for the lean amine cooler, 800 $W/(m^2 \cdot K)$ for the reboiler, and 1000 $W/(m^2 \cdot K)$ for the condenser [19]. This research assumes standard shell and tube heat exchangers.

For the pump and fan, the duty is used as the dimensioning parameter; however, the mass flow is also used in the Aspen In-Plant cost estimator. The fan and pumps are designed with an adiabatic efficiency of 75% in Aspen HYSYS.

3.3. Capital Cost Estimation

In this study, the capital cost (CAPEX) estimation is done by the Enhanced Detailed Factor (EDF) method. This approach is based on installation factors for each item of process equipment. According to Ali et al. [14], there are numerous advantages of using the EDF method. This method performs well and gives an accurate estimate in the early-stage cost estimate.

The cost for each piece of equipment was calculated with the use of the Aspen In-Plant Cost-estimator (v.12) software, which gives the price in Euro (€) for the year 2019 (1st Quarter), and the default location is Rotterdam in Netherland. The exact location of Rotterdam is assumed in this evaluation. To calculate the cost for each piece of equipment, an installation factor sheet made by Nils Henrik Eldrup is used [15]. This sheet includes all the installation factors for the equipment, and it requires that the equipment cost is calculated in carbon steel (CS). Each equipment component will have a total installation factor, which is the sum of all sub-factors (direct costs, engineering, administration, commissioning, and contingency).

Except for the flue gas fan, which is built of carbon steel, all equipment is supposed to be made of stainless steel (SS316). For welded and rotating equipment, the material factors to convert costs in SS316 to CS are 1.75 and 1.30, respectively.

The installed cost factors were estimated by using the values reported in the detailed factor table [15] for the year 2020. This implies that the equipment cost must first be converted to cost data for 2020. Then, the EDF method will be applied for finding the total installed cost. Finally, the total installed price must be adjusted for inflation from 2020 to 2021. The cost-inflation index that has been used is tabulated in Table 2 [20]. The currency will be converted to Norwegian kroner (NOK) with the factor of 9.8, which is taken from early October 2021 from Norges Bank [21].

Table 2. Cost-inflation indexes: 2019–2021 [20].

Year	Cost-Inflation Index
2019	289
2020	301
2021	317

3.4. Operating Cost Estimation

The operational and maintenance cost (OPEX) contributes to a significant part of the total costs. It is common practice to divide OPEX into fixed and variable costs. The fixed costs are maintenance costs and operating labor costs. The maintenance costs are usually assumed as a percentage of the equipment installation cost (EIC) in the range of 2% to 6%, which in this study considered 3%. The operating labor costs depend on the number of workers and the operating hours throughout the year. Variable costs are mainly utility costs such as raw materials, electricity, cooling water, steam, solvents, and other consumables. The yearly cost of the mentioned utilities must be calculated [14]. In Table 3, the OPEX assumptions and specifications are presented [15].

Table 3. OPEX assumptions and specifications [15].

Item	Symbol	Unit	Value
Operating lifetime	n	(Years)	25 [1]
Operating hours p	-	(h/year)	8000
Discount rate	r	(%)	8
Exchange rate	-	(NOK/EUR)	9.8
Electricity cost	-	(EUR/kWh)	0.06
Steam cost	-	(EUR/kWh)	0.015
Cooling water cost	-	(EUR/m^3)	0.022
Water process cost	-	(EUR/m^3)	0.203
MEA cost	-	(EUR/m^3)	1516
Maintenance cost	-	(EUR/year)	3% of CAPEX
Operator cost	-	(EUR/year)	80,414 ($\times$6 operators)
Engineer cost	-	(EUR/year)	156,650 (1 engineer)

[1] 2 years construction + 23 years operation.

4. Methods for Optimization

For the economic evaluation, a sensitivity analysis of our configuration is conducted. A series of case studies are performed to investigate the effect of the different variables on the cost. The variables are the packing height, absorber efficiency, and lean/rich heat exchanger minimum temperature difference. To compare the various project alternatives and arrive at the optimum solution, it is necessary to compute the yearly CO_2 capturing cost for each alternative, where this amount is estimated from Equation (1) [14,15].

$$\text{Annual capturing cost of } CO_2 = (\text{Total annual cost})/(\text{Mass of captured } CO_2) \qquad (1)$$

$$\text{Total annual cost} = \text{Annualized CAPEX} + \text{Annualized OPEX} \qquad (2)$$

To calculate the annualized CAPEX, the operational lifetime and the interest rate needs to be known.

$$\text{Annualized CAPEX (euro/year)} = \text{CAPEX}/(\text{Annualized factor}) \qquad (3)$$

$$\text{Annualized factor} = \sum_{i=1}^{n} \left[\frac{1}{(1+r)^n} \right] \qquad (4)$$

where n is the operative lifetime (for one year construction) and r is the interest rate.

The equipment's size will differ when parameters are changed in order to perform a sensitivity analysis on the base case. A comparison between the EDF method and the cost-to-capacity method, also known as the Power Law, will be applied to the equipment. The idea of the Power Law is that the change in equipment's size or performance is not necessarily linear to the costs but that cost is a function of capacity raised to an exponential factor. This can be expressed by Equation (5) [22].

$$(\text{Cost of B})/(\text{Cost of A}) = ((\text{Capacity of B})/(\text{Capacity of A}))^e \tag{5}$$

where e is an exponential size factor that typically varies in the order of magnitude from 0.35 to 1.70 depending on the type of equipment [23]; for this study, it is assumed that the exponential factor is 1.0 for the absorber and desorber column, and for the rest of the equipment, it is considered as a factor of 0.65.

4.1. Absorber Packing Height

For this case study, the stages have been adjusted from 18 to 24 stages to find the lowest CO_2 captured cost (NOK/kg CO_2). It is assumed that the pressure drop in the absorber is a function of the number of stages and correlated with a factor of 1 kPa per stage. This will also affect the needed pressure increase from the fan. This is the same approach used by Kallevik; however, he used 0.94 kPa per stage [6]. The absorber efficiency, the lean/rich heat exchanger minimum temperature difference, and lean amine composition have been kept constant through the case study. However, to keep them constant, the lean amine feed and the inlet temperature of the desorber had to be adjusted.

4.2. Removal Efficiency

To explore economic performance while modifying the overall removal efficiency, a case study was conducted. First, the CO_2 captured cost (NOK/kg CO_2) was evaluated for two different efficiency cases (85% and 95%) as well as the base case (90%). Then, to find a more accurate case, by changing the lean amine flow rate, efficiency was adjusted from 82% to 89%, and for each case, the CO_2 captured cost was calculated. Here, the number of stages in absorber and desorber, the composition of recycling amine solvent, and the minimum temperature approach in the lean/rich heat exchanger remained the same as the base case.

4.3. The Lean/Rich Heat Exchanger Minimum Temperature Approach

Another case study was conducted to look into the economic performance of the lean/rich heat exchanger when the degree of heat recovery was changed. In the minimum approach temperature, CO_2 captured cost was calculated for four cases ($\Delta T_{min} = 5.5\,°C$, $\Delta T_{min} = 7.5\,°C$, $\Delta T_{min} = 12.5\,°C$, and $\Delta T_{min} = 15\,°C$) as well as the base case ($\Delta T_{min} = 10\,°C$). According to the result obtained from these cases, the study was expanded on extra cases from 11 °C to 17 °C. Assuming the constant temperature for the reboiler outlet is equal to 120 °C and changing the temperature of the outlet-rich amine from the lean/reach heat exchanger, the ΔT_{min} was adjusted for each case. For a particular total CO_2 removal efficiency, all flue gas and absorption column parameters were kept constant throughout the investigation, and the same was done for the rate and composition of lean amine flow.

4.4. Approach to Size Factor's Exponent Derivation

The methodology for deriving an exponent factor assumes a non-linear relationship between the size and the equipment cost—cost is a function of size raised to an exponent. However, a linear relationship appears by applying the natural logarithm to the size and the price. Therefore, the individual exponent can be derived by taking the natural logarithm of a data set of size and cost, making a data plot of the results, and then performing a linear regression from the data plot. The slope of the line represents the exponent, and the R-squared indicates how close the linear regression is to matching the data set [22]. To derive the individual equipment exponent factors, two sets of data from the Aspen

In-Plant Cost Estimator was used. The first data set is from the "the absorber packing height"-variation, and the second is from "the CO_2-efficiency and the ΔT_{min}"-variations.

5. Results and Discussion

5.1. Base Case Evaluation

Figure 3 illustrates the equipment cost in the CO_2 capture plant for the base case study. Total equipment cost is about 480 MNOK, and it is evident that the absorber is the most expensive equipment, with more than 40% of the total cost. It is worth mentioning that about 57% of the absorber's cost is the packing cost. The lean/rich heat exchanger and reboiler are the other expensive types of equipment, contributing 29% and 13% of the total cost, respectively.

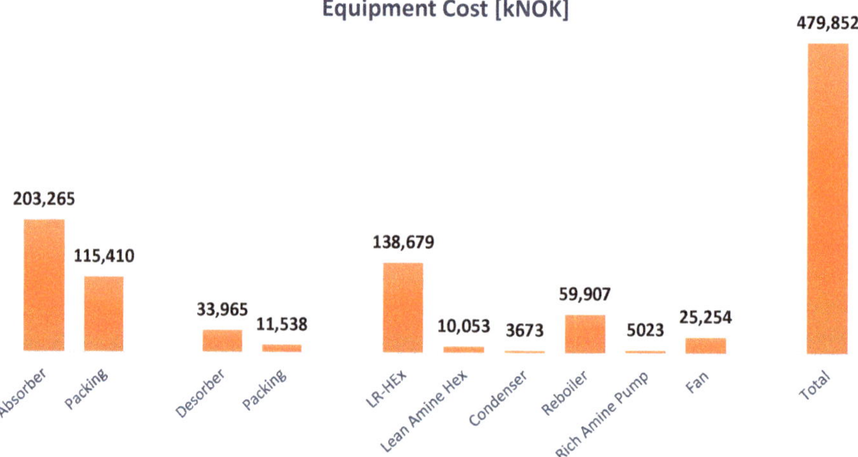

Figure 3. Equipment cost for CO_2 capture plant.

Figure 4 shows the total OPEX for the base case study. With a cost of more than 61 MNOK per year, steam is the most expensive utility for this plant. This is about 60% of the total OPEX (105.5 MNOK). For each step of the analysis, the steam consumption has been calculated and compared. As shown in Figure 4, maintenance, MEA, and electricity are the other expensive cost components beside steam.

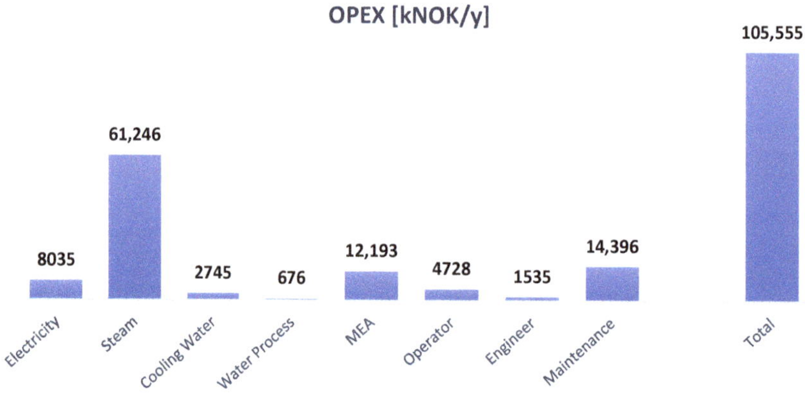

Figure 4. Operation cost for the base case study.

5.2. Minimum Temperature Approach (ΔT_{min}) in Lean/Rich Heat Exchanger

This study is a trade-off between the area in the lean/rich heat exchanger and the requirements for external utility. When ΔT_{min} is changed, the area of the heat exchanger is changed. The most important trade-off is between the capital cost affected by the change in the size of the lean/rich heat exchanger and the operational cost involved in the difference in the steam consumption.

The calculation of the sensitivity analysis for the lean/rich heat exchanger minimum temperature approach (ΔT_{min}) is shown in Figure 5. The graph compares CO_2 captured cost (NOK/t CO_2) between the EDF- and Power Law method. According to Figure 5, the optimum ΔT_{min} is equal to 12.5 °C when the Power Law method has been used, while with the EDF method, the lowest cost achieved is with a $\Delta T_{min} = 15$ °C. For further evaluation, other sensitivity analyses where the minimum temperature approach ranged from 11 °C to 17 °C were conducted and are shown in Figure 6.

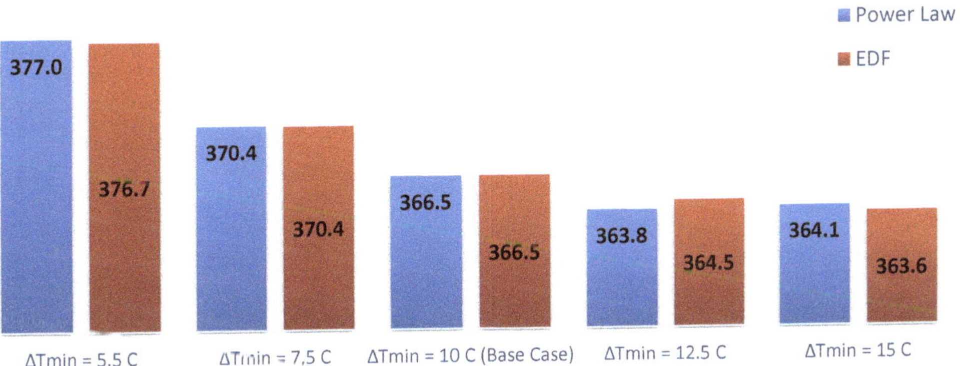

Figure 5. Sensitivity analyses for minimum temperature approach, ΔT_{min} (first optimization).

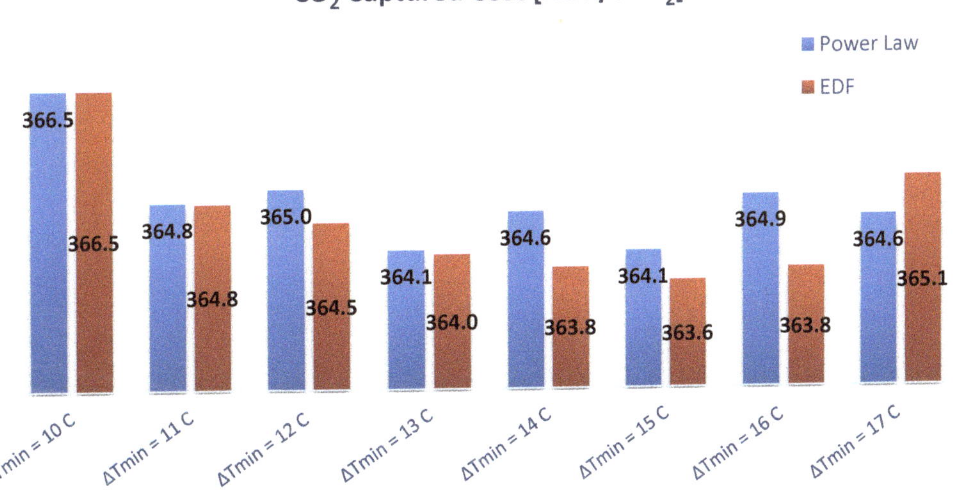

Figure 6. Sensitivity analysis for minimum temperature approach, ΔT_{min} (second optimization).

Figure 6 shows that 15 °C is the calculated optimal minimum temperature approach for the two procedures, in which CO_2 captured cost for the EDF- and Power Law method are 363.6 NOK/t CO_2 and 364.1 NOK/t CO_2, respectively. However, a similar result is seen for the Power Law method at ΔT_{min} = 13 °C. The results obtained from the EDF method are much more stable than the Power Law and clearly show that the lowest CO_2 captured cost has occurred at ΔT_{min} = 15 °C (Figure 6). In terms of the Power Law method, the same exponent value (0.65) has been adopted for all types of equipment and in all various circumstances and capacities in this study. This could be a reason for the instability in the results of this method.

Figure 7 indicates the consumption of steam as the most expensive utility in different cases by evaluating the duty of reboiler per kilogram CO_2 captured. Obviously, by increasing the minimum temperature approach in the lean/rich heat exchanger, the amount of steam consumed in the reboiler increased. The steam consumption for ΔT_{min} optimum calculated in this study is about 3820 kJ/kg.

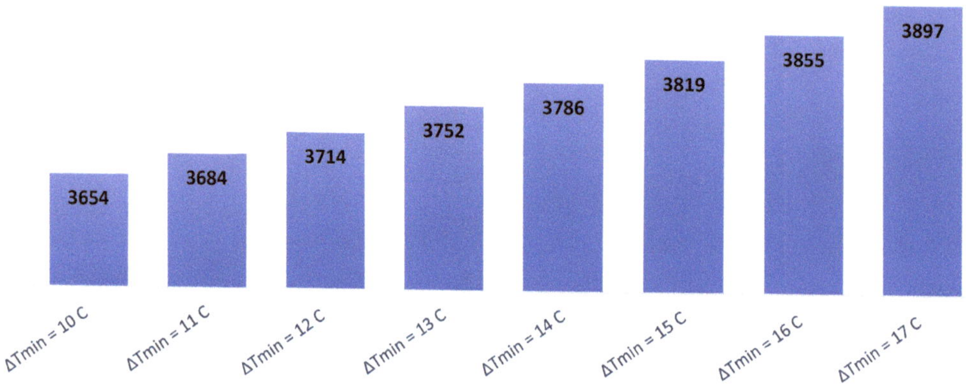

Figure 7. Steam consumption for ΔT_{min} sensitivity analyses (second optimization).

5.3. CO_2 Removal Efficiency

First, for finding the optimum efficiency, by adjusting the flow rate of amine solvent circulation, CO_2 captured costs for 85% and 95% efficiency were calculated and compared to the base case (90% efficiency). Figure 8 shows almost similar results for both the EDF and the Power Law methods. It seems that in this study, 85% is the optimal case. After this analysis, extra removal efficiencies alternatives were evaluated to obtain the exact optimum efficiency (Figure 9).

Figure 9 compares the results of efficiencies from 82% to 89% in both the EDF- and Power Law methods. According to the graph, 87% is the optimum calculated efficiency in both approaches. The CO_2 captured cost is 360 NOK/t and 360.8 NOK/t for the EDF- and Power Law methods, respectively.

There is an instability in the results shown in Figure 9 for both the EDF- and Power Law method. Although the graph indicates almost the same trend for both scenarios, the reason for discrepancies between the obtained results for both models could be the selection of the same exponent in the Power Law method for all equipment in various capacities, similar to what was explained in the previous section for the minimum temperature approach. Additionally, uncertainties in calculations and simulations could be another reason for instability in the results.

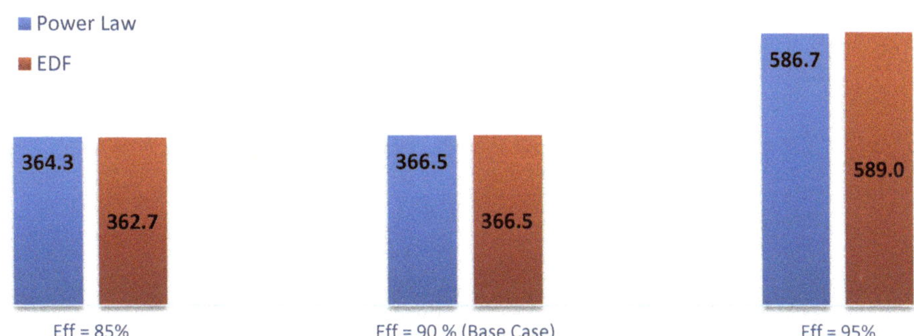

Figure 8. Sensitivity analyses for CO$_2$ capture efficiency (first optimization).

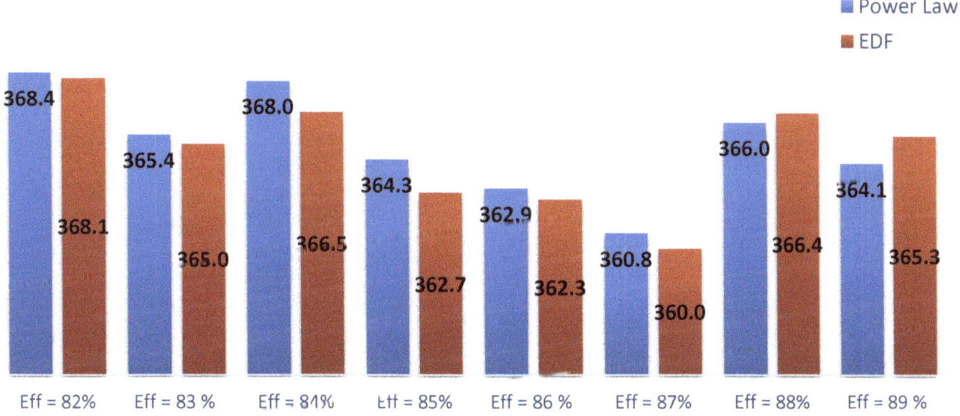

Figure 9. Sensitivity analyses for CO$_2$ capture efficiency (second optimization).

The total annual cost and amount of CO$_2$ captured in a year are demonstrated in Figure 10. This proves that by increasing the efficiency from 82% to 90%, all other items, including CAPEX, OPEX, total annual cost, and CO$_2$ captured, gradually go up, and the results are entirely stable. So, the variation in the results shown in Figure 9 could be due to CO$_2$ captured cost dependency on the total annual cost and the amount of CO$_2$ captured simultaneously, which cause some uncertainty in results for some cases, especially 83% or 84% efficiency.

The steam consumption in the reboiler has been compared for various efficiencies in Figure 11. It shows almost identical steam consumption for removal efficiency from 82% to 89% in the range of 3472 to 3603 kJ/kg CO$_2$ captured, while the calculated duty of the reboiler for 95% efficiency is about 6324 kJ per kilogram CO$_2$ captured. This is almost twice as much as other cases, so this could be a significant reason why the CO$_2$ captured cost in this efficiency is much higher than in other cases.

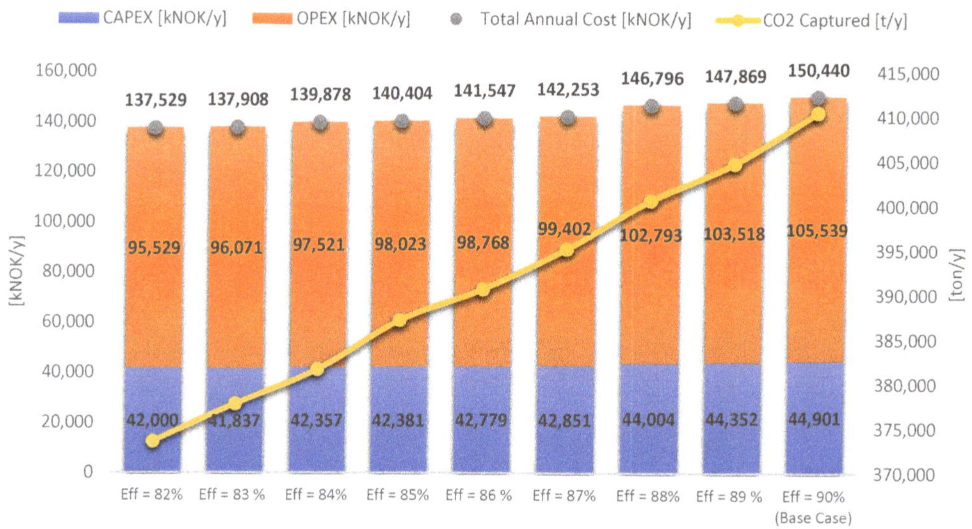

Figure 10. Total annual cost and CO_2 captured for optimum efficiency analysis.

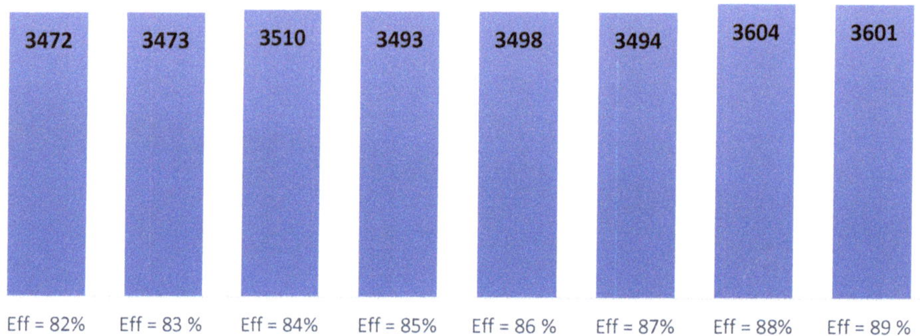

Figure 11. Steam consumption in sensitivity analyses for optimum efficiency (second optimization).

5.4. Absorber Packing Height

This study is a trade-off between amine flow rate and the number of stages in the absorber: the solvent flow rate increases when the number of absorber stages decreases, and vice versa. The significant capital costs affected by this change are the size of the absorber and the heat transfer area in the lean/rich heat exchanger. The cost of the lean/rich heat exchanger is affected due to the change in amine flow. The primary operational costs affected by the change are fan power consumption, steam consumption, and amine consumption. The fan power consumption varies because the pressure is adjusted to simulate the pressure loss over the packing. The steam consumption is affected due to the amine flow change, and the amine consumption is probably affected by various temperatures and amine flow in the absorber.

Figure 12 indicates the change in CO_2 captured cost (NOK/t CO_2) by adjusting the number of stages from 18 to 24 stages for the EDF- and Power Law method. Some variations between the two approaches can be observed, but no significant trend. There is also an indication of a global calculated minimum at stage number 19 and a local calculated minimum at stage number 22.

CO$_2$ Captured Cost [NOK/t CO$_2$]

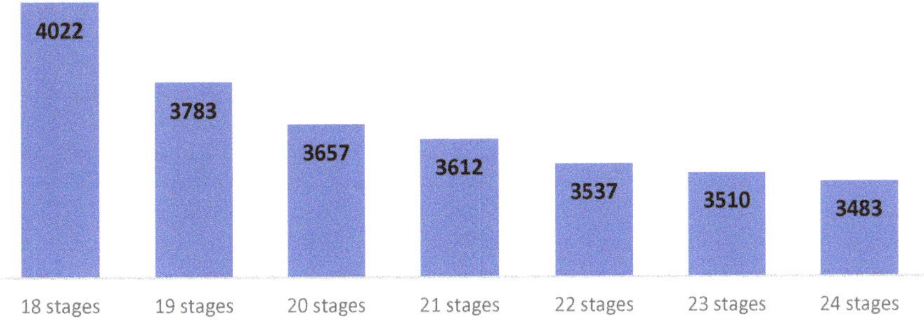

Figure 12. Sensitivity analyses for absorber packing height.

Figure 13 shows the steam consumption (kJ/kg CO$_2$) at different absorber stages. When the number of stages increases the flow rate of the amine solvent decreases, so a declining trend in reboiler duty can be observed, with the largest decline from stage 18 to 20 and a gradually flattening curve after that. The steam represents the most considerable expense. Even though its rate of change declines gradually with the increment of stages, the largest decline happens from 18 to 19 and the second largest from 19 to 20. By only considering the calculated numbers, the local minimum at 22 could be due to the significant decline in steam consumption from stage 21 to 22. However, the most likely explanation is either instability in the simulations or uncertainties in the calculations.

Steam consumption [kJ/kg CO$_2$]

Figure 13. Steam consumption in sensitivity analyses for absorber packing height.

5.5. Size factor's Exponent Derivation

In Table 4, the derived exponents for the different types of equipment can be observed. The exponents are derived using the data set from the sensitivity analysis of "the absorber packing height". It is worth noticing the high R-squared value for all the types of equipment except the fan.

Table 4. Exponent derivation results from the sensitivity analysis of "the absorber packing height" data set.

Equipment	Material	Capacity Unit	Size	Cost (kEUR)	Data Points	Exponent Factor	R^2
Lean HEx	SS316	m^2	322.7	177.3	9	0.71	0.960
Lean/rich HEx	SS316	m^2	8258.1	3311.6	9	1.07	0.992
Condenser	SS316	m^2	66.9	48.6	9	0.7	0.985
Reboiler	SS316	m^2	3050.4	1430.2	9	0.86	0.981
Pump	SS316	kW	26.4	60.9	9	0.75	0.989
Fan	CS	kW	1020.3	528.7	9	0.23	0.937

Table 5 indicates the derived exponents for the different types of equipment, using the sensitivity analysis of "the CO_2-efficiency and the ΔT_{min}" data set. It is worth noticing that this data set does not include the fan. Moreover, the R-squared is high value for all the types of equipment except the condenser.

Table 5. Exponent derivation results from the sensitivity analysis of "the CO_2-efficiency and the ΔT_{min}" data set.

Equipment	Material	Capacity Unit	Size	Cost (kEUR)	Data Points	Exponent Factor	R^2
Lean HEx	SS316	m^2	251.2	150.8	20	0.77	0.972
Lean/rich HEx	SS316	m^2	5890.3	2344.1	20	1.00	0.999
Condenser	SS316	m^2	78.1	57.8	20	0.67	0.943
Reboiler	SS316	m^2	3056.7	1396.3	20	0.98	0.997
Pump	SS316	kW	25.8	60.2	20	1.02	0.999

The R-squared value measures how close the data are to the fitted regression line. An R-squared value of 1 is an ideal approximation, and 0 indicates that the model explains none of the data points [24]. The R-squared values in Tables 4 and 5 suggest that all exponents—for both data sets and the given size range, equipment type and material—are over 94%. This indicates that when using the exact specifications and scope, all the exponents are usable and a better choice than the average.

The lean/rich heat exchanger, the reboiler, and the pump suggest that all the exponents for both data sets are over 98% and are almost ideal. This indicates that assuming the same specifications and size range, using these exponents with the Power law, will practically give the same result as the EDF method.

The exponents for the lean heat exchanger, the lean/rich heat exchanger, and the condenser are similar for both data sets. This indicates that the exponents can be used outside the size range of the given data sets, and when choosing which one from the given data set, the R-squared value and the proximity to the size range need to be considered.

The exponents for the reboiler and the pump deviate more for both data sets than the other equipment but indicate a value above the average exponent. This suggests that the exponents should be used with more care outside the given size range.

The exponent for the fan is only obtained from the "absorber height" data set.

The exponents presented in Tables 4 and 5 are derived for the given size range, material, and equipment. The data are obtained from the Aspen In-Plant Cost Estimator and not the actual price. Additionally, the data sets are small. Therefore, it is essential to warrant a level of caution when using the exponents. However, most published exponents from the industry do not support specific items. Therefore, using the derived exponents will probably give a more precise result than the general exponent.

5.6. Comparison with Earlier Reports

The predicted cost of 360–380 NOK (36–38 Euro) per 1 ton CO_2 captured is low when compared to more detailed cost estimates on CO_2 capture cost reported in the literature, which show more than 50 Euro/t [14,15]. A probable reason for this could be the assumption of no pre-treatment, such as inlet gas purification or cooling, and no post-treatment, such as compression, transport, or storage of CO_2. Additionally, it is more than the calculated cost of 180–190 NOK/t (18–19 Euro/t), which Øi et al. have evaluated [13].

A comparison between the results of this study and the other literature is shown in Table 6. All these studies have used a 29–30 (W%) MEA solvent in their simulations. The concentration of inlet CO_2 varies due to variation in plant's flue gases (NGCC power plant and Cement).

Table 6. Comparison of simulation results with literature.

Study	CO_2 Capture Rate (%)	CO_2 Concentration (mol%)	ΔT_{min} (°C)	Absorber Packing Height (m)	Reboiler Duty (kJ/kg)
This work (Base Case)	90	7.5	10	20	3654
Ali et al. [14]	90	22–28	10	15	3970
Aromada et al. [3]	85	3.73	10	20	3600
Øi et al. [13]	90	17.8	10	12	3500
Amrollahi et al. [25]	90	3.8	8.5	13	3740
Sipöcz et al. [26]	90	4.2	10	26.9 *	3930
Nwaoha et al. [27]	90	11.5	10	22 (36 Stages)	3860

* Not defined whether it is packing height or total column height.

5.6.1. Absorber Stages

A similar study in Kallevik's master thesis [6] is for 90% removal efficiency and finds the global calculated minimum captured cost in 19 stages. Moreover, the hot utility requirement decreased from 4.02 to 3.51 MJ/kg CO_2 as the stages increased from stages 18 to 23. This is almost identical to Kallevik's findings, ranging from 3.88 to 3.51 MJ/kg CO_2 from stages 18 to 23 [6].

Øi et al. [13] calculated the global minimum at 12 m absorber packing height for 90% cleaning efficiency with a CO_2-content of 17.8 mole%. This is a significant difference compared to the result in this report. When performing this sensitivity study, it was impossible to get the absorber to converge at 90% cleaning efficiency with fewer than 16 stages. This is similar to the Kalleviks study [6], where it only achieved a 90% removal rate between 14 and 23 stages. Husebye et al. argue that an increase in CO_2 concentration results in a decline in cost—especially between 2.5 and 10%. This happens because of higher CO_2 transfer, which needs less solvent [28]—therefore, one could have expected a lower number of stages.

5.6.2. Minimum Temperature Approach in the Lean/Rich Heat Exchanger

The results obtained for the EDF method clearly show that at ΔT_{min} equal to 15 °C, the lowest CO_2 captured cost has been calculated, corresponding to Aromada et al.'s article [15]. In terms of the Power Law method, the calculated optimum minimum temperature approach in this study is ΔT_{min} = 13 °C, similar to Kallevik [6] and Øi [13] studies. The cost can probably be reduced by changing to another type of heat exchanger. Øi points out in his PhD [7] that a plate heat exchanger may reduce the cost and will probably reduce the optimum ΔT_{min}. According to Øi [7], a low ΔT_{min} will reduce steam and reboiler duty. On the other hand, a high ΔT_{min} will reduce the heat exchanger cost.

5.6.3. Total Efficiency

From the first optimization in this study, 85% is the optimal case, which corresponds to what is reported in Øi et al.'s article [13] and Kallevik's thesis [6]. Moreover, Ali [14] calculated the minimum CO_2 captured cost by changing the efficiency from 85% to 90%, in which the optimal case was reported as 87%. In this study, by evaluating the extra alternatives from 82% to 90%, the calculated optimum CO_2 capture efficiency corresponds to 87%.

5.6.4. Reboiler Duty

The duty of the reboiler for 90% CO_2 removal efficiency and calculated optimum $\Delta T_{min} = 15\,°C$ is about 3820 kJ/kg CO_2 captured, which almost corresponds to the values reported in previous works [11]. In Ali et al. [14], this figure for 90% efficiency and $\Delta T_{min} = 10°C$ is about 3970 kJ/kg, while for 85% efficiency and $\Delta T_{min} = 15\,°C$, it is more than 4000 kJ/kg. In this study, the calculated reboiler duties at ΔT_{min} equal to 10 °C for 90% and 85% efficiency are 3654 kJ/kg and 3493 kJ/kg, respectively. For 85% efficiency and $\Delta T_{min} = 10\,°C$, Øi [11], Ali [14], and Aromada [3] reported in their studies the calculated reboiler duty to 3300 kJ/kg, 3900 kJ/kg and 3600 kJ/kg, respectively.

6. Conclusions

This work presented a cost estimation of an amine-based CO_2 capture process for a waste burning facility at Klemetsrud in Norway. The CAPEX and OPEX calculations were done by performing simulations in Aspen HYSYS using available data in Fortum's FEED report. For the base case, the total cost was calculated to 37.5 EUR per ton CO_2 captured, and energy consumption in the reboiler was calculated to 3654 kJ/kg.

The sensitivity analysis for the CO_2 removal efficiency is a trade-off between the quantity of CO_2 captured and the expense of a higher amine flow rate, which resulted in larger lean/rich heat exchanger, reboiler, and steam usage. This study resulted in 87% as the optimum calculated efficiency in the EDF- and Power Law methods. There the CO_2 captured costs were 36.7 EUR/t and 36.8 EUR/t, respectively. The reboiler duty was around 3500 kJ per kg CO_2 captured in the parameter assessment.

Sensitivity analysis of the ΔT_{min} in the lean/rich heat exchanger is a trade-off between the area of the lean/rich heat exchanger and the required exterior utility. This analysis showed an optimum ΔT_{min} of 15 °C from both the EDF- and Power Law methods, with a CO_2 removal cost of 37 EUR/t CO_2. However, the Power Law approach also yielded a comparable result at $\Delta T_{min} = 13\,°C$. In addition, the reboiler's average energy usage was around 3800 kJ per kg CO_2 captured.

The sensitivity analysis for the absorber packing height looks at the relationship between the amine flow rate and the number of absorber stages: as the number of absorber stages lowers, the solvent flow rate rises. The size of the absorber and the heat transfer area in the lean/rich heat exchanger are two key capital expenses that are influenced by this shift. Fan power consumption, steam consumption, and amine consumption are the three main operating costs affected by this adjustment. The fan power consumption fluctuates since the pressure is modified to compensate for pressure loss across the packing. Due to the amine flow modification, both steam and amine consumption are affected. This study resulted in 19 m as the optimum packing height. The CO_2 removal costs for the EDF and Power Law techniques are 37.3 EUR/t and 37.1 EUR/t CO_2 captured, respectively. In addition, the reboiler uses 3780 kJ of energy to extract 1 kg of CO_2 from the flue gas.

A new size of factor exponent in the Power Law approach has been proposed by considering the calculated cost from the EDF-method. The exponents have different values that are specific for different equipment. The derivation on average resulted in the following exponents: for the lean heat exchanger 0.74, for the lean/rich heat exchanger 1.03, for the condenser 0.68, for the reboiler 0.92, for the pump 0.88, and for the fan 0.23.

Even though the optimization in this work only decreases the CO_2 capture cost by about 3% compared to the base case, the total improvement is significant because the total cost is very large.

Author Contributions: Conceptualization, L.E.Ø. and S.A.A.; methodology, S.A.A., S.S. and B.W.-S.; software, S.S., F.F. and B.W.-S.; validation, L.E.Ø., S.V. and S.S.; formal analysis, L.E.Ø., S.S. and S.V.; investigation, S.S., S.V., F.F. and B.W.-S.; resources, L.E.Ø.; data curation, S.S., S.V. and B.W.-S.; writing—original draft preparation, S.S.; writing—review and editing, S.S., S.K. and L.E.Ø.; visualization, S.S. and L.E.Ø.; supervision, L.E.Ø. and S.K.; project administration, S.V. and B.W.-S. All authors have read and agreed to the published version of the manuscript.

Funding: This research received no external funding.

Institutional Review Board Statement: Not applicable.

Informed Consent Statement: Not applicable.

Data Availability Statement: Not applicable.

Conflicts of Interest: The authors declare no conflict of interest.

References

1. Causes of Climate Change. Available online: https://ec.europa.eu/clima/climate-change/causes-climate-change_en (accessed on 26 October 2021).
2. IEA—International Energy Agency, IEA. Available online: https://www.iea.org (accessed on 26 October 2021).
3. Aromada, S.A.; Øi, L. Simulation of improved absorption configurations for CO_2 capture. In Proceedings of the 56th Conference on Simulation and Modelling (SIMS 56), Linköping, Sweden, 7–9 October 2015; Linköping University Electronic Press: Linköping, Sweden, 2015.
4. Jinadasa, M.H.W.N.; Jens, K.-J.; Halstensen, M. Process Analytical Technology for CO_2 Capture. In *Carbon Dioxide Chemistry, Capture and Oil Recovery*; Karamé, I., Shaya, J., Srour, H., Eds.; InTech: London, UK, 2018. [CrossRef]
5. Patel, H.A.; Byun, J.; Yavuz, C.T. Carbon Dioxide Capture Adsorbents: Chemistry and Methods. *ChemSusChem* **2017**, *10*, 1303–1317. [CrossRef] [PubMed]
6. Kallevik, O.B. Cost Estimation of CO_2 Removal in HYSYS. Master's Thesis, Høgskolen i Telemark, Porsgrunn, Norway, 2010.
7. Øi, L.E. Removal of CO_2 from Exhaust Gas. Ph.D. Thesis, Telemark University College, Porsgrunn, Norway, 2012.
8. Rubin, E.S.; Short, C.; Booras, G.; Davison, J.; Ekstrom, C.; Matuszewski, M.; McCoy, S. A proposed methodology for CO_2 capture and storage cost estimates. *Int. J. Greenh. Gas Control* **2013**, *17*, 488–503. [CrossRef]
9. Van der Spek, M.; Roussanaly, S.; Rubin, E.S. Best practices and recent advances in CCS cost engineering and economic analysis. *Int. J. Greenh. Gas Control* **2019**, *83*, 91–104. [CrossRef]
10. Luo, X. Process Modelling, Simulation and OPTIMISATION of Natural Gas Combined Cycle Power Plant Integrated with Carbon Capture, Compression and Transport. Ph.D. Thesis, School of Engineering, The University of Hull, Hull, UK, 2016.
11. Øi, L.E.; Bråthen, T.; Berg, C.; Brekne, S.K.; Flatin, M.; Johnsen, R.; Moen, I.G.; Thomassen, E. Optimization of configurations for amine based CO_2 absorption using Aspen HYSYS. *Energy Procedia* **2014**, *51*, 224–233. [CrossRef]
12. Roussanaly, S.; Lindqvist, K.; Anantharaman, R.; Jakobsen, J. A Systematic Method for Membrane CO_2 Capture Modeling and Analysis. *Energy Procedia* **2014**, *63*, 217–224. [CrossRef]
13. Øi, L.E.; Eldrup, N.; Aromada, S.; Haukås, A.; Helviglda Hæstad, J.; Lande, A.M. Process Simulation, Cost Estimation and Optimization of CO_2 Capture using Aspen HYSYS. In Proceedings of the 61st International Conference of Scandinavian Simulation, Virtual Conference, Oulu, Finland, 22–24 September 2020.
14. Ali, H.; Eldrup, N.H.; Normann, F.; Skagestad, R.; Øi, L.E. Cost Estimation of CO_2 Absorption Plants for CO2 Mitigation–Method and Assumptions. *Int. J. Greenh. Gas Control* **2019**, *88*, 10–23. [CrossRef]
15. Aromada, S.A.; Eldrup, N.H.; Øi, L.E. Capital cost estimation of CO_2 capture plant using Enhanced Detailed Factor (EDF) method: Installation factors and plant construction characteristic factors. *Int. J. Greenh. Gas Control* **2021**, *110*, 103394. [CrossRef]
16. FEED-Study-Report-DG3_redacted_version_03-2.pdf. Available online: https://ccsnorway.com/wp-content/uploads/sites/6/2020/07/FEED-Study-Report-DG3_redacted_version_03-2.pdf (accessed on 26 October 2021).
17. Fagerheim, S. Process Simulation of CO_2 Absorption at TCM Mongstad. Master's Thesis, University of South-Eastern Norway, Porsgrunn, Norway, 2019.
18. Iliuta, I.; Larachi, F. Modeling and simulations of NO_x and SO_2 seawater scrubbing in packed-bed columns for marine applications. *Catalysts* **2019**, *9*, 489. [CrossRef]
19. Ali, H. Techno-Economic Analysis of CO_2 Capture Concepts. Ph.D. Thesis, University of South-Eastern Norway, Porsgrunn, Norway, 2019.
20. ClearTax, Cost Inflation Index FY 2021-22—Overview, Calculation, Benefits & Examples. Available online: https://cleartax.in/s/cost-inflation-index (accessed on 28 October 2021).
21. Exchange Rates. Available online: https://www.norges-bank.no/en/topics/Statistics/exchange_rates/ (accessed on 28 October 2021).
22. Baumann, C. Cost-to-Capacity Method: Applications and Considerations. *M&TS J.* **2014**, *30*, 49–56.
23. Smith, R. *Chemical Process Design and Integration*; John Wiley & Sons Ltd.: Chichester, UK, 2005; pp. 17–19.
24. "What Is R-Squared?", Investopedia. Available online: https://www.investopedia.com/terms/r/r-squared.asp (accessed on 6 November 2021).
25. Amrollahi, Z.; Ystad, P.A.M.; Ertesvåg, I.S.; Bolland, O. Optimized process configurations of post-combustion CO_2 capture for natural-gas-fired power plant–Power plant efficiency analysis. *Int. J. Greenh. Gas Control* **2012**, *8*, 1–11. [CrossRef]
26. Sipöcz, N.; Tobiesen, A.; Assadi, M. Integrated modelling and simulation of a 400 MW NGCC power plant with CO_2 capture. *Energy Procedia* **2011**, *4*, 1941–1948. [CrossRef]

27. Nwaoha, C.; Beaulieu, M.; Tontiwachwuthikul, P.; Gibson, M.D. Techno-economic analysis of CO2 capture from a 1.2 million MTPA cement plant using AMP-PZ-MEA blend. *Int. J. Greenh. Gas Control* **2018**, *78*, 400–412. [CrossRef]
28. Husebye, J.; Brunsvold, A.L.; Roussanaly, S.; Zhang, X. Techno Economic Evaluation of Amine based CO_2 Capture: Impact of CO_2 Concentration and Steam Supply. *Energy Procedia* **2012**, *23*, 381–390. [CrossRef]

processes

Article

Di- and Mono-Rhamnolipids Produced by the *Pseudomonas putida* PP021 Isolate Significantly Enhance the Degree of Recovery of Heavy Oil from the Romashkino Oil Field (Tatarstan, Russia)

Liliya Biktasheva, Alexander Gordeev, Svetlana Selivanovskaya and Polina Galitskaya *

Institute of Environmental Sciences, Kazan Federal University, 18 Kremlyovskaya St., 420008 Kazan, Russia; biktasheval@mail.ru (L.B.); drgor@mail.ru (A.G.); svetlana.selivanovskaya@kpfu.ru (S.S.)
* Correspondence: gpolina33@yandex.ru; Tel.: +7-903-340-40-23

Abstract: Around the globe, only 30–50% of the amount of oil estimated to be in reservoirs ("original oil in place") can be obtained using primary and secondary oil recovery methods. Enhanced oil recovery methods are required in the oil processing industry, and the use of microbially produced amphiphilic molecules (biosurfactants) is considered a promising efficient and environmentally friendly method. In the present study, biosurfactants produced by the *Pseudomonas putida* PP021 isolate were extracted and characterized, and their potential to enhance oil recovery was demonstrated. It was found that the cell-free biosurfactant-containing supernatant decreased the air–water interface tension from 74 to 28 mN m^{-1}. Using TLC and FTIR methods, the biosurfactants produced by the isolate were classified as mono- and di-rhamnolipid mixtures. In the isolates' genome, the genes *rhlB* and *rhlC*, encoding enzymes involved in the synthesis of mono- and di-rhamnolipids, respectively, were revealed. Both genes were expressed when the strain was cultivated on glycerol nitrate medium. As follows from the sand-packed column and core flooding simulations, biosurfactants produced by *P. putida* PP021 significantly enhance the degree of recovery, resulting in additional 27% and 21%, respectively.

Keywords: enhanced oil recovery; biosurfactant; mono- and di-rhamnolipids; sand-packed column simulation of oil recovery; core flooding simulation of oil recovery

Citation: Biktasheva, L.; Gordeev, A.; Selivanovskaya, S.; Galitskaya, P. Di- and Mono-Rhamnolipids Produced by the *Pseudomonas putida* PP021 Isolate Significantly Enhance the Degree of Recovery of Heavy Oil from the Romashkino Oil Field (Tatarstan, Russia). *Processes* **2022**, *10*, 779.
https://doi.org/10.3390/pr10040779

Academic Editors: Venko N. Beschkov and Konstantin Petrov

Received: 10 February 2022
Accepted: 13 April 2022
Published: 15 April 2022

1. Introduction

Oil recovery, which is the extraction of crude oil from an oil field, can be subdivided into primary, secondary, and tertiary, or enhanced. Primary oil recovery occurs due to natural pressure in an oil reservoir using simple mechanical pumps, while secondary recovery techniques include artificial creation of additional pressure in the oil reservoir using injection water. Around the globe, only 30–50% of the amount of oil estimated to be in reservoirs ("original oil in place") can be obtained using primary and secondary oil recovery methods [1]. Enhanced oil recovery methods usually include injection of steam (thermal methods), gas, or chemicals to decrease oil viscosity. These methods enable the extraction of 10–40% more oil over the initial amount [2]. Recently, alternative methods with higher efficacy, lower cost, and lower environmental impact for enhanced oil recovery have been required.

The Romashkino oil field situated in Tatarstan (Russia) has a large oil field area, a small formation dip, many reservoir layers, a wide oil-water transition zone, and a complicated sedimentary environment. It has been exploited since 1948, and the oil recovery degree since then has decreased from 53% to 45%. Because of its high oil viscosity (2.6–4.5 mPa·s) and high sulfur and wax contents (1.3 and 3.2%, respectively), the Romashkino oil field urgently needs the development and implementation of new efficient methods for enhancing oil recovery [3,4].

Biosurfactants are amphiphilic microbial products that may be used for enhanced oil recovery since they can reduce crude oil–water interfacial tension (IFT) and alter emulsification and wettability. Biosurfactants possess lower toxicity and higher biodegradability than chemical surfactants. Due to their relatively lower efficacy (in comparable concentrations) and higher costs of production, biosurfactants are still not widely used in large-scale oil production or in the remediation of polluted sites [5–7]. However, new biosurfactant producers and new biosurfactant types with improved characteristics are often being discovered, providing good prospects for the future commercial use of biosurfactants for enhanced oil recovery, considering the growing costs of chemical biosurfactants and the growing awareness of their negative impact [5,8–12]. It should be noted that there is a lack of publications concerning specific aspects of biosurfactant usability for enhanced oil recovery, e.g., information available about the recovery degree of heavy oils in complicated sedimentary environments such as in the Romashkino oil field situated in Tatarstan, Russia [13].

The hydrophobic component of biosurfactants usually consists of long-chain fatty acids or saturated or unsaturated hydrocarbons, while the hydrophilic component is made up of organic acids, alcohols, or other carbohydrates [14,15]. Thus, according to their chemical structure, biosurfactants are classified into glycolipids, lipopeptides, phospholipids, neutral lipids, substituted fatty acids, and polysaccharides [16,17].

Biosurfactant producers, which include yeast, fungi, and mainly bacteria, originate from oil-polluted aquatic and soil environments as well as from the rhizosphere or other plant-associated locations [16,18,19]. Biosurfactants are produced by microbes for different goals, e.g., for transportation across membranes, for protection from predators, parasites and competitors, for interaction with hosts, for the formation of biofilms, and for changes in the properties of environments and nutrient sources, such as wetting and metal sequestration [12,20,21]. In environments containing hydrocarbons, microbes produce biosurfactants to provide interactions between cell physiology, the cell surface, and hydrocarbons that are substrates for the cell. Thus, low molecular mass biosurfactants are used to solubilize poorly soluble hydrocarbons by means of micelles and aggregate formation, while high molecular mass biosurfactants emulsify the hydrocarbons in water medium, and they may be bound to the cell wall, modifying the membrane, and enabling hydrocarbons to pass across the wall. It has been reported that some types of biosurfactants stimulate the growth of their producing strains, playing a vital role in the interaction between microbes and their environments. However, many mechanisms of biosurfactant synthesis and many goals that microbes achieve using biosurfactants have not yet been determined [8].

Rhamnolipids are well-studied biosurfactants due to their high emulsifying activity as well as high yields [22–25]. In large-scale bioreactors, from 1 to 10–12 and in some cases to 30–50 g L^{-1} rhamnolipids may be obtained utilizing different types of oils as a carbon source [26–29]. Structurally, rhamnolipids consist of one or two rhamnoses and one or two β-hydroxyl fatty acids of different carbon chain lengths (C$_8$-C$_{10}$, C$_{10}$-C$_{10}$, C$_{10}$-C$_{12}$, C$_{10}$-C$_{12:1}$ etc.). It has been reported that rhamnolipids with different structures possess different activities; thus, mono-rhamnolipids have been demonstrated to be more inhibitory towards plant fungal pathogens, and their emulsifying activity is higher than that of di-rhamnolipids [9,30–32].

Rhamnolipids are produced by some representatives of *Pseudomonas*, *Dietzia*, and other bacterial genera, while the pathogenic *P. aeuroginosa* is the most well studied efficient producer [32–34]. Non-pathogenic *Pseudomonas* species such as *P. putida*, *P. citronellolis*, *P. cepacian*, and other *Pseudomonas* spp. are less mentioned as potential rhamnolipid producers in the scientific literature [35–39]. Mono-rhamnolipid synthesis in bacteria involves the rhamnotransferase RhlB, while dTDP-L-rhamnose and β-hydroxyl fatty acids are used as precursors. Di-rhamnolipids are synthesized using dTDP-L-rhamnose, and mono-rhamnolipids are precursors involving rhlC rhamnotransferase. In addition, rhamnolipid production is controlled by genetic regulation of the *rmlBDAC* and *rhlAB* operons. It was demonstrated that the transformation of *Pseudomonas* strains with plasmids carrying *RhlB* and *RhlC* encoding genes enabled them to produce mono- and di-rhamnolipids respectively [32]. Interestingly, in most publications cited above, rhamnolipids of *Pseudomonas* origin are comprehensively characterized in terms of their chemical composition and ability to emulsify oil, collapse drops, inhibit pathogens, and alter the interface tension but not in terms of their ex situ capacity to enhance oil recovery, in particular heavy oil.

However, taking into account the promising abilities of rhamnolipids as surfactants produced by *Pseudomonas*, they are good candidates to be used in microbially enhanced oil recovery. The objective of the present study was to obtain and characterize biosurfactants produced by *P. putida* PP021 isolated from oil-polluted soil and to investigate their potential to enhance the recovery of heavy crude oil from the Romashkino oil field using two ex situ techniques: sand-packed column and core flooding.

2. Materials and Methods

Biosurfactant producing strain *P. putida* PP021 was isolated from an old oil spill situated in the production area of the Romashkino oil field (Tatarstan, Russia). The strain was identified on the basis of 16S rRNA gene sequencing and stored in the laboratory museum of the Institute of Environmental Sciences of Kazan Federal University (Russia).

To estimate the biosurfactant production activity, the strain was cultivated in glycerol nitrate medium at 35 °C and 180 rpm for 1–6 days. Each day, the optical density of the growing culture was measured at 600 nm, and the cell-free supernatant was obtained by centrifugation at 8000 rpm for 10 min. Interfacial tension was measured using a KS20 (Kruess, Hamburg, Germany) tensiometer using the Dui Nui method.

In order to obtain the biosurfactants, the cell-free supernatant obtained as described above was pH-adjusted using 2 N HCl, incubated overnight at 4 °C and centrifuged at 10,000 rpm and 4 °C for 20 min. The precipitated fraction was dissolved in a chloroform:methanol (2:1, v/v) mixture, and the crude biosurfactant mixture was obtained using a rotary evaporator under vacuum. The yield of the acid precipitated fraction of the biosurfactants (APF) was estimated to be 9.8 g from 1 L of cell-free supernatant.

Characterization of biosurfactants was conducted using two methods—thin-layer chromatography (TLC) and Fourier transform infrared spectroscopy (FTIR).

Separation of different rhamnolipid fractions was performed by employing TLC. To assess the composition of the biosurfactants obtained, the crude mixture was dissolved in CHCl3/CH3OH (1:1 v/v) and spotted on a silica gel plate (G60, Merck, Darmstadt, Germany). The TLC solvent was a mixture of chloroform:methanol:ammonia solution (65:35:5, $v/v/v$). The Rf values of the spots obtained were calculated from the TLC plate exposed to UV light. To detect sugars, lipids, and free amino groups in the TLC spots, the following chromogenic agents were used: (i) 100 mL of acetic acid supplemented with 2 mL of sulfuric agent and 1 mL of p-anisaldehyde, (ii) iodine vapor, and (iii) 1% ninhydrin reagent. After staining, the plates were incubated at 110 °C for 10 min for spot development. Mono- and di-rhamnolipid standards (Sigma Aldrich, Burlington, MA, USA, 99% pure) were used to identify different fractions of rhamnolipids on the TLC plates [40].

To determine the type and structure of biosurfactants, Fourier transform infrared spectroscopy was performed using LUMOS I (BRUKER, Billerica, MA, USA) for the APF. The spectra were collected from 400 to 4000 wavenumbers (cm^{-1}).

The ability of the bacterial strain to synthesize mono- and di-rhamnolipids was analyzed using real-time PCR with specific primers for the *rhlB* and *rhlC* genes, respectively (forward and reverse, 5′-3′: GCCCACGACCAGTTCGAC and CATCCCCCTCCCTATGAC, CCGAAGCTTATGAGCGGCCTGTTCCACT and CTTGGAATTCCCGGAAGCTACGGACG). DNA was extracted from the pellet of the overnight culture after centrifugation using the FastDNA™ SPIN Kit for Soil (MP Biomedicals) according to the manufacturer's instructions. Real-time PCR was carried out with SYBR Green in a total volume of 20 μL that contained 1X PCR buffer, 0.2 mM dNTPs, 1 μM of each primer, 2 mM MgCl₂, 1.25 U Taq DNA polymerase, and 1 μM DNA. Amplification was carried out in a CFX 96 (Bio-Rad Laboratories, Inc., Hercules, California, USA) thermocycler using the following program: denaturation at 94 °C for 2 min, 35 cycles of 94 °C for 30 s, 58 °C for 1 min, extension at 72 °C for 1.5 min, and final extension at 72 °C for 7 min [9,41,42]. To analyze the expression levels of the *rhlB* and *rhlC* genes, RNA was extracted from the isolate and converted to cDNA.

Before RNA extraction, the strain was cultivated (i) in glycerol nitrate medium at 35 °C and 180 rpm for 6 days and (ii) in LB medium under the same conditions. RNA was extracted using the RNeasy PowerSoil Total RNA Kit and RNase-Free DNase Set (Qiagen, Hilden, Germany). RNA was converted to cDNA using the Transcriptor High Fidelity cDNA Synthesis Kit (Roche, Basel, Switzerland). cDNA was further used in the real-time PCR procedure described above. The efficiency of each primer set was calculated in Bio–Rad CFX 96 software using a calibration curve obtained from 10-fold serial dilutions of cDNA. Relative expression levels were calculated with the Ct values obtained and corresponding primer efficiencies as described by Pfaffl and co-authors [43].

The potential of APF to enhance tertiary oil recovery was investigated in two types of experiments: sand column experiments and core flooding experiments. Crude oil for both experiments was obtained from the oil well number 2313 of the Romashkino oil field (Tatarstan, Russia) in June 2021. The crude oil had the following characteristics: density— 0.91 g cm^{-1}, viscosity—3.7 mPa·s, wax content—3.2%. In both experiments, a solution of APF in brine (NaCl, 5 wt%) with a concentration of 100 mg L^{-1} was used.

For the column experiment, twenty grams of soil mixture (sieved agricultural soil/ sieved sand 1:1 *w:w*) was packed into a glass column with a diameter of 1.5 cm and a length of 25 cm. The soil contained 32%, 26%, and 42% of clay, silt, and sand, respectively, the content of organic matter was 3.1 g kg^{-1} and the pH was 7.1. The clean river sand was air-dried and sieved with a 2 mm mesh. Eight centimeters of the column on the top remained empty to enable the addition of liquids. The column was filled stepwise with brine until the volumes of added and released water had equalized. The difference between the total volumes of added and exceeded water was assumed to be the pore volume (PV). Furthermore, crude oil was forced to pass through the column, and the volume of brine released from it was assumed to be the initial oil in place. The column was incubated for 1 day at room temperature, and 6 PV of brine was added to the column in a stepwise manner to simulate secondary oil recovery. The total amount of oil released from the column after brine flooding was registered. Furthermore, 6 PV of APF solution in brine was added to the top of the column, and the column was incubated for 3 days at room temperature. An additional 10 PV of brine was added stepwise to the column to simulate enhanced oil recovery. The total volume of oil released from the column after APF solution and the second brine addition was registered. The rates of secondary and tertiary oil recovery relating to the initial oil in place amount were calculated as percentages [44].

For the core flooding experiment, the core was obtained from the producing reservoir of the Romashkino oil field located in Tatarstan (Russia). It was cleaned in a Soxhlet apparatus (BUCHI, Flawil, Switzerland) using a chloroform:methanol mixture (75:25) and dried at 80 °C for 24 h. The cleaned core was weighed and fixed in the core holder of the flooding system presented in Figure 1. The fixed core was vacuumed for 2 h to remove pore water and air. Then, it was filled with brine at a flow rate of 0.5 mL min^{-1} for 12 h, and the weight of the saturated core was determined. The difference between the two core weights before and after saturation reflected the PV of the core (PV). Subsequently, crude oil was injected into the core at a flow rate of 0.3 mL min^{-1} for 24 h, and the volume of brine displaced by the crude oil was registered to calculate the original oil in place amount. Afterward, the core was incubated for 3 days. To simulate the process of water flooding, brine was injected at a rate of 0.5 mL min^{-1} until no more oil flowed out from the core, and the volume of oil was determined in the flooding process to estimate the secondary oil recovery. To simulate tertiary oil recovery, APF solution in brine (approximately 6 PV) was added to the core at a rate of 0.3 mL min^{-1} until no more oil flowed out, and the volume of oil was registered in the flooding process [45,46]. All manipulations with the core were conducted at 37.8 °C, as this temperature was reported to be typical for the Romashkino oil field [3].

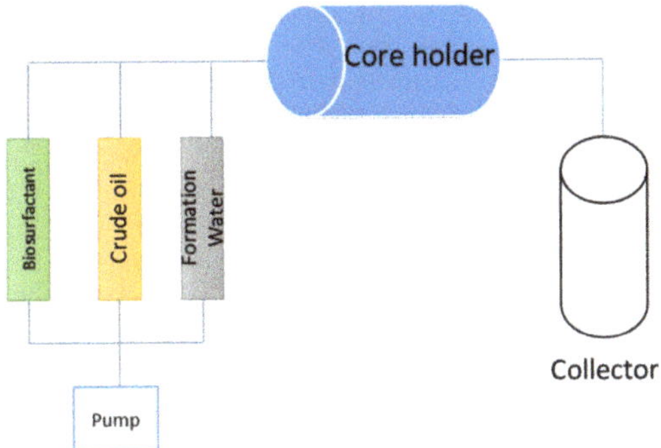

Figure 1. Scheme of core flooding equipment.

Air–water tension assessment, and the determination of gene expression was conducted in four independent cultures analyzed in triplicate. Corresponding figures contain the means and standard deviations. The qPCR results were analyzed using Student's *t* test, and other results were analyzed using the Mann–Whitney U test. All statistical analyses were performed at a significance level of 0.05.

3. Results

3.1. Air–Water Interface Tension

Supernatant obtained in the process of *P. putida* PP021 incubation led to alteration of the air–water surface tension (Figure 2a). Immediately after bacterial inoculation, the surface tension of the supernatant was estimated to be 73.8 mN m^{-1}, while on the 3rd and following days after inoculation, it decreased to ~27 mN m^{-1}. The results obtained are in line with those presented in the scientific literature for other efficient biosurfactant producers of *Pseudomonas* origin and suggest the high potential of *P. putida* PP021 to produce biosurfactants for enhancing oil recovery [36,47]. It should be noted that the optical density of bacteria reached a plateau only on the 5th day of incubation (Figure 2b). At least three reasons might explain this absence of correlation between the bacterial density and the activity of biosurfactants observed. First, bacteria could stop producing biosurfactants; second, the ratio between different types of biosurfactants (that possess different properties) could change; and third, surfactant molecule saturation could be reached at the interface [9,40]. Regardless, the relative alteration of surface tension per unit of bacterial density, or per cost of incubation duration, is an important economic characteristic for the commercial production of biosurfactants.

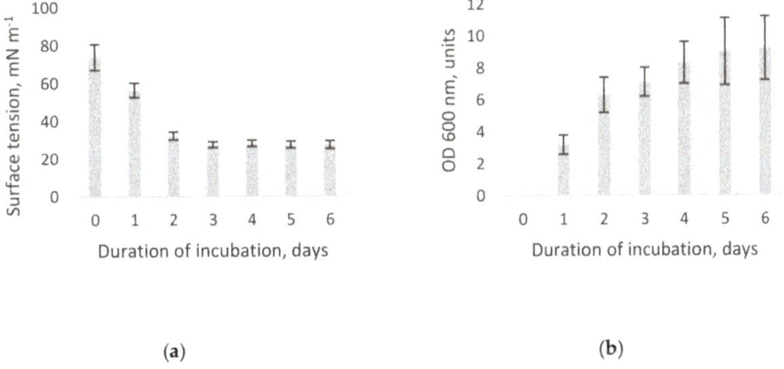

Figure 2. Surface tension of (**a**) and optical density (**b**) of the *P. putida* PP021 cultural medium.

3.2. TLC and FTIR

The biosurfactants produced by *P. putida* PP021 were extracted and purified to characterize their chemical structure. After purification, TLC was conducted to separate different surfactants. Exposure of the TCL plate to UV revealed three main spots (Figure 3a) with retention factors (Rf) of 0.79, 0.74, and 0.36. Further, the TLC plate was treated with iodine vapor, ninhydrine, and anisaldehyde to reveal the basic chemical nature of the biosurfactants (Figure 3b–d). The results suggest that proteins were absent, while carbohydrates and lipids were present in two of three spots (with Rf = 0.74 and Rf = 0.36); therefore, these two spots contain biosurfactants belonging to glycolipids. Indeed, bacteria of the *Pseudomonas* genus are reported to be able to produce biosurfactants from the glycolipid class that contain one or two rhamnoses and fatty acids of different C-chain lengths [10,30]. The two spots therefore may be considered mono- and di-rhamnolipids.

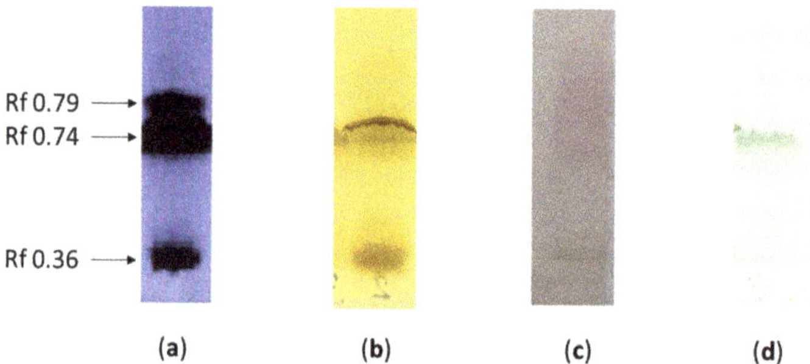

Figure 3. Characterization of the chemical structure of the biosurfactant using TLC: (**a**) spots under UV light, (**b**) spots after iodine staining, (**c**) spots after ninhydrin staining, (**d**) spots after anisaldehyde staining.

These results were confirmed by analysis of the FTIR spectrum of the APF (Figure 4). Hydroxyl groups identified by symmetric O–H stretching referred to carbohydrate fragments, the region after 1648 cm^{-1} to 1316 cm^{-1} corresponded to groups characteristic of carbonyl groups in unsaturated aliphatic carboxylic acids, a weak C=O ester signal at 1724 cm^{-1} and fan stretching vibration for CH$_2$. The observed peaks correspond to the spectra described by other authors for rhamnolipids [48–50].

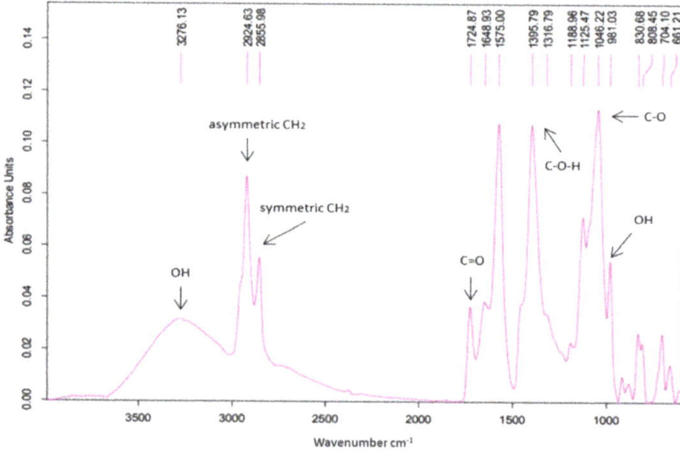

Figure 4. FTIR spectrum of the APF.

3.3. Presence and Relative Expression of Genes Encoding Rhamnotransferase Synthesis

The results of chemical characterization of the surfactants produced by *P. putida* PP021 were confirmed using molecular biology methods. The presence and expression of genes encoding enzymes involved in the rhamnolipid synthesis were analyzed using RT–PCR and reverse transcription RT–PCR techniques (Figure 5). It was revealed that both the *rhlB* and *rhlC* genes, encoding rhamnosyltransferase 1, which catalyzes dTDP-L-rhamnose, and β-hydroxy fatty dTDP-L-rhamnose accessions, which compose mono-rhamnolipis, and rhamnosyltransferase 2, which catalyzes dTDP-L-rhamnose and mono-rhamnolipid accessions, were present in the genome of the isolate (data not shown). Both genes were also expressed when the isolate was cultivated in glycerol nitrate medium for 6 days. The relative expression level of the *rhlB* gene was higher than that of *rhlC*. Indeed, rhamnosyltransferase 1 is involved in the synthesis of both mono- and di-rhamnolipids, while rhamnosyltransferase 2 is involved only in di-rhamnolipid synthesis [9]. We assume that this confirms the presence of both types of rhamnolipids in the cell-free supernatants of the isolate. Interestingly, the relative expression levels of the *rhlB* and *rhlC* genes were significantly higher ($p < 0.05$) when the isolate was grown in a medium with glycerol as a sole carbon source than when it was grown in the nutrient-rich LB medium. This confirms that rhamnotransferases 1 and 2 are inducible but not constitutive enzymes and suggests that the yield of rhamnolipids produced by the *P. putida* PP021 strain can be increased by optimization of the incubation conditions.

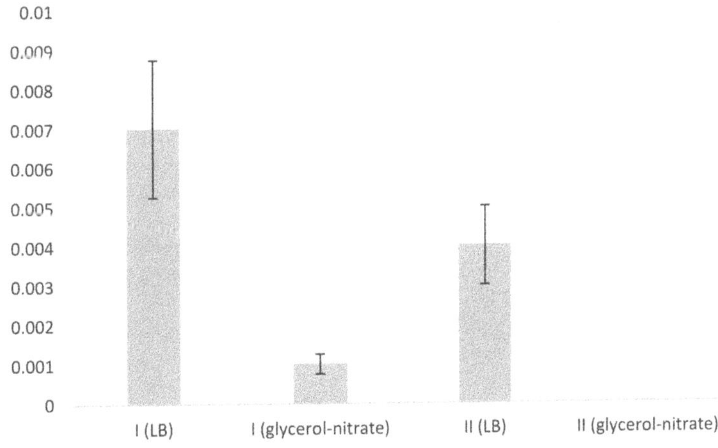

Figure 5. Relative expression of the *rhlB*-encoding rhamnotrasferase 1 (I) and *rhlC*-encoding rhamno-transferase 2 (II) genes in *P. putida* PP021 grown on LB and glycerol nitrate media.

3.4. Enhanced Oil Recovery

The efficacy of the APF to enhance oil recovery was assessed using two simulation methods: sand-packed columns incubated at room temperature and sandstone cores incubated at 37.8 °C, which corresponds to the average temperature in the oil reservoir number 2313. The sand-packed column method has the advantage of reproducibility and the possibility to conduct many experimental variants at the same time (e.g., with different biosurfactant concentrations or with different types of surfactants) and therefore to compare them correctly. The disadvantage of this method lies in the lower surface area of the column's material in comparison with the original core and the consequent exaggerated value of the enhanced oil recovery. In contrast, the use of a natural reservoir core in the reservoir temperature conditions allows practically relevant values of enhanced oil recovery to be obtained. However, it does not allow the comparison of several methods of enhancement or to conduct experiments in many replicates to obtain statistically significant results [44–46]. Thus, the combination of the two simulation methods might overcome disadvantages of both of them and provide relevant conclusions.

In the sand-packed column experiment, brine was forced to pass through the column, and using the difference of added and excessive brine, the PV was determined to be 6.21 mL. After crude oil was forced into the column with the brine-saturated pores, a volume of brine excessed from it (3.81 mL). This volume was assumed to be the amount of original oil in place. Original oil in place occupied approximately 61% of the PV. Later on, the degrees of oil recovery were calculated in relation to this amount. The secondary oil recovery simulated by flooding of the column with brine resulted in 1.03 mL of oil (27%). The enhanced oil recovery simulated by treatment with biosurfactant solution led to the elution of 1.83 mL of oil (i.e., an additional 48%).

In the core flooding experiment, the PV was estimated to be 12.67 mL, and the original oil in place amount was 6.30 mL. Thus, the original oil in place occupied approximately 50%, which is lower than that in the column experiment. Presumably, sand and soil particles of the artificial column had larger surface areas than the original core, and therefore, more oil was absorbed there on the particles and in the pores. As shown in Figure 6, secondary oil recovery increased intensively, shortly after brine injection. Furthermore, the recovery slowed down, and after the injection of 2.5 PV, the recovery degree remained stable independent of brine injection. The secondary oil recovery degree in the core flooding experiment was estimated to be 40–42% of the amount of original oil in place, which was significantly higher than that in the column experiment. Presumably, the oil in the core was contained in the pores and was not as bound to particles as in the column experiment. The addition of the APF led to a 20–22% additional increase in oil recovery. This result can be assessed as efficient compared with reports of other authors, especially considering the type of crude oil that was used in the experiment—heavy oil—which has high wax and sulfur contents [5]. The mechanism of biosurfactant efficacy lies in its ability to alter the interface tension since it increased the capillary number, which in turn lowered the residual oil saturation [51]. The result of tertiary oil recovery enhancement obtained in this study is also efficient compared with conventional chemical flooding agents such as polymers, surfactants, composites, and alkali. Thus, chemical flooding is considered to be efficient if over 10% of the original oil in place is recovered in addition to the initial water flooding [52].

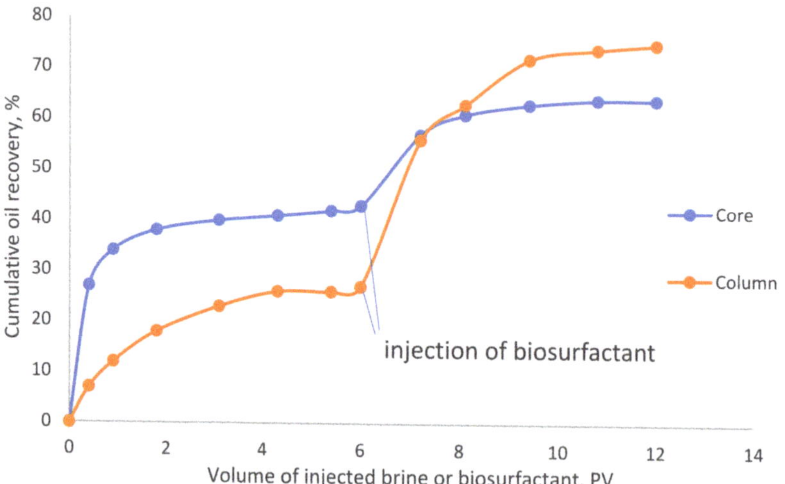

Figure 6. Secondary (before biosurfactant injection) and enhanced (after biosurfactant injection) oil recovery as estimated in sand-packed column and core flooding experiments.

4. Conclusions

It can be concluded that biosurfactants produced by the newly isolated strain *P. putida* PP021 contain compounds belonging to the di- and mono-rhamnolipid groups. The cell-free supernatant obtained from the cultural medium after 6 days of cultivation of the isolate in the glycerol nitrate medium decreased the air–water interface tension from 73.8 to 27.3 mN m^{-1}. The APF produced by the isolate enhanced the degree of recovery of heavy crude oil from the Romashkino oil fields (Russia), by 48% and 21% as observed in the oil in sand-packed column and core flooding experiments, respectively.

Author Contributions: Conceptualization, P.G. and S.S.; methodology, L.B.; validation, A.G.; formal analysis, P.G.; investigation, L.B. and A.G.; resources, L.B.; data curation, S.S.; writing—original draft preparation, S.S.; writing—review and editing, P.G.; visualization, A.G.; supervision, S.S.; funding acquisition, P.G. All authors have read and agreed to the published version of the manuscript.

Funding: This work was supported by the Ministry of Science and Higher Education of the Russian Federation under agreement No. 075-15-2020-931 within the framework of the development program for a world-class Research Center "Efficient development of the global liquid hydrocarbon reserves".

Institutional Review Board Statement: Not applicable.

Informed Consent Statement: Not applicable.

Data Availability Statement: Data is contained within the article.

Conflicts of Interest: The authors declare no conflict of interest. The funders had no role in the design of the study; in the collection, analyses, or interpretation of data; in the writing of the manuscript, or in the decision to publish the results.

References

1. Shibulal, B.; Al-Bahry, S.N.; Al-Wahaibi, Y.M.; Elshafie, A.E.; Al-Bemani, A.S.; Joshi, S.J. Microbial enhanced heavy oil recovery by the aid of inhabitant spore-forming bacteria: An insight review. *Sci. World J.* **2014**, *2014*, 309159. [CrossRef] [PubMed]
2. Alikhlalov, K.; Dindoruk, B. Conversion of cyclic steam injection to continous steam injection. In Proceedings of the Proceedings-SPE Annual Technical Conference and Exhibition, Denver, CO, USA, 30 October–2 November 2011.
3. Lu, L.; Liu, Z.; Liu, H.; Yan, Y. Study on Technical Measures of Romashkino Oil Field after Entering Ultra-High Water Cut Stage. *Engineering* **2013**, *5*, 622–628. [CrossRef]
4. Galitskaya, P.; Gumerova, R.; Ratering, S.; Schnell, S.; Blagodatskaya, E.; Selivanovskaya, S. Oily waste containing natural radionuclides: Does it cause stimulation or inhibition of soil bacterial community? *J. Plant Nutr. Soil Sci.* **2015**, *178*, 825–833. [CrossRef]
5. Liu, Q.; Niu, J.; Liu, Y.; Li, L.; Lv, J. Optimization of lipopeptide biosurfactant production by Bacillus licheniformis L20 and performance evaluation of biosurfactant mixed system for enhanced oil recovery. *J. Pet. Sci. Eng.* **2022**, *208*, 109678. [CrossRef]
6. Gumerova, R.K.; Galitskaya, P.Y.; Badrutdinov, O.R.; Selivanovskaya, S.Y. Changes of hydrocarbon and oil fractions contents in oily waste treated by different methods of bioremediation. *Neft. Khozyaystvo-Oil Ind.* **2013**, 118–120.
7. Galitskaya, P.Y.; Gumerova, R.K.; Selivanovskaya, S.Y. Bioremediation of oil waste under field experiment. *World Appl. Sci. J.* **2014**, *30*, 1694–1698. [CrossRef]
8. Alsayegh, S.Y.; Al Disi, Z.; Al-Ghouti, M.A.; Zouari, N. Evaluation by MALDI-TOF MS and PCA of the diversity of biosurfactants and their producing bacteria, as adaption to weathered oil components. *Biotechnol. Rep.* **2021**, *31*, e00660. [CrossRef]
9. Zhao, F.; Yuan, M.; Lei, L.; Li, C.; Xu, X. Enhanced production of mono-rhamnolipid in Pseudomonas aeruginosa and application potential in agriculture and petroleum industry. *Bioresour. Technol.* **2021**, *323*, 124605. [CrossRef]
10. Ambust, S.; Das, A.J.; Kumar, R. Bioremediation of petroleum contaminated soil through biosurfactant and Pseudomonas sp. SA3 amended design treatments. *Curr. Res. Microb. Sci.* **2021**, *2*, 100031. [CrossRef]
11. Johnson, P.; Trybala, A.; Starov, V.; Pinfield, V.J. Effect of synthetic surfactants on the environment and the potential for substitution by biosurfactants. *Adv. Colloid Interface Sci.* **2021**, *288*, 102340. [CrossRef]
12. Markande, A.R.; Patel, D.; Varjani, S. A review on biosurfactants: Properties, applications and current developments. *Bioresour. Technol.* **2021**, *330*, 124963. [CrossRef]
13. Nazina, T.N.; Pavlova, N.K.; Tatarkin, Y.V.; Shestakova, N.M.; Babich, T.L.; Sokolova, D.S. Development of the biotechnology for the enhancement of oil recovery from carbonate petroleum reservoirs of the tatarstan republic region. *Actual Probl. Oil Gas* **2012**, *6*, 16.
14. Manga, E.B.; Celik, P.A.; Cabuk, A.; Banat, I.M. Biosurfactants: Opportunities for the development of a sustainable future. *Curr. Opin. Colloid Interface Sci.* **2021**, *56*, 101514. [CrossRef]

15. Rudakova, M.A.; Galitskaya, P.Y.; Selivanovskaya, S.Y. Biosurfactants: Current application trends. *Uchenye Zap. Kazan. Univ. Seriya Estestv. Nauk.* **2021**, *163*, 177–208. [CrossRef]
16. Rani, M.; Weadge, J.T.; Jabaji, S. Isolation and Characterization of Biosurfactant-Producing Bacteria From Oil Well Batteries with Antimicrobial Activities Against Food-Borne and Plant Pathogens. *Front. Microbiol.* **2020**, *11*, 64. [CrossRef]
17. Nalini, S.; Parthasarathi, R. Optimization of rhamnolipid biosurfactant production from Serratia rubidaea SNAU02 under solid-state fermentation and its biocontrol efficacy against Fusarium wilt of eggplant. *Ann. Agrar. Sci.* **2018**, *18*, 108–115. [CrossRef]
18. Marchut-Mikolajczyk, O.; Drożdżyński, P.; Pietrzyk, D.; Antczak, T. Biosurfactant production and hydrocarbon degradation activity of endophytic bacteria isolated from *Chelidonium majus* L. *Microb. Cell Fact.* **2018**, *17*, 171. [CrossRef]
19. Sachdev, D.P.; Cameotra, S.S. Biosurfactants in agriculture. *Appl. Microbiol. Biotechnol.* **2013**, *97*, 1005–1016. [CrossRef]
20. Banerjee, S.; Ghosh, U. Production, purification and characterization of biosurfactant isolated from Bacillus oceanisediminis H2. *Mater. Today Proc.* **2021**. [CrossRef]
21. Wu, J.J.; Chou, H.P.; Huang, J.W.; Deng, W.L. Genomic and biochemical characterization of antifungal compounds produced by Bacillus subtilis PMB102 against Alternaria brassicicola. *Microbiol. Res.* **2021**, *251*, 126815. [CrossRef]
22. Pacwa-Płociniczak, M.; Płaza, G.A.; Piotrowska-Seget, Z.; Cameotra, S.S. Environmental applications of biosurfactants: Recent advances. *Int. J. Mol. Sci.* **2011**, *12*, 633–654. [CrossRef] [PubMed]
23. Shi, J.; Chen, Y.; Liu, X.; Li, D. Rhamnolipid production from waste cooking oil using newly isolated halotolerant Pseudomonas aeruginosa M4. *J. Clean. Prod.* **2021**, *278*, 123879. [CrossRef]
24. Müller, M.M.; Kügler, J.H.; Henkel, M.; Gerlitzki, M.; Hörmann, B.; Pöhnlein, M.; Syldatk, C.; Hausmann, R. Rhamnolipids-Next generation surfactants? *J. Biotechnol.* **2012**, *162*, 366–380. [CrossRef] [PubMed]
25. Henkel, M.; Hausmann, R. Diversity and Classification of Microbial Surfactants. In *Biobased Surfactants*; AOCS Press: Urbana, IL, USA, 2019.
26. Robineau, M.; Le Guenic, S.; Sanchez, L.; Chaveriat, L.; Lequart, V.; Joly, N.; Calonne, M.; Jacquard, C.; Declerck, S.; Martin, P.; et al. Synthetic Mono-Rhamnolipids Display Direct Antifungal Effects and Trigger an Innate Immune Response in Tomato against Botrytis Cinerea. *Molecules* **2020**, *25*, 3108. [CrossRef]
27. Rocha, V.A.L.; de Castilho, L.V.A.; de Castro, R.P.V.; Teixeira, D.B.; Magalhães, A.V.; Gomez, J.G.C.; Freire, D.M.G. Comparison of mono-rhamnolipids and di-rhamnolipids on microbial enhanced oil recovery (MEOR) applications. *Biotechnol. Prog.* **2020**, *64*, e2981. [CrossRef]
28. Kronemberger, F.d.A.; Borges, C.P.; Freire, D.M.G. Fed-Batch Biosurfactant Production in a Bioreactor. *Int. Rev. Chem. Eng.* **2010**, *2*, 513–518.
29. Raza, Z.A.; Khan, M.S.; Khalid, Z.M. Evaluation of distant carbon sources in biosurfactant production by a gamma ray-induced Pseudomonas putida mutant. *Process Biochem.* **2007**, *42*, 686–692. [CrossRef]
30. Gaur, V.K.; Sharma, P.; Sirohi, R.; Varjani, S.; Taherzadeh, M.J.; Chang, J.S.; Yong Ng, H.; Wong, J.W.C.; Kim, S.H. Production of biosurfactants from agro-industrial waste and waste cooking oil in a circular bioeconomy: An overview. *Bioresour. Technol.* **2022**, *343*, 126059. [CrossRef]
31. Tahzibi, A.; Kamal, F.; Assadi, M.M. Improved production of rhamnolipids by a Pseudomonas aeruginosa mutant. *Iran. Biomed. J.* **2004**, *8*, 25–31.
32. Tiso, T.; Zauter, R.; Tulke, H.; Leuchtle, B.; Li, W.J.; Behrens, B.; Wittgens, A.; Rosenau, F.; Hayen, H.; Blank, L.M. Designer rhamnolipids by reduction of congener diversity: Production and characterization. *Microb. Cell Fact.* **2017**, *16*, 225. [CrossRef]
33. Cha, M.; Lee, N.; Kim, M.; Kim, M.; Lee, S.-J. Heterologous production of Pseudomonas aeruginosa EMS1 biosurfactant in Pseudomonas putida. *Bioresour. Technol.* **2008**, *99*, 2192–2199. [CrossRef]
34. Poonguzhali, P.; Rajan, S.; Parthasarathi, R.; Srinivasan, R.; Kannappan, A. Optimization of biosurfactant production by Pseudomonas aeruginosa using rice water and its competence in controlling Fusarium wilt of Abelmoschus esculentus. *S. Afr. J. Bot.* **2021**, *in press, corrected proof.* [CrossRef]
35. Ángeles, M.T.; Refugio, R.V. In situ biosurfactant production and hydrocarbon removal by Pseudomonas putida CB-100 in bioaugmented and biostimulated oil-contaminated soil. *Braz. J. Microbiol.* **2013**, *44*, 595–605. [CrossRef]
36. Janek, T.; Lukaszewicz, M.; Krasowska, A. Identification and characterization of biosurfactants produced by the Arctic bacterium Pseudomonas putida BD2. *Colloids Surf B Biointerfaces* **2013**, *110*, 379–386. [CrossRef]
37. Ray, M.; Kumar, V.; Banerjee, C. Kinetic modelling, production optimization, functional characterization and phyto-toxicity evaluation of biosurfactant derived from crude oil biodegrading Pseudomonas sp. IITISM 19. *J. Environ. Chem. Eng.* **2022**, *10*, 107190. [CrossRef]
38. Da Silva, R.C.; Luna, J.M.; Rufino, R.D.; Sarubbo, L.A. Ecotoxity of the formulated biosurfactant from Pseudomonas cepacia CCT 6659 and application in the bioremediation of terrestrial and aquatic environments impacted by oil spills. *Process Saf. Environ. Prot.* **2021**, *154*, 338–347. [CrossRef]
39. Koutinas, M.; Kyriakou, M.; Kostas, A.; Hadjicharalambous, M.; Kaliviotis, E.; Pasias, D.; Kazamias, G.; Varavvas, C.; Vyrides, I. Enhanced biodegradation and valorization of drilling wastewater via simultaneous production of biosurfactants and polyhydroxyalkanoates by Pseudomonas citronellolis SJTE-3. *Bioresour. Technol.* **2021**, *340*, 125679. [CrossRef]

40. Sun, W.; Cao, W.; Jiang, M.; Saren, G.; Liu, J.; Cao, J.; Ali, I.; Yu, X.; Peng, C.; Naz, I. Isolation and characterization of biosurfactant-producing and diesel oil degrading Pseudomonas sp. CQ2 from Changqing oil field, China. *RSC Adv.* **2018**, *8*, 39710–39720. [CrossRef]

41. Agarwal, A.; Liu, Y. Remediation technologies for oil-contaminated sediments. *Mar. Pollut. Bull.* **2015**, *101*, 483–490. [CrossRef]

42. Galitskaya, P.; Biktasheva, L.; Blagodatsky, S.; Selivanovskaya, S. Response of bacterial and fungal communities to high petroleum pollution in different soils. *Sci. Rep.* **2021**, *11*, 18. [CrossRef]

43. Pfaffl, M.W. A new mathematical model for relative quantification in real-time RT-PCR. *Nucleic Acids Res.* **2001**, *29*, 45e. [CrossRef]

44. Zargar, A.N.; Lymperatou, A.; Skiadas, I.; Kumar, M.; Srivastava, P. Structural and functional characterization of a novel biosurfactant from Bacillus sp. IITD106. *J. Hazard. Mater.* **2022**, *423*, 127201. [CrossRef] [PubMed]

45. Aboelkhair, H.; Diaz, P.; Attia, A. Biosurfactant production using Egyptian oil fields indigenous bacteria for microbial enhanced oil recovery. *J. Pet. Sci. Eng.* **2022**, *208*, 109601. [CrossRef]

46. Liu, Q.; Niu, J.; Yu, Y.; Wang, C.; Lu, S.; Zhang, S.; Lv, J.; Peng, B. Production, characterization and application of biosurfactant produced by Bacillus licheniformis L20 for microbial enhanced oil recovery. *J. Clean. Prod.* **2021**, *307*, 127193. [CrossRef]

47. Tuleva, B.K.; Ivanov, G.R.; Christova, N.E. Biosurfactant Production by a New Pseudomonas Putida Strain. *Z. Nat. C* **2002**, *57*, 356–360. [CrossRef]

48. Deepika, K.V.; Raghuram, M.; Bramhachari, P.V. Rhamnolipid biosurfactant production by Pseudomonas aeruginosa strain KVD-HR42 isolated from oil contaminated mangrove sediments. *Afr. J. Microbiol. Res.* **2017**, *11*, 218–231. [CrossRef]

49. Leitermann, F.; Syldatk, C.; Hausmann, R. Fast quantitative determination of microbial rhamnolipids from cultivation broths by ATR-FTIR Spectroscopy. *J. Biol. Eng.* **2008**, *2*, 13. [CrossRef]

50. Sabturani, N.; Latif, J.; Radiman, S.; Hamzah, A. Spectroscopic analysis of rhamnolipid produced by produced by Pseudomonas aeruginosa UKMP14T. *Malays. J. Anal. Sci.* **2016**, *20*, 31–43. [CrossRef]

51. Al-Anssari, S.; Arain, Z.-U.-A.; Shanshool, H.A.; Ali, M.; Keshavarz, A.; Iglauer, S.; Sarmadivaleh, M. Effect of Nanoparticles on the Interfacial Tension of CO_2-Oil System at High Pressure and Temperature: An Experimental Approach. In Proceedings of the SPE Asia Pacific Oil & Gas Conference and Exhibition, Virtual, 17–19 November 2020.

52. Bai, Y.; Wang, F.; Shang, X.; Lv, K.; Dong, C. Microstructure, dispersion, and flooding characteristics of intercalated polymer for enhanced oil recovery. *J. Mol. Liq.* **2021**, *340*, 117235. [CrossRef]

Article

Improving Epoxy Resin Performance Using PPG and MDI by One-Step Modification

Yong Wen [1,2,*], Xudong Liu [2] and Lang Liu [1]

[1] State Key Laboratory of Chemistry and Utilization of Carbon Based Energy Resources, College of Chemistry, Xinjiang University, Urumqi 830017, China; liulang@xju.edu.cn
[2] School of Civil Engineering and Architecture, Xinjiang University, Urumqi 830047, China; 230208258@seu.edu.cn
* Correspondence: wenyong_9731@126.com

Abstract: The toughening modification of epoxy resin by polyurethane prepolymer (PU) can effectively solve the disadvantage of high brittleness in its application. In this study, a convenient way to toughen epoxy resins was explored, and the monomers PPG and MDI for the synthesis of polyurethane prepolymers were used for a one-step modification of epoxy resins. The test results of viscosity and elongation at break showed that P-M reduced the viscosity of the epoxy resin and improved the toughness. Especially when the content of P-M was 25%, the elongation at the break of the modified EP reached 196.56%. From a thermogravimetric and pyrolysis kinetic analysis, the P-M modification had better thermal stability than the PU modification. These findings have particular implications for the toughening and engineering applications of epoxy resins.

Keywords: epoxy resin; modified; viscosity; thermal decomposition; kinetic

1. Introduction

Epoxy resin (EP) has been widely used in coatings, adhesives, aviation, construction materials, composite materials, and photoelectric materials, due to its high bonding strength, mechanical strength, excellent chemical resistance, electrical insulation, and dimensional stability [1–5]. However, the shortcomings of epoxy resin such as high brittleness and easy cracking also greatly limit its application in the industrial field. The low toughness and poor crack resistance of EP caused by highly cross-linked structures has always been a problem and research direction for scholars to solve [6,7]. Consequently, many efforts have been focused on improving the toughness of EP in recent years.

In order to improve the toughness of EP, second components such as nanoparticles [8,9], fibers [10,11], and polymers [12,13] are used as toughening modifiers or even composite modification. Yang et al. [8] used the prepared regenerated cellulose and nanocarbon dioxide hybrid material as a modifier to increase the tensile strength and impact toughness of epoxy resin by 38% and 40%, respectively. Wang et al. [7] used liquid γ-aminopropyl triethoxysilane (APTES)-modified Sm2O3 nanoparticles to improve the mechanical properties of epoxy resins, and the peak elongation at break of the composite was only 0.68%. Although the modification of epoxy resin by nanomaterials can improve its flexibility to a certain extent, the problems of dispersion of nanoparticles in epoxy resin and interfacial incompatibility have not been solved. On the other hand, the uneconomical nature of nanoparticles also precludes them from being widely used in engineering.

Polymers are also often used to improve the toughness of epoxy resins, [14] of which polyurethanes have been widely studied due to their good flexibility [15]. Since Frisch first reported polyurethane (PU)-modified EP [16], many related works have been widely studied due to the superior flexibility and excellent mechanical properties of interpenetrating network polymers (IPNs) [17,18]. Bakar and Kostrzewa synthesized PUs with different isocyanates and polyols and discussed the effect of PU types on the toughness of

Citation: Wen, Y.; Liu, X.; Liu, L. Improving Epoxy Resin Performance Using PPG and MDI by One-Step Modification. *Processes* **2022**, *10*, 929. https://doi.org/10.3390/pr10050929

Academic Editor: Chi-Min Shu

Received: 21 February 2022
Accepted: 3 May 2022
Published: 7 May 2022

EP. Experiment results indicated that an IPN was formed through a graft reaction between the isocyanate-terminated PU and EP, greatly improving the toughness and reducing the glass transition temperature of the EP [19–23]. Li's research results showed that there was a graft reaction between the polymer network I formed by epoxy resin with a curing agent and polymer network II obtained via PUP reactions with a chain extender and cross-linker [24–26]. Dharmalingam Sivanesan et al. prepared polycaprolactone polyols with different molecular weights and chain lengths, and then prepared polyurethane-modified epoxy resins with hexamethylene diisocyanate. The results showed that the addition of polyurethanes and EP formed cross-linking bonds, thereby increasing the tensile strength from 63 MPa to 81 MPa [27]. Studies by many scholars have proved that due to the cross-linking effect of polyurethane and EP, the phenomenon of phase separation is avoided in the system, thereby improving the mechanical properties.

In previous studies, the modification of EP usually required two stages, namely the preparation of polyurethane and the blending with epoxy resin. In this paper, the steps of preparing the polymer were omitted. Moreover, the research on the toughening and modification of epoxy resin by PPG and MDI, the monomer for preparing the polyurethane prepolymer, has rarely been reported. In this study, epoxy resin was modified with MDI and PPG, compared with the traditional polyurethane prepolymer modification, and the effect of the modifier addition on the mechanical properties of epoxy resin was discussed. In addition, the viscosity, initial setting time, and thermal stability of the modified epoxy resin were also discussed in this study; the thermal decomposition kinetic parameters were analyzed by Kissinger and Flynn–Wall–Ozawa (F–W–O) methods, respectively.

2. Experimental Section

2.1. Materials

4,4′-Diphenylmethane diisocyanate (MDI) and polypropylene glycol (PPG Mw = 1000 g mol^{-1}) were supplied by Jining Baiyi Chemical Co. (Shandong, China). PPG was an industrial product that was dehydrated under a vacuum at 120 °C for 2 h before use. Epoxy resin (EP), with an epoxy value of 0.51 mol/100 g, was purchased from Guangzhou Qian Chemical Co., Ltd. (Guangzhou, China). The curing agent (model 421, amine value was 380 ± 20 mg KOH/g) was obtained from Boyu Biological Engineering Co., Ltd. (Shandong, China).

2.2. Preparation of Polyurethane Prepolymer

Under vacuum conditions, the PPG was added dropwise into the preweighed MDI using a peristaltic pump at a speed of 30 r/min (n (NCO): n (OH) = 1.5). Then, the reaction temperature was kept at 70 °C for 3.5 h and it yielded PUP. The viscosity of the PUP sample was 13,400 mPa·s, and the isocyanate content measured by the di-n-butylamine back drop method is 2.96%.

2.3. Preparation of Modified Epoxy Resin

EP, modified with different contents of PPG and MDI (n (NCO): n (OH) = 1.5), 0, 10, 20, 25, and 30 wt %, was prepared. First, PPG and MDI were added into EP and stirred at room temperature for 15 min. Then, the curing agent (33 wt % of EP) was added and stirred for 15 min. The mixture was poured into a mold coated with vacuum silicone grease and placed in a vacuum oven at 80 °C for 2 h to remove air bubbles. Next, it was demolded after curing for 24 h at room temperature, and the obtained materials were denoted as E-(P-M X%), where X was the amount of PPG and MDI.

For comparison, PPG and MDI in the above system were replaced with the as-synthesized PUP, and the other conditions were the same as the above procedure, leading to the PUP-modified EP with 0, 10, 20 25 and 30 wt %, which is designated as E-(PUP X%). The specific steps of sample preparation were shown in Figure 1.

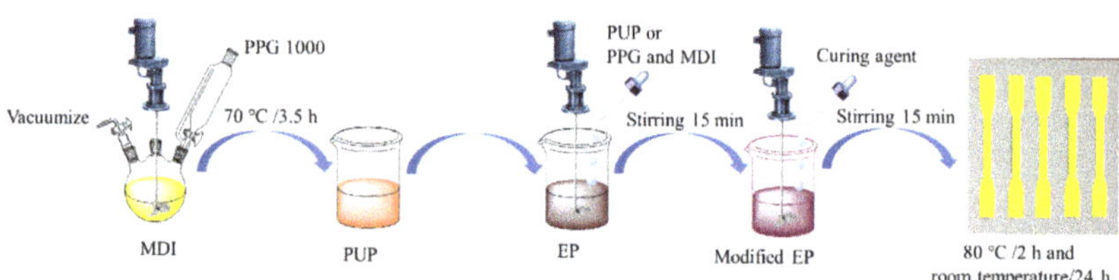

Figure 1. Preparation of samples.

2.4. Testing Methods

The VERTEX 70 (Brook, Germany) was used for the Fourier transform infrared spectroscopy (FT-IR) to test the sample structure with IR spectra from 400 to 4000 cm^{-1}.

The viscosity test of the sample was carried out according to GB/T 2974-2013, using the NDJ-1F rotational viscometer of Shanghai Changji Geological Instrument Co., Ltd., the rotor model was no. 29, and the speed was 10 r/min. The time from when the sample was measured to when the viscosity reached 60,000 mpa·s was selected as the initial setting time. We took the average value of three parallel experiments for each sample.

The tensile strength and elongation at break were tested on the FR-100C electronic universal testing machine (Farui, China) according to the GB/T 2567-2008 standard, and the tensile rate was 2 mm/min until the samples broke. The values of tensile strength and elongation at break were the average values of 3 samples. The tensile strength of the sample was calculated as Equation (1):

$$\sigma_t = \frac{F}{b \times h} \tag{1}$$

In the formula, σt is the tensile strength of the sample (MPa); F is the maximum load that the sample bears (N); b is the width of the sample (mm); h is the thickness of the sample (mm).

The elongation at break of the sample was calculated Equation (2):

$$e = \frac{\Delta L}{L_0} \times 100\% \tag{2}$$

where e is the elongation at break; L_0 is the gauge length of the sample (mm); ΔL is the gauge length elongation at break (mm).

Additionally, the tensile strength and elongation at break were tested on the FR-100C electronic universal testing machine (Farui, China) according to the GB/T 2567-2008 standard. Thermal stability was monitored on an STA 7300 (Hitachi, Japan) from room temperature to 600 °C, with heating rates of 5, 10, 15, 20 and 25 °C/min under a nitrogen atmosphere.

2.5. Kinetic Analysis of Thermal Decomposition

The thermal degradation process of EP can be expressed as Equation (3) when only considering whole chemical changes:

$$A(s) \rightarrow B(s) + C(g) \tag{3}$$

The conversion rate is defined as Equation (4):

$$\alpha = \frac{m_0 - m_t}{m_0 - m_1} \tag{4}$$

where m_0 and m_1 are the initial mass of the sample and the final mass after reaction, respectively, and m_t represents the mass of the sample at time t.

The decomposition rate $d\alpha/dt$ is a function of the conversion rate α and temperature T, as Equation (5):

$$\frac{d\alpha}{dt} = k(T)f(\alpha) \tag{5}$$

where k is the reaction rate constant. This term is a function of temperature T and can be expressed by the Arrhenius formula, as Equation (6):

$$k = A\exp(-\frac{E_a}{RT}) \tag{6}$$

where A is the frequency factor (s^{-1}), E_a is the activation energy of the decomposition reaction (KJ mol^{-1}), R is the general gas constant ($R = 8.314$ J mol^{-1} K^{-1}), and T is the reaction temperature (K).

Substituting Equations (6) and (5), we obtain Equation (7):

$$\frac{d\alpha}{dt} = A\exp(-\frac{E_a}{RT})f(\alpha) \tag{7}$$

For a constant heat rate $\beta = dT/dt$ thermal degradation system, the kinetic parameters can be obtained from the overall rate equation, as Equation (8):

$$\frac{d\alpha}{dT} = \frac{A}{\beta}\exp(-\frac{E_a}{RT})f(\alpha) \tag{8}$$

Several methods of calculating kinetic parameters can be obtained through various deductions of Equation (8). In this paper, we calculated the kinetic parameters of the modified EP by adopting the classic Kissinger (Equation (9)) and Flynn–Wall–Ozawa (Equation (10)) equations [28,29]. The advantage of the Kissinger method is that Ea can be obtained without understanding the thermal degradation reaction mechanism. Moreover, the Flynn–Wall–Ozawa (F–W–O) method is also known as the integral method, and it is very suitable for explaining the kinetic parameters under complex thermal degradation reactions.

$$\ln(\frac{\beta}{T_p^2}) = -\frac{E_a}{RT_p} + \ln\frac{AR}{E_a} \tag{9}$$

$$\ln(\beta) = \lg\frac{AE_a}{RG(\alpha)} - 2.315 - 0.4567\frac{E_a}{RT} \tag{10}$$

3. Results and Discussion

3.1. FT-IR of P-M-Modified Epoxy Resin

Figure 2 presents the FT-IR spectra of modifying agents and E-(P-M X%). The characteristic peaks at 3400 and 913 cm^{-1} are stretching vibrations of the hydroxyl and epoxy groups in the EP, respectively, and peaks at 2271 and 1523 cm^{-1} are stretching vibrations of the N=C=O groups and the benzene rings in the MDI. The vibrational peaks of C–O–C and –OH in the PPG are at 1108 and 3475 cm^{-1}, respectively. In the PUP, the characteristic peaks at 2271 and 1735 cm^{-1} are NCO and urethane groups, respectively.

In the infrared spectra of E-(P-M X%), it can be clearly seen that the N=C=O characteristic peak near 2270 cm^{-1} in the MDI disappears, and a new characteristic peak that belongs to C=O in carbamate appears near 1730 cm^{-1}. The disappearance of the N=C=O groups may be due to the reaction with a curing agent containing more reactive amino groups (Equation (11)). The IR results prove that during the E-(P-M X%), PPG and MDI reacted to form carbamate, and a part of the MDI was consumed by the curing agent.

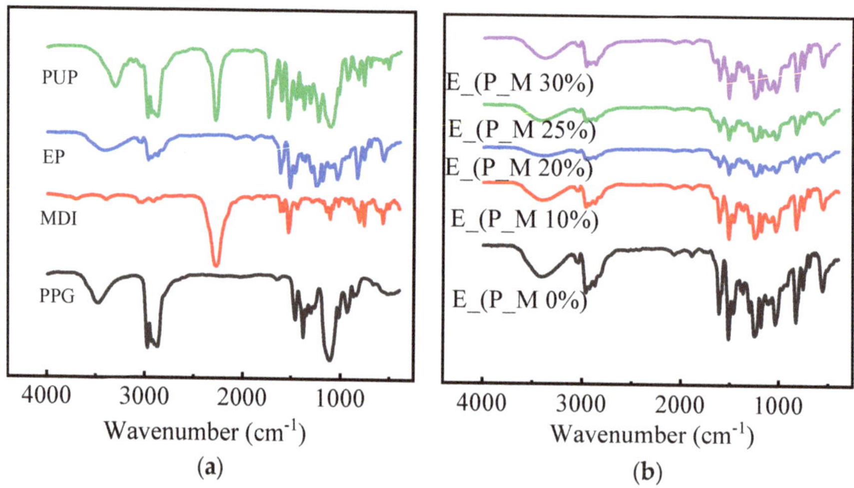

Figure 2. FT-IR spectra of modifying agents and E-(P-M X%). (**a**) Modifiers and pure EP, (**b**) Modified Epoxy Resin.

$$-N{=}C{=}O \quad + \quad -NH_2 \quad \longrightarrow \quad -\overset{H}{N}-\overset{\overset{\textstyle O}{\|}}{C}-\overset{H}{N}- \tag{11}$$

3.2. Properties of Modified Epoxy Resin

The viscosity of EP has a significant effect on its processing ability and the properties of cured samples. The lower viscosity facilitates operability during later construction [30]. Figure 3 shows the viscosity and initial setting time results for the PUP-M and P-M samples. In the viscosity test, two extremes can be seen, the viscosity of the PUP-M sample increases sharply with the increase of the amount of this modifier, because the prepared PUP itself has a higher viscosity due to its larger molecular weight. The epoxy resin itself has a large viscosity base, so it shows a viscosity trend, while the PPG and MDI monomers as modifiers are just the opposite, because PPG and MDI as monomers have a very low viscosity. Adding it to the epoxy resin acts as a diluent to a certain extent, so the higher the amount of monomer added, the lower the viscosity of the system.

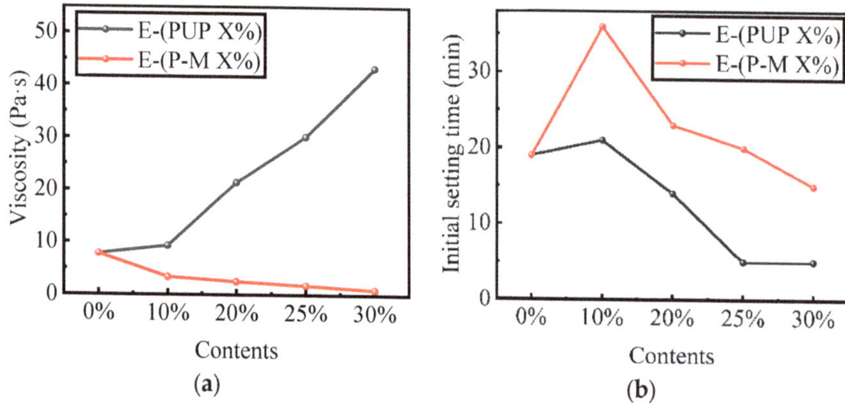

Figure 3. Viscosity and initial setting time of the modified EP. (**a**) Viscosity, (**b**) initial setting time.

Notably, the initial setting time of the two modified EPs first increases and then decreases, but the reason for this trend is different. With the increase in P-M content, the active NCO group increases, the reaction becomes more intense, and the viscosity rises faster, so the initial setting time is short, while the initial viscosity of the PUP modification system increases with the rise in PUP content. When the PUP content is 30%, the initial viscosity of the system is 43,274 mPa·s, which dramatically shortens the time for the system's viscosity to reach 60,000 mPa·s (i.e., the initial setting time). At the same dosage, the initial setting time of the P-M modification is always slightly higher than that of the PUP. In practice, this means that the P-M modification provides a longer operational time.

In conclusion, the viscosity of the system can be greatly reduced during the modification of P-M, and the operating time estimated from the initial setting time can also be adjusted by the amount of P-M added.

Figure 4 represents the tensile strength and elongation at break of the PUP-M and P-M samples. The tensile strength showed a decreasing trend in both modified samples, but the mechanism was different. In the process of the P-M modification, an infrared analysis revealed that the MDI also consumed part of the curing agent 421 when the PPG reacted to generate PUP, which resulted in the incomplete curing of the EP itself when the amount of modifier was too high. However, a decrease in tensile strength of the PUP-M samples has been reported in the literature due to the formation of alkanone structures by the epoxy groups in PUP and EP with NCO end groups, [31] which led to the decrease in tensile strength.

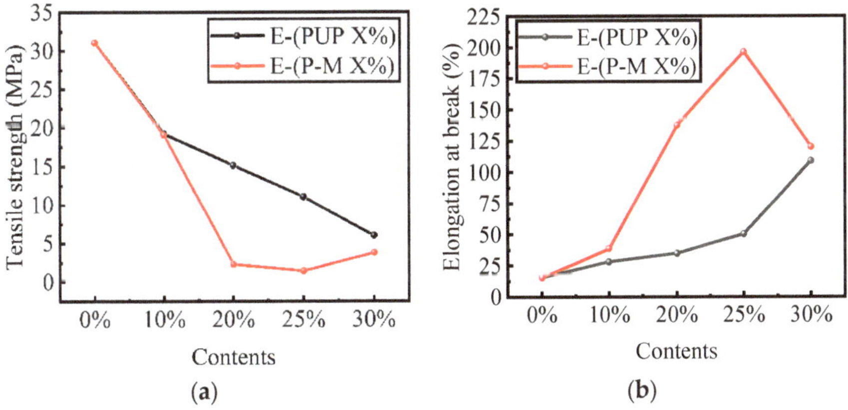

Figure 4. Tensile strength and elongation at break of the modified EP. (**a**) Tensile strength, (**b**) elongation at break.

The P-M modification showed a larger elongation at break, and for (EP-P-M 25%), the elongation at break achieved a maximum of 196.56%. It can be clearly seen that both modification methods can improve the elongation at the break of EP to a certain extent. This is because both modifiers contain the generated PUP, and PPG as a flexible segment in PUP can improve the sample's flexibility, which has been reported in many papers [32].

The elongation at the break of P-M was higher than that of PUP. During the modification process of P-M, the curing agent 421 consumed a part of the MDI, which caused free PPG to form a continuous collision polymer due to the dehydration and condensation of a gunshot during the curing process of EP. The result of the infrared analysis confirmed this conjecture. Obviously, the chain polymer of free PPG dehydration–condensation in the system is beneficial to the flexibility of the system. With the increase of P-M content, the late resin hardly has any strength, so the elongation at break shows a decreasing trend at that time.

3.3. Thermal Properties

Figure 5 displays the TG (weight loss) and DTG (weight derivative) curves of EP-(P-M X%) and EP-(PUP X%). The initial decomposition temperature (Ti) of unmodified EP is 325.96 °C, and the weight loss reached a maximum at 368.83 °C. EP-(PUP X%) slightly reduces the initial decomposition temperature of EP, while EP-(P-M X%) obviously has a higher Ti than EP-(PUP X%), which indicates that EP-(P-M X%) modification makes the polymer have higher crosslinking density, which is attributed to the MDI in the modification process not only reacted with PPG to form PUP but also grafted into EP through the reaction with curing agent. EP-(PUP X%) exhibits lower thermal stability mainly due to the presence of more relatively weak urethane linkages in the system [33]. However, this phenomenon of reduced thermal stability was not found in the P-M modification process, because in addition to the grafting of PUP to EP, there was also a part of MDI to EP to form oxazolidinones.

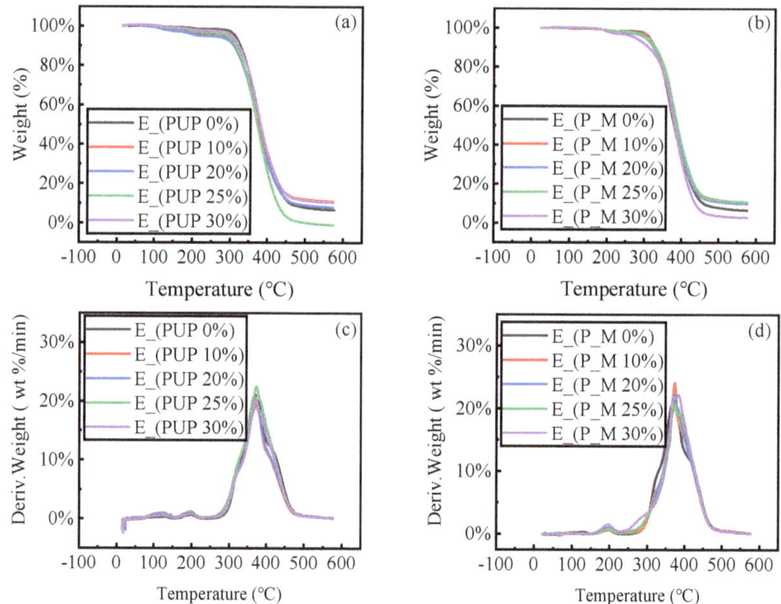

Figure 5. TG and DTG of the modified EP. (**a**) TG of PUP modified EP, (**b**) TG of P-M modified EP, (**c**) DTG of PUP modified EP, (**d**) DTG of P-M modified EP.

3.4. Thermal Decomposition Kinetic

Kissinger Method

For the Kissinger method, Ea was determined by the slope of the linear fit of ln (β/Tp2) to 1/Tp with varying P-M contents (Figure 6). The thermal degradation kinetic parameters are listed in Table 1. Consequently, there was a linear relationship for different P-M contents, signifying the reliability of the Kissinger method under this system. The Ea value of the unmodified EP was 58.97 KJ·mol^{-1}; E-(P-M 20%) and E-(P-M 25%) increased by 11.32 KJ·mol^{-1} and 6.47 KJ·mol^{-1}, respectively, compared to the unmodified EP, suggesting that the material needs to absorb more heat during thermal degradation, i.e., it has a higher thermal stability. The Ea value calculated by Kissinger method was basically consistent with the Ti results shown by thermal stability, which both proved that the EP-(P-M X%) had a higher thermal stability. When Ea/RT < 10, the value of Ea calculated by the Kissinger method is highly reliable (i.e., the error is less than 5%) [34].

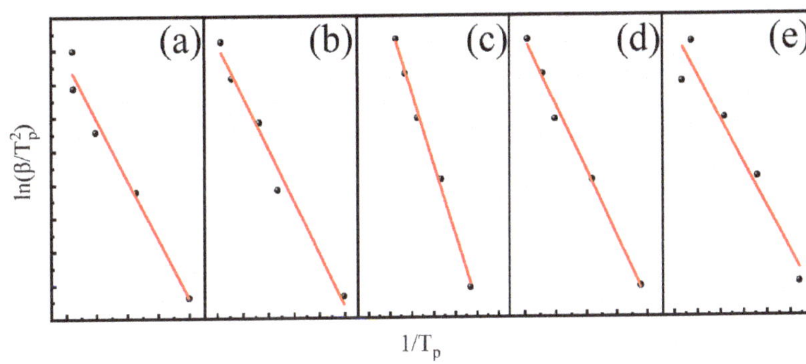

Figure 6. Relationship between ln (β/Tp) and 1/Tp in thermal degradation. (**a**) EP-(P-M 0%), (**b**) EP-(P-M 10%), (**c**) EP-(P-M 20%), (**d**) EP-(P-M 25%), (**e**) EP-(P-M 30%).

Table 1. The thermal degradation kinetic parameters for the Kissinger method.

P-M Contents	R^2	Slope	Intercept	E_a (KJ·mol^{-1})
EP-(P-M 0%)	0.9765	-7093.42	10.49	58.97
EP-(P-M 10%)	0.9822	-5933.12	7.07	49.33
EP-(P-M 20%)	0.9960	-8454.20	13.86	70.29
EP-(P-M 25%)	0.9920	-7870.63	12.21	65.44
EP-(P-M 30%)	0.9392	-5211.02	4.94	43.32

For the F–W–O method, Ea was determined by the slope of the linear fit of lg β to $1/T_p$ with different P-M contents (Figure 7). Unlike the Kissinger method, E_a of the thermal degradation system at each moment can be calculated directly. In this paper, E_a was calculated with a 10 to 80% conversion rate (increments of 10%), and the results are shown in Table 2. E_a was highly dependent on the conversion rate due to the multistage reactivity of the modified EP during thermal degradation [26]. Moreover, Ea gradually increased from low to high conversion because the degradation reactions occurred at various stages in the thermal degradation process. The initial stage was primarily involved in the evaporation of water, so Ea was low.

Table 2. F–W–O method to calculate the thermal degradation kinetic parameters.

α	Ea (KJ·mol^{-1})				
	EP-(P-M 0%)	EP-(P-M 10%)	EP-(P-M 20%)	EP-(P-M 25%)	EP-(P-M 30%)
10%	42.92	49.09	50.81	41.90	34.44
20%	51.34	51.43	57.83	56.73	52.05
30%	56.03	52.45	57.85	59.37	55.58
40%	57.58	53.84	58.06	61.46	57.11
50%	58.48	55.79	58.64	58.58	57.38
60%	59.97	56.61	58.89	58.01	57.81
70%	62.54	59.51	62.06	60.08	59.93
80%	67.34	63.51	66.21	63.24	63.32
Average	57.03	55.28	58.79	57.42	54.70

The Ea obtained by the F–W–O method did not differ much from that of the Kissinger method because the Kissinger method does not involve the conversion rate during the degradation process. The thermal degradation kinetic parameters obtained by the two methods proved that the P-M modification did not significantly reduce the material's thermal stability like the PUP modification.

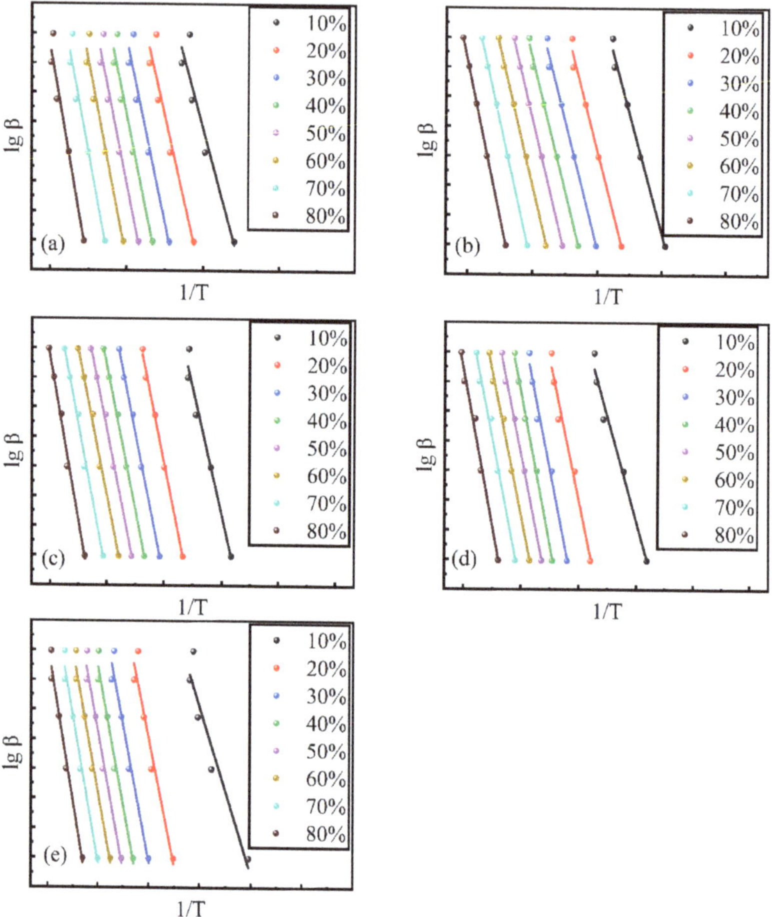

Figure 7. Relationship between lgβ and 1/Tp in thermal degradation. (**a**) EP-(P-M 0%), (**b**) EP-(P-M 10%), (**c**) EP-(P-M 20%), (**d**) EP-(P-M 25%), (**e**) EP-(P-M 30%).

4. Conclusions and Future Work

This study laid a theoretical foundation for the facile preparation of EP with good flexibility. The modification of EP with MDI and PPG was used to compare with conventional PUP modification. Due to the low viscosity of the added monomer, which acts as a diluent in the system, the viscosity was obviously reduced from 7820 mPa·S to 1034 mPa·S. Interestingly, as the monomer content increased, the highly reactive NCO group increased. The increase of the group made the viscosity of the system rise sharply in the later stage. The one-step modification of EP by MDI and PPG greatly improved the flexibility, and a maximum value of 196.56% was obtained when the dosage was 25%. This indicated that both modification methods reduced the tensile strength. In addition, the thermogravimetric test and thermal degradation kinetics results both proved that the one-step modified EP had better thermal stability.

This kind of P-M-EP with a good flexibility has great application prospects in industry, such as road repair materials and concrete crack repair. Of course, the application performance in practical engineering, such as the ability to repair cracks, will be systematically reported in follow-up studies.

Author Contributions: Conceptualization, Y.W.; methodology, Y.W.; software, Y.W.; validation, Y.W. and L.L.; formal analysis, X.L.; investigation, X.L.; resources, Y.W.; data curation, X.L.; writing—original draft preparation, X.L.; writing—review and editing, L.L.; supervision, Y.W. and L.L.; project administration, Y.W.; funding acquisition, Y.W. All authors have read and agreed to the published version of the manuscript.

Funding: This research was funded by the National Natural Science Foundation of China (51768068) and the TIANSHAN young talent project of Xinjiang (2018Q014).

Institutional Review Board Statement: Not applicable.

Informed Consent Statement: Not applicable.

Data Availability Statement: Not applicable.

Conflicts of Interest: The authors declare that they have no known competing financial interests or personal relationships that could have appeared to influence the work reported in this paper.

References

1. Vilčáková, J.; Lenka, K.; Marek, J.; Moučka, R.; Vicha, R.; Sedlacik, M.; Kovalcik, A.; Machovsky, M.; Zazantseva, N. Enhanced Charpy impact strength of epoxy resin modified with vinyl-terminated polydimethylsiloxane. *J. Appl. Polym. Sci.* **2018**, *135*, 45720. [CrossRef]

2. Ma, H.; Xin, Z.; Feifei, J.; Tsai, S.-B. A Study on Curing Kinetics of Nano-Phase Modified Epoxy Resin. *Sci. Rep.* **2018**, *8*, 3045. [CrossRef]

3. Ricciardi, M.R.; Papa, I.; Langella, A.; Langella, T.; Lopresto, V.; Antonucci, V. Mechanical properties of glass fibre composites based on nitrile rubber toughened modified epoxy resin. *Compos. Part B Eng.* **2018**, *139*, 259–267. [CrossRef]

4. Tan, Y.; Shao, Z.-B.; Yu, L.-X.; Long, J.-W.; Qi, M.; Chen, L.; Wang, Y.-Z. Piperazine-modified ammonium polyphosphate as monocomponent flame-retardant hardener for epoxy resin: Flame retardance, curing behavior and mechanical property. *Polym. Chem.* **2016**, *7*, 3003–3012. [CrossRef]

5. Zhang, Y.; Shi, C.; Qian, X.; Jing, J.; Jin, L. DOPO/Silicon/CNT Nanohybrid Flame Retardants: Toward Improving the Fire Safety of Epoxy Resins. *Polymers* **2022**, *14*, 565. [CrossRef]

6. Yahyaie, H., Morteza, E.; Hamed-Vakili, T.; Mafi, E.R. Toughening mechanisms of rubber modified thin film epoxy resins. *Prog. Org. Coat.* **2013**, *76*, 286–292. [CrossRef]

7. Wang, H.; Qunying, H.; Yutao, Z. Influences of Modified Sm_2O_3 on Thermal Stability, Mechanical and Neutron Shielding Properties of Aminophenol Trifunctional Epoxy Resin. *Polymers* **2022**, *14*, 638. [CrossRef]

8. Yang, J.; Hengxu, W.; Xiaohuan, L.; Fu, S.; Song, P. A nano-TiO_2/regenerated cellulose biohybrid enables simultaneously improved strength and toughness of solid epoxy resins. *Compos. Sci. Technol.* **2021**, *212*, 108884. [CrossRef]

9. Mao, D.; Chen, J.; Ren, L.; Zhang, K., Yuen, M.M.F.; Zeng, X.; Sun, R.; Xu, J.-B.; Wong, C.-P. Spherical core-shell Al@Al_2O_3 filled epoxy resin composites as high-performance thermal interface materials. *Compos. Part A Appl. Sci. Manuf.* **2019**, *123*, 260–269. [CrossRef]

10. Misumi, J.; Oyama, T. Low viscosity and high toughness epoxy resin modified by in situ radical polymerization method for improving mechanical properties of carbon fiber reinforced plastics. *Polymer* **2018**, *156*, 1–9. [CrossRef]

11. Wang, W.; Zhou, G.; Yu, B.; Peng, M. New reactive rigid-rod aminated aromatic polyamide for the simultaneous strengthening and toughening of epoxy resin and carbon fiber/epoxy composites. *Compos. Part B Eng.* **2020**, *197*, 108044. [CrossRef]

12. Zhang, J.; Mi, X.; Chen, S.; Xu, Z.; Zhang, D.; Miao, M.; Wang, J. A bio-based hyperbranched flame retardant for epoxy resins. *Chem. Eng. J.* **2020**, *381*, 122719. [CrossRef]

13. Liu, X.-F.; Liu, B.-W.; Luo, X.; Guo, D.-M.; Zhong, H.-Y.; Chen, L.; Wang, Y.-Z. A novel phosphorus-containing semi-aromatic polyester toward flame retardancy and enhanced mechanical properties of epoxy resin. *Chem. Eng. J.* **2020**, *380*, 122471. [CrossRef]

14. Kou, Y.; Zhou, W.; Li, B.; Dong, L.; Duan, Y.-E.; Hou, Q.; Liu, X.; Cai, H.; Chen, Q.; Dang, Z.-M. Enhanced mechanical and dielectric properties of an epoxy resin modified with hydroxyl-terminated polybutadiene. *Compos. Part A Appl. Sci. Manuf.* **2018**, *114*, 97–106. [CrossRef]

15. Xu, Y.; Luo, J.; Liu, X.; Liu, R. Polyurethane modified epoxy acrylate resins containing ε-caprolactone unit. *Prog. Org. Coat.* **2020**, *141*, 105543. [CrossRef]

16. Frisch, H.-L.; Frisch, K.-C.; Klempner, D. Glass Transitions of Topologically Interpenetrating Polymer Networks. *Polym. Eng. Sci.* **1974**, *14*, 646–650. [CrossRef]

17. Chen, S.; Zhang, X.; Wang, Q.; Wang, T. Physical Properties of Micro Hollow Glass Bead Filled Castor Oil-Based Polyurethane/Epoxy Resin IPN Composites. *J. Macromol. Sci. Part B* **2017**, *56*, 161–169. [CrossRef]

18. Huo, L.; Wang, D.; Liu, H.; Jia, P.; Gao, J. Cytoxicity, dynamic and thermal properties of bio-based rosin-epoxy resin/ castor oil polyurethane/carbon nanotubes bio-nanocomposites. *J. Biomater. Sci. Polym. Ed.* **2016**, *27*, 1100–1114. [CrossRef]

19. Bakar, M.; Duk, R.; Przybyłek, M.; Kostrzewa, M. Mechanical and Thermal Properties of Epoxy Resin Modified with Polyurethane. *J. Reinf. Plast. Compos.* **2008**, *28*, 2107–2118. [CrossRef]

20. Bakar, M.; Hausnerova, B.; Kostrzewa, M. Effect of diisocyanates on the properties and morphology of epoxy/polyurethane interpenetrating polymer networks. *J. Thermoplast. Compos. Mater.* **2013**, *26*, 1364–1376. [CrossRef]
21. Bakar, M.; Kostrzewa, M.; Pawelec, Z. Preparation and properties of epoxy resin modified with polyurethane based on hexamethylene diisocyanate and different polyols. *J. Thermoplast. Compos. Mater.* **2014**, *27*, 620–631. [CrossRef]
22. Kostrzewa, M.; Hausnerova, B.; Bakar, M.; Dalka, M. Property evaluation and structure analysis of polyurethane/epoxy graft interpenetrating polymer networks. *J. Appl. Polym. Sci.* **2011**, *122*, 1722–1730. [CrossRef]
23. Kostrzewa, M.; Hausnerova, B.; Bakar, M.; Siwek, E. Effects of various polyurethanes on the mechanical and structural properties of an epoxy resin. *J. Appl. Polym. Sci.* **2011**, *119*, 2925–2932. [CrossRef]
24. Li, Z.-H.; Huang, Y.-P.; Ren, D.-Y.; Zheng, Z.-Q. Structural characteristics and properties of polyurethane modified TDE-85/MeTHPA epoxy resin with interpenetrating polymer networks. *J. Cent. South Univ. Technol.* **2008**, *15*, 305–308. [CrossRef]
25. Li, Z.-H.; Huang, Y.-P.; Ren, D.-Y.; Zheng, Z.-Q. Structural characteristics and properties of PU-modified TDE-85/MeTHPA epoxy resin. *J. Cent. South Univ. Technol.* **2007**, *14*, 753–758. [CrossRef]
26. Ren, D.-Y.; Li, X.-H.; Li, Z.-H. Influence Factors Study on the Properties of Epoxy Resin Modified by Polyurethane. *Adv. Mater. Res.* **2012**, *472–475*, 1937–1940. [CrossRef]
27. Sivanesan, D.; Kim, S.; Jang, T.W.; Kim, H.J.; Song, J.; Seo, B.; Lim, C.-S.; Kim, H.-G. Effects of flexible and rigid parts of ε-caprolactone and tricyclodecanediol derived polyurethane on the polymer properties of epoxy resin. *Polymer* **2021**, *237*, 124374. [CrossRef]
28. Othman, M.B.H.; Khan, A.; Ahmad, K.; Zakaria, M.R.; Ullah, F.; Akil, H.M. Kinetic investigation and lifetime prediction of Cs–NIPAM–MBA-based thermo-responsive hydrogels. *Carbohydr. Polym.* **2016**, *136*, 1182–1193. [CrossRef]
29. Díaz, E.G.; González, E.; García, M.F.; Asensio, I.A. Kinetic Study of the Pyrolysis of Canary Pine: The Relationship between the Elemental Composition and the Kinetic Parameters. *Ind. Eng. Chem. Res.* **2018**, *57*, 9094–9101. [CrossRef]
30. Ma, S.; Liu, X.; Fan, L.; Jiang, Y.; Cao, L.; Tang, Z.; Zhu, J. Synthesis and Properties of a Bio-Based Epoxy Resin with High Epoxy Value and Low Viscosity. *ChemSusChem* **2014**, *7*, 555–562. [CrossRef]
31. Jia, M.; Hadjichristidis, N.; Gnanou, Y.; Feng, X. Polyurethanes from Direct Organocatalytic Copolymerization of p-Tosyl Isocyanate with Epoxides. *Angew. Chem. Int. Ed. Engl.* **2021**, *60*, 1593–1598. [CrossRef] [PubMed]
32. Peng, Y.J.; He, X.; Wu, Q.; Sun, P.-C.; Wang, C.-J.; Liu, X.-Z. A new recyclable crosslinked polymer combined polyurethane and epoxy resin. *Polymer* **2018**, *149*, 154–163. [CrossRef]
33. Koradiya, S.B.; Adroja, P.-P.; Patel, J.P.; Ghumara, R.Y.; Parsania, P.H. Curing, Spectral and Thermal Study of Epoxy Resin of Bisphenol-C and Its Polyester Polyols Based Polyurethanes. *Polym. Plast. Technol. Eng.* **2012**, *51*, 1545–1549. [CrossRef]
34. Mothé, C.G.; de Freitas, J.S. Lifetime prediction and kinetic parameters of thermal decomposition of cashew gum by thermal analysis. *J. Therm. Anal. Calorim.* **2018**, *131*, 397–404. [CrossRef]

Article

Thermochemical Analysis of a Packed-Bed Reactor Using Finite Elements with FlexPDE and COMSOL Multiphysics

Sebastian Taco-Vasquez [1,*], César A. Ron [1], Herman A. Murillo [1], Andrés Chico [1] and Paul G. Arauz [2]

[1] Department of Chemical Engineering, Escuela Politécnica Nacional, Quito 170517, Ecuador; cesar.ron@epn.edu.ec (C.A.R.); herman.murillo@epn.edu.ec (H.A.M.); andres.chico@epn.edu.ec (A.C.)
[2] Department of Mechanical Engineering, Universidad San Francisco de Quito, Quito 170901, Ecuador; parauz@usfq.edu.ec
* Correspondence: sebastian.taco@epn.edu.ec

Abstract: This work presents the thermochemical analysis of a packed-bed reactor via multi-dimensional CFD modeling using FlexPDE and COMSOL Multiphysics. The temperature, concentration, and reaction rate profiles for methane production following the Fischer–Tropsch (F-T) synthesis were studied. To this end, stationary and dynamic differential equations for mass and heat transfer were solved via the finite element technique. The transport equations for 1-D and 2-D models using FlexPDE consider dispersion models, where the fluid and the catalyst can be treated as either homogeneous or heterogenous systems depending on the gradient extents. On the other hand, the 3-D model obtained in COMSOL deems the transport equations incorporated in the Porous Media module. The aim was to compare the FlexPDE and COMSOL models, and to validate them with experimental data from literature. As a result, all models were in good agreement with experimental data, especially for the 2-D and 3-D dynamic models. In terms of kinetics, the predicted reaction rate profiles from the COMSOL model and the 2-D dynamic model followed the temperature trend, thus reflecting the temperature dependence of the reaction. Based on these findings, it was demonstrated that applying different approaches for the CFD modeling of F-T processes conducts reliable results.

Keywords: multi-dimensional CFD modeling; Fischer-Tropsch synthesis; reactor modeling; solid catalysts; heat transfer; kinetics

Citation: Taco-Vasquez, S.; Ron, C.A.; Murillo, H.A.; Chico, A.; Arauz, P.G. Thermochemical Analysis of a Packed-Bed Reactor Using Finite Elements with FlexPDE and COMSOL Multiphysics. *Processes* **2022**, *10*, 1144. https://doi.org/10.3390/pr10061144

Academic Editor: Blaž Likozar

Received: 18 May 2022
Accepted: 4 June 2022
Published: 7 June 2022

1. Introduction

Most processes employ beds for the solid phase in the reactor design for fluids flowing through solid catalysts. There are fluidized beds that allow for working at temperatures distributed uniformly alongside the bed and allow the catalyst to regenerate. However, their operation's high costs and complexity make fluidized beds suitable only for being employed regarding a few industrial processes. The other option is to use fixed beds (also called packed beds), whose advantage, other than being cheaper, is that separation of phases after the operation is more effortless [1]. These reactors must control main parameters, such as pressure drop and heat transfer between reactants and catalysts. Single-packed beds are usually employed for adiabatic processes. Still, when there are highly exothermic reactions, shell and tubes packed beds (multi-tubular) are preferred due to their high heat transfer. In the packed-bed reactor diameter, bigger diameters will worsen the heat transfer, and hot spots will occur. In contrast, a small diameter allows for a better heat transfer and reduces the temperature gradient between the center and the reactor wall. Although fluidized beds can have five to ten times greater heat transfer than a packed-bed reactor, allowing isothermal conditions [2], other drawbacks are related to catalyst losses because of catalyst transportation between the reactor and regenerator.

Consequently, it is impossible to have a single large diameter tube packed with a catalyst. Instead, the reactor should be made up of several tubes. At the industrial scale, multitube packed-bed reactors have been shown to have some advantages, including a behavior close to plug flow conditions, improved catalyst retention, catalyst separation not being required, and the feasibility of the process to be scaled up [3]. Still, heat removal is the primary consideration in the design of packed-bed reactors for highly exothermic reactions. It is desired to operate these beds within a range of temperatures to optimize productivity and selectivity [4].

In this context, the tube diameter for exothermic reactions is generally between two and 5 cm [5], and the tube-to-particle diameter is about eight depending on the application [1]. Smaller particles tend to decrease the bed's radial thermal conductivity, causing an excessive pressure drop; however, larger particles cause high intraparticle mass transfer resistances, reducing the heat removal from the catalyst to the fluid. This increases the risk of thermal instability of the particle, also showing hot spots inside the catalyst that can damage it [6]. The pressure drop in a porous medium is crucial within a reacting channel, and it affects the heat transfer coefficient, as it must be modified. Darcy's law is one option to study the existing pressure drop in a porous medium. Accordingly, there are different approaches to link Darcy's law to heat transfer equations (e.g., by the Nusselt number). Shah et al. [7] performed a simulation for an alumina-based nanofluid flowing through a porous cavity. Herein, a model was used that associates the Nusselt number with the Darcy's parameter, which depends on the medium permeability and the channel length.

Efficient heat removal is associated with the catalytic material. For instance, a V-Mo-based catalyst conducting the maleic anhydride production from benzene oxidation at 350 °C was conducted in a multi-tubular reactor with two cm-tubes. Unfortunately, heat removal was not as efficient as expected even when using small tubes, as temperature increments of 100 °C were observed in the shell, which can be transmitted to the tubes [1]. In this context, some authors have investigated catalyst geometries and other strategies to achieve better performance on heat transfer. Visconti et al. [8] and Asalieva et al. [9] prepared cobalt-based catalysts onto metallic supports for the Fischer–Tropsch (F-T) reaction using different geometries and wash-coating methodologies. It resulted in reduced hot spots reaching more isothermal conditions.

The Fischer–Tropsch reaction (ΔHr = -165 kJ/mol$_{CO}$) is one study case of highly exothermic reactions, which entails the syngas (i.e., a CO/H$_2$ mixture) conversion to long-chain hydrocarbons towards fuel production [10]. The reaction is sensitive to temperature, so a controlled heat transfer inside the reactor is critical to keep selectivity at reasonable levels, as it can be hindered at high temperatures [11].

Modeling and simulation determine design parameters and process operating conditions to optimize their performance and keep the system operating within a safe range. The most critical parameters, apart from kinetics, are those that control radial heat transfer from the packed bed to the cooling tube [12]. Due to the substantial difference between heat transfer parameters measured under reactive and non-reactive conditions reported in the literature, choosing the best correlations that lead to the most accurate evaluation of these parameters is essential.

The existing literature about the modeling of the F-T reaction addresses different approaches. Recently, computational fluid dynamics (CFD) has been performed for modeling this process. Mohammad et al. [13] provided remarkable insights into the CFD modeling of F-T synthesis using microchannel microreactors. These authors investigated the effect of different parameters focusing mainly on the channel design. For example, they found that the maximum temperature observed in the channel can be reduced by increasing the channel width. On the other hand, the configuration of the channels was also studied, resulting in lower productivity losses when setting the channels in a series. However, the authors suggest further analysis of heat removal mechanisms.

Abusrafa et al. [14] developed a 2-D model for comparing a microfibrous F-T reactor to a conventional packed-bed reactor. The thermal performance of both reactors on conversion and hydrocarbon selectivity was assessed. The authors also discussed the scale-up potential of this novel reactor, suggesting that CFD modeling can be considered the first approach to enhance the conventional F-T technologies.

Chabot et al. [15] studied a 2-D pseudo-homogenous model to determine the thermal behavior over cobalt supported on alumina, concluding that liquid hydrocarbon selectivity was decreased when isothermal conditions were not achieved. Na et al. [16] modeled a 2-D axisymmetric microchannel reactor and proposed a genetic algorithm for multi-objective optimization using a cobalt-based catalyst. Their findings indicated that catalyst zone division could increase productivity and decrease temperature increments.

On the other hand, the pressure drops in CFD simulations for porous medium (e.g., considering Darcy's law) is often considered only in the axial direction showing viscous and inertial resistances. Pavlišič et al. [17] used a modified Darcy's law, incorporating the kinetic energy of the fluid to the inertial term in a laminar flow regime for packed-bed reactors. Their simulation results suggest a channeling strategy to decrease induced pressure differences. Nevertheless, Darcy's law is valid for fluid mainly influenced by viscous effects and negligible at higher velocities. In this case, other correlations such as Ergun and Forchheimer can predict the pressure drop as a function of velocity where some terms are added to correct the mentioned drawbacks [18,19].

The modeling of F-T synthesis must also consider the kinetics equations. In this regard, Rai et al. [20] proposed a power law-based kinetic model for microchannel and packed-bed reactors, assuming different reaction orders. The activation energy was lower for the microchannel reactor and reached >92% of CO conversion; however, a lower temperature gradient was found for the packed-bed reactor. In most cases, reaction orders for CO and H_2 were found to be positive, but there was a case following a negative order for CO, which leads to a process inhibition due to the CO absorption by the catalyst.

This paper proposed a multi-dimensional CFD modeling of the F-T reaction in a packed-bed reactor considering temperature and concentration gradients for methane production from syngas as a first approach. Experimental validation was also performed to assess the validity of the proposed models, as indicated in our previous work [21]. According to Deutschmann [22], if the concentration and temperature gradients between the gas and the catalyst are small, a homogeneous model can be used. In contrast, if the gradients are significant, a model for each phase must be applied, considering the mass and heat transport between the fluid phase and the catalyst bed. For instance, when using Ansys Fluent, heat transfer for heterogeneous catalysis is commonly simulated by the non-equilibrium thermal model, as Pessoa et al. [23] address the CFD simulation of packed bed fermentation bioreactor. For the present study, the equations describing the catalyst bed's mass, energy, and reaction transfer were incorporated into the models. 1-D models (along the packed-bed length) can be used when the temperature and concentration gradients on the radial axis are small; otherwise, a 2-D model is used. In this regard, 1-D and 2-D models were developed using FlexPDE. In addition, a 3-D model was attained using COMSOL Multiphysics. Finally, all models were validated with experimental data to assess the accuracy of the proposed models. In the case of the FlexPDE models, dispersion models (i.e., modified transport equations using a dimensional numbers like the Peclet number) were applied to perform the corresponding simulations, whereas the COMSOL model employs the default transport equations defined by the available modules for porous media modeling. As modeling of the F-T synthesis has been previously addressed in the literature, this work intended to further investigate different CFD software to establish the reliability of the available approaches to describe the transport of species in porous media.

2. Model Description

For multitube packed-bed reactors, all tubes should ideally behave the same; this implies that the inlet conditions, wall temperature, concentrations, and temperature profiles are the same. Therefore, we have chosen one cylindrical tube (as indicated in Figure 1) to perform our analysis over a packed-bed reactor. The two spatial variables will be the radial direction (r) and the axial or longitudinal direction (z). The angular direction was not considered for this simulation since there is no variation in concentration or temperature on the angular axis.

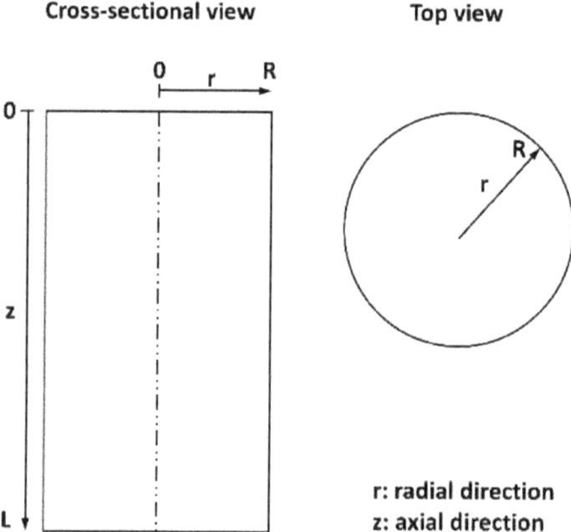

Figure 1. A cross-sectional and top view of a single cylindrical packed-bed reactor.

In this section, the four relevant packed-bed models were discretized: (a) 2-dimensional stationary model; (b) the 1-dimensional dynamic-state model; (c) 2-dimensional dynamic model; and (d) 3-dimensional dynamic model. For models (a), (b), and (c), the governing differential equations were derived and solved using FlexPDE software. On the other hand, model (d) was implemented and solved using COMSOL Multiphysics.

The transport equations, boundary conditions, and the experimental data to validate the proposed models were collected from Skaare [24].

2.1. 2-D Stationary Model

It was initially determined that heat and mass transfer are steady-state, as shown in Equations (1) and (2). The temperature $\theta = T/T_0$, and the axes $z' = z/L$ and $r' = r/R$ were discretized.

$$0 = -\frac{\partial \theta}{\partial z'} + \frac{1}{Pe'_{hz}}\frac{\partial^2 \theta}{\partial z'^2} + \frac{1}{Pe'_{hr}}\left[\frac{1}{r'}\frac{\partial}{\partial r'}\left(r'\frac{\partial \theta}{\partial r'}\right)\right] + \frac{L}{u\rho_g C_{pg}T_0}\rho_b\frac{(1-\varepsilon_0)}{(1-\varepsilon_0)}(\Delta H)r_v \quad (1)$$

$$0 = \frac{\partial y_i}{\partial z'} + \frac{1}{Pe'_{mz}}\frac{\partial^2 y_i}{\partial z'^2} + \frac{1}{Pe'_{mr}}\frac{1}{r'}\frac{\partial}{\partial r'}\left(r'\frac{\partial y'}{\partial r'}\right) + \frac{L}{u_o C_{0i}}\rho_b\frac{(1-\varepsilon_0)}{(1-\varepsilon_0)}r_{vi} \quad (2)$$

The boundary conditions are listed from Equations (3)–(8).

$$\frac{\partial \theta}{\partial r} = \frac{\partial y_i}{\partial r} = 0 \text{ When } r = 0, \text{ for all } z \quad (3)$$

$$-\frac{\partial \theta}{\partial r'} = \frac{UcR}{\lambda_{cr}}(\theta - \theta_c) \text{ When } r = R, \text{ for all } z \quad (4)$$

$$\frac{\partial y_i}{\partial r'} = 0 \text{ When } r = R, \text{ for all } z \quad (5)$$

$$\frac{\partial\theta}{\partial z} = Pe'_{hz}(\theta - 1) \text{ When } z = 0, \text{ for all } r \tag{6}$$

$$\frac{\partial y_i}{\partial z} = Pe'_{mz}(y_i - 1) \text{ When } z = 0, \text{for all } r \tag{7}$$

$$\frac{\partial\theta}{\partial z} = \frac{\partial y_i}{\partial z} = 0 \text{ When } z = L, \text{ for all } r \tag{8}$$

2.2. 1-D Dynamic Model

Likewise, heat and mass equations are shown in Equations (9) and (10), respectively, for dynamic state.

$$\frac{L\varepsilon}{u(3600)} + \frac{L\,cp_s\rho_s(1-\varepsilon)}{u\rho_f cp_f(3600)}\frac{\partial\theta}{\partial t} =$$
$$-\frac{\partial\theta}{\partial z'} + \frac{1}{Pe'_{hz}}\frac{\partial^2\theta}{\partial z'^2} + \frac{L}{u\rho_g C_{pg}T_0}\rho_b\frac{(1-\varepsilon)}{(1-\varepsilon_0)}(\Delta H)r_v + \frac{4U_c L}{d_t u\rho_g C_{pg}}(\theta - \theta_c) \tag{9}$$

$$\frac{L\varepsilon}{u(3600)}\frac{\partial y_i}{\partial t} = \frac{\partial y_i}{\partial z'} + \frac{1}{Pe'_{mz}}\frac{\partial^2 y_i}{\partial z'^2} + \frac{L}{uC_{0i}}\rho_b\frac{(1-\varepsilon)}{(1-\varepsilon_0)}r_{vi} \tag{10}$$

The boundary conditions are listed from Equations (11)–(13).

$$\frac{\partial\theta}{\partial z} = Pe'_{hz}(\theta - 1) \text{ When } z = 0, \text{ for all } r \tag{11}$$

$$\frac{\partial y_i}{\partial z} = Pe'_{mz}(y_i - 1) \text{ When } z = 0, \text{ for all } r \tag{12}$$

$$\frac{\partial\theta}{\partial z} = \frac{\partial y_i}{\partial z} = 0 \text{ When } z = L, \text{ for all } r \tag{13}$$

2.3. 2-D Dynamic Model

The heat and mass transfer equations are represented by Equations (14) and (15), respectively, for the 2-dimensional dynamic model.

$$\frac{L\varepsilon}{u(3600)} + \frac{L\,cp_s\rho_s(1-\varepsilon)}{u\rho_f cp_f(3600)}\frac{\partial\theta}{\partial t} =$$
$$-\frac{\partial\theta}{\partial z'} + \frac{1}{Pe'_{hz}}\frac{\partial^2\theta}{\partial z'^2} + \frac{1}{Pe'_{hr}}\left[\frac{1}{r'}\frac{\partial}{\partial r'}\left(r'\frac{\partial\theta}{\partial r'}\right)\right] + \frac{L}{u\rho_g C_{pg}T_0}\rho_b\frac{(1-\varepsilon)}{(1-\varepsilon_0)}(\Delta H)r_v \tag{14}$$

$$\frac{L\varepsilon}{u(3600)}\frac{\partial y_i}{\partial t} = \frac{\partial y_i}{\partial z'} + \frac{1}{Pe'_{mz}}\frac{\partial^2 y_i}{\partial z'^2} + \frac{1}{Pe'_{mr}}\frac{1}{r'}\frac{\partial}{\partial r'}\left(r'\frac{\partial y'}{\partial r'}\right) + \frac{L}{uC_{0i}}\rho_b\frac{(1-\varepsilon)}{(1-\varepsilon_0)}r_{vi} \tag{15}$$

The boundary conditions are listed from Equations (16)–(21).

$$\frac{\partial\theta}{\partial r} = \frac{\partial y_i}{\partial r} = 0 \text{ When } r = 0, \text{ for all } z \tag{16}$$

$$-\frac{\partial\theta}{\partial r'} = \frac{U_c R}{\lambda_{cr}}(\theta - \theta_c) \text{ When } r = R, \text{ for all } z \tag{17}$$

$$\frac{\partial y_i}{\partial r'} = 0 \text{ When } r = R, \text{ for all } z \tag{18}$$

$$\frac{\partial\theta}{\partial z} = Pe'_{hz}(\theta - 1) \text{ When } z = 0, \text{ for all } r \tag{19}$$

$$\frac{\partial y_i}{\partial z} = Pe'_{mz}(y_i - 1) \text{ When } z = 0, \text{ for all } r \tag{20}$$

$$\frac{\partial\theta}{\partial z} = \frac{\partial y_i}{\partial z} = 0 \text{ When } z = L, \text{ for all } r \tag{21}$$

The procedure to calculate the heat transfer coefficient ("Uc") is detailed in the Supplementary Material in Equations (S16)–(S25). As previously mentioned, the models for this work were simulated for methane synthesis, which is expressed by Equation (22).

$$CO + 3H_2 \rightarrow CH_4 + H_2O \tag{22}$$

The model was studied for CO conversion and CH_4 formation, requiring only the calculation of concentration profiles of CO and CH_4. The rate of H_2 consumption will depend on the rates of CO consumption and CH_4 formation. Moreover, it will depend on the carbon number distribution of the ethane fraction (following the guidelines of Rodemerck et al. [25]) according to the Schulz–Flory–Anderson distribution. The parameter will describe it for propane and the C_3/C_2 ratio known for ethane. These parameters were considered constant, irrespective of the rector's position and time. The selectivity of hydrocarbons was calculated based on the reaction rate of propane according to the Schulz–Flory–Anderson distribution, which depends on the reaction rate expressions of CH_4, and CO. Selectivity parameters are dependent on the temperature and the partial pressure of CO and H_2. In this regard, the reaction rate expressions are found in the Supplementary Material in Equations (S1)–(S16).

The relative simplicity of the equation proposed by Yates and Satterfield [26] was chosen as the basis of calculation for the rate of CO consumption (see Equation (23)) because it showed good correlations with the experimental data. The same relationship was recommended for calculating methane formation, as indicated in Equation (24).

$$-r_{co} = \frac{(1 - \beta_{co}t)A'_{co}e^{Eco/RT_0\left(\frac{1-\theta}{\theta}\right)}y_{co}y_{H_2}}{(1 + K_{co}y_{co})^2} \tag{23}$$

$$r_{CH_4} = \frac{(1 - \beta_{CH4}t)A'_{CH_4}e^{E_{CH_4}/RT_0\left(\frac{1-\theta}{\theta}\right)}y_{CO}^{\frac{1}{2}}y_{H_2}}{(1 - K_{co}y_{co})^2} \tag{24}$$

Once the equations of state and the reaction stoichiometry had been determined, the different equations were entered into FlexPDE. The mesh was defined with 1200 nodes along two dimensions: radial (r) and axial (z).

2.4. 3-D Dynamic Model

A 3-D representation of 1/8 of one cylindrical packed-bed reactor was drawn in AutoCAD and imported into COMSOL Multiphysics. Given that COMSOL renders results for the entire cylindrical geometry using axial symmetry, only 1/8 of the entire geometry was utilized to reduce the computational calculation cost of the simulation. The Brinkman equation for fluid transport in porous media (Brinkman Equations module, br) [27] was utilized to solve the flowing fluid in the reactor. To determine the concentration profiles, the advection–dispersion–reaction equation (Transport of Diluted Species in Porous Media module, tds; and Reacting Flow Diluted Species module, rfd) [27] was incorporated into the model to represent the mass transfer processes of reactants and products. The heat transfer equation (Heat Transfer in Porous Media module, het) [27] was also incorporated into the model to determine the temperature profile.

The physicochemical parameters for each module were the same as those utilized in the above-described models. Once the modules were set up, a mesh size analysis was performed to determine the element size that minimizes the Mean-Absolute-Error (MAE) relative to the experimental data, as indicated in the Supplementary Material in Table S1. The mesh type that minimized the MAE was the "Extremely fine". The 3-dimensional geometry was discretized to 2400 elements and 765 nodes for this mesh.

In this section, the physicochemical parameters of the system to be studied and its components are presented. It included fluid, species, packed-bed and catalyst, heat and mass transfer coefficients, and reaction coefficients. All these values are required to define the models. Tables 1 and 2 show the experimental parameters used in the model extracted from Skaare [24].

Table 1. The reactor and catalyst dimensions and physicochemical properties.

Properties	Value
L (m)	1.5
R (m)	12.5×10^{-3}
R_o (m)	17.5×10^{-3}
ρ_b (kg/m^3)	0.89×10^3
ρ_p (kg/m^3)	1.59×10^3
ε_0	0.44
ε_p	0.485
d_p (m)	3.3×10^{-3}
C_{ps} (J/(kg K))	965
λ_p (W/(m K))	0.4
α_w (W/(m^{3}K))	430
λ_w (W/(m K))	18.0

Table 2. The fluid and gas conditions, dimensionless numbers, and kinetic parameters considered in simulations.

Fluid Properties		Dimensionless Numbers		Masic Peclets		Kinetic Parameters	
T_o (K)	498	Re_p	119	$1/Pe'_{mz}CO$	0.00111	K_{CO}	1.2
P_T (MPa)	1	Pr	0.57	$1/Pe'_{mz}H_2$	0.00112	E_{CO}	110
P_{N2} (Mpa)	0.77	Sc_{CO}	0.8945	$1/Pe'_{mz}CH_4$	0.00111	E_{CH4}	140
u_o (m/s)	0.16	Sc_{H2}	0.2451	$1/Pe'_{mr}CO$	2.54486	A'_{CO}	0.0041
Co_{N2} (mol/m^3)	186	Sc_{CH4}	0.8155	$1/Pe'_{mr}H_2$	2.74323	A'_{CH4}	0.0006
Co_{CO} (mol/m^3)	17 4			$1/Pe'_{mr}CH_4$	2.55211	β_{CO}	0.0043
Co_{H2} (mol/m^3)	38.2					β_{CH4}	0.0097
Physical properties of gas mixture		Distribution parameters		Caloric Peclets		Thermal conductivity	
ρ_g (kg/m^3)	5.77	α	0.69	$1/Pe'_{hz}$	0.00196	λ_{cr} (W/(mK))	0.6172
C_{pg} (J/(kg/K))	1333	γ	0.88	$1/Pe'_{hr}$	4.81445	Uc (W/(m^{2}K))	525.12
μ_g (Ns/m^2)	2.56×10^{-5}						
λ_g (W/(mK))	0.05972						

3. Results and Discussion

3.1. Analysis and Validation of the 2-D Stationary Model and the 3-D Dynamic Model

Figure 2a shows the CO concentration of the stationary 2-dimensional model (Flex-PDE), dynamic 3-dimensional model (COMSOL), and the experimental data. Since there is no gradient of experimental concentrations through the radial axis, the profiles in the center of the bed were compared with the experimental one. For the stationary 2-dimensional model, we can observe good agreement between simulation and experimentation at distances close to the entrance of the bed in the axial direction. However, differences are observed as we move further away from the entrance in the axial direction. This is because the simulated data for the stationary model do not consider the variation in the catalyst activity, which decreases over time. This activity is proportionally related to the reaction rate [28], which will be lower.

Consequently, the simulated data for the stationary model present lower concentrations than the experimental ones; by not considering the decrease in catalyst activity, the conversion of CO increases, as shown in the previous figure. For example, the simulated CO concentration at the reactor outlet is 9 mol/m^3, while the experimental one is kept at 11.5 mol/m^3. For the dynamic 3-D model, the simulation agrees with the experimental data, following the same trend and showing a maximum discrepancy of ~15%, which is in line with Shin et al. [29].

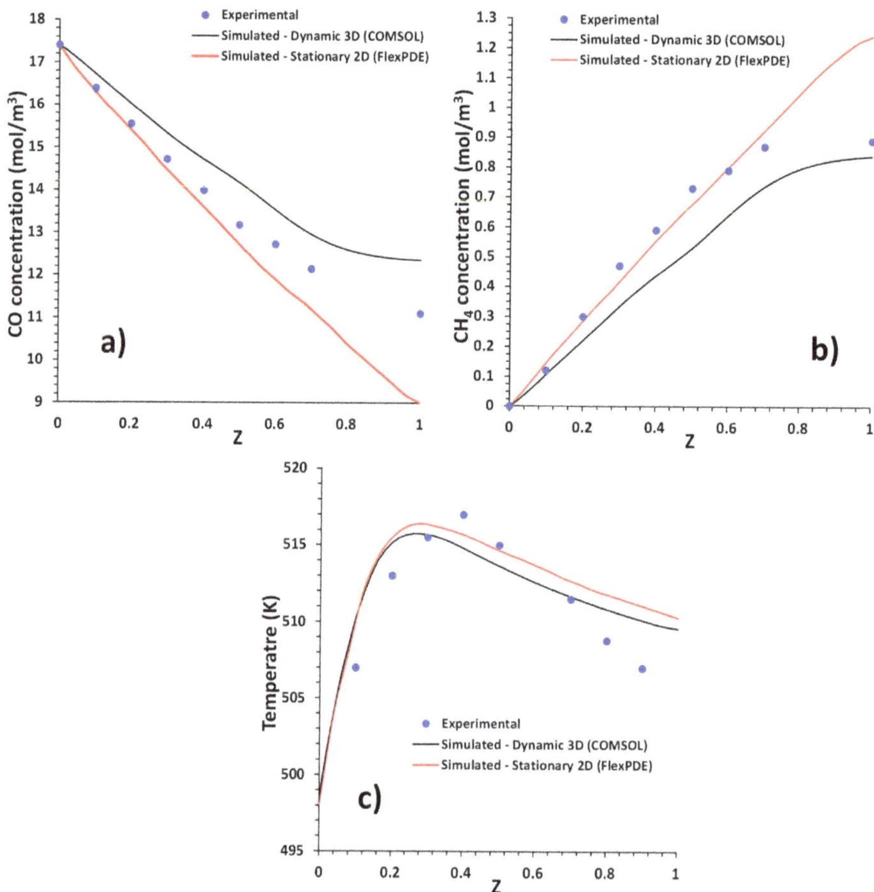

Figure 2. The simulated and experimental data for the 2-D stationary and the 3-D dynamic models: (a) CO concentration, (b) CH_4 concentration, and (c) Temperature.

On the other hand, Figure 2b shows the simulated and experimental CH_4 concentrations in the axial direction of the reactor. For the stationary 2-D model, CH_4 simulated concentration is in good agreement with the experimental data at a distance close to the entrance of the bed in the axial direction. Likewise, differences are observed as we move further away from the entrance in the axial direction. Again, this difference between the experimental and simulated concentration increases because the stationary model does not consider catalyst activity changes over time. For example, the CH_4 concentration at the reactor outlet is 0.9 mol/m^3, while the simulated one is 1.4 mol/m^3. This is because the reaction rate is higher when the catalyst activity factor is not considered since this factor is less than 1 [30]. Consequently, for a product such as CH_4, the simulated concentration will be greater than the experimental concentration since the former has a higher reaction rate, thus showing a more significant transformation of reactants to products [31]. For the dynamic 3-dimensional model, the simulation agrees with the experimental data, following the same trend and showing a maximum discrepancy of ~15%, also in concordance with Shin et al. [29].

The simulated and experimental temperature ©n the reactor center shows these results in Figure 2c. We can see that the simulated data differ from the experimental data by ~5 degrees. The difference between simulated and experimental values is due to two reasons: temperature measurements will depend on the heat capacity of the thermocouple that will affect the temperature data collection and the difficulty of taking the data for such a small radius of tube (1.25 cm). The experimental and simulated data have the same trend, as shown in Figure 2c, making the model valid. Furthermore, due to the tendency of the experimental data to present a maximum temperature point near the bed entrance, the theory proves that the reaction rate will be higher, which was also indicated by Irani et al. [32].

3.2. Analysis and Validation of the 1-D Dynamic Model and the 3-D Dynamic Model

Figure 3 includes the simulation results of the dynamic 1-dimensional model. Figure 3a compares the experimental and simulated values considering the 1-dimensional model. It is observed that the simulated concentration values are higher than the experimental ones, which means that the CO did not react enough in the simulation. This is because the reaction rate is lower for the one-dimensional model than for the two-dimensional one, which is the one that is closest to reality. CO did not react enough in the simulation because the one-dimensional model takes the radial average temperature to calculate the reaction rate (in line with Kuncharam and Dixon [33]).

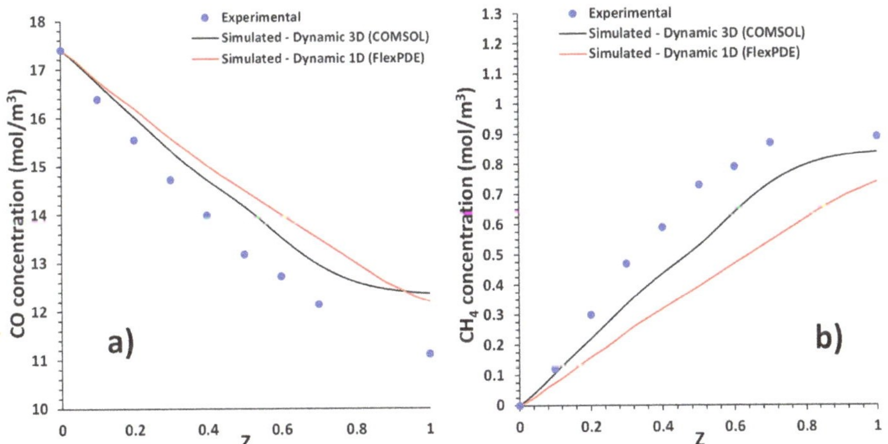

Figure 3. The simulated and experimental data for the 1-D dynamic and the 3-D dynamic models: (**a**) CO concentration, and (**b**) CH$_4$ concentration.

In contrast, the two-dimensional model takes the reaction rate in two dimensions. Therefore, the one-dimensional model will present less conversion than the two-dimensional model. Consequently, it can be concluded that the two-dimensional model presents better results and is the one that is closer to reality than the one-dimensional one.

Figure 3b shows the CH$_4$ concentration profile in the axial direction for the 1-dimensional model. The CH$_4$ concentration is lower for the simulation relative to the experiments. The experimental CH$_4$ concentration at the reactor outlet was 0.9 mol/m^3, while the simulated one was 0.7 mol/m^3. Both are different because not enough CO reacted to produce CH$_4$ to concentrations close to those observed experimentally.

3.3. Analysis and Validation of the 2-D Dynamic Model and the 3-D Dynamic Model

Figure 4 includes the simulation results for the dynamic 2-dimensional model. Figure 4a compares the simulated and experimental CO mol/m^3 concentrations. Both the dynamic 2-dimensional and the 3-dimensional models agree with the experimental data. It is observed that when taking a dynamic model, the approximation of simulated and experimental data improves, as also explained by Mandić et al. [34]. This is mainly because the activity of the catalyst is considered over time. Thus, it loses its activity and causes the reaction rate to slow down, which is closer to reality. The experimental CO concentration at the reactor outlet was 11.5 mol/m^3, while the simulated one was 10.7 mol/m^3.

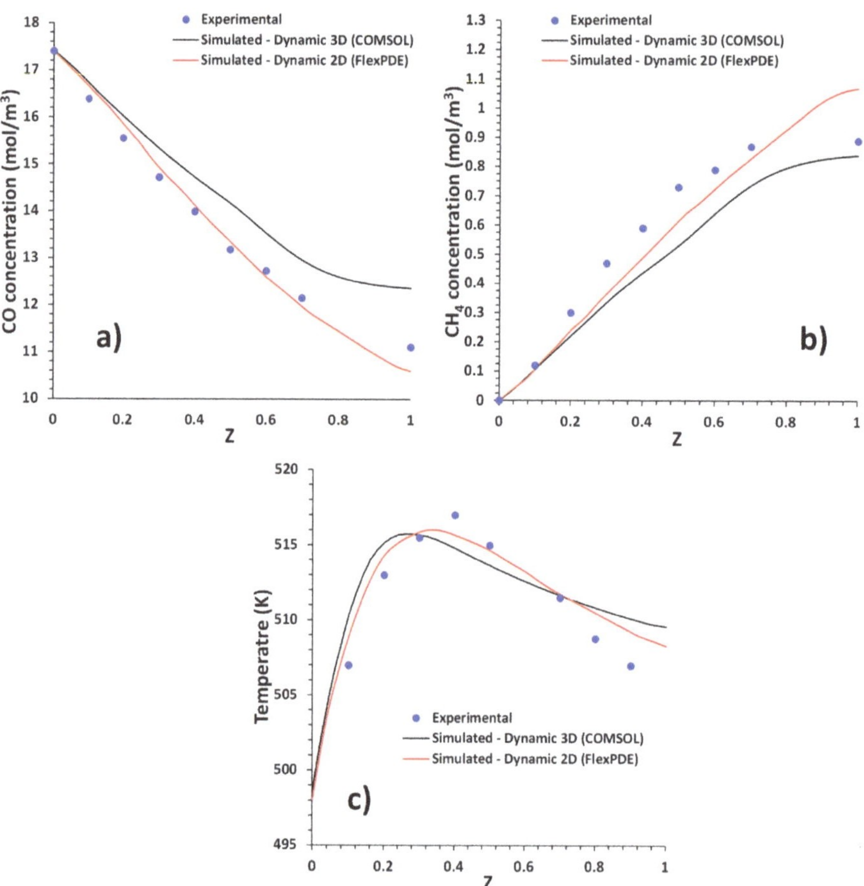

Figure 4. The simulated and experimental data for the 2-D dynamic and the 3-D dynamic models: (**a**) CO concentration, (**b**) CH$_4$ concentration, and (**c**) Temperature.

Figure 4b shows the comparison of simulated and experimental CH$_4$ concentrations. Both the dynamic 2-dimensional and the 3-dimensional models agree with the experimental data. The experimental CH$_4$ concentration at the reactor outlet was 0.9 mol/m^3, while the simulated one was 1.1 mol/m^3. Figure 4 shows that the dynamic 2-dimensional and the dynamic 3-dimensional models agree with the experimental data.

Figure 5 shows the temperature variation in the radial and axial directions for the dynamic 2-dimensional (Figure 5a) and the dynamic 3-dimensional model (Figure 5b). On the other hand, Figure 6 shows the reaction rate of CO in the radial and axial directions for both the dynamic 2-dimensional (Figure 6a) and the dynamic 3-dimensional model (Figure 6b). Figure 5a,b follow the same trend as Figure 6a,b. This reflects that increased/decreased temperatures imply increased/decreased reaction rates. The geometry built and utilized to construct the dynamic 3-dimensional model in COMSOL is available in the Supplementary Material in Figures S1–S4.

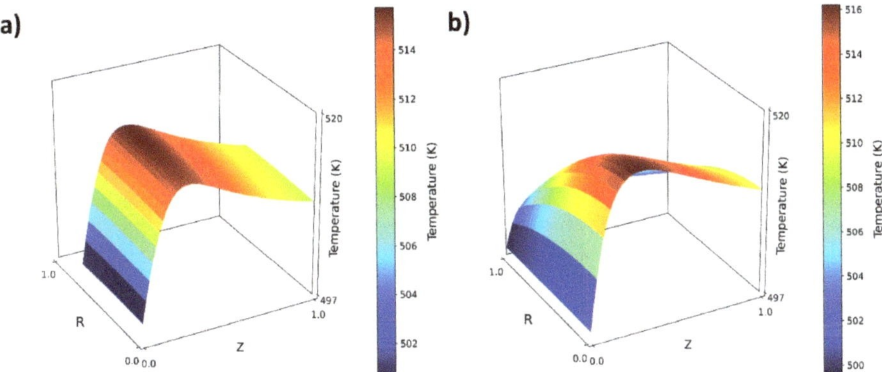

Figure 5. Bed temperature for (**a**) 3-dimensional dynamic model (COMSOL), and (**b**) 2-dimensional dynamic model (FlexPDE).

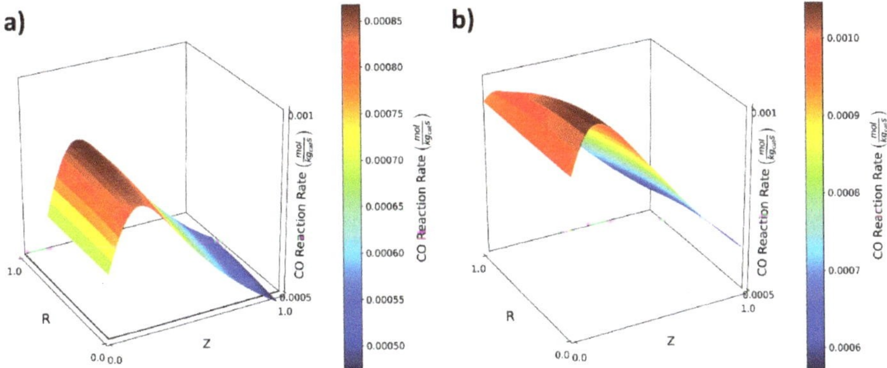

Figure 6. The CO Reaction Rate (mol/kg$_{cat}$ s) for (**a**) 3-dimensional dynamic model (COMSOL), and (**b**) 2-dimensional dynamic model (FlexPDE).

4. Conclusions

In this work, the thermochemical analysis of a cylindrical packed-bed reactor was carried out using the finite element methodology. Different CFD software such as FlexPDE and COMSOL Multiphysics was employed. Several approaches considering stationary and dynamic simulations and a multi-dimensional study were proposed for methane production using the F-T synthesis. The 1-D and 2-D simulations performed with FlexPDE and the 3-D simulation from COMSOL were compared to experimental data resulting in an excellent agreement, validating the concentrations of CO and CH$_4$ and the temperature profiles. In the axial direction, both stationary and dynamic schemes described the experimental phenomena accurately.

On the other hand, surface responses for temperature and CO reaction rate were developed to observe the dependence of axial and radial directions. In this regard, although similar trends were observed when comparing FlexPDE and COMSOL models, there were few differences, which can be attributed to the different types of transport equations employed. For instance, FlexPDE models were built based on dispersion models using the Peclet number, which seems to be more sensitive to heat transfer as the temperature was not as stable as the COMSOL model.

Since a comparison between different CFD modeling software is proposed in this work, it can be suggested that it would be interesting to consider Ansys Fluent for further studies. This software also provides some heat transfer models for porous media, like the non-equilibrium thermal model, which focuses on the fluid and the catalyst treated separately. As aforementioned, in real applications, both phases are not supposed to be at thermal equilibrium, so more equations must be solved in these cases.

Modeling reactors using heterogenous catalysts can be divided into (a) transport phenomena and (b) kinetic modeling. These two areas interact to predict temperature profiles and concentrations in space and time. This study uses COMSOL and FlexPDE to model the heat and mass transfer and the kinetics of the reactor. This manuscript uses F-T as a reaction example which has simple reaction kinetics; however, this study can serve as starting point to model more complex reactions and will allow readers to use it as a basis for more detailed models.

Supplementary Materials: The following supporting information can be downloaded at: https://www.mdpi.com/article/10.3390/pr10061144/s1.

Author Contributions: Conceptualization, S.T.-V. and C.A.R.; methodology, S.T.-V., C.A.R., H.A.M., A.C. and P.G.A.; software, S.T.-V., C.A.R., H.A.M., A.C. and P.G.A.; validation, S.T.-V., C.A.R., H.A.M.; formal analysis, S.T.-V. and C.A.R.; investigation, S.T.-V., C.A.R. and H.A.M.; funding acquisition, S.T.-V.; writing—original draft preparation, S.T.-V., C.A.R. and H.A.M.; writing—review and editing, S.T.-V., A.C. and P.G.A., supervision, S.T.-V., project administration, S.T.-V. and P.G.A. All authors have read and agreed to the published version of the manuscript.

Funding: This research was funded by Escuela Politécnica Nacional, via the project PIMI 16-07. The APC was funded by Escuela Politécnica Nacional.

Data Availability Statement: Not applicable.

Acknowledgments: We acknowledge the support of Escuela Politécnica Nacional, via the project PIMI 16-07.

Conflicts of Interest: The authors declare no conflict of interest.

Nomenclature

A'_i	Modified exponential factor for species i in the Arrhenius expression	
a_v	External surface area per unit volume of the bed	
C_i	Concentration of species i in the fluid phase	mol/m^3
C_{0i}	Initial concentration of the species i	mol/m^3
C_{si}	Concentration of species i in the solid phase	mol/m^3
Co_{si}	Initial concentration on the surface	mol/m^3
c_{pg}	Heat capacity of the fluid phase	$J/(kgK)$
c_{ps}	Heat capacity of the fluid phase	$J/(kgK)$
d_p	Equivalent particle diameter,	m
d_t	Tube diameter	m
E_i	Activation energy for species i	J/mol
ΔH_i	Enthalpy of reaction	J
h_c	Coefficient of heat transfer between solid and fluid phase	W/m^3K
k_c	Mass transfer coefficient between solid and fluid phase	m/s
k_p	Reaction constant for the Ficher-Tropsch polymerization	
k_t	Reaction constant for the determination of Ficher-Tropsch	
L	Bed length	m
Nu	Nusselt number	
Nu_w	Nusselt number	
P	Pressure	MPa
Po	Initial Pressure	MPa
P_i	Partial pressure for species i	MPa
Pe_h	Peclet number of heat	
$Pe°_h$	Molecular Peclet number of heat based on particle diameter	
Pe_{hv}	Radial effective heat peclet number based on particle diameter	
Pe_{hz}	Axial effective heat peclet number based on particle diameter	

Pe'_{hv}	Radial effective Peclet number of heat based on bed dimensions	
Pe'_{hz}	Axial effective Peclet number of heat based on bed dimensions	
Pe_{mv}	Radial effective Peclet number of mass based on particle diameter	
Pe_{mz}	Effective axial Peclet number of mass based on particle diameter	
Pe'_{mr}	Radial effective Peclet number of mass based on bed dimensions	
Pe'_{mz}	Radial effective Peclet number of mass based on bed dimensions	
Pr	Prandlt number	
R	Internal radius of tube	m
R_0	External radius of tube	m
R'	Ratio of the internal radius of the tube to the particle diameter	
Re_p	Reynolds number based on particle diameter	
R_t	Radius of tube	m
r	Radial coordinate	m
r'	Dimensionless Radial coordinate	
r_n	Reaction rate of formation of n alkanes	
r_{vi}	Reaction rate of species i	mol/(kgcat s)
$(-\Delta H)_{rv}$	Total heat generation by reaction	W/kg cat
S_n	Selectivity of n-alkanes	
Sc	Schmidt number	
Sh	Sherwood number	
T	Fluid temperature	K
T_0	Initial fluid temperature	K
T_s	Solid Temperature	K
To_s	Solid initial temperature	K
T_c	Cooling liquid temperature	K
T_w	Wall temperature	K
t	Time	h
U	Total Heat Transfer Coefficient for one dimension,	W/(m^{2}K)
U_c	Total Heat Transfer Coefficient	W/(m^{2}K)
u	Velocity, m/s	
u_0	Average speed	m/s
V_p	Volume of the particle	m
X	Axial dimensionless coordinate z/L in the graphs	
Y	Radial dimensionless r/R coordinate in graphs	
y_i	Dimensionless concentration of species i, C_i/C_{io}	
z	Axial coordinate	m
z'	Dimensionless Axial coordinate z/L	

Greek Letters

α	Probability constant for the alkane chain to grow	
α_w	Heat Transfer coefficient through the wall	
βi	Deactivation coefficient for species i	1/h
γ	C2/C3 ratio	
ε	Local porosity	
ε_0	Average porosity	
ε_p	Porosity of the catalyst particle	
η	Effectiveness Factor	
λ^0_c	Thermal conductivity due to conduction	W/(mK)
λ^f_c	Thermal conductivity due to convection	W/(mK)
λ_{cr}	Mean radial thermal conductivity of the bed	W/(mK)
λ_{cz}	Mean axial thermal conductivity of the bed	W/(mK)
λ_f	Thermal conductivity of the fluid phase	W/(mK)
λ_p	Thermal conductivity through the particle	W/(mK)
λ_w	Thermal conductivity of the bed wall	W/(mK)
μ_g	Fluid Viscosity	Ns/m^2
θ	Adimensional Temperature T/To	
θ_c	Dimensionless temperature of the cooling fluid Tc/To	
ρ_b	Density of the catalyst based on the volume of the bed	kg/m^3
ρ_g	Density of the fluid phase	kg/m^3
ρ_p	Density of the catalyst particle	kg/m^3

References

1. Perry, R.; Green, D.; Maloney, J. *Perry's Chemical Engineers' Handbook*, 7th ed.; McGraw-Hill: Madrid, Spain, 2001; ISBN 84-481-3008-1.
2. Cocco, R.; Karri, S.; Knowlton, T. Introduction to Fluidization. *Chem. Eng. Prog.* **2014**, *110*, 21–29.
3. Fratalocchi, L.; Groppi, G.; Visconti, C.G.; Lietti, L.; Tronconi, E. Packed-POCS with Skin: A Novel Concept for the Intensification of Non-Adiabatic Catalytic Processes Demonstrated in the Case of the Fischer-Tropsch Synthesis. *Catal. Today* **2022**, *383*, 15–20. [CrossRef]
4. Merino, D.; Sanz, O.; Montes, M. Effect of the Thermal Conductivity and Catalyst Layer Thickness on the Fischer-Tropsch Synthesis Selectivity Using Structured Catalysts. *Chem. Eng. J.* **2017**, *327*, 1033–1042. [CrossRef]
5. Calverley, E.M.; Witt, P.M.; Sweeney, J.D. Reactor Runaway Due to Statistically Driven Axial Activity Variations in Graded Catalyst Beds: Loading from Pre-Measured Single Tube Aliquots. *Chem. Eng. Sci.* **2013**, *90*, 170–178. [CrossRef]
6. Philippe, R.; Lacroix, M.; Dreibine, L.; Pham-Huu, C.; Edouard, D.; Savin, S.; Luck, F.; Schweich, D. Effect of Structure and Thermal Properties of a Fischer–Tropsch Catalyst in a Fixed Bed. *Catal. Today* **2009**, *147*, S305–S312. [CrossRef]
7. Shah, Z.; Kumam, P.; Ullah, A.; Khan, S.N.; Selim, M.M. Mesoscopic Simulation for Magnetized Nanofluid Flow within a Permeable 3D Tank. *IEEE Access* **2021**, *9*, 135234–135244. [CrossRef]
8. Visconti, C.G.; Tronconi, E.; Lietti, L.; Groppi, G.; Forzatti, P.; Cristiani, C.; Zennaro, R.; Rossini, S. An Experimental Investigation of Fischer–Tropsch Synthesis over Washcoated Metallic Structured Supports. *Appl. Catal. A Gen.* **2009**, *370*, 93–101. [CrossRef]
9. Asalieva, E.; Gryaznov, K.; Kulchakovskaya, E.; Ermolaev, I.; Sineva, L.; Mordkovich, V. Fischer–Tropsch Synthesis on Cobalt-Based Catalysts with Different Thermally Conductive Additives. *Appl. Catal. A Gen.* **2015**, *505*, 260–266. [CrossRef]
10. Fratalocchi, L.; Visconti, C.G.; Groppi, G.; Lietti, L.; Tronconi, E. Intensifying Heat Transfer in Fischer-Tropsch Tubular Reactors through the Adoption of Conductive Packed Foams. *Chem. Eng. J.* **2018**, *349*, 829–837. [CrossRef]
11. Chandra, V.; Vogels, D.; Peters, E.A.J.F.; Kuipers, J.A.M. A Multi-Scale Model for the Fischer-Tropsch Synthesis in a Wall-Cooled Packed Bed Reactor. *Chem. Eng. J.* **2021**, *410*, 128245. [CrossRef]
12. Dixon, A.G.; Van Dongeren, J.H. The Influence of the Tube and Particle Diameters at Constant Ratio on Heat Transfer in Packed Beds. *Chem. Eng. Process. Process Intensif.* **1998**, *37*, 23–32. [CrossRef]
13. Mohammad, N.; Chukwudoro, C.; Bepari, S.; Basha, O.; Aravamudhan, S.; Kuila, D. Scale-up of High-Pressure F-T Synthesis in 3D Printed Stainless Steel Microchannel Microreactors: Experiments and Modeling. *Catal. Today* **2021**, *397–399*, 182–196. [CrossRef]
14. Abusrafa, A.E.; Challiwala, M.S.; Wilhite, B.A.; Elbashir, N.O. Thermal Assessment of a Micro Fibrous Fischer Tropsch Fixed Bed Reactor Using Computational Fluid Dynamics. *Processes* **2020**, *8*, 1213. [CrossRef]
15. Chabot, G.; Guilet, R.; Cognet, P.; Gourdon, C. A Mathematical Modeling of Catalytic Milli-Fixed Bed Reactor for Fischer–Tropsch Synthesis: Influence of Tube Diameter on Fischer Tropsch Selectivity and Thermal Behavior. *Chem. Eng. Sci.* **2015**, *127*, 72–83. [CrossRef]
16. Na, J.; Kshetrimayum, K.S.; Lee, U.; Han, C. Multi-Objective Optimization of Microchannel Reactor for Fischer-Tropsch Synthesis Using Computational Fluid Dynamics and Genetic Algorithm. *Chem. Eng. J.* **2017**, *313*, 1521–1534. [CrossRef]
17. Pavlišič, A.; Ceglar, R.; Pohar, A.; Likozar, B. Comparison of Computational Fluid Dynamics (CFD) and Pressure Drop Correlations in Laminar Flow Regime for Packed Bed Reactors and Columns. *Powder Technol.* **2018**, *328*, 130–139. [CrossRef]
18. Amiri, L.; Ghoreishi-Madiseh, S.A.; Hassani, F.P.; Sasmito, A.P. Estimating Pressure Drop and Ergun/Forchheimer Parameters of Flow through Packed Bed of Spheres with Large Particle Diameters. *Powder Technol.* **2019**, *356*, 310–324. [CrossRef]
19. Khan, S.; Selim, M.M.; Gepreel, K.A.; Ullah, A.; Ikramullah; Ayaz, M.; Mashwani, W.K.; Khan, E. An Analytical Investigation of the Mixed Convective Casson Fluid Flow Past a Yawed Cylinder with Heat Transfer Analysis. *Open Phys.* **2021**, *19*, 341–351. [CrossRef]
20. Rai, A.; Anand, M.; Farooqui, S.A.; Sibi, M.G.; Sinha, A.K. Kinetics and Computational Fluid Dynamics Study for Fischer–Tropsch Synthesis in Microchannel and Fixed-Bed Reactors. *React. Chem. Eng.* **2018**, *3*, 319–332. [CrossRef]
21. Oñate, W.; Maldonado, S.; Taco, S.; Caiza, G. Computational Fluid Dynamic Simulation with Experimental Validation in Turbine Pipeline. *Int. J. Recent Technol. Eng.* **2019**, *8*, 927–931.
22. Deutschmann, O. Modeling of the Interactions Between Catalytic Surfaces and Gas-Phase. *Catal. Lett.* **2014**, *145*, 272–289. [CrossRef]
23. Pessoa, D.R.; Finkler, A.T.J.; Machado, A.V.L.; Mitchell, D.A.; de Lima Luz, L.F. CFD Simulation of a Packed-Bed Solid-State Fermentation Bioreactor. *Appl. Math. Model.* **2019**, *70*, 439–458. [CrossRef]
24. Skaare, S. Reaction and Heat Transfer in a Wall-Cooled Fixed Bed Reactor, University of Trondheim, The Norwegian Institute of Technology, Laboratory of Industrial Chemistry, [Thesis Submitted for the Doctor of Engineering Degree]. 1993. Available online: http://www.fischer-tropsch.org/DOE/DOE_reports/95717546/de95717546_toc.htm (accessed on 17 May 2022).
25. Rodemerck, U.; Holeňa, M.; Wagner, E.; Smejkal, Q.; Barkschat, A.; Baerns, M. Catalyst Development for CO_2 Hydrogenation to Fuels. *ChemCatChem* **2013**, *5*, 1948–1955. [CrossRef]
26. Yates, I.C.; Satterfield, C.N. Intrinsic Kinetics of the Fischer-Tropsch Synthesis on a Cobalt Catalyst. *Energy Fuels* **2002**, *5*, 168–173. [CrossRef]
27. *Introduction to the Porous Media Flow Module User's Guide, COMSOL Multiphysics®v. 5.5*; COMSOL AB: Stockholm, Sweden, 2019; pp. 6–38.
28. Vervloet, D.; Kapteijn, F.; Nijenhuis, J.; Van Ommen, J.R. Fischer–Tropsch Reaction–Diffusion in a Cobalt Catalyst Particle: Aspects of Activity and Selectivity for a Variable Chain Growth Probability. *Catal. Sci. Technol.* **2012**, *2*, 1221–1233. [CrossRef]

29. Shin, M.S.; Park, N.; Park, M.J.; Jun, K.W.; Ha, K.S. Computational Fluid Dynamics Model of a Modular Multichannel Reactor for Fischer–Tropsch Synthesis: Maximum Utilization of Catalytic Bed by Microchannel Heat Exchangers. *Chem. Eng. J.* **2013**, *234*, 23–32. [CrossRef]
30. Guettel, R.; Turek, T. Comparison of Different Reactor Types for Low Temperature Fischer–Tropsch Synthesis: A Simulation Study. *Chem. Eng. Sci.* **2009**, *64*, 955–964. [CrossRef]
31. Moazami, N.; Wyszynski, M.L.; Mahmoudi, H.; Tsolakis, A.; Zou, Z.; Panahifar, P.; Rahbar, K. Modelling of a Fixed Bed Reactor for Fischer–Tropsch Synthesis of Simulated N2-Rich Syngas over Co/SiO$_2$: Hydrocarbon Production. *Fuel* **2015**, *154*, 140–151. [CrossRef]
32. Irani, M.; Alizadehdakhel, A.; Pour, A.N.; Proulx, P.; Tavassoli, A. An Investigation on the Performance of a FTS Fixed-Bed Reactor Using CFD Methods. *Int. Commun. Heat Mass Transf.* **2011**, *38*, 1119–1124. [CrossRef]
33. Kuncharam, B.V.R.; Dixon, A.G. Multi-Scale Two-Dimensional Packed Bed Reactor Model for Industrial Steam Methane Reforming. *Fuel Process. Technol.* **2020**, *200*, 106314. [CrossRef]
34. Mandić, M.; Dikić, V.; Petkovska, M.; Todić, B.; Bukur, D.B.; Nikačević, N.M. Dynamic Analysis of Millimetre-Scale Fixed Bed Reactors for Fischer-Tropsch Synthesis. *Chem. Eng. Sci.* **2018**, *192*, 434–447. [CrossRef]

Article

Renewable Polymers Derived from Limonene

Roman Aleksandrovich Lyubushkin, Natalia Igorevna Cherkashina *, Dar'ya Vasil'yevna Pushkarskaya, Dar'ya Sergeyevna Matveenko, Alexander Sergeevich Shcherbakov and Yuliya Sergeevna Ryzhkova

Department of Theoretical and Applied Chemistry, Belgorod State Technological University Named after V.G. Shukhov, 308012 Belgorod, Russia
* Correspondence: cherkashina.ni@bstu.ru

Abstract: Renewable natural and synthetic basic substances can be used to produce biodegradable polymers. Several methods of the polymerization of terpene limonene have been evaluated. The polymerization methods evaluated are radical polymerization, cationic polymerization and thiol-ene polymerization. The free-radical polymerization of limonene with azobisisobutyronitrile (AIBN) as an initiator was carried out. The cationic polymerization of limonene was carried out using AlCl$_3$ as a catalyst. The copolymerization of limonene with mercaptoethanol, 2-mercaptoethyl ether without an initiator and with an AIBN initiator was studied and it was also shown that polymerization can proceed spontaneously. The resulting compounds were investigated by NMR and FTIR spectroscopy. The values of the molecular weight characteristics of the samples obtained are presented, such as: number-average molecular weight, hydrodynamic radius and characteristic viscosity, depending on the method of production. The coefficients α (molecular shape) in the Mark–Kuhn–Houwink equation are determined according to the established values of the characteristic viscosity. According to the values obtained, the AC molecules in solution have parameters α 0.14 to 0.26, which corresponds to a good solvent and the molecular shape-dense coil.

Keywords: bio-based elastomers; azobisisobutyronitrile; copolymerization; thiol-ene-click; (S)-(−)-limonen

Citation: Lyubushkin, R.A.; Cherkashina, N.I.; Pushkarskaya, D.V.; Matveenko, D.S.; Shcherbakov, A.S.; Ryzhkova, Y.S. Renewable Polymers Derived from Limonene. *ChemEngineering* **2023**, *7*, 8. https://doi.org/10.3390/chemengineering7010008

Academic Editor: George Z. Papageorgiou

Received: 2 November 2022
Revised: 26 December 2022
Accepted: 13 January 2023
Published: 17 January 2023

1. Introduction

Polymers are used in everyday life due to their wide range of chemical, mechanical, thermal and electro-physical properties [1,2]. By now, the problem of using renewable resources to meet needs without creating harmful effects on human health and the environment is quite acute. Nature provides mankind with a rich polymer material on the basis of which useful products can be obtained. Their production requires a non-renewable, depleting petroleum feedstock, which has significant energy costs during processing and produces toxic waste. Renewable natural and synthetic basic substances can be used to produce biodegradable polymers. There are principles for the selection of suitable monomers and methods for modifying natural origin polymers in order to impart the desired properties. One of these substances is limonene [3–5]. Limonene is found in many essential oils and is an adaptable chemical raw material [6]. As a component of turpentine, limonene can be obtained as a by-product from over 300 different plants [7–9]. Limonene makes up 92–97% of citrus rind oils and its global production, around 400,000 tons per year, is growing [10–13].

One of the promising areas for the application of limonene is its copolymerization to obtain new functional polymers [14–16].

Many scientific studies have been devoted to the polymerization of limonene [17–19]. It was shown in [20] that D-limonene faces difficulties during polymerization, which lead to low monomer conversion and molecular weight. To obtain poly (limonene) with a higher molecular weight, it is necessary to simultaneously reduce the concentrations of the monomer and initiator.

To increase the degree of polymerization, it is possible to use the photochemical radical polymerization of limonene at low temperatures using a combination of type II photoinitiators and alkyl halide initiators [21].

One of the main problems hindering the extension of the scope of application of modified polymers from renewable sources of biological origin is the low efficiency and selectivity of condensation reactions. In this context, there has been genuine interest in the development of new conjugation methods, collectively called click chemistry. Among the attractive features of click-chemistry methods are soft condensation conditions, high efficiency and selectivity. A special place among these processes is occupied by radical polymerization with reversible chain transfer (RCT) according to the addition–fragmentation mechanism due to tolerance to the functional groups of monomers and soft conditions [22–24]. This process is based on the use of sulphur-containing compounds of the general structure Z-C (=S) S-R. Under the right conditions of synthesis, macromolecules are formed whose α- and ω-end groups are determined by the chemical nature of the RCT agent used (R- and Z-C (=S) S-, respectively). The use of RCT agents, in which the R SH and C=C groups, allows such macromolecules to be further used as blanks for click reactions. Terpene-based monomers can then be used to synthesize various polymers.

This type of reaction can be carried out under soft conditions by simply mixing thiol and olefin substrates with terminal olefins. It has previously been demonstrated that thiol-ene systems can lead to extremely homogeneous glasses, elastomers and adhesives [25].

This article presents the results of a study of terpene limonene polymerization using radical polymerization, cationic polymerization and thiol-ene polymerization methods.

2. Materials and Methods

The following reagents were used to carry out the polymerization: toluene (99.5%, for synthesis), methanol (for analysis, Sigma Aldrich, Saint Louis, MO, USA), Limonene (b.p. = 176–177 °C, $[\alpha]20 = +113 \pm 2$, d = 0.8411, Sigma Aldrich), Azobisisobutyronitrile (98%, Fluka), twice reprecipitated from methanol, $AlCl_3$ (anhydrous, 99.5%, Sigma Aldrich), 2-mercaptoethanol (99.0%, Sigma Aldrich), 2-Mercaptoethyl ether (95%, Sigma Aldrich).

The results for determining the structure of the resulting compounds are obtained using measuring instruments: VERTEX 70 FTIR spectrometer (Bruker, Ettlingen, Germany); AVANCE III HD NMR spectrometer (400 MHz, Bruker, Ettlingen, Germany).

Molecular mass characteristics: weight-average molecular weight (Mw), number-average molecular weight (Mn) and polydispersity of the samples were determined by gel permeation chromatography using an Agilent 1260 Infinity II Multi-Detector GPC/SEC System (Agilent Technologies, Santa Clara, CA, USA) chromatograph with triple detection. Separation was carried out on a PLgel Mixed-E column designed for the analysis of oligomers and low molecular weight polymers using tetrahydrofuran stabilized with 250 ppm ionol as mobile phase. The column was calibrated using polydisperse polystyrene standards (Agilent, USA). The eluent feed rate is 1 mL/min; the sample volume is 100 µL. The alkyd resin samples are dissolved in tetrahydrofuran at a concentration of 5 mg/mL and held for 12 h until completely dissolved. All samples are filtered through a 0.22 µm polytetrafluoroethylene membrane filter Millipore (Merck KGaA, Darmstadt, Germany) after complete dissolution. Data are collected and processed using Agilent GPC/SEC MDS software.

3. Results and Discussion

3.1. Radical Polymerization

The free-radical polymerization of limonene with azobisisobutyronitrile (AIBN) as an initiator was carried out. The process of the radical polymerization of limonene involves the following steps:

I.   Initiation (Figure 1a);

II.   Chain propagation (Figure 1b);

III. Interruption (Figure 1c).

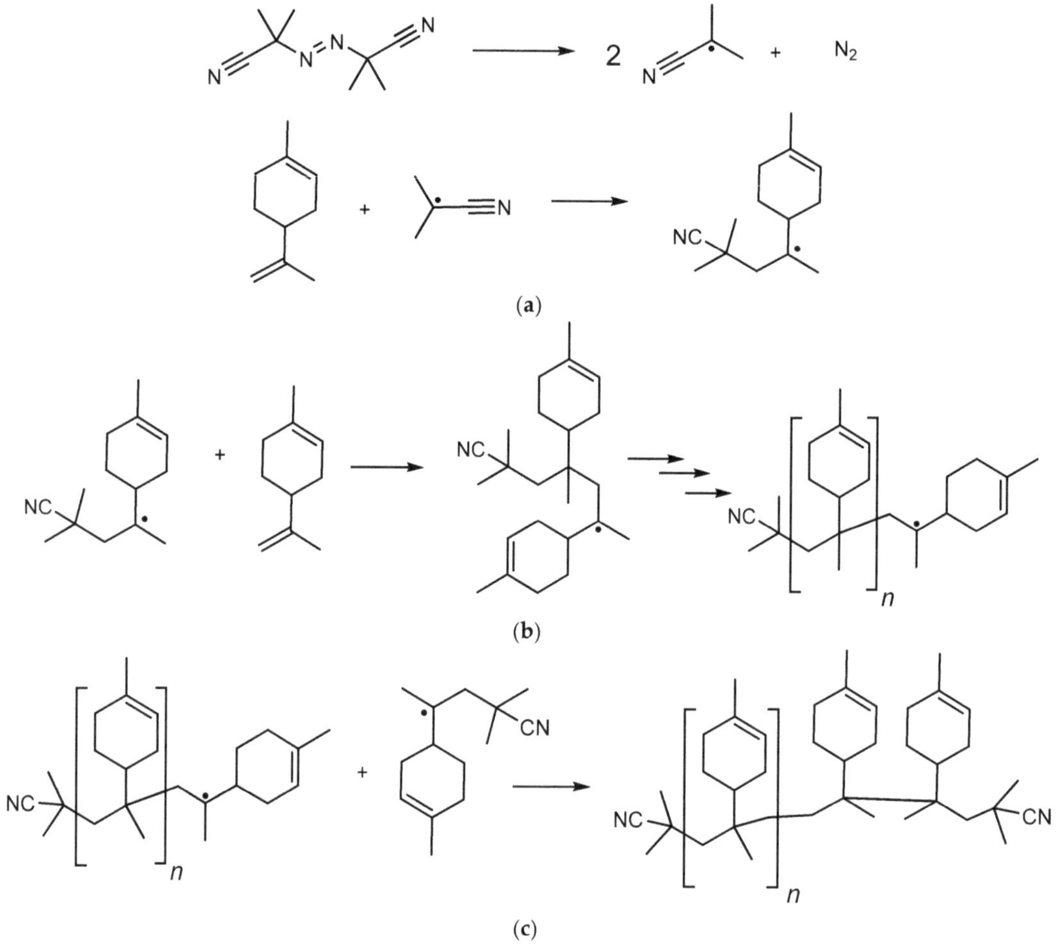

Figure 1. Reaction schemes of various stages of limonene radical polymerization: (**a**) initiation; (**b**) chain growth; (**c**) break.

No polymer precipitation was observed when the crude reaction mixture was precipitated with cooled methanol. Figure 2 shows the ^{1}H NMR spectrum of the resulting product; there are no characteristic peaks of the polylimonene, and there is no decrease in the peaks of the inner double bond a (5.4 ppm) and the outer double bond b (4.7 ppm); i.e., the spectrum is identical to that of the initial monomer. The peak around 1.4–1.7 ppm corresponds to six protons in the limonene methyl groups (C–H).

The homopolymerization of monoterpenes by the free-radical mechanism is difficult due to the presence of allylic C-H bonds. The reason for this effect is that the chain transfer to the allylic monomer leads not only to the decay of the growing macroradical, but also to a complete interruption of the kinetic chain due to the abstraction of a hydrogen atom in the α-position to the allylic bond with the formation of a resonantly stabilized allylic radical. Limonene has two unconjugated double bonds due to steric effects and different radical stability [26,27]; it is difficult to homopolymerize limonene during free-radical polymerization due to destructive chain transfer caused by more reactive allyl hydrogen [28] (Figure 3).

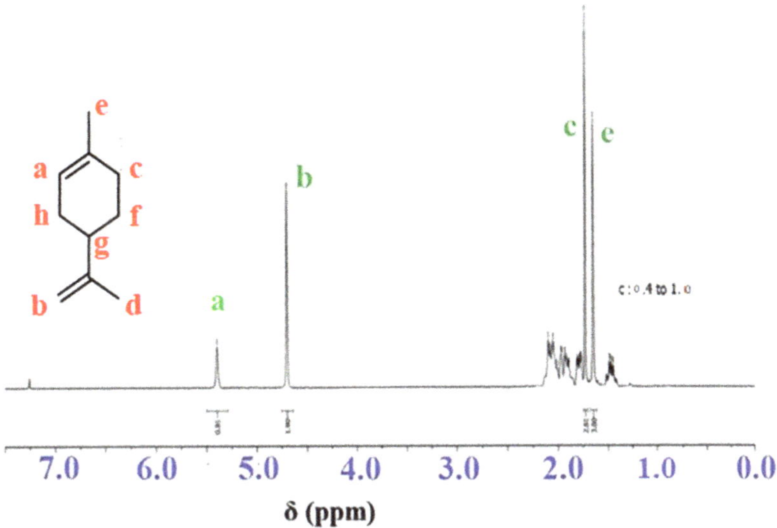

Figure 2. ^{1}H-NMR spectrum of the free radical polymerization reaction mixture of limonene with AIBN as initiator.

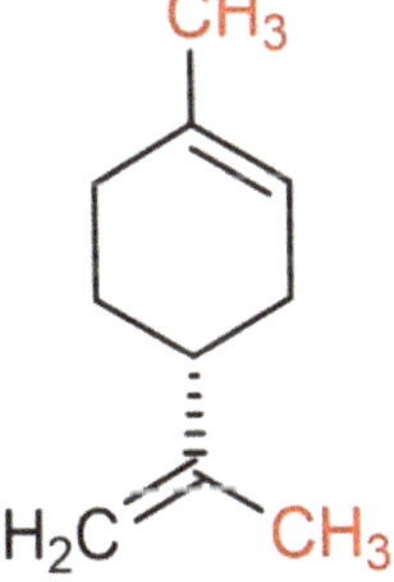

Figure 3. Alyl carbons in the structure of limonene.

This so-called degradation chain transfer to monomer or autoinhibition is the main reason for the low molecular weight of polymers and their inhibitory effect in polymerization processes derived from allyl monomer products.

3.2. Cationic Polymerization

The cationic polymerization of limonene using $AlCl_3$ as a catalyst was carried out according to the following scheme. Equal parts by weight of limonene and dried toluene were cooled down to 0 °C or −10 °C in a nitrogen atmosphere in a round bottom flask. A 5% wt. % $AlCl_3$ equivalent to limonene was added to the solution. Then, the temperature was slowly increased from 0 °C to 50 °C in intervals of 20 min. ^{1}H-NMR analysis was used to monitor the reactions.

The $AlCl_3$ catalyst was removed by stirring with 0.1 M HCl until the orange color disappeared and the organic phase was washed four times with 0.1 M NaOH and twice with deionized water. The organic phase was then dried with magnesium sulphate and the toluene was evaporated. The resulting reaction products were dissolved in tetrahydrofuran and precipitated with cold methanol and dried.

Cationic polymerization, as with radical polymerization, is a chain reaction consisting of four stages: preinitiation, initiation, propagation and interruption (Figure 4).

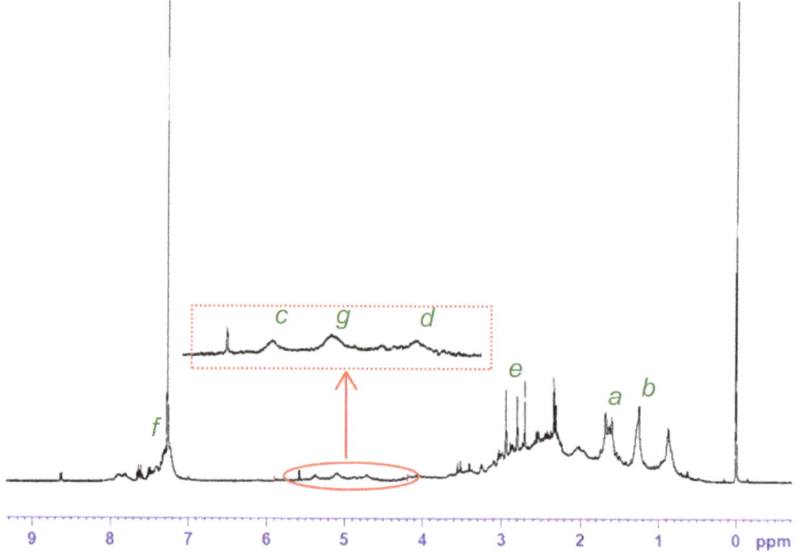

Figure 4. Cationic polymerization of limonene.

The ^{1}H NMR spectrum of the resulting product is shown in Figure 5. It can be seen that the peak of about 1.4–1.7 ppm corresponds to six protons in the methyl groups of limonene (C–H). It is noted that this peak also appears at about 0.8 m.d. for polylimonene.

Figure 5. ^{1}H-NMR spectrum of the product of cationic polymerization of limonene with AlCl$_3$ as initiator.

In addition, in the poly (limonene) spectrum, a peak is observed at ~1.2 m.d. corresponding to the protons of the methylene group (-CH$_2$-), which is not observed in the monomer spectrum, which indicates the break of the double bond and is an indication of the polymerization of limonene. It is also noted that the peaks related to the endocyclic and exocyclic double bonds of limonene are located at 5.4 (one proton) and 4.7 ppm (two protons), respectively. The ratio of peak intensities of exocyclic/endocyclic double bonds is 2:1, which is lower, indicating that the exocyclic double bonds of limonene mainly react through an additional radical reaction, confirming the result previously mentioned regarding the polymerization process.

It is expected that only these exocyclic double bonds will participate in polymerization, whereas endocyclic double bonds will remain in the polymer chain. However, a small amount of unreacted exocyclic double bond of limonene was also observed, with a peak of about 4.7 ppm, indicating that in addition to the limonene link included as a result of propagation, a different type of limonene link that contains endocyclic and exocyclic alkene groups is formed in the process. This result indicates that there was a case of destructive chain transfer caused by allyl hydrogen. Limonene contains more than one allyl hydrogen, and all of them can become reactive centers for this parallel reaction. Although the highly substituted tertiary allyl radical usually represents the highest degree of stability, in the case of limonene, the formation of a primary allyl radical is more likely due to less steric difficulties, and it is the dominant compound formed during the chain transfer reaction. Given that these allyl radicals are not prone to propagate, they eventually reproduce with each other or, more likely, with radicals in the propagation chain near the end of the process. Despite this expected behavior during the polymerization of limonene, it is extremely important to make sure that the intensity of these peaks is significantly reduced, which may indicate a decrease in the frequency of these chain transfer reactions. It is necessary to highlight some new peaks that have appeared in the spectrum. The peak is about 3 ppm. It is characteristic of a proton bound to an alcohol bond of the initiator (R-OH), which indicates the presence of part of its structure in the polymer. It is worth noting that the peak at 5.0 ppm refers to the myrcene links included in the polymer. Myrcene is the main impurity found in limonene; due to similar boiling points, limonene and myrcene are not easily separated.

Based on this analysis, it was possible to estimate the conversion of the monomer at about 10%.

The FTIR spectra of limonene (Figure 6) show an absorption band of valence vibrations at 1640 cm^{-1} corresponding to the exocyclic double bond, which disappears during the polymerization process.

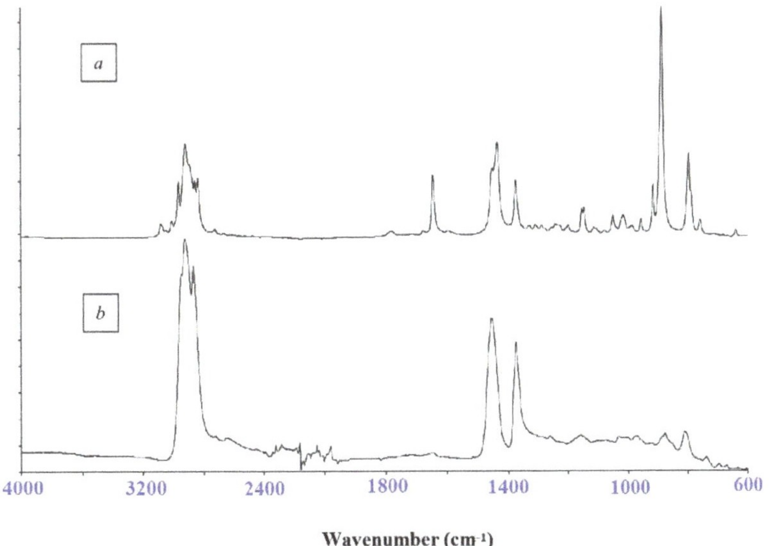

Figure 6. The FTIR spectrum of limonene (**a**) and cationic homopolymerization product (**b**).

3.3. Thiol-Ene Polymerization

The copolymerization of limonene with mercaptoethanol, 2-mercaptoethyl ether without an initiator and with an AIBN initiator was studied and it was also shown that the polymerization could proceed spontaneously.

The reactivity of the endocyclic and exocyclic double bond in the limonene structure is different, resulting in different monomers depending on the reaction time (Figure 7). The reaction of 2-mercaptoethanol and limonene, obtained at room temperature after 24 h, gives the addition products at the exocyclic bond. After 48 h, both the double bonds of limonene react with thiol to form a difunctional monomer with end hydroxyl groups, most suitable for polycondensation. Increasing the reaction temperature to 60 °C results in a polymer.

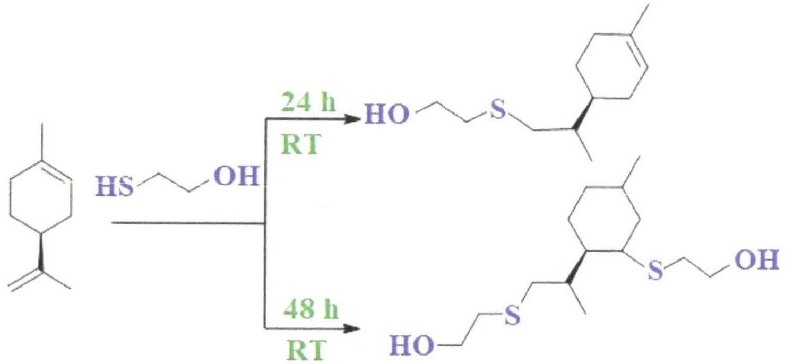

Figure 7. Reaction of 2-mercaptoethanol and limonene for 24 and 48 h at room temperature.

A similar situation was observed with 2-mercaptoethyl ether; polymerization could not be achieved (Figure 8).

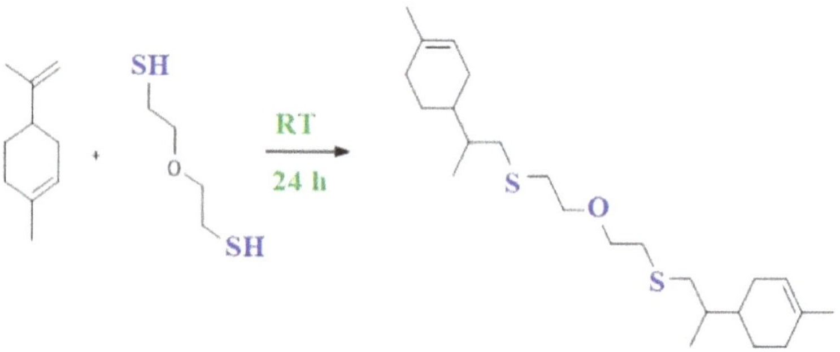

Figure 8. Scheme of addition of mercaptoethyl ether to limonene at room temperature.

The copolymerization of limonene with 2-mercaptoethyl ether using AIBN as an initiator was carried out to obtain high molecular weight products. These reactions follow a free-radical chain mechanism and generally lead to anti-Markovnik products. The initially formed thiol radical attacks limonene, forming a carbon radical. The carbon radical then reacts with the thiol molecule to form the final product and a new thiol radical, thereby lengthening the radical chain (Figure 9). Since this involves the cleavage of the S-H bond, the overall rate of reaction will strongly depend on the thiol structure as well as the lifetime of the thiol. The unpaired electron on the sulphur atom is in the highest binding orbital. This determines their electrophilic properties, and the electron affinity of thiol radicals is higher than that of hydroxyl HO-. As a result, thiol radicals recombine with each other more easily than hydroxyl radicals.

Limonene and 2-mercaptoethyl ether were mixed and stirred in a nitrogen atmosphere for 15 min. After 15 min, AIBN, 1 wt. % was added. (per monomer). The mixture was lowered into a preheated oil bath and left to polymerize for three hours. ^{1}H-NMR analysis was carried out after polymerization. The mixture was then dissolved in tetrahydrofuran and added drop by drop to the cold methanol. The polymer was collected by decanting the solvent.

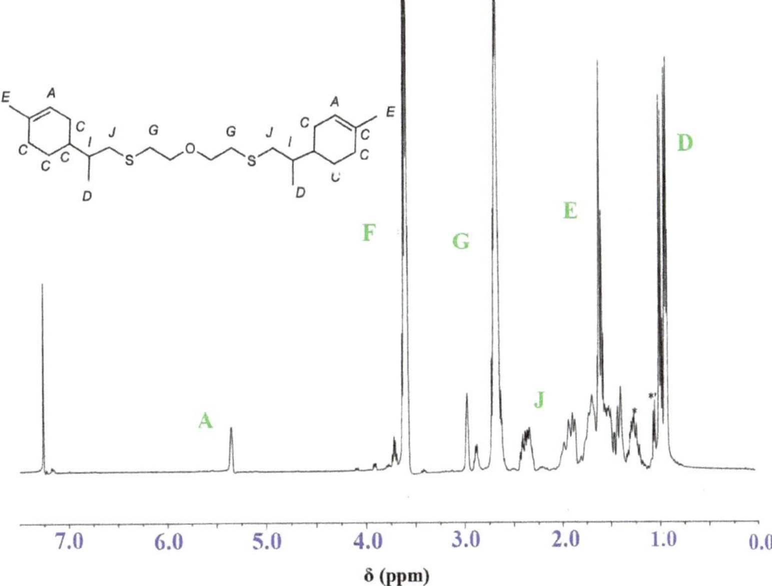

Figure 9. Scheme of the reaction of copolymerization of limonene with 2-mercaptoethyl ether using AIBN as an initiator.

The ^{1}H-NMR spectra of the sample after precipitation are shown in Figure 10. It can be seen that the exocyclic double bond signals have disappeared.

Figure 10. ^{1}H-NMR spectrum of the polymerization product of limonene and 2-mercaptoethyl ether with AIBN as initiator.

The number-average molecular weight (Mn) of the synthesized samples was determined using an Agilent 1260 Infinity II Multi-Detector GPC/SEC System chromatograph. The data obtained are presented in Table 1.

Table 1. The value of the molecular-weight characteristics of the samples depending on the synthesis.

Sample	Number Average Molecular Weight Value Mn (g/mol)	Hydrodynamic Radius, Rhn (nm)	Intrinsic Viscosity, IVn (dL/g)	Parameters of the Mark–Kuhn–Houwink Equation	
				α	K
Limonene/AIBN	1500	3.05	0.222692	0.29	1664.71
Limonene AlCl₃	2486	4.33	0.399749	0.08	18,981.87
Limonene/2-Mercapto-ethyl ether	3100	4.37	0.395570	0.13	11,453.21
Limonene/2-Mercapto-ethyl ether/AIBN	5400	5.12	0.370520	0.17	6523.38

Two main regions can be distinguished on the curves of the molecular mass distribution of the synthesized samples (Figure 11): a range of about 1000 g/mol, in which a mixture of monomeric and primary dimeric condensation products is identified, and with an increase in the duration of the process, the intensity of this region decreases significantly; a high-molecular region of 10,000–10,000,000 g/mol corresponding to oligomeric compounds. The coefficients α (molecular shape) in the Mark–Kuhn–Houwink equation are determined according to the established values of the characteristic viscosity. According to the values obtained, the AC molecules in solution have parameters α 0.14 to 0.26, which correspond to a good solvent and the molecular shape-dense coil.

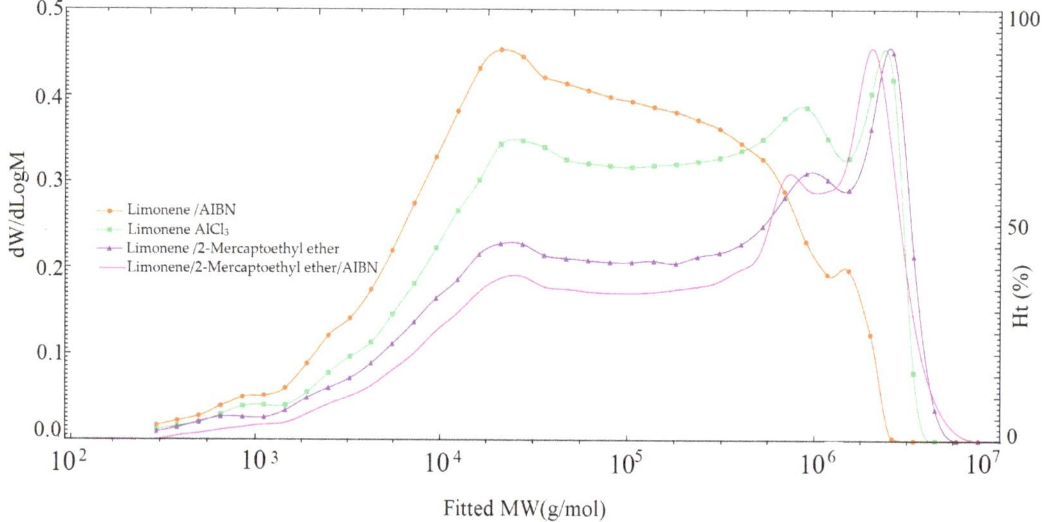

Figure 11. Molecular mass distribution of the synthesized samples.

4. Conclusions

The samples obtained by thiol-ene polymerization have the highest molecular weight; homopolymerization in the presence of aluminum chloride gives a product with a low molecular weight. The free radical polymerization of limonene with AIBN does not give the possibility to produce polylimonene; however, as a general conclusion it is shown that limonene can polymerize and thus has the properties necessary for the subsequent creation of materials based on it.

Further research will be aimed at increasing the yield of the limonene polymerization reaction and effective synthesis methods, as well as the search for practically important copolymers of limonene.

Author Contributions: Conceptualization, R.A.L. and N.I.C.; methodology, validation, D.V.P. and D.S.M.; formal analysis, N.I.C.; investigation, A.S.S. and Y.S.R.; writing—original draft preparation, R.A.L.; writing—review and editing, R.A.L. and Y.S.R.; supervision, N.I.C.; funding acquisition, N.I.C. All authors have read and agreed to the published version of the manuscript.

Funding: This work was realized using the equipment of the High Technology Center at BSTU named after V.G. Shukhov, the framework of the State Assignment of the Ministry of Education and Science of the Russian Federation, project No. FZWN-2021-0015.

Conflicts of Interest: The authors declare no conflict of interest. The funders had no role in the design of the study; in the collection, analyses, or interpretation of data; in the writing of the manuscript, or in the decision to publish the results.

References

1. Bakshi, P.R.; Londhe, V.Y. Widespread applications of host-guest interactive cyclodextrin functionalized polymer nanocomposites: Its meta-analysis and review. *Carbohydr. Polym.* **2020**, *242*, 116430. [CrossRef] [PubMed]
2. Ogay, V.; Mun, E.A.; Kudaibergen, G.; Baidarbekov, M.; Kassymbek, K.; Zharkinbekov, Z.; Saparov, A. Progress and Prospects of Polymer-Based Drug Delivery Systems for Bone Tissue Regeneration. *Polymers* **2020**, *12*, 2881. [CrossRef] [PubMed]
3. Ishmuratov, G.Y.; Legostaeva, Y.V.; Botsman, L.P.; Nasibullina, G.V.; Muslukhov, R.R.; Kazakov, D.V.; Tolstikov, G.A. Ozonolytic transformations of (S)-(−)-limonene. *Russ. J. Org. Chem.* **2012**, *48*, 18–24. [CrossRef]
4. Martínez, Q.H.; Paez-Mozo, E.A.; Martínez, O.F. Selective Photo-epoxidation of (R)-(+)- and (S)-(−)-Limonene by Chiral and Non-Chiral Dioxo-Mo(VI) Complexes Anchored on TiO$_2$-Nanotubes. *Top Catal.* **2021**, *64*, 36–50. [CrossRef]
5. Alexandrino, T.D.; Madureira de Medeiros, T.D.; Ruiz, A.L.T.G.; Favaro, D.C.; Pastore, G.M.; Bicas, J.L. Structural properties and evaluation of the antiproliferative activity of limonene-1,2-diol obtained by the fungal biotransformation of R-(+)- and S-(−)-limonene. *Chirality* **2022**, *34*, 887–893. [CrossRef] [PubMed]
6. Corma, A.; Iborra, S.; Velty, A. Chemical routes for the transformation of biomass into chemicals. *Chem. Rev.* **2007**, *107*, 2411–2502. [CrossRef] [PubMed]
7. Panwar, D.; Panesar, P.S.; Chopra, H.K. Recent Trends on the Valorization Strategies for the Management of Citrus By-products. *Food Rev. Int.* **2021**, *37*, 91–120. [CrossRef]
8. Burdock, G.A. *Fenaroli's Handbook of Flavour Ingredients*, 3rd ed.; CRC Press: Boca Raton, FL, USA, 1995.
9. Oliveira, R.M.A.; Henriques, J.D.O.; Sartoratto, A.; Maciel, M.R.W.; Martinez, P.F.M. Evaluation of Limonene in sugarcane wax extraction. *Sustain. Chem. Pharm.* **2022**, *27*, 100657. [CrossRef]
10. Santiago, B.; Moreira, M.T.; Feijoo, G.; González-García, S. Identification of environmental aspects of citrus waste valorization into D-limonene from a biorefinery approach. *Biomass Bioenergy* **2020**, *143*, 105844. [CrossRef]
11. Ozturk, B.; Winterburn, J.; Gonzalez-Miquel, M. Orange peel waste valorisation through limonene extraction using bio-based solvents. *Biochem. Eng. J.* **2019**, *151*, 107298. [CrossRef]
12. Negro, V.; Mancini, G.; Ruggeri, B.; Fino, D. Citrus waste as feedstock for bio-based products recovery: Review on limonene case study and energy valorization. *Bioresour. Technol.* **2016**, *214*, 806–815. [CrossRef] [PubMed]
13. Braddock, R.J. *Handbook of Citrus By-Products and Processing Technology*; John Wiley & Sons: New York, NY, USA, 1999.
14. Wulf, G.; Mayer, B.; Lommatzsch, U. Plasma Co-Polymerization of HMDSO and Limonene with an Atmospheric Pressure Plasma Jet. *Plasma* **2022**, *5*, 44–59. [CrossRef]
15. Ren, S.; Trevino, E.; Dubé, M.A. Copolymerization of Limonene with n-Butyl Acrylate. *Macromol. React. Eng.* **2015**, *9*, 339–349. [CrossRef]
16. Zhang, Y.; Dubé, M.A. Copolymerization of 2-Ethylhexyl Acrylate and D-Limonene. *Polym. Technol. Eng.* **2015**, *54*, 499–505. [CrossRef]
17. Singh, A.; Kamal, M. Synthesis and characterization of polylimonene: Polymer of an optically active terpene. *J. Appl. Polym. Sci.* **2012**, *125*, 1456–1469. [CrossRef]
18. Mija, A.; Louisy, E.; Lachegur, S.; Khodyrieva, V.; Martinaux, P.; Olivero, S.; Michelet, V. Limonene dioxide as a building block for 100% bio-based thermosets. *Green. Chem.* **2021**, *23*, 9855–9859. [CrossRef]
19. Louisy, E.; Khodyrieva, V.; Olivero, S.; Michelet, V.; Mija, A. Use of Limonene Epoxides and Derivatives as Promising Monomers for Biobased Polymers. *ChemPlusChem* **2022**, *87*, e202200190. [CrossRef]
20. Coelho, F.M.; Vieira, R.P. Synthesis of Renewable Poly(limonene): A Kinetic Modeling Study to Improve the Polymerization. *Braz. Arch. Biol. Technol.* **2020**, *63*, e20200022. [CrossRef]
21. de Oliveira, E.R.M.; Vieira, R.P. Synthesis and Characterization of Poly(limonene) by Photoinduced Controlled Radical Polymerization. *J. Polym. Environ.* **2020**, *28*, 2931–2938. [CrossRef]

22. Moad, G.; Chiefari, J.; Chong, Y.K.; Krstina, J.; Mayadunne, R.T.A.; Postma, A.; Rizzardo, E.; Thang, S.H. Living free radical polymerization with reversible addition—Fragmentation chain transfer (the life of RAFT). *Polym. Int.* **2000**, *49*, 993–1001. [CrossRef]
23. Moad, C.L.; Moad, G. Fundamentals of reversible addition–fragmentation chain transfer (RAFT). *Chem. Teach. Int.* **2021**, *3*, 3–17. [CrossRef]
24. Destarac, M. On the Critical Role of RAFT Agent Design in Reversible Addition-Fragmentation Chain Transfer (RAFT) Polymerization. *Polym. Rev.* **2011**, *51*, 163–187. [CrossRef]
25. Carrodeguas, L.P.; Chen, T.T.D.; Gregory, G.L.; Sulley, G.S.; Williams, C.K. High elasticity, chemically recyclable, thermoplastics from bio-based monomers: Carbon dioxide, limonene oxide and ε-decalactone. *Green Chem.* **2020**, *22*, 8298–8307. [CrossRef]
26. Killops, K.L.; Campos, L.M.; Hawker, C.J. Robust, Efficient, and Orthogonal Synthesis of Dendrimers via Thiol-ene "Click" Chemistry. *J. Am. Chem. Soc.* **2008**, *130*, 5062–5064. [CrossRef]
27. Hoyle, C.E.; Lee, T.Y.; Roper, T. Thiol–enes: Chemistry of the past with promise for the future. *J. Polym. Sci. Part A* **2004**, *42*, 5301–5338. [CrossRef]
28. Baimark, Y.; Rungseesantivanon, W.; Prakymoramas, N. Synthesis of flexible poly(l-lactide)-b-polyethylene glycol-b-poly(l-lactide) bioplastics by ring-opening polymerization in the presence of chain extende. *e-Polymers* **2020**, *20*, 423–429. [CrossRef]

Article

Electrochemical Synthesis-Dependent Photoelectrochemical Properties of Tungsten Oxide Powders

Anastasia Tsarenko [1], Mikhail Gorshenkov [2], Aleksey Yatsenko [1], Denis Zhigunov [3], Vera Butova [4], Vasily Kaichev [5] and Anna Ulyankina [1,*]

[1] Research Institute "Nanotechnologies and New Materials", Platov South-Russian State Polytechnic University (NPI), 346428 Novocherkassk, Russia; tsarenkoanasteisha@yandex.ru (A.T.); alexyats-npi@yandex.ru (A.Y.)

[2] Department of Physical Materials Science, National University of Science & Technology (MISIS), 119049 Moscow, Russia; mvgorshenkov@gmail.com

[3] Center for Photonics and Quantum Materials, Skolkovo Institute of Science and Technology, 121205 Moscow, Russia; dmzhigunov@physics.msu.ru

[4] The Smart Materials Research Institute, Southern Federal University, 344090 Rostov-on-Don, Russia; butovav86@gmail.com

[5] Department of Investigation of Catalysts, Boreskov Institute of Catalysis, 630090 Novosibirsk, Russia; vvk@catalysis.ru

* Correspondence: anya-barbashova@yandex.ru

Abstract: A rapid, facile, and environmentally benign strategy to electrochemical oxidation of metallic tungsten under pulse alternating current in an aqueous electrolyte solution was reported. Particle size, morphology, and electronic structure of the obtained WO_3 nanopowders showed strong dependence on electrolyte composition (nitric, sulfuric, and oxalic acid). The use of oxalic acid as an electrolyte provides a gram-scale synthesis of WO_3 nanopowders with tungsten electrochemical oxidation rate of up to 0.31 $g \cdot cm^{-2} \cdot h^{-1}$ that is much higher compared to the strong acids. The materials were examined as photoanodes in photoelectrochemical reforming of organic substances under solar light. WO_3 synthesized in oxalic acid is shown to exhibit excellent activity towards the photoelectrochemical reforming of glucose and ethylene glycol, with photocurrents that are nearly equal to those achieved in the presence of simple alcohol such as ethanol. This work demonstrates the promise of pulse alternating current electrosynthesis in oxalic acid as an efficient and sustainable method to produce WO_3 nanopowders for photoelectrochemical applications.

Keywords: tungsten oxide; electrochemical synthesis; pulse alternating current; photoelectrochemical reforming

Citation: Tsarenko, A.; Gorshenkov, M.; Yatsenko, A.; Zhigunov, D.; Butova, V.; Kaichev, V.; Ulyankina, A. Electrochemical Synthesis-Dependent Photoelectrochemical Properties of Tungsten Oxide Powders. *ChemEngineering* **2022**, *6*, 31. https://doi.org/10.3390/chemengineering6020031

Academic Editor: Alírio E. Rodrigues

Received: 9 March 2022
Accepted: 13 April 2022
Published: 15 April 2022

1. Introduction

Photoelectrochemical (PEC) water splitting provides a great opportunity to produce environmentally friendly H_2 fuel by direct conversion of solar energy to establish a CO_2 zero emission society. The efficiency of PEC water splitting reaction, as a matter of fact, is quite low due to the recombination of the photogenerated charge carriers [1]. The enhanced efficiency of hydrogen evolution can be reached by introducing various sacrificial electron donor reagents in water for the irreversible reaction with h^+ from photogenerated electron-hole pairs [2].

The most used sacrificial agents for photo- and phoelectrochemical reforming of organic materials are simple alcohols, such as methanol, ethanol, etc. [3]. However, enormous amounts of waste containing glycerol and ethylene glycol are generated by industries, resulting in a serious environmental problem [4]. Moreover, involvement of biomass-derived compounds offers a feasible approach for both sustainable H_2 production and biomass valorization with renewable solar energy [5]. Lately, a PEC cell functioning in the presence of an organic sacrificial agent was referenced as a photoactivated fuel cell (PFC) [6]. PFC consists of a semiconductor photoanode that performs an organic oxidation reaction under illumination and a cathode where H_2 may be formed as a result of water or H^+ reduction. Although all PFC components can affect its performance, the nature of semiconductor electrode has an essential role in its functioning.

Up until now, a wide range of semiconductors have been explored as photoanodic materials, including metal oxides, nitrides, and sulfides. Most of the PEC studies have been mainly focused on metal oxides based photoanodes due to their low cost and environmental friendliness [7].

Tungsten oxide (WO_3) is considered as one of the most earth-abundant and promising candidates for the photoanode in PEC [8,9]. On the one hand, its relatively high absorption of solar light (up to 480 nm) due to a small band gap of 2.5–2.8 eV and its deep valence band position perfectly match the thermodynamic energy requirements needed to drive water/organics oxidation. On the other hand, WO_3 exhibits high stability in harsh acidic environment, a moderate hole diffusion length (~150 nm) compared to other common semiconductor oxides, and better electron transport (12 $cm^2 \cdot V^{-1} \cdot s^{-1}$) compared to TiO_2 (0.3 $cm^2 \cdot V^{-1} \cdot s^{-1}$) [10]. The photoelectrochemical efficiency of WO_3 is determined by its structure-dependent properties. The most widely used approach for the fabrication of diverse WO_3 nanostructures is wet-chemical synthesis. However, there is still a great challenge to control over their properties using a facile and green method [11]. Most of the employed techniques are complex, and thus less suitable for industrial applications. Recently, electrochemical synthesis of nanosized metal oxides has been considered as a sustainable and economically attractive method to produce highly efficient catalysts [12,13]. Electrosynthesis demonstrates many advantages compared with other physical and chemical methods such as low temperature, aqueous media, simple operation, etc. The most works have been focused on the anodization of tungsten as a valve metal to fabricate WO_3 thin films on the metal substrate [14,15]. However, there is a sense to develop electrochemical synthesis of WO_3 powders with controllable characteristics for various applications. There have been several reports about the production of WO_3 powder particles by an electrochemical route. Particularly, the preparation of $WO_3 \cdot 2H_2O$ nanoplatelets by anodization of tungsten under breakdown conditions was described elsewhere [16]. However, a study on the photoelectrochemical properties of WO_3 powders prepared in different electrolytes under pulse alternating current (PAC) is not reported in the literature yet. In our previous works, the effect of PAC synthesis conditions (current density and electrolytic media) on structural and functional properties of semiconductor materials such as ZnO [17] and CuO_x [18] was reported.

In the current study we demonstrate that WO_3 nanopowders can be successfully synthesized by safe, fast, and scalable electrochemical method using pulse alternating current. The influence of electrolytic media (oxalic, sulfuric, and nitric acid) on structural characteristics and electronic structure as well as photoelectrochemical properties of WO_3 nanoparticles was studied. These results highlight a novel electrochemical strategy to fabricate WO_3 powders for photoelectrochemical applications.

2. Materials and Methods

2.1. Preparation of WO_3 Powders

WO_3 nanopowders were synthesized by electrochemical method under pulse alternating current using different aqueous electrolyte solutions such as oxalic acid (0.5 M), sulfuric acid (0.5 M), and nitric acid (0.5 M). Electrochemical synthesis was executed at a current density ratio (j_a:j_c) of 3:3 $A \cdot cm^{-2}$ for 1 h with tungsten (W) foils (0.1 mm thickness) as the anode and cathode at a constant stirring speed of 200 rpm. The constant temperature of electrolyte was maintained using a cooling jacket. After synthesis the obtained powders were separated, washed with ethanol until neutral pH was achieved and dried. The as-prepared nanopowders were annealed in air using a muffle furnace at 500 °C for 3 h.

2.2. Characterization of WO₃ Powders

Morphologies of the powders obtained during tungsten electrochemical oxidation under pulse alternating current were characterized using a scanning electron microscope (SEM), JEOL, with an acceleration voltage of 30 kV and a JEOL JEM-2100 transmission electron microscope (TEM) operating at 200 kV. X-ray diffraction (XRD) measurements were performed to confirm the crystal structure using an ARL X'TRA X-ray diffractometer (Thermo Fisher Scientific Inc., Waltham, MA, USA) with Cu Kα radiation (λ = 1.5406 Å). The Rietveld refinement of the XRD data has been done using the JAVA based software namely Materials Analysis Using Diffraction (MAUD). Raman spectra were recorded on a DRX Raman microscope (Thermo Fisher Scientific Inc.) using a 532 nm laser for excitation. UV–vis reflectance spectra (UV-vis DRS) were recorded on a UV-2600 UV–visible spectrophotometer (Shimadzu) with an integrating sphere attachment. BET analysis was performed using a Micromeritics ASAP 2020 Physisorption BET instrument. Surface elemental composition and electronic valence band spectra were examined using X-ray photoelectron spectroscopy (XPS). The spectra were recorded on an X-ray photoelectron spectrometer (SPECS Surface Nano Analysis GmbH) using non-monochromatic Al Kα radiation (hv = 1486.6 eV). To analyze the presence of impurities, survey spectra were taken using the pass energy of 50 eV. The valence band, C1s, O1s, and W4f spectra, were obtained using the pass energy of 20 eV. The CasaXPS software was used to analyze the XPS spectra. Charge correction was performed by setting the C1s peak at 284.8 eV from adventitious hydrocarbon.

2.3. Photoelectrochemical Measurements

Fabrication of WO₃ photoanode was performed through a drop-casting method. First, 50 mg of the WO₃ nanopowder dispersed and ultrasonicated for 30 min in 2 mL of isopropanol was dropped onto a fluorine-doped tin oxide (FTO) glass substrate. Prior to drop casting, the FTO substrate was cleaned and ultrasonicated with acetone and then water. The fabricated WO₃ film was then dried at room temperature, annealed in air at 500 °C for 30 min, and used as the working electrode.

Photoelectrochemical properties of the WO₃/FTO photoanode were studied in a three-electrode electrochemical cell equipped with a quartz window using a P-45X potentiostat-galvanostat system (Elins, Russia) where Ag/AgCl (in 3.5 M KCl) and platinum wire were used as the reference and counter electrodes, respectively. The geometrical surface area of WO₃/FTO photoanode was equal to 1.0 cm². The measurements were performed in 0.5 M aqueous H₂SO₄ solution as the electrolyte. The impact of various sacrificial reagents on the photoelectrochemical activity of WO₃ was studied using the aqueous electrolyte, which contained 0.5 M H₂SO₄ and 5 mM of glycerol, ethylene glycol, glucose, and ethanol. A solar simulator equipped with a 500 W Xenon lamp was employed as the light source (AM 1.5G). The simulated solar illumination was adjusted to 1 sun (100 mW·cm⁻²). Linear sweep voltammograms (LSV) were recorded from 0 to 1.6 V vs. Ag/AgCl at a fixed scan rate of 10 mV·s⁻¹ with chopping light. The photocurrent density-time transients were measured at 0.8 V vs. Ag/AgCl. Open circuit potential (OCP) of the prepared photoanodes was determined in the dark and light illumination. Applied bias photon-to-current efficiency (ABPE) was calculated using the following equation:

$$\text{ABPE (100\%)} = J \cdot \frac{1.23 - E_{RHE}}{P_{light}} \cdot 100\%$$

where J is the photocurrent density, E_{RHE} is the potential versus RHE, and P_{light} is the simulated solar light intensity.

The electrochemical impedance measurements were carried out with a three-electrode system in 0.5 M H₂SO₄ electrolyte under the same illumination conditions in the frequency range of 0.1 Hz–50 kHz. Electrochemically active surface area (ECSA) determination was performed at scan rates of 10–50 mV s⁻¹ in 0.5 M H₂SO₄.

3. Results and Discussion

3.1. Characterization of the Powders

To investigate the microstructural characteristics of the products obtained during tungsten electrochemical oxidation under pulse alternating current, electron microscopy was employed. Figure 1a–f show the SEM images of the powders formed in the presence of nitric (WO_3-NA), oxalic (WO_3-OA), and sulfuric (WO_3-SA) acids and annealed at 500 °C. It is evident that the electrolyte has a great influence on the morphology and size of the prepared powders. In the presence of HNO_3, the product of W electrochemical oxidation is mainly composed of the large and well-formed plates with a length of 400–600 nm and thickness of 30–60 nm mixing with smaller nanoparticles with a side length of ~30 nm (Figure 1a,d). Similar morphology was obtained by authors [19] during W anodization at high temperature in strongly acidic electrolyte without the addition of complexing agent. Using H_2SO_4 as an electrolyte led to the disappearance of large plates (Figure 1b,e,h). The irregular nanoparticles (20–30 nm in size) are mainly present. However, a small number of particles with a length size of ~100 nm is still observed. The sample prepared in oxalic acid is consisted of the predominantly round-shaped particles with their side lengths distributed in the range of ~10–30 nm (Figure 1c,f,i). TEM, HRTEM, and the corresponding fast-Fourier-transform (FFT) images were demonstrated in Figure 1g–i.

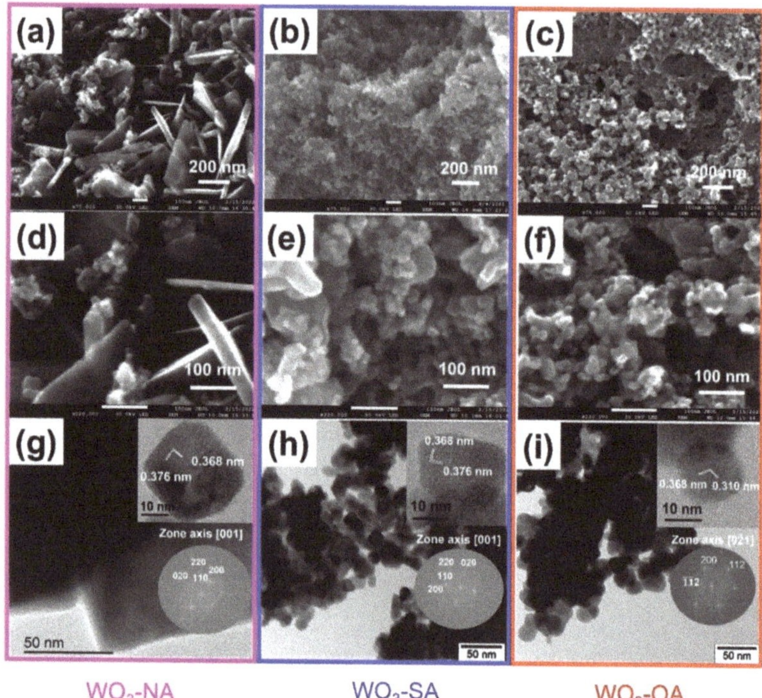

Figure 1. SEM (**a–f**), TEM (**g–i**), and corresponding HRTEM and FFT images (inset) of the products of tungsten electrochemical oxidation under pulse alternating current in nitric (**a,d,g**), sulfuric (**b,e,h**), and oxalic (**c,f,i**) acids.

For WO_3-NA and WO_3-SA samples, the measured lattice spacing of d = 0.376 nm and d = 0.368 nm can be assigned to the (020) and (200) crystal planes of the monoclinic WO_3 structure, respectively. For WO_3-OA, the lattice spacings of d = 0.361 and d = 0.310 can be assigned to the (200) and (112) planes of the monoclinic WO_3, respectively. The monoclinic phase was found to be the most stable and suitable for photoelectrochemical applications [20].

To examine the influence of electrolyte on the composition and crystalline structure of the products obtained during W electrochemical oxidation under pulse alternating current X-ray diffraction and Raman spectroscopy were used for additional characterization. Based on XRD results, the phase analysis of the obtained products is rather difficult due to non-Bragg scattering phenomena together with extremely broadened peaks. The overlapping of reflections is reported to be typical for such samples [21]. However, the XRD patterns of all the samples could be well fitted to the monoclinic WO_3 crystal phase with space group of P 2_1/n (ICSD # 80056) [21] that is consistent with the HRTEM data. Figure 2a shows the experimental patterns obtained for the three samples, along with the simulated diffractograms using the MAUD software. The crystallographic parameters obtained for all samples from the Rietveld refinements are reported in Table S1. The fitting data demonstrate the variation of WO_3 lattice parameters that can be due to morphological changes of the prepared powders [22]. The average crystallite size of the particles calculated using XRD data according to Debye-Scherer equation [23] was found to be 37.2, 18.6, and 13.3 nm for WO_3-NA, WO_3-SA, and WO_3-OA, respectively.

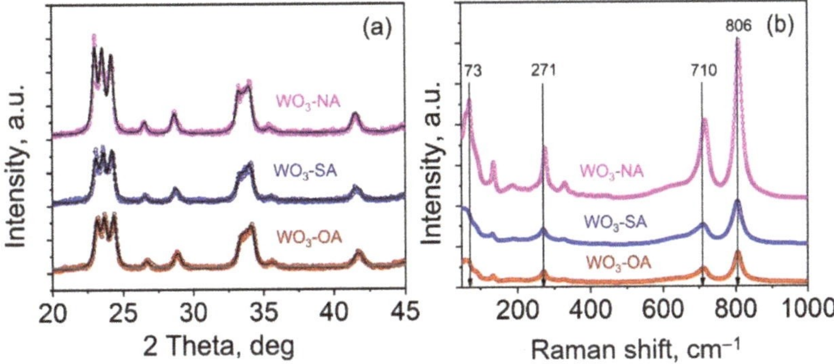

Figure 2. XRD patterns (**a**) and Raman spectra (**b**) of products obtained by electrochemical oxidation of tungsten under pulse alternating current using oxalic acid (WO_3-OA), sulfuric acid (WO_3-SA), and nitric acid (WO_3-NA). The measured diffraction data are shown by circles, and the calculated are shown by solid lines in black.

Raman spectroscopy is an effective way to study subtle change in the crystal structure due to high sensitivity to the phase transition and the structure defects [24]. The Raman spectra (Figure 2b) of all samples exhibited vibrations in three regions at 900–600, 400–200, and below 200 cm^{-1} which corresponds to the W–O stretching, W–O deformation, and lattice modes of the monoclinic WO_3 crystal phase, respectively. All the Raman spectra have been fitted by Lorentzian components (Figure S1). The bands at 271, 710, and 806 cm^{-1} were identified as the three strongest modes of monoclinic tungsten oxide. The Raman bands at 271 and 327 cm^{-1} correspond to O–W–O bending modes. The two sharp peaks at 710 and 806 cm^{-1} both result from O–W^{6+}–O stretching modes in the WO_6 octahedral unit [25].

The significant increase in intensity of the O–W–O Raman vibrations of WO_3-NA sample is most probably correlated with the degree of crystallinity and the particle size of the powders. The higher peak intensity reflects a considerable improvement in the structural order of the powders in terms of bond length and angle of the O–W–O bonding [26]. Some authors demonstrated that the weak and broad peaks in Raman spectra could be attributed to the local lattice imperfections, revealing the absence of partial O atoms in the WO_{3-x} structure [27]. Moreover, a blue shift of the peaks is evident when nitric acid is used for the WO_3 synthesis that can be also due to the improved crystallinity of the material or the crystallite size effect [26,28].

The surface elemental compositions of the synthesized WO_3 nanopowders were studied by X-ray photoelectron spectroscopy. The data are shown in Figures S2 and S3. The XPS scan (Figure S2a) indicates the presence of W, O, and C as main elements in the WO_3 samples. The C1s spectra in Figure S2b can be deconvoluted into three peaks at 284.8, 286.3–286.5, and 288.8–288.9 eV, and are due to C-C/C-H, C-OH/C-O-C, and O=C−C bonds, respectively. For all tested samples, the W4f spectra was approximated well by two spin-orbital $W4f_{7/2}$-$W4f_{5/2}$ doublets (Figure 3a). The main doublet with the $W4f_{7/2}$ binding energy of 35.5 eV corresponds to W^{6+} species and a less intense doublet with the $W4f_{7/2}$ binding energy of 33.8–33.9 eV can be probably ascribed to W^{5+} species [29,30]. The content of W^{5+} species in all samples calculated using the XPS data is 3–4%. The O1s peak can be approximated by three components (Figure 3b). The peak at the highest binding energy (530.2–530.4 eV) derives from the oxygen in the strong W–O bonds of WO_3. The O1s peak at 531.5–531.7 eV corresponds to the oxygen from the surface hydroxyl groups. The O1s peak at 532.9–533.1 eV is assigned to the oxygen from water molecules physiosorbed on the surface of oxide [29,31,32]. Furthermore, the nature of electrolytes used during synthesis exerts negligible influence on the chemical states of W and O. To determine the relative band edges of the WO_3 crystals, the valence band spectra were recorded and shown in Figure 3c. WO_3 synthesized in oxalic and sulfuric acids have a deeper VB maximum by 0.1 eV than WO_3 prepared in nitric acid, suggesting a stronger oxidation power of photoexcited VB holes in the former [20].

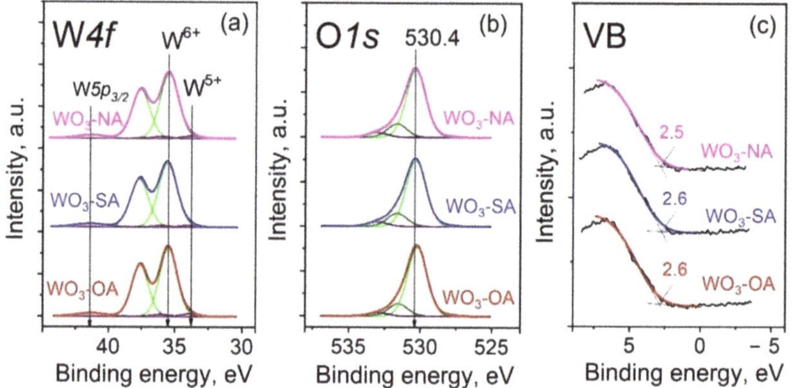

Figure 3. Deconvoluted high-resolution XPS spectra of W4f core level (**a**) and O1s core levels (**b**). XPS valence band edge plot of WO_3 prepared in different electrolytes (**c**).

Figure S3 shows the nitrogen adsorption–desorption isotherms of all samples prepared using different acidic electrolytes. According to the IUPAC classification, WO_3 nanoparticles exhibit type IV nitrogen physisorption isotherm with type H3 hysteresis loop, resulting from pore structures giving wedge-shaped or groove mesopores of 2–12 nm in diameter presumably formed by aggregates with plate-like nanoparticles [33]. The inset table in Figure S3 shows the corresponding SSA and pore volume values. It can be observed that SSA of WO_3-SA and WO_3-NA samples are quite similar (11.4 and 14.2 m^2/g, respectively). The reduction in surface area, which is probably due to a higher agglomeration effect, is more pronounced for WO_3-OA (7.58 m^2/g).

The UV-vis diffuse reflectance spectra of the WO_3 samples are shown in Figure 4a. It is evident that WO_3-OA sample exhibits enhanced visible light absorption in the range 460–800 nm compared to WO_3-SA and WO_3-NA. The more efficient harvesting of visible light is commonly ascribed to the possible modification in the crystal structure [34,35]. Generally, the photoactivity of the semiconductor materials is defined by the band gap energy. The optical band gap (E_g) of the prepared materials was extrapolated via the Kubelka-Munk method and extracted from the plot of $(F(R) \times h\nu)^{1/2}$ versus E ($h\nu$). The E_g values WO$_3$-OA, WO$_3$-SA, and WO$_3$-NA were 2.53, 2.55, and 2.62 eV, respectively (Figure 4b). The band gap values are consistent with reported ones for monoclinic WO$_3$ synthesized using chemical routes [22,35]. The difference in band gap can originate from the different morphology and orientation of crystal facets. It was reported earlier that electronic band structure of WO$_3$ can be tuned by formation of crystals with different morphology, such as quasi-cubic and rectangular sheet [20].

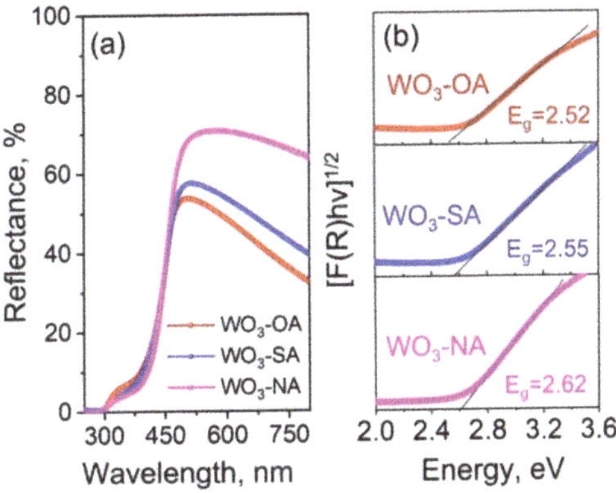

Figure 4. Diffuse reflectance (**a**) and band gap determination curves (**b**) of WO$_3$ prepared using different electrolytes.

3.2. Possible Formation Mechanism

The oxidation of W electrode during anodic pulse takes place indicating the fast formation of a compact isolating WO$_3$ layer according to the following reaction (1) [36]:

$$W + 3H_2O \rightarrow WO_3(s) + 6H^+ + 6e^- \tag{1}$$

Although WO$_3$ is thermodynamically stable in acidic solutions, this layer can undergo electrical-field-induced dissolution (2) by the action of protons, resulting in WO$_2{}^{2+}$ species formation.

$$WO_3 + 2H^+ \rightarrow WO_2^{2+} + 2H_2O \tag{2}$$

Due to its great affinity to oxygen, tungsten forms a metal/metal oxide/metal ion irreversible system. The chemical nature of the oxide layer on the W substrate is generally influenced by pH, temperature, concentration, presence of complexing agents, etc.

At high pH, electrochemical dissolution of the tungsten substrate through the oxide is also possible according to the reaction (3):

$$W + 2H_2O \rightarrow WO_2^{2+} + 4H^+ + 4e^- \tag{3}$$

The highest electrochemical oxidation rate of tungsten (up to 0.31 g·cm^{-2}·h^{-1}) was observed in the oxalic acid solution compared to nitric and sulfuric acids (Figure S4). Moreover, an increase in the oxalic acid concentration resulted in a higher dissolution rate of the WO$_3$ layer. These results imply that OA promoted the active dissolution of the substrate via complexation, forming soluble complexes between oxalate and tungsten released to the electrolyte. The formation of WO$_3$ films by anodization through a dissolution-precipitation mechanism in the presence of citric acid as an complexing agent was previously shown [9].

Upon reaching supersaturation conditions, the soluble species WO_2^{2+} precipitates on the electrode surface [37].

$$WO_2^{2+} + 2H_2O \rightarrow WO_3 \cdot H_2O + 2H^+ \tag{4}$$

The formation of orthorhombic tungsten oxide monohydrate ($WO_3 \cdot H_2O$ space group *Pmnb*, No. 62, Z = 4, a = 5.2477 Å, b = 10.7851 Å, c = 5.1440 Å) [38] as a product of W oxidation under pulse alternating current was confirmed by XRD for all used electrolytes (Figure S5). The electrolytic media influence the crystallinity of the as-prepared $WO_3 \cdot H_2O$. The lowest crystallinity confirmed by low and wide peaks in the XRD pattern was observed in the presence of oxalic acid. It can be explained by the decrease in the ordering effect in oxalic acid, which resulted in the formation of less compact structures [39].

During cathodic pulse, the hydrogen ions are reduced to hydrogen gas molecules:

$$2H^+ + 2e^- \rightarrow H_2 \tag{5}$$

Hydrogen bubbles, after reaching some critical size, detach from the electrode and move to the electrolyte/air interface and promote the removal of the already formed $WO_3 \cdot H_2O$ particles from the electrode surface [40]. Thereby, application of pulse alternating current (sequence of anodic and cathodic pulses with pauses between them) results in formation of microporous multi-layer film of $WO_3 \cdot H_2O$ which has micro-scale wicking channels (Figure S6). No WO_3 powder is observed in the electrolyte solution when an anodic pulse current of the same density is used.

3.3. Photoelectrochemical Performance

PEC properties of WO_3 nanoparticles synthesized in different electrolytes were evaluated by various electrochemical techniques in an aqueous electrolyte solution (H_2SO_4, 0.5 M). LSV curves under chopped illumination were recorded for all samples to illustrate the electrochemical synthesis-dependent photoelectrochemical properties of the prepared powders. H_2SO_4 was chosen as a supporting electrolyte considering the advantages of acidic media for the proton exchange membrane and the hydrogen evolution process. The anodic current for all samples in the dark is small as expected for semiconductor materials. The anodic current densities under simulated solar light are seen to follow the order WO_3-OA > WO_3-SA > WO_3-NA (Figure 5a). The WO_3-OA photoanode provided a substantial current density of 0.24 mA/cm^2 at 0.8 V vs. Ag/AgCl in the 0.5 M H_2SO_4 supporting electrolyte, as can be seen from transient photocurrent measurement (Figure 5b).

To evaluate intrinsic potential of PEC system for the electrode/electrolyte interfacial hole injection rate, the OCP of the prepared photoanodes was measured under light irradiation and in the dark. As shown in Figure S7a, d(OCP)/d(time) vs. time curve of WO_3-OA shows a sharper decay profile that indicates more efficient electron-hole separation. In comparison, d(OCP)/d(time) vs. time profile was recorded for WO_3-OA sample in the H_2SO_4/ethanol solution (Figure S7b). The adding of ethanol results in a faster decay profile compared with that in the H_2SO_4 solution due to suppressing the carrier trapping recombination [41]. Thereby, the most crucial thing was to understand how a given photoanode material responses to the different oxygenated species (sacrificial agent). For this reason, chopped-light LSV curves were recorded in aqueous solutions (5 mM) of glycerol, ethylene glycol, glucose, and ethanol (Figure 5c).

Adding of all sacrificial agents results in the extensive current increase under light irradiation. The highest value was obtained in the presence of ethanol. Raptis et al. [6] also observed the highest current in the presence of ethanol using WO_3 photoanodes. However simple alcohols such as ethanol are not ideal biomass-derived feedstocks for PEC reforming. In Figure 5e, the increase in photocurrent for sacrificial agent solution is provided. This enhancement can determine the selectivity towards photoelectrochemical reforming of organic compounds [42]. Excluding glycerol, increases in photocurrent were found to vary between 63 and 87%. Thereby, glucose as a biomass-derived feedstock can be oxidized at WO_3 photoanodes synthesized under pulse alternating current with current densities comparable to simple alcohols. The phenomenon of significant photocurrent enhancement can be attributed to the so-called current doubling effect at WO_3 photoelectrodes. It can be observed when two electrons are produced in the presence of semiconductor nanoparticles for one photon absorbed. In the case of ethanol, the following reactions (6) and (7) may occur [43]:

$$CH_3CH_2OH + h^+ \rightarrow CH_3CH_2O^\cdot + H^+ \tag{6}$$

$$CH_3CH_2O^\cdot \rightarrow CH_3CHO + H^+ + e^- \tag{7}$$

The applied bias photon-to-current efficiency (ABPE) is then carried out to quantitatively evaluate the performance of the electrodes (Figure S8). The maximum ABPE was calculated to be 0.2% at 0.8 V vs. RHE in the presence of H_2SO_4/ethanol solution.

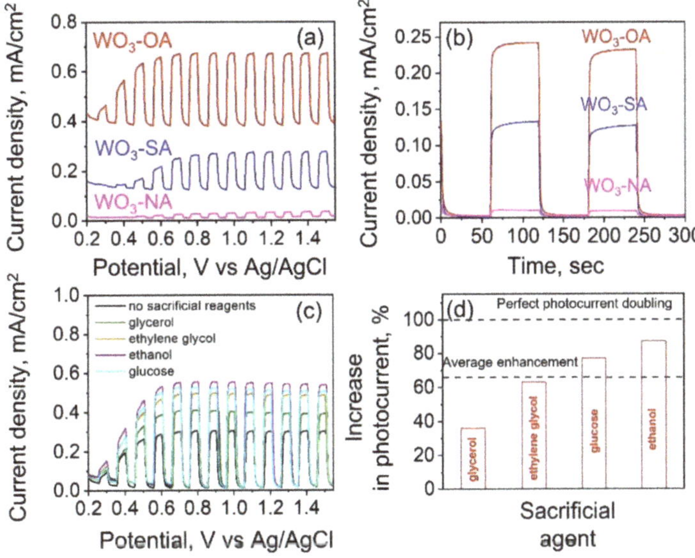

Figure 5. LSV curves (**a**) and transient photocurrent response at 0.8 V vs. Ag/AgCl (**b**) under chopped solar illumination in 0.5 M H_2SO_4; LSV curves (**c**) and increase in photocurrent (**d**) for WO_3-OA sample in 5 mM sacrificial agent solution compared with 0.5 M H_2SO_4 solution.

For a better understanding of differences in photoelectrochemical performance of WO_3, the electrochemical active surface area (ECSA) was estimated from the electrochemical double layer capacitance (C_{dl}) measurements. The non-Faradaic region was typically a 0.1 V window about OCP, and all measured currents in this region were assumed to be due to double-layer charging [44,45]. Figure S9a–c show the CV of WO_3 samples prepared in different electrolytes. The linear slope of capacitive current versus scan rate could be used to calculate the C_{dl} of the samples. The large slope represents the large ECSA. C_{dl} values followed the order WO_3-NA < WO_3-SA < WO_3-OA (Figure S9d) which are in accordance with the observed photoelectrochemical activity. The increased C_{dl} value indicated an enhanced electrochemically active surface area, which is beneficial for photoelectrochemical performance.

Electrochemical impedance spectroscopy (EIS) studies were carried out to investigate the charge transport behavior of the photoanode materials prepared in different electrolytes. Figure S10 presents the Nyquist impedance plots of the WO_3-OA, WO_3-SA, and WO_3-NA samples under illumination. The impedance spectra can be divided into a semicircle arc in the high frequency region and a straight line in the low frequency region. Equivalent circuit model was proposed to illustrate corresponding electrochemical system. R_s characterizes the ohmic resistance of electrolyte solution and FTO glass resistance. R_s was found to be approximately the same for all samples due to the similar experimental conditions (electrolyte and photoelectrochemical cell) [46]. CPE_{dl} represents a capacitance of double layer. R_{ct} is the transfer charge resistance at the electrode/electrolyte interface. Decrease in R_{ct} value indicates a lower resistance due to more efficient charge transfer and separation of photo generated electron–hole pairs. The estimated values of R_S and R_{ct} are 33.0, 32.6, 31.4, and 55.1, 69.3, 84.2 Ω, respectively, for WO_3-OA, WO_3-SA, and WO_3-NA. Z_W is the Warburg impedance which describes the diffusion transport in the electrolyte solution. In addition, the straight line obtained in the low frequency region is vertically more linear for WO_3-OA compared to other samples, indicating that the sample prepared in oxalic acid has a lower diffusion resistance.

4. Conclusions

In summary, we have described a simple, industrially applicable, and environmentally friendly synthesis of WO_3 nanopowders by electrochemical oxidation of tungsten under pulse alternating current in acidic electrolytes. Investigation of electrolyte composition (oxalic, sulfuric, and nitric acid) was performed to establish a relation between the characteristics of the obtained materials and their photoelectrochemical performance. We demonstrated that using nitric acid as an electrolyte results in formation of the mixture of plates with a length of 400–600 nm and nanoparticles with a side length of approximately 30 nm. Irregular and round-shaped nanoparticles with a side length of 10–30 nm were obtained in sulfuric and oxalic acids, respectively. The high tungsten oxidation rate of up to 0.31 g cm^{-2} h^{-1} was achieved in the oxalic acid solution. The maximum photocurrent density was found to be ~0.25 mA/cm^2 and 0.55 mA/cm^2 at 0.8 V vs. Ag/AgCl in H_2SO_4 and H_2SO_4/ethanol solution, respectively. Glucose is found to be a biomass-derived feedstock which can be oxidized at electrochemically synthesized WO_3 photoanodes with photocurrents comparable to ethanol. The enhanced PEC activity of tungsten oxide prepared in oxalic acid could be attributed to the optimal morphological, electronic, and charge-transfer properties. This work demonstrates the promise of gram-scale electrosynthesis in oxalic acid as an efficient and sustainable method to produce WO_3 nanopowders for photoelectrochemical applications.

Supplementary Materials: The following supporting information can be downloaded at: https://www.mdpi.com/article/10.3390/chemengineering6020031/s1, Table S1: Fitting data of XRD patterns of WO_3 samples using MAUD software; Figure S1: Lorentzian curve fitting for Raman spectra of WO_3-OA (a), WO_3-SA (b) and WO_3-NA (c) samples; Figure S2: Nitrogen adsorption–desorption isotherms of WO_3-OA (a), WO_3-SA (b) and WO_3-NA (c) samples; Figure S3: Effect of electrolyte concentration on tungsten electrochemical oxidation rate using different electrolytes; Figure S4: XRD patterns of the as-prepared powders by electrochemical oxidation of tungsten under pulse alternating current in comparison with the reference pattern of tungstite $WO_3 \cdot H_2O$; Figure S5: SEM image of the film structure obtained on tungsten electrode during electrochemical oxidation of W in oxalic acid under of pulse alternating current; Figure S6: OCP decay derivative of d(OCP)/d(Time) vs. time curves of WO_3 samples in H_2SO_4 (a) and in H_2SO_4/ethanol (b) solution; Figure S8: Applied bias photon-to-current efficiency (ABPE, %) as a function of applied potential for WO3-OA sample; Figure S9: CVs at different scan rates in the potential range of 0.52−0.72 V vs. the RHE (nonfaradaic region) in 0.5 M H_2SO_4 solution of (a) WO_3-OA, (b) WO_3-SA and (c) WO_3-NA; (d) Capacitive currents measured at 0.62 V vs. the RHE as a function of scan rates; Figure S10: Nyquist plots of a WO_3 samples obtained in different electrolytes. The inset shows the equivalent circuit used for fitting the data with the resistances R_s and R_{ct}, the Warburg impedance Z_w, the constant phase element CPE_{dl} and the capacitance C_l.

Author Contributions: Conceptualization, A.U.; Investigation, A.T., M.G., A.Y., D.Z., V.B. and V.K.; Methodology, A.U.; Writing—Original draft, A.T. and A.U.; Writing—Review & editing, A.U. All authors have read and agreed to the published version of the manuscript.

Funding: This research was funded by RUSSIAN SCIENCE FOUNDATION, grant number 21-79-00079.

Institutional Review Board Statement: Not applicable.

Informed Consent Statement: Not applicable.

Data Availability Statement: Not applicable.

Acknowledgments: The TEM study was carried out on the equipment of the Center Collective Use "Materials Science and Metallurgy" which was purchased with the financial support of the Russian Federation represented by the Ministry of Education and Science (No. 075-15-2021-696). The XPS study was conducted using the equipment of the Center of Collective Use "National Center of Catalyst Research". The authors also thank the Shared Research Center "Nanotechnologies" of Platov South-Russian State Polytechnic University (NPI) for XRD investigations.

Conflicts of Interest: The authors declare no conflict of interest.

References

1. Momeni, M.M.; Akbarnia, M.; Ghayeb, Y. Preparation of S–W-codoped TiO$_2$ nanotubes and effect of various hole scavengers on their photoelectrochemical activity: Alcohol series. *Int. J. Hydrogen Energy* **2020**, *45*, 33552–33562. [CrossRef]
2. Kurenkova, A.Y.; Markovskaya, D.V.; Gerasimov, E.Y.; Prosvirin, I.P.; Cherepanova, S.V.; Kozlova, E.A. New insights into the mechanism of photocatalytic hydrogen evolution from aqueous solutions of saccharides over CdS-based photocatalysts under visible light. *Int. J. Hydrogen Energy* **2020**, *45*, 30165–30177. [CrossRef]
3. Markovskaya, D.V.; Zhurenok, A.V.; Kurenkova, A.Y.; Kremneva, A.M.; Saraev, A.A.; Zharkov, S.M.; Kozlova, E.A.; Kaichev, V.V. New titania-based photocatalysts for hydrogen production from aqueous-alcoholic solutions of methylene blue. *RSC Adv.* **2020**, *10*, 34137–34148. [CrossRef]
4. López, C.R.; Melián, E.P.; Ortega Méndez, J.A.; Santiago, D.E.; Doña Rodríguez, J.M.; González Díaz, O. Comparative study of alcohols as sacrificial agents in H$_2$ production by heterogeneous photocatalysis using Pt/TiO$_2$ catalysts. *J. Photochem. Photobiol. A Chem.* **2015**, *312*, 45–54. [CrossRef]
5. Liu, X.; Wei, W.; Ni, B.-J. Photocatalytic and Photoelectrochemical Reforming of Biomass. In *Solar-to-Chemical Conversion*; WILEY-VCH: Weinheim, Germany, 2021; pp. 389–417.
6. Raptis, D.; Dracopoulos, V.; Lianos, P. Renewable energy production by photoelectrochemical oxidation of organic wastes using WO$_3$ photoanodes. *J. Hazard. Mater.* **2017**, *333*, 259–264. [CrossRef]
7. Wang, Y.; Zhang, J.; Balogun, M.S.; Tong, Y.; Huang, Y. Oxygen Vacancy-Based Metal Oxides Photoanodes in Photoelectrochemical Water Splitting. *Mater. Today Sustain.* **2022**, *18*, 100118. [CrossRef]
8. Shandilya, P.; Sambyal, S.; Sharma, R.; Mandyal, P.; Fang, B. Properties, optimized morphologies, and advanced strategies for photocatalytic applications of WO$_3$ based photocatalysts. *J. Hazard. Mater.* **2022**, *428*, 128218. [CrossRef]
9. Zhang, J.; Salles, I.; Pering, S.; Cameron, P.J.; Mattia, D.; Eslava, S. Nanostructured WO$_3$ photoanodes for efficient water splitting via anodisation in citric acid. *RSC Adv.* **2017**, *7*, 35221–35227. [CrossRef]
10. Shabdan, Y.; Markhabayeva, A.; Bakranov, N.; Nuraje, N. Photoactive Tungsten-Oxide Nanomaterials for Water-Splitting. *Nanomaterials* **2020**, *10*, 1871. [CrossRef]
11. Dutta, V.; Sharma, S.; Raizada, P.; Thakur, V.K.; Khan, A.A.P.; Saini, V.; Asiri, A.M.; Singh, P. An overview on WO$_3$ based photocatalyst for environmental remediation. *J. Environ. Chem. Eng.* **2021**, *9*, 105018. [CrossRef]
12. Izzi, M.; Sportelli, M.C.; Ditaranto, N.; Picca, R.A.; Innocenti, M.; Sabbatini, L.; Cioffi, N. Pros and Cons of Sacrificial Anode Electrolysis for the Preparation of Transition Metal Colloids: A Review. *ChemElectroChem* **2020**, *7*, 386–394. [CrossRef]
13. Gao, D.; Li, H.; Wei, P.; Wang, Y.; Wang, G.; Bao, X. Electrochemical synthesis of catalytic materials for energy catalysis. *Chin. J. Catal.* **2022**, *43*, 1001–1016. [CrossRef]
14. Zhang, T.; Paulose, M.; Neupane, R.; Schaffer, L.A.; Rana, D.B.; Su, J.; Guo, L.; Varghese, O.K. Nanoporous WO$_3$ films synthesized by tuning anodization conditions for photoelectrochemical water oxidation. *Sol. Energy Mater. Sol. Cells* **2020**, *209*, 110472. [CrossRef]
15. Fernández-Domene, R.M.; Sánchez-Tovar, R.; Lucas-Granados, B.; Roselló-Márquez, G.; García-Antón, J. A simple method to fabricate high-performance nanostructured WO$_3$ photocatalysts with adjusted morphology in the presence of complexing agents. *Mater. Des.* **2017**, *116*, 160–170. [CrossRef]
16. Wu, S.; Li, Y.; Chen, X.; Liu, J.; Gao, J.; Li, G. Fabrication of WO$_3$·2H$_2$O nanoplatelet powder by breakdown anodization. *Electrochem. Commun.* **2019**, *104*, 106479. [CrossRef]
17. Ulyankina, A.; Molodtsova, T.; Gorshenkov, M.; Leontyev, I.; Zhigunov, D.; Konstantinova, E.; Lastovina, T.; Tolasz, J.; Henych, J.; Licciardello, N.; et al. Photocatalytic degradation of ciprofloxacin in water at nano-ZnO prepared by pulse alternating current electrochemical synthesis. *J. Water Process Eng.* **2021**, *40*, 101809. [CrossRef]

18. Ulyankina, A.; Leontyev, I.; Maslova, O.; Allix, M.; Rakhmatullin, A.; Nevzorova, N.; Valeev, R.; Yalovega, G.; Smirnova, N. Copper oxides for energy storage application: Novel pulse alternating current synthesis. *Mater. Sci. Semicond. Process.* **2018**, *73*, 111–116. [CrossRef]

19. Sadek, A.Z.; Zheng, H.; Breedon, M.; Bansal, V.; Bhargava, S.K.; Latham, K.; Zhu, J.; Yu, L.; Hu, Z.; Spizzirri, P.G.; et al. High-Temperature Anodized WO_3 Nanoplatelet Films for Photosensitive Devices. *Langmuir* **2009**, *25*, 9545–9551. [CrossRef]

20. Xie, Y.P.; Liu, G.; Yin, L.; Cheng, H.-M. Crystal facet-dependent photocatalytic oxidation and reduction reactivity of monoclinic WO_3 for solar energy conversion. *J. Mater. Chem.* **2012**, *22*, 6746–6751. [CrossRef]

21. Pokhrel, S.; Birkenstock, J.; Dianat, A.; Zimmermann, J.; Schowalter, M.; Rosenauer, A.; Ciacchi, L.C.; Mädler, L. In situ high temperature X-ray diffraction, transmission electron microscopy and theoretical modeling for the formation of WO_3 crystallites. *CrystEngComm* **2015**, *17*, 6985–6998. [CrossRef]

22. Desseigne, M.; Dirany, N.; Chevallier, V.; Arab, M. Shape dependence of photosensitive properties of WO_3 oxide for photocatalysis under solar light irradiation. *Appl. Surf. Sci.* **2019**, *483*, 313–323. [CrossRef]

23. Efkere, H.İ.; Gümrükçü, A.E.; Özen, Y.; Kınacı, B.; Aydın, S.Ş.; Ates, H.; Özçelik, S. Investigation of the effect of annealing on the structural, morphological and optical properties of RF sputtered WO_3 nanostructure. *Phys. B Condens. Matter* **2021**, *622*, 413350. [CrossRef]

24. Li, Y.; Tang, Z.; Zhang, J.; Zhang, Z. Defect Engineering of Air-Treated WO_3 and Its Enhanced Visible-Light-Driven Photocatalytic and Electrochemical Performance. *J. Phys. Chem. C* **2016**, *120*, 9750–9763. [CrossRef]

25. Jin, B.; Wang, J.; Xu, F.; Li, D.; Men, Y. Hierarchical hollow WO_3 microspheres with tailored surface oxygen vacancies for boosting photocatalytic selective conversion of biomass-derived alcohols. *Appl. Surf. Sci.* **2021**, *547*, 149239. [CrossRef]

26. Vargas-Consuelos, C.I.; Seo, K.; Camacho-López, M.; Graeve, O.A. Correlation between Particle Size and Raman Vibrations in WO_3 Powders. *J. Phys. Chem. C* **2014**, *118*, 9531–9537. [CrossRef]

27. Li, Y.H.; Liu, P.F.; Pan, L.F.; Wang, H.F.; Yang, Z.Z.; Zheng, L.R.; Hu, P.; Zhao, H.J.; Gu, L.; Yang, H.G. Local atomic structure modulations activate metal oxide as electrocatalyst for hydrogen evolution in acidic water. *Nat. Commun.* **2015**, *6*, 8064. [CrossRef]

28. Abbaspoor, M.; Aliannezhadi, M.; Tehrani, F.S. Effect of solution pH on as-synthesized and calcined WO_3 nanoparticles synthesized using sol-gel method. *Opt. Mater.* **2021**, *121*, 111552. [CrossRef]

29. Colton, R.J.; Rabalais, J.W. Electronic structure to tungsten and some of its borides, carbides, nitrides, and oxides by x-ray electron spectroscopy. *Inorg. Chem.* **1976**, *15*, 236–238. [CrossRef]

30. Fleisch, T.H.; Mains, G.J. An XPS study of the UV reduction and photochromism of MoO_3 and WO_3. *J. Chem. Phys.* **1982**, *76*, 780–786. [CrossRef]

31. Ramana, C.V.; Vemuri, R.S.; Kaichev, V.V.; Kochubey, V.A.; Saraev, A.A.; Atuchin, V.V. X-ray Photoelectron Spectroscopy Depth Profiling of La_2O_3/Si Thin Films Deposited by Reactive Magnetron Sputtering. *ACS Appl. Mater. Interfaces* **2011**, *3*, 4370–4373. [CrossRef]

32. Al-Kandari, H.; Al-Kharafi, F.; Al-Awadi, N.; El-Dusouqui, O.M.; Katrib, A. Surface electronic structure–catalytic activity relationship of partially reduced WO_3 bulk or deposited on TiO_2. *J. Electron Spectrosc. Relat. Phenom.* **2006**, *151*, 128–134. [CrossRef]

33. Cheshme Khavar, A.H.; Moussavi, G.; Yaghmaeian, K.; Mahjoub, A.R.; Khedri, N.; Dusek, M.; Vaclavu, T.; Hosseini, M. A new Ru(ii) polypyridyl complex as an efficient photosensitizer for enhancing the visible-light-driven photocatalytic activity of a TiO_2/reduced graphene oxide nanocomposite for the degradation of atrazine: DFT and mechanism insights. *RSC Adv.* **2020**, *10*, 22500–22514. [CrossRef]

34. Liu, Q.; Wang, F.; Lin, H.; Xie, Y.; Tong, N.; Lin, J.; Zhang, X.; Zhang, Z.; Wang, X. Surface oxygen vacancy and defect engineering of WO_3 for improved visible light photocatalytic performance. *Catal. Sci. Technol.* **2018**, *8*, 4399–4406. [CrossRef]

35. Wang, L.; Wang, Y.; Cheng, Y.; Liu, Z.; Guo, Q.; Ha, M.N.; Zhao, Z. Hydrogen-treated mesoporous WO_3 as a reducing agent of CO_2 to fuels (CH_4 and CH_3OH) with enhanced photothermal catalytic performance. *J. Mater. Chem. A* **2016**, *4*, 5314–5322. [CrossRef]

36. Anik, M.; Cansizoglu, T. Dissolution kinetics of WO_3 in acidic solutions. *J. Appl. Electrochem.* **2006**, *36*, 603–608. [CrossRef]

37. Fernández-Domene, R.M.; Sánchez-Tovar, R.; Segura-Sanchís, E.; García-Antón, J. Novel tree-like WO_3 nanoplatelets with very high surface area synthesized by anodization under controlled hydrodynamic conditions. *Chem. Eng. J.* **2016**, *286*, 59–67. [CrossRef]

38. Seifollahi Bazarjani, M.; Hojamberdiev, M.; Morita, K.; Zhu, G.; Cherkashinin, G.; Fasel, C.; Herrmann, T.; Breitzke, H.; Gurlo, A.; Riedel, R. Visible Light Photocatalysis with c-WO_{3-x}/WO_3 × H_2O Nanoheterostructures In Situ Formed in Mesoporous Polycarbosilane-Siloxane Polymer. *J. Am. Chem. Soc.* **2013**, *135*, 4467–4475. [CrossRef]

39. Pham, N.L.; Luu, T.L.A.; Nguyen, H.L.; Nguyen, C.T. Effects of acidity on the formation and adsorption activity of tungsten oxide nanostructures prepared via the acid precipitation method. *Mater. Chem. Phys.* **2021**, *272*, 125014. [CrossRef]

40. Cherevko, S.; Kulyk, N.; Chung, C.-H. Pulse-reverse electrodeposition for mesoporous metal films: Combination of hydrogen evolution assisted deposition and electrochemical dealloying. *Nanoscale* **2012**, *4*, 568–575. [CrossRef]

41. Tang, R.; Wang, L.; Zhang, Z.; Yang, W.; Xu, H.; Kheradmand, A.; Jiang, Y.; Zheng, R.; Huang, J. Fabrication of MOFs' derivatives assisted perovskite nanocrystal on TiO_2 photoanode for photoelectrochemical glycerol oxidation with simultaneous hydrogen production. *Appl. Catal. B Environ.* **2021**, *296*, 120382. [CrossRef]

42. Esposito, D.V.; Forest, R.V.; Chang, Y.; Gaillard, N.; McCandless, B.E.; Hou, S.; Lee, K.H.; Birkmire, R.W.; Chen, J.G. Photoelectrochemical reforming of glucose for hydrogen production using a WO$_3$-based tandem cell device. *Energy Environ. Sci.* **2012**, *5*, 9091–9099. [CrossRef]
43. Kalamaras, E.; Lianos, P. Current Doubling effect revisited: Current multiplication in a PhotoFuelCell. *J. Electroanal. Chem.* **2015**, *751*, 37–42. [CrossRef]
44. McCrory, C.C.L.; Jung, S.; Peters, J.C.; Jaramillo, T.F. Benchmarking Heterogeneous Electrocatalysts for the Oxygen Evolution Reaction. *J. Am. Chem. Soc.* **2013**, *135*, 16977–16987. [CrossRef] [PubMed]
45. Nayak, A.K.; Verma, M.; Sohn, Y.; Deshpande, P.A.; Pradhan, D. Highly Active Tungsten Oxide Nanoplate Electrocatalysts for the Hydrogen Evolution Reaction in Acidic and Near Neutral Electrolytes. *ACS Omega* **2017**, *2*, 7039–7047. [CrossRef]
46. Yoo, S.J.; Lim, J.W.; Sung, Y.-E.; Jung, Y.H.; Choi, H.G.; Kim, D.K. Fast switchable electrochromic properties of tungsten oxide nanowire bundles. *Appl. Phys. Lett.* **2007**, *90*, 173126. [CrossRef]

 *chemengineering*

Article

Evaluating Electrochemical Properties of Layered Na$_x$Mn$_{0.5}$Co$_{0.5}$O$_2$ Obtained at Different Calcined Temperatures

Le Minh Nguyen [1], Van Hoang Nguyen [1,2,*], Doan My Ngoc Nguyen [1,3], Minh Kha Le [2,3], Van Man Tran [1,2,3] and My Loan Phung Le [1,2,3]

1 Applied Physical Chemistry Laboratory (APCLAB), VNUHCM-University of Science, Ho Chi Minh City 700000, Vietnam

2 Viet Nam National University Ho Chi Minh City (VNUHCM), Ho Chi Minh City 700000, Vietnam

3 Department of Physical Chemistry, Faculty of Chemistry, VNUHCM-University of Science, Ho Chi Minh City 700000, Vietnam

* Correspondence: nvhoang@hcmus.edu.vn

Abstract: P-type layered oxides recently became promising candidates for Sodium-ion batteries (NIBs) for their high specific capacity and rate capability. This work elucidated the structure and electrochemical performance of the layered cathode material Na$_x$Mn$_{0.5}$Co$_{0.5}$O$_2$ (NMC) with x~1 calcined at 650, 800 and 900 °C. XRD diffraction indicated that the NMC material possessed a phase transition from P3- to P2-type layered structure with bi-phasic P3/P2 at medium temperature. The sodium storage behavior of different phases was evaluated. The results showed that the increased temperature improved the specific capacity and cycling stability. P2-NMC exhibited the highest initial capacity of 156.9 mAh·g^{-1} with capacity retention of 76.2% after 100 cycles, which was superior to the initial discharge capacity of only 149.3 mAh·g^{-1} and severe capacity fading per cycle of P3-NMC, indicating high robust structure stability by applying higher calcination temperature. The less stable structure also contributed to the fast degradation of the P3 phase at high current density. Thus, the high temperature P2 phase was still the best in sodium storage performance. Additionally, the sodium diffusion coefficient was calculated by cyclic voltammetry (CV) and demonstrated that the synergic effect of the two phases facile the sodium ion migration. Hard carbon | | P2-NMC delivered a capacity of 80.9 mAh·g^{-1} and 63.3% capacity retention after 25 cycles.

Keywords: calcinated temperature; electrochemical performance; sodium-ion batteries; Na$_x$Mn$_{0.5}$Co$_{0.5}$O$_2$; P-type layered structure

Citation: Nguyen, L.M.; Nguyen, V.H.; Nguyen, D.M.N.; Le, M.K.; Tran, V.M.; Le, M.L.P. Evaluating Electrochemical Properties of Layered Na$_x$Mn$_{0.5}$Co$_{0.5}$O$_2$ Obtained at Different Calcined Temperatures. *ChemEngineering* **2023**, 7, 33. https://doi.org/10.3390/chemengineering7020033

Academic Editor: Lucun Wang

Received: 23 February 2023
Revised: 25 March 2023
Accepted: 31 March 2023
Published: 10 April 2023

1. Introduction

The planet warming due to the greenhouse gases exhausted from burning fossil fuels is the primary motivation for switching to renewable energy sources and electricity. In this context, electrochemical energy storage technologies such as rechargeable batteries, which are large storage capacity, easy to install and low cost, are crucial to adapt to the interruption of energy sources. For a long time, lithium-ion batteries (LIBs) have been the dominant power sources for various electronic and electric devices serving every aspect of modern life. Recently, the expansion of the electric vehicles market required more lithium-ion cells to produce enormous energy to ensure proper travel distance. Despite the large number of lithium-ion cells consumed worldwide and growing steadily every year, the predicted shortfall of lithium resources is still a massive challenge for the stability of LIBs uses and the downfall of its price [1]. Therefore, alternative power sources that lower costs and use more abundant and sustainable elements are ever needed. Many researchers were interested in LIB analogs using alkaline and earth elements such as sodium, potassium, aluminum and magnesium. Among them, sodium-ion batteries (SIBs) gained high maturity and were poised to commercial. Sodium-ion batteries' price will be lower than their lithium counterparts because of the abundance of raw materials, including sodium and metal elements [2]. However, the batteries' currently low energy density, about half of LIBs, limits their applications primarily to household and stationary markets.

Improvement in component engineering, especially cathode electrodes, is expected to improve the energy density of SIB. The aim of achieving high energy density SIBs could be gained by using layered structure oxides that conventionally possess high specific capacity. Sodium transition oxides typically crystallize in O3-type and P2-type layered structures, as named by Delmas and co-workers [3], in which the sodium ions were accommodated at octahedral and prismatic sites, respectively. P2-type materials give good reversible capacity, rate performance and longer cycle life compared to O3-type ones due to higher sodium diffusibility [4–10]. The unitary compounds such as Na_xMnO_2 and Na_xCoO_2 containing only one transition metal in the repeating $(MO_2)_n$ sheets were intensively investigated [11,12], but their poor performance prevented them from being considered as cathode materials for SIBs. Introducing extra ions through substitution and doping was verified to improve the electrochemical properties via synergistic effect. Up to now, abundant multicomponent materials with P2-type layered structures were evaluated during the development of cathode materials for SIBs [4–10].

Among several combinations built for sodium insertion cathode such as Fe-Mn, Ni-Mn, Ni-Mn-Co, Fe-Ni-Mn, Mn-Co incorporation is attractive for its specific capacity and rate capability [13–21]. For instance, $P2-Na_{2/3}[Mn_{0.8}Co_{0.2}]O_2$ illustrated excellent electrochemical performance with a capacity of about 175 mAh·g^{-1} at 0.1C, and over 90% of its initial capacity remained after 300 cycles at 0.1C and 10C [22]. For sodium-deficient layered cathodes, P3 and P2 were low and high-temperature phases, respectively [23]. Typically, Chen and co-workers reported the preparation of $P2-Na_{0.67}Co_{0.5}Mn_{0.5}O_2$ and $P3/P2-Na_{0.66}Co_{0.5}Mn_{0.5}O_2$ at 950 °C and 800 °C, respectively, through a facile and straightforward sol-gel route [19,24]. The P2-phase material delivered a high discharge capacity of 147 mAh·g^{-1} at 0.1C rate and excellent cyclic stability with nearly 100% capacity retention over at least 100 cycles at 1C [19]. The P2/P3-phase material exhibited an impressively higher discharge capacity of 156.1 mAh·g^{-1} at 1C in the voltage range of 1.5–4.3 V [24]. It was noticed that a change in synthesis temperatures affected the macro and nanostructure of the cathode materials related to the electrochemical properties. The two phases intergrowth cathodes showed the improvement in electrochemical performance but the proper ratio of the two phases was not considered.

In this work, we evaluated the structure and the electrochemical properties relationship of $Na_xMn_{0.5}Co_{0.5}O_2$ (NMC) obtained at 650, 800 and 900 °C. We started from a sodium/transition element ratio of unity that was the same as the O3-type phases to lower the mixture's temperature and improve the diffusion in solid state reaction. XRD analysis of the samples indicated that the phase transformation from P3 to P2 phase with intermediate P3/P2 phase coexist as the calcination temperature increased. The electrochemical evaluation suggested that high temperature was better for this material to maximize performance in SIBs due to its own structure stability. Additionally, the synergic effect of two-phase coexist in the P3/P2-NMC enhanced the sodium migration. Full-cell was made using P2-NMC, as the cathode delivered the initial discharge capacity of 80.9 mAh·g^{-1} and remained 63.3% of that value after 25 cycles.

2. Materials and Methods

$NaMn_{0.5}Co_{0.5}O_2$ (NMC) was prepared by the conventional solid-state reaction. The starting precursor mixture including Na_2CO_3 (Merck, >99%), $MnCO_3$ (Sigma-Aldrich, >99%) and $Co(NO_3)_2 \cdot 6H_2O$ (Sigma-Aldrich, >99%) in a Na/Mn/Co ratio of 2.1:1:1, respectively, was mixed with 1 mL distilled water to make a homogeneous slurry. The slurry was stored for about 30 min to stable and was then heated at 500 °C for 12 h in the air to decompose before being grounded to obtain precursor powder. The next heat treatment step was calcinating the precursor bronze between 650 °C and 900 °C for 12 h in the Ar atmosphere to gain the final materials. In this step, the materials were removed from the furnace at calcinated temperature to cool rapidly in the air and, then, transferred to an

Ar-filled glovebox (GP-Campus, Jacomex). The final samples were named after the calcined temperatures, which were NMC-650, NMC-800 and NMC-900.

The crystal structure of the synthesized NMC samples was examined using X-ray diffraction performed on D8 Advanced (Bruker) diffractometer equipped with a CuKα X-ray source (λ = 0.15418 nm). The measurement was conducted in a 2θ angle range from $10°$ to $70°$ with $0.02°$/step/0.25 s scan rate. In addition, the lattice parameters were obtained from Rietveld refinement using Material Studio (version 2017).

The Raman spectrum was collected on Xplora One Raman Spectrometer (Horiba) equipped with Ar$^+$ laser with a wavelength of 532 nm. The morphologies of NMC materials were evaluated using FE-SEM S-4800 (Hitachi) scanning electron microscope with energy-dispersive X-ray spectroscopy (EDS) coupled optionally to explore the chemical composition and elements distribution.

The cathode electrode was prepared in the Ar-filled glovebox. The as-prepared NMC material, conducting carbon C65 and poly(vinylidene fluoride-co-hexafluoropropylene) (PVDF-HFP) binder in a weight ratio of 80:15:5, respectively, was mixed with N-methyl-2-pyrrolidone (NMP) solvent under magnetic stirring for at least 2 h to make a slurry. Then, the slurry was coated onto technical Al foil, dried at 80 °C in the glovebox before being placed in a vacuum oven at 100 °C for 12 h. Finally, the foil was moved to the glovebox and was punched into round pieces with a diameter of 12.7 mm and mass loading of 2–3 mg·cm^{-2}.

For the anode, commercial hard carbon (Kureha Battery Materials, Tokyo, Japan) (90%), carbon C65 (5%) and carboxylmethyl cellulose (CMC) (10% solution in water) were mixed under magnetic stirring for at least 2 h to obtain a homogeneous slurry. The slurry was coated on a Cu foil, dried at 100 °C for 12 h in the vacuum oven and was then punched into round electrodes with 12.7 mm-diameter and mass loading of 1 mg·cm^{-2}.

Coin cells CR2032 were used in electrochemical properties characterization. The cells were assembled in the Ar-filled glovebox with the cathode film as working electrode, a circular Na foil as reference and counter electrodes, two Whatman glass fibers (GF/F) as separator, and the electrolyte was a solution of 1 M NaClO$_4$ in propylene carbonate (PC) with 2% fluoroethylene carbonate (FEC) additive. The full-cell was assembled using coin cell CR2032 with the same structure as above, but hard carbon was used as the counter electrode.

Electrochemical testing was controlled by multi-channel MPG-2 and VSP battery cycler (Biologic, France). Cyclic voltammetry (CV) was conducted with a scan rate of 20–100 μV·s^{-1} between 1.5 and 4.5 V. Charge–discharge performance was processed in Galvanostatic mode with different C-rate (1C corresponding to a nominal specific capacity of 150 mAh·g^{-1}). Full-cell was cycled with a current density of 1C (that was calculated based on the cathode) between 0.1 and 4.5 V. Electrochemical impedance spectroscopy (EIS) was measured by applying an alternating potential with an amplitude of 10 mV in the frequency range from 1 MHz to 100 mHz at equilibrium potential.

For the XRD analysis to evaluate the structure during charging, the cells were charged by a current density of C/20 for 5 cycles and were then rested for at least 3 h before being opened in the glovebox to obtain the cathode films. The cathode films were washed in dimethyl carbonate (DMC) at least three times before being dried in a hot plate at 100 °C and sealed. The, they were subjected to XRD analysis on D8 Advance (Bruker) coupled with a CuKα source (λ = 1.5814 Å) and the scan rate was $0.02°$/step/0.25 s.

3. Results

Figure 1 shows the thermal behavior of the precursor between room temperature and 1000 °C in the dry air. It was observed in the TGA curve that the precursor exhibited several mass losses, which started from about 80 °C and lasted until the end of the investigated temperature range. The DSC curve displayed a sloping curve, which demonstrated endothermal reactions. The first mass loss of about 2% with a slope at roughly 83 °C in the TGA curve was of water desorption, corresponding to the first peak in the DSC curve.

Additionally, the acceleration of mass loss was observed as starting from roughly 200, 550 and 750 °C, corresponding to the continued decomposition of residual anions/metal salts, oxygen removal, and sodium evaporation under an oxidation atmosphere. During this time, the solid-state reaction between the components was also activated, and the reconstruction of the precursor happened. It was noticed in the TGA that the decomposition seemed to finish at around 800 °C, where no significant mass loss was observed.

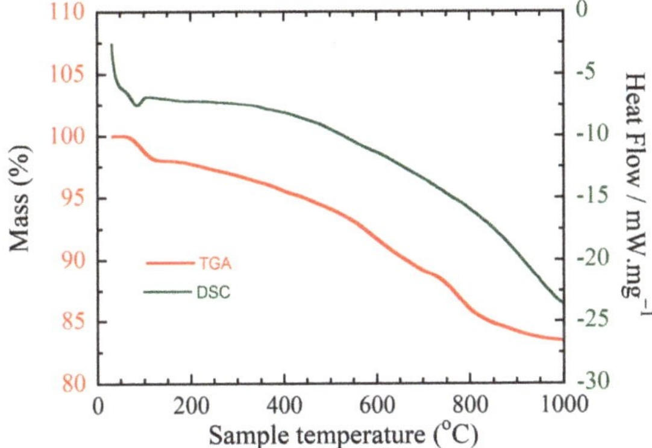

Figure 1. TG/DSC curves of the precursors.

The XRD of the samples at 650 °C, 800 °C and 900 °C (Figure 2a–c) revealed the structure evolution from P3 to P2 during the temperature increase and that the mass loss on TGA might have come with structure reconstruction. For instance, the P3-type layered structure was observed in the sample at 650 °C, while the second phase P2 was seen in the samples above 800 °C. The formation of P3 phase was at a lower temperature than the previously studied P3-phase cathodes that calcined at 700/750 °C [25–27]. The P3 structure obtained Cm space group (PDF#00-054-0839). On the other hand, the P2-type layered structure with P6$_3$/mmc space group (PDF#01-070-3726) was gained in the sample NMC-900. The presence of P2 phase at a high temperature of 900 °C coincided with previous studies because the phase was more stable at high temperatures than P3 phase [23,28,29].

The Rietveld refinement results for structure analysis of the materials shown in Figure 2a–c revealed a good match between the simulation and the experiment patterns with a depleted different curve. The reasonably low R_{wp} (11.23, 9.38, and 6.64% for the calcinated temperature of 650, 800 and 900 °C, respectively) showed that the higher temperature exhibited less mismatch due to higher crystallinity. In addition, it was observed that almost all the samples were high purity because of the absence of unidentified peaks. The Rietveld refinement results also indicated that the percentage of the P2 phase was slightly higher compared to the P3 phase when calcined at 800 °C. Table 1 lists the lattice parameters of different phases in the materials. The calculated results of a, c axes and the volume V did not possess significant differences in the same phases.

The Raman spectra for the three samples with different temperatures shown in Figure 2d were similar to those obtained from Na$_{2/3}$Co$_{2/3}$Mn$_{2/3}$O$_2$ [30], but the peaks shifted forward due to the difference in sodium concentration and element composition. The spectra presented two peaks at 630 and 581 cm^{-1}, which were attributed to the deformation and symmetric stretching vibrations (A$_{1g}$ and E$_g$, respectively) of metal-oxygen (Mn/Co–O) bonds of Mn/CoO$_6$ octahedral in layered Na-metal transition oxides [30,31]. The peaks at 488 and around 800 cm^{-1} could be explained by impurity (other phase structure and metal oxides) or sublattice. It was induced that the P3 and P2 layered structure of NMC material were probably distinguishable by Raman spectra, because of the different between the spectra of the samples calcined at 650 and 900 °C. Namely, the peak splitting at 488 cm^{-1} and the peak shifting from 800 cm^{-1} to high wavelength were seen in the spectra of the 900 °C sample, which were expected to differ the P2 phase from the P3 phase.

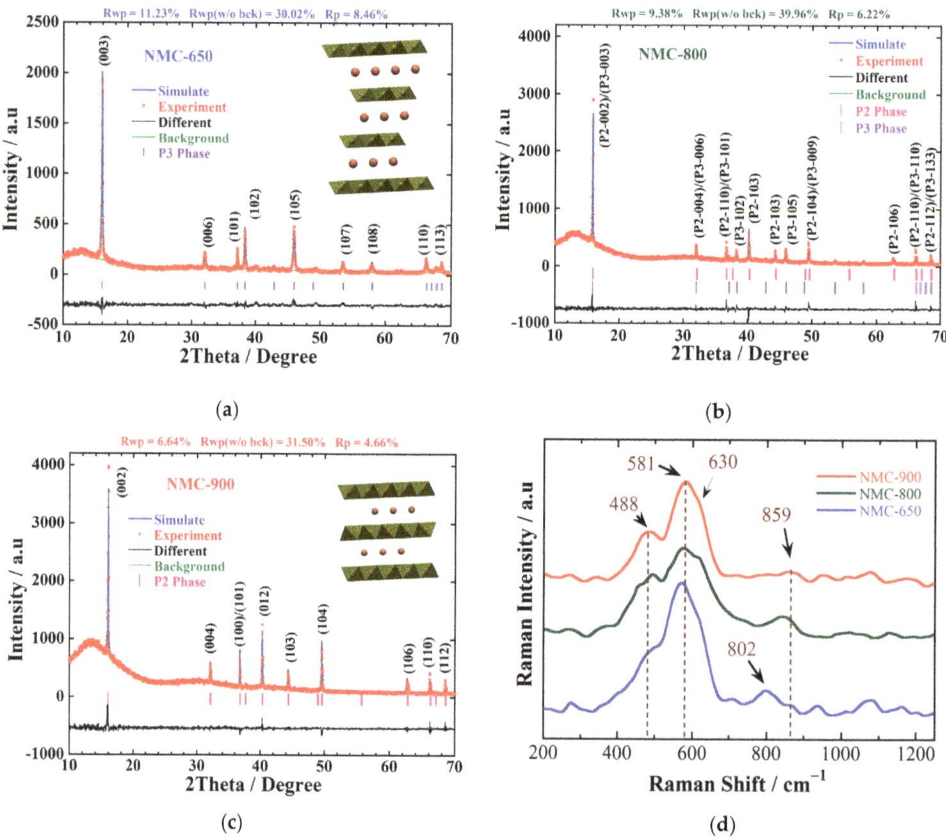

Figure 2. Rietveld refinement results of the NMC materials obtained at various calcined temperatures (**a**) NMC-650, (**b**) NMC-800 and (**c**) NMC-900, (**d**) Raman spectra of the synthesized materials. Inset of (**a**,**c**) shows the schematic of crystal structures of P3 and P2-type layered oxide, respectively.

Table 1. Crystallographic parameters of $Na_xMn_{0.5}Co_{0.5}O_2$ materials prepared at different calcination temperatures of 650, 800 and 900 °C.

Sample ID	Notation	Space Group	Lattice Parameters			Weight Percentage (Rietveld)
			a (Å)	c (Å)	V (Å³)	
NMC-650	P3	Cm	2.831	16.530	114.730	100.00
NMC-800	P2	P63/mmc	2.823	11.192	77.258	58.60
	P3	Cm	2.827	16.735	115.799	41.40
NMC-900	P2	P63/mmc	2.825	11.178	77.279	100.00

SEM images of P2- and P3-NMC (Figure 3) showed irregular-shaped polygonal particles with a size of a few micrometers made up of the aggregation of flaky particles with smaller dimensions. The sample morphologies were almost similar, but the particle size of P2-NMC was relatively larger than P3-NMC due to higher calcined temperature. EDS images of element contribution confirmed the presence of all essential elements in the samples.

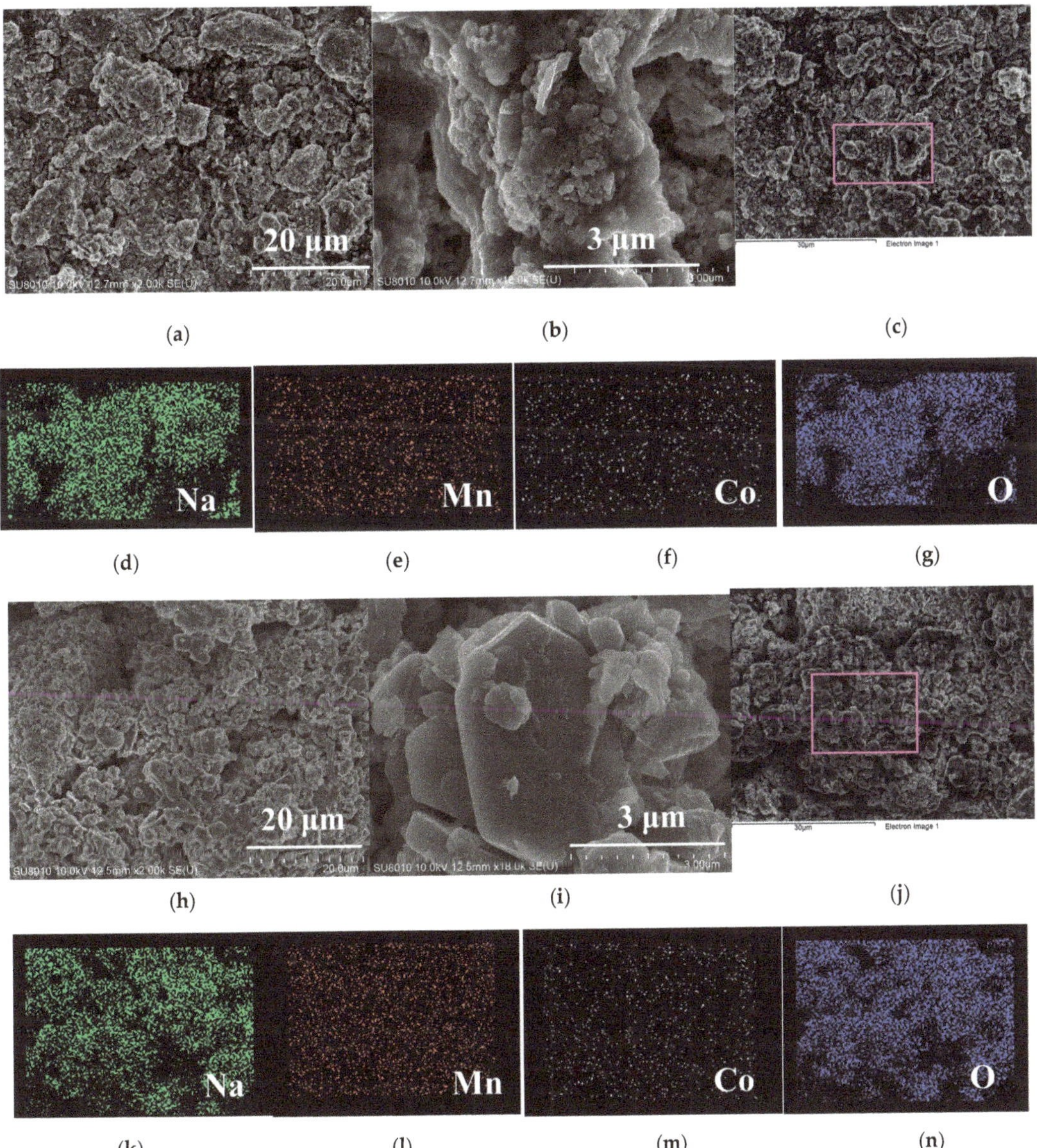

Figure 3. (**a–c**) SEM images of NMC material calcined at 650 °C and (**d–g**) EDX mapping results of area in the SEM image (**c**); (**h–j**) SEM images of NMC material calcined at 800 °C and (**k–n**) EDX mapping results of area in the SEM image (**j**).

The electrochemical properties characterization of the materials is presented hereafter. CV curves of the synthesized samples are given in Figure 4a. Typically, the CV curves showed many peak couples corresponding to several successive phase transitions and structure evolution during sodiation/desodiation. For the voltage range 1.5 to 4.5 V, the CV curves could be divided into three regions at about 2.7 and 4.3 V knots. The first and the second regions had peaks that mainly contributed to Mn^{4+}/Mn^{3+} and Co^{4+}/Co^{3+} redox couples, respectively, while the last one contributed by Co^{4+}/Co^{3+} redox reaction and decomposition of the electrolyte when being charged up to 4.5 V. Additionally, the CV curve of the sample calcined at 900 °C was smoother, and the peak couples were rather more reversible. It could be noticed that the CV curve of NMC-800 was a superposition of the others because both P3 and P2 phases existed. In general, the CV feature of P-type layered NMC was almost the same, regardless of the difference in the crystal lattice, probably because they had the same redox species and composition.

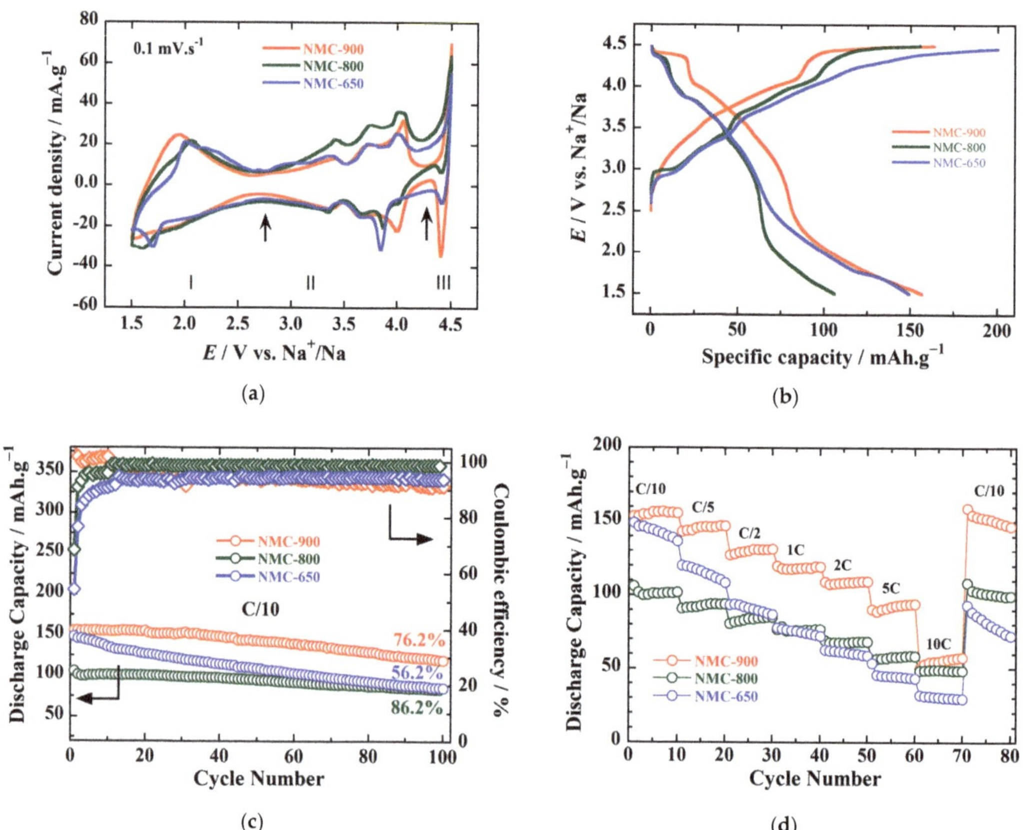

Figure 4. (**a**) Cyclic voltammogram of NMC materials in 1.5–4.5 V voltage windows at a scan rate of 100 µV·s^{-1}. (**b**) The first charge–discharge voltage profiles, (**c**) evolution of capacity and Coulombic efficiency with cycle number of the material cycled in a voltage range of 1.5–4.5 V and (**d**) rate capability in the same voltage limit.

The galvanostatic charge/discharge curves of the initial cycle are shown in Figure 4b. All the samples showed characteristic voltage profiles that were rather alike with relatively high discharge capacities, coinciding with the electrochemical behavior shown in CV curves. However, the samples NMC-650 and NMC-800 exhibited high irreversible capacity in the first cycle, resulting in low Coulombic efficiency with the value of about 74.2% and 68.5% for NMC-650 and NMC-800, respectively. It could be noticed that NMC-800, with a complex composition including P3 and P2 phase, presented an even higher irreversible capacity than the P3 phase and P2 phase components. Meanwhile, the sample NMC-900 exceptionally displayed a nearly reversible charge/discharge cycle with Coulombic efficiency of about 96.6%. In general, the P-phases with sodium deficiency typically displayed high Coulombic efficiency above 100% [4,8,19]. Thus, the materials obtaining P2 phase with a high sodium content of 1, such as the O3 phase, gained high Coulombic efficiency due to the excess Na^+ ions that were possible due to the electrolyte.

Moreover, the redox couple Co^{4+}/Co^{3+} was activated when charged to a high cutoff voltage and contributed to the large irreversible charge capacity of the samples NMC-650 and NMC-800. For the sample NMC-900, the huge sodium loss via evaporation due to high calcined temperature reduced the Coulombic efficiency. As a result, the initial discharge capacity of 149.3, 106.2 and 156.9 mAh·g^{-1} were obtained for the sample NMC-650, NMC-800 and NMC-900, respectively; hence, the Coulombic efficiency were 74.6, 68.4 and 89.8%, respectively.

Figure 4c shows the capacity and Coulombic efficiency of the samples during long-term charge–discharge. It was observed that the samples NMC-800 and NMC-900 attained stable cycling that slightly changed in capacity for each cycle. The sample NMC-900 remained a capacity higher than the others, with a capacity retention of 76.2% after 100 cycles. On the other hand, the capacity of NMC-650 reduced gradually and the capacity retention was about 56.2% after 100 cycles. The sample NMC-800 had a two-phase component, but capacity retention was the best of the three, with a value of 86.2% after 100 cycles. Additionally, the NMC-800 and NMC-900 samples possessed a slightly capacity increase in the first dozen cycles that could be assigned to the increase in active surface and active site of sodium that were mentioned elsewhere [32,33] and obviously contributed to the higher capacity retention of these materials. This could be related to the enhanced electrolyte permeability and the reconstruction of the electrode and the interface.

The rate capability of the samples is illustrated in Figure 4d. It was observed that the three samples displayed acceptable rate capability. Of the three, the samples NMC-650 and NMC-800 exhibited good capacity retention in which the capacities dropped steadily and were negligible when the current density increased from C/10 to 10C. Meanwhile, the capacity of NMC-900 fell significantly at the high current density of above 5C. However, when the current density returned to C/10, only NMC-650 showed a sharp decrease in capacity compared to the initial value at the same current density. Furthermore, both NMC-900 and NMC-800 possessed the capacity increase in the first few cycles at different C-rates. The extended capacity could contribute to the high capacity recovery that was even higher than the capacity at C/10 at the beginning when the current returned to C/10 after cycling at different C-rates.

The long-term charge–discharge at higher current density was applied to characterize the stability of electrodes (Figure 5). Generally, the capacities of the three samples were stable during the cycle with low-capacity decay that coincided with the rate capability test. It was observed that the capacity retention at the high rate was better because of shortened retention time in high voltage regions and the contribution of capacity from the high voltage redox couple [21]. NMC-650 displayed insufficient capacity and capacity retention compared to the other two in all the test conditions, while NMC-900 performed the best. Additionally, the Coulombic efficiency of approximately 100% was gained by all the samples at various rates.

Figure 6a,b show the XRD patterns of the bare electrodes NMC-650 and NMC-800 and the electrodes at the fifth cycle. Compared with the XRD of the bare NMC-650 and NMC-800 shown in Figure 2a,b, respectively, it could be seen that the samples retained the original structure and phase component, revealing the good structure stability. Furthermore, the fading of high redox couples during the first several cycles seemed not to be correlated with the appearance of a new phase or structure changes. Thus, the results demonstrated that P-type layers displayed high structural stability. Our previous work also indicated that NMC-900 could remain a P2-type layered structure after 50 cycles of charging-discharging [21]. Typically, P3 phase exhibits phase transformation P3-O3 near the end of discharge [34–36], while P2 phase remains its structure during the cycling [19,37]. Further work will be carried out to analyze the phase transition of the different phases during the cycling.

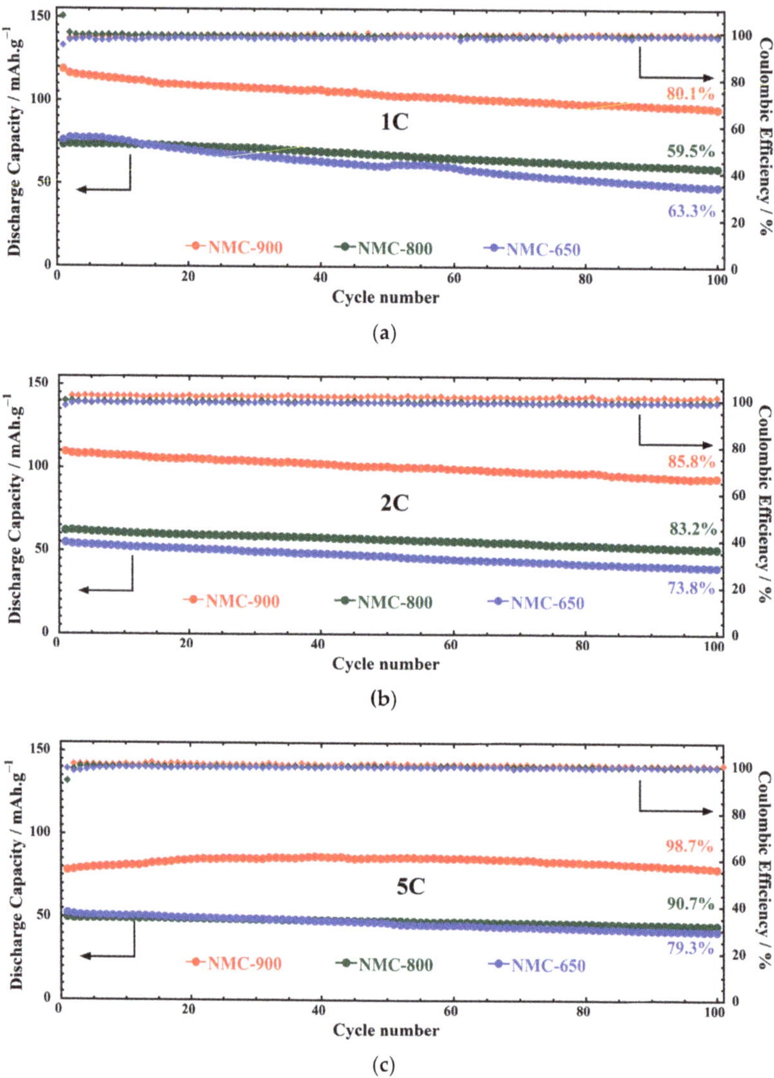

Figure 5. Electrochemical performances of NMC cathodes upon cycle tests in the range of 1.5–4.5 V at (**a**) 1C, (**b**) 2C and (**c**) 5C, respectively.

EIS spectra of the electrodes were also collected at the beginning of the cycling test and at the fifth cycle as shown in Figure 6c,d. The EIS spectra showed a semicircle at high and medium frequencies of the charge transfer resistance and one huge semicircle at low frequency due to ion diffusion in the porous electrode [38]. Typically, the high frequency charge transfer process could be simulated by a single Randles circuit including a capacitance and a resistance in series, while the low frequency semicircle could be simulated by a series combination of capacitances and resistors [39]. The equivalent circuit used to simulate the charge transfer resistance is shown in Figure 6e, showing two charge transfer processes which, in turn, occurred on the cathode interface film (R_1) and cathode electrolyte interface (R_2). Table 2 provides the value of resistance parameters obtained after simulation.

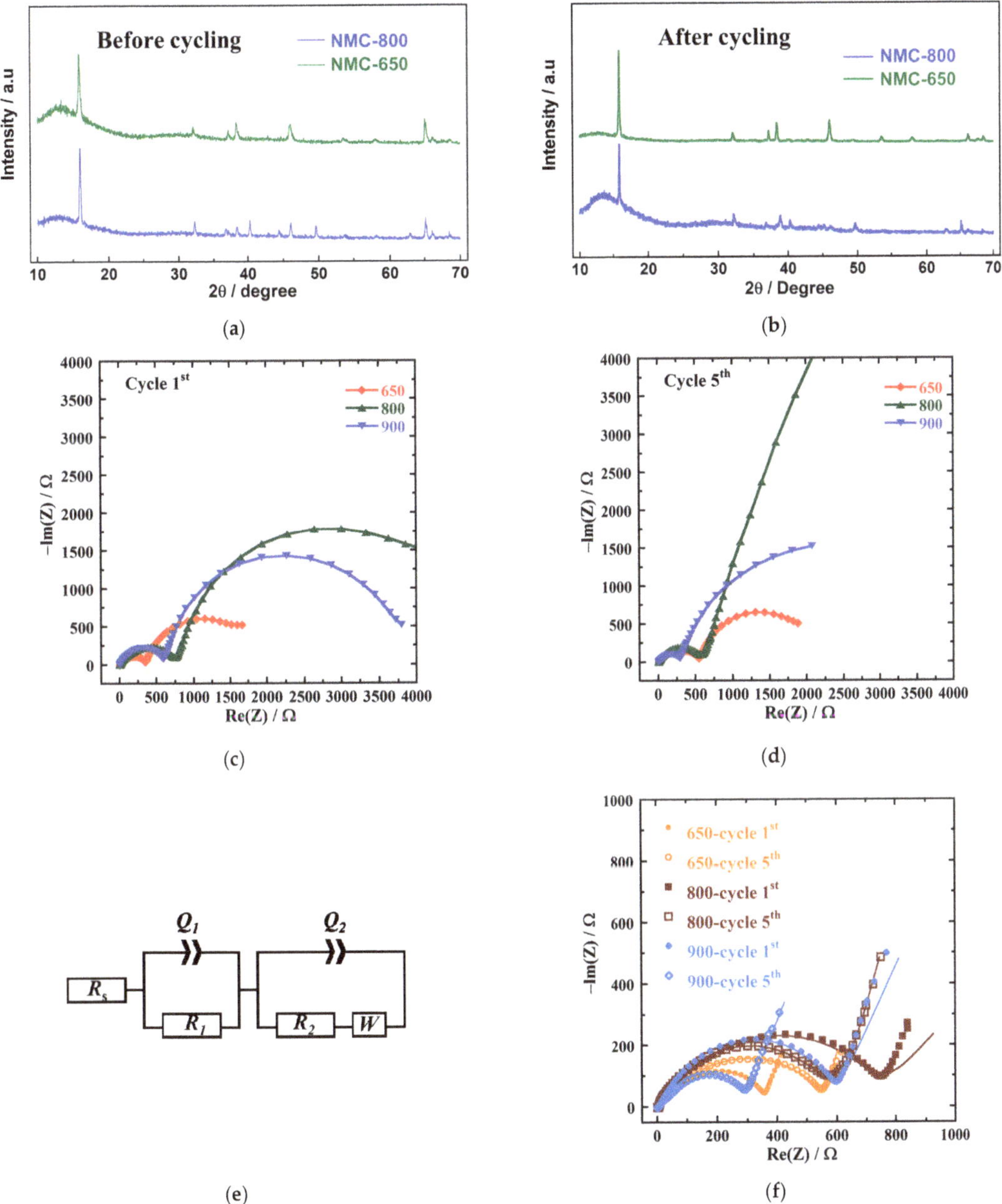

Figure 6. XRD patterns of the electrodes obtained (**a**) before and (**b**) after cycling. EIS spectra of the electrode NMC-650, NMC-800 and NMC-900 obtained at (**c**) the first and (**d**) the fifth cycles. (**e**) Equivalent circuit that use to fit the EIS results and (**f**) Nyquist plot for the electrodes with a frequency range from 10^6 to 0.1 Hz. The measurement data are shown by the dots and the fitting curves are presented by the solid lines.

The performance of P2-NMC was also evaluated in full-cell configuration using hard carbon as the anode (Figure 8). The mass ratio between the cathode and the anode of 1.4 was chosen for the highest capacity. As displayed in Figure 8a, the voltage profile of the full-cell was the combination of the voltage profile of the cathode and the anode. The first charge capacity of about 151.4 mAh·g^{-1} was obtained and was comparable to the corresponding value of the material delivered in the half-cell on the first cycle and was much higher than the discharge capacity of only 80.9 mAh·g^{-1}. This resulted in the low Coulombic efficiency of 51.9%. The irreversible capacity of hard carbon anode leads to the low Coulombic efficiency, so that presodiation techniques were widely used to optimize the performance of full-cells [43–45]. For the following cycles, the Coulombic efficiency varied around 92% and the capacity gradually declines were seen as the results of side reactions of the electrolyte on the electrodes due to wide operation voltage. The capacity retention at 25th cycle was 63.3%.

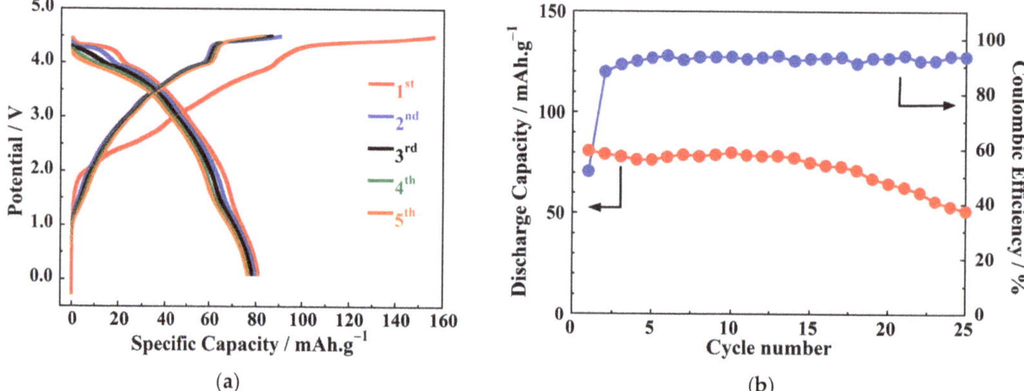

Figure 8. Cycling performance of hard carbon ‖ NMC-900 full-cell. (a) The voltage profile and (b) evolution of capacity and Coulombic efficiency with cycle number.

4. Discussion

The P-type layered structures exist within 650–900 °C, given the high thermal stability of the P-phase in the Na-Mn-Co system. Several mechanisms were explored considering the capacity fading of high voltage materials, including intragranular and lattice fracture caused by lattice change during phase transition, surface structural change and irreversible lattice oxygen release [46]. The results indicated that the phase structure had tie connection to the cycling stability that was observed to be improved by elevated temperature. Figure 7e,f show that the P3 phase could have a facile environment for sodium ion conductivity, but poor capacity retention compared to the P2 phase. Additionally, as could be seen on the first cycle (Figure 4b), the sample at low temperature NMC-650 and NMC-800 exhibited a severe side reaction on charge at high voltage plateaus that contributed mainly from the decomposition of the electrolyte and may be assisted by the high active surface and the stability of the electrode surface in the electrolyte. This occurred repeatedly in consecutive cycles, leading to the low Coulombic efficiency of the NMC-650 sample (Figure 4c). The performance of the hard carbon ‖ P2-NMC full-cell was still far below to the half-cell, emphasizing the important role of presodiation techniques of the anode.

For comparison, the electrochemical performance of some electrode materials at high voltage range are listed in Table 3. The Mn-Co system delivered lower specific capacity than other Mn-based systems, but good cycling stability. This demonstrated the role of Co in stabilizing the structure and enhancing the electrochemical performance of the electrode materials as previously reported [47,48].

The two-phase intergrowth cathodes were reported to improve the rate capability and specific capacity contributed by the synergic effect of both components [24,41,49–51].

This work revealed that the P3/P2 integration in NMC-800 was beneficial to capacity retention, but the capacity was lower than the single-phase components, which could be associated, respectively, with the robust intrinsic structure and the severe side reaction of the electrolyte that induced more sodium lost during the solid electrolyte interface (SEI) formation, resulting in relatively higher SEI and charge transfer resistance over others (Figure 6c–e). Further works should be taken to understand the microstructure and surface structure of the bi-phasic electrode to better understand the sodium storage performance.

Table 3. Comparison of the electrochemical performance of electrode materials at high cutoff voltage.

	Voltage Range	Specific Capacity $(\text{mah}\cdot\text{g}^{-1})$/Current Density	Capacity Retention/Cycle Number	Refs
P2-$Na_{0.7}Cu_{0.2}Fe_{0.2}Mn_{0.6}O_2$	1.5–4.5 V	~190 (C/10)	~69% (30)	[52]
P2-$NaMn_{0.5}Fe_{0.5}O_2$	2.0–4.3 V	185 (C/10)	60% (50)	[53]
P2-$Na_{2/3}Ni_{1/3}Mn_{2/3}O_2$	2.0–4.5 V	133 (~C/10)	42% (50)	[54]
P2-$Na_{0.66}Mn_{0.9}Mg_{0.1}O_2$	2.0–4.5 V	152 (~C/5)	71% (100)	[55]
P3-$Na_{0.9}Ni_{0.5}Mn_{0.5}O_2$	1.5–4.5 V	102 (1C)	78% (500)	[56]
P2-$NaCo_{0.5}Mn_{0.5}O_2$ (This work)	1.5–4.5 V	149 (C/10)	76% (100)	

5. Conclusions

In summary, $Na_xMn_{0.5}Co_{0.5}O_2$ possessed phase transition from P3- to P2-type layered structure when calcinated at 650 and 900 °C with bi-phasic P3/P2 coexisted in a medium temperature of 800 °C. The electrochemical properties examinations indicated that the higher calcinated temperature improved the crystallinity and strongly contributed to the cycling stability and performance. For instance, the sample NMC-900 exhibited the highest specific capacity and capacity retention at high rates that were superior to the others. Meanwhile, the bi-phase P3/P2 NMC-800 delivered the lowest specific capacity, but showed the highest diffusion coefficient, contributing to the high-rate capability and structure stability during sodium intercalation/deintercalation. The results demonstrated that single phases were better for $Na_xMn_{0.5}Co_{0.5}O_2$ material and P2 phase was preferred in terms of electrochemical performance. Further works should be conducted to understand the electrochemical performance of the bi-phasic P3/P2 NMC-800 that probably benefits and improves the performance of Na-Mn-Co electrode system.

Author Contributions: L.M.N., suggested the idea and methods, conducted the experiment and prepared the manuscript; D.M.N.N., conducted the experiment and prepared the manuscript; V.H.N., suggested the idea and methods and edited and submitted the manuscript; M.K.L., V.M.T. and M.L.P.L., lecture contribution to complete the manuscript. All authors have read and agreed to the published version of the manuscript.

Funding: This research was funded by University of Science, Viet Nam National University Ho Chi Minh City, under grant number HH 2021-03.

Informed Consent Statement: Not applicable.

Data Availability Statement: The data that support the findings of this study are available from the authors, upon reasonable request.

Conflicts of Interest: The authors declare no conflict of interest.

References

1. Palomares, V.; Serras, P.; Villaluenga, I.; Hueso, K.B.; Carretero-González, J.; Rojo, T. Na-ion batteries, recent advances and present challenges to become low cost energy storage systems. *Energy Environ. Sci.* **2012**, *5*, 5884. [CrossRef]
2. Pan, H.; Hu, Y.-S.; Chen, L. Room-temperature stationary sodium-ion batteries for large-scale electric energy storage. *Energy Environ. Sci.* **2013**, *6*, 2338. [CrossRef]
3. Delmas, C.; Fouassier, C.; Hagenmuller, P. Structural classification and properties of the layered oxides. *Phys. BC* **1980**, *99*, 81–85. [CrossRef]
4. Yuan, D.; Hu, X.; Qian, J.; Pei, F.; Wu, F.; Mao, R.; Ai, X.; Yang, H.; Cao, Y. P2-type $Na_{0.67}Mn_{0.65}Fe_{0.2}Ni_{0.15}O_2$ cathode material with high-capacity for sodium-ion battery. *Electrochim. Acta* **2014**, *116*, 300–305. [CrossRef]
5. Kang, W.; Yu, D.Y.W.; Lee, P.-K.; Zhang, Z.; Bian, H.; Li, W.; Ng, T.-W.; Zhang, W.; Lee, C.-S. P2-type $Na_xCu_{0.15}Ni_{0.20}Mn_{0.65}O_2$ cathodes with high voltage for high-power and long-life sodium-ion batteries. *ACS Appl. Mater. Interfaces* **2016**, *8*, 31661–31668. [CrossRef]
6. Lee, D.H.; Xu, J.; Meng, Y.S. An advanced cathode for Na-ion batteries with high rate and excellent structural stability. *Phys. Chem. Chem. Phys.* **2013**, *15*, 3304. [CrossRef]
7. Li, Y.; Yang, Z.; Xu, S.; Mu, L.; Gu, L.; Hu, Y.-S.; Li, H.; Chen, L. Air-stable copper-based P2-$Na_{7/9}Cu_{2/9}Fe_{1/9}Mn_{2/3}O_2$ as a new positive electrode material for sodium-ion batteries. *Adv. Sci.* **2015**, *2*, 1500031. [CrossRef]
8. Yabuuchi, N.; Kajiyama, M.; Iwatate, J.; Nishikawa, H.; Hitomi, S.; Okuyama, R.; Usui, R.; Yamada, Y.; Komaba, S. P2-type $Na_x[Fe_{1/2}Mn_{1/2}]O_2$ made from earth-abundant elements for rechargeable Na batteries. *Nat. Mater.* **2012**, *11*, 512–517. [CrossRef]
9. Sun, X.; Jin, Y.; Zhang, C.-Y.; Wen, J.-W.; Shao, Y.; Zang, Y.; Chen, C.-H. $Na[Ni_{0.4}Fe_{0.2}Mn_{0.4-x}Ti_x]O_2$: A cathode of high capacity and superior cyclability for Na-ion batteries. *J. Mater. Chem. A* **2014**, *2*, 17268–17271. [CrossRef]
10. Vassilaras, P.; Toumar, A.J.; Ceder, G. Electrochemical properties of $NaNi_{1/3}Co_{1/3}Fe_{1/3}O_2$ as a cathode material for Na-ion batteries. *Electrochem. Commun.* **2014**, *38*, 79–81. [CrossRef]
11. Mendiboure, A.; Delmas, C.; Hagenmuller, P. Electrochemical intercalation and deintercalation of Na_xMnO_2 bronzes. *J. Solid State Chem.* **1985**, *57*, 323–331. [CrossRef]
12. Delmas, C.; Braconnier, J.; Fouassier, C.; Hagenmuller, P. Electrochemical intercalation of sodium in Na_xCoO_2 bronzes. *Solid State Ion.* **1981**, *3–4*, 165–169. [CrossRef]
13. Carlier, D.; Cheng, J.H.; Berthelot, R.; Guignard, M.; Yoncheva, M.; Stoyanova, R.; Hwang, B.J.; Delmas, C. The P2-$Na_{2/3}Co_{2/3}Mn_{1/3}O_2$ phase: Structure, physical properties and electrochemical behavior as positive electrode in sodium battery. *Dalton. Trans.* **2011**, *40*, 9306. [CrossRef]
14. Manikandan, P.; Heo, S.; Kim, H.W.; Jeong, H.Y.; Lee, E.; Kim, Y. Structural characterization of layered $Na_{0.5}Co_{0.5}Mn_{0.5}O_2$ material as a promising cathode for sodium-ion batteries. *J. Power Sources* **2017**, *363*, 442–449. [CrossRef]
15. Yang, P.; Zhang, C.; Li, M.; Yang, X.; Wang, C.; Bie, X.; Wei, Y.; Chen, G.; Du, F. P2-$NaCo_{0.5}Mn_{0.5}O_2$ as a positive electrode material for sodium-ion batteries. *ChemPhysChem* **2015**, *16*, 3408–3412. [CrossRef]
16. Bucher, N.; Hartung, S.; Franklin, J.B.; Wise, A.M.; Lim, L.Y.; Chen, H.-Y.; Weker, J.N.; Toney, M.F.; Srinivasan, M. P2-$Na_xCo_yMn_{1-y}O_2$ (y = 0, 0.1) as cathode materials in sodium-ion batteries-effects of doping and morphology to enhance cycling stability. *Chem. Mater.* **2016**, *28*, 2041–2051. [CrossRef]
17. Wang, X.; Tamaru, M.; Okubo, M.; Yamada, A. Electrode properties of P2-$Na_{2/3}Mn_yCo_{1-y}O_2$ as cathode materials for sodium-ion batteries. *J. Phys. Chem. C* **2013**, *117*, 15545–15551. [CrossRef]
18. Chen, X.; Zhou, X.; Hu, M.; Liang, J.; Wu, D.; Wei, J.; Zhou, Z. Stable layered P3/P2 $Na_{0.66}Co_{0.5}Mn_{0.5}O_2$ cathode materials for sodium-ion batteries. *J. Mater. Chem. A* **2015**, *3*, 20708–20714. [CrossRef]
19. Zhu, Y.-E.; Qi, X.; Chen, X.; Zhou, X.; Zhang, X.; Wei, J.; Hu, Y.; Zhou, Z. A P2-$Na_{0.67}Co_{0.5}Mn_{0.5}O_2$ cathode material with excellent rate capability and cycling stability for sodium ion batteries. *J. Mater. Chem. A* **2016**, *4*, 11103–11109. [CrossRef]
20. Xu, X.; Ji, S.; Gao, R.; Liu, J. Facile synthesis of P2-type $Na_{0.4}Mn_{0.54}Co_{0.46}O_2$ as a high capacity cathode material for sodium-ion batteries. *RSC Adv.* **2015**, *5*, 51454–51460. [CrossRef]
21. Le Nguyen, M.; Van Nguyen, H.; Ghosh, N.; Garg, A.; Van Tran, M.; Le, P. High-voltage performance of P2-$Na_xMn_{0.5}Co_{0.5}O_2$ layered cathode material. *Int. J. Energy Res.* **2021**, *46*, 5119–5133. [CrossRef]
22. Konarov, A.; Kim, H.J.; Voronina, N.; Bakenov, Z.; Myung, S.-T. P2-$Na_{2/3}MnO_2$ by co incorporation: As a cathode material of high capacity and long cycle life for sodium-ion batteries. *ACS Appl. Mater. Interfaces* **2019**, *11*, 28928–28933. [CrossRef] [PubMed]
23. Hwang, J.-Y.; Myung, S.-T.; Sun, Y.-K. Sodium-ion batteries: Present and future. *Chem. Soc. Rev.* **2017**, *46*, 3529–3614. [CrossRef] [PubMed]
24. Chen, X.; Song, J.; Li, J.; Zhang, H.; Tang, H. A P2/P3 composite-layered cathode material with low-voltage decay for sodium-ion batteries. *J. Appl. Electrochem.* **2021**, *51*, 619–627. [CrossRef]
25. Shi, Y.; Zhang, Z.; Jiang, P.; Gao, A.; Li, K.; Zhang, Q.; Sun, Y.; Lu, X.; Cao, D.; Lu, X. Unlocking the potential of P3 structure for practical Sodium-ion batteries by fabricating zero strain framework for Na^+ intercalation. *Energy Storage Mater.* **2021**, *37*, 354–362. [CrossRef]
26. Zhang, L.; Wang, J.; Li, J.; Schuck, G.; Winter, M.; Schumacher, G.; Li, J. Preferential occupation of Na in P3-type layered cathode material for sodium ion batteries. *Nano Energy* **2020**, *70*, 104535. [CrossRef]
27. Zhou, Y.-N.; Wang, P.-F.; Zhang, X.-D.; Huang, L.-B.; Wang, W.-P.; Yin, Y.-X.; Xu, S.; Guo, Y.-G. Air-stable and high-voltage layered p3-type cathode for sodium-ion full battery. *ACS Appl. Mater. Interfaces* **2019**, *11*, 24184–24191. [CrossRef]

28. Lei, Y.; Li, X.; Liu, L.; Ceder, G. Synthesis and stoichiometry of different layered sodium cobalt oxides. *Chem. Mater.* **2014**, *26*, 5288–5296. [CrossRef]
29. Han, M.H.; Acebedo, B.; Gonzalo, E.; Fontecoba, P.S.; Clarke, S.; Saurel, D.; Rojo, T. Synthesis and electrochemistry study of P2- and O3-phase $Na_{2/3}Fe_{1/2}Mn_{1/2}O_2$. *Electrochim. Acta* **2015**, *182*, 1029–1036. [CrossRef]
30. Sendova-Vassileva, M.; Stoyanova, R.; Carlier, D.; Yoncheva, M.; Zhecheva, E.; Delmas, C. Raman spectroscopy study on $Na_{2/3}Mn_{1-x}Fe_xO_2$ oxides. *Adv. Sci. Technol.* **2010**, *74*, 60–65.
31. Qu, J.F.; Wang, W.; Chen, Y.; Li, G.; Li, X.G. Raman spectra study on nonstoichiometric compound Na_xCoO_2. *Phys. Rev. B* **2006**, *73*, 092518. [CrossRef]
32. Wang, C.; Yan, J.; Li, T.; Lv, Z.; Hou, X.; Tang, Y.; Zhang, H.; Zheng, Q.; Li, X. A coral-like FeP@NC anode with increasing cycle capacity for sodium-ion and lithium-ion batteries induced by particle refinement. *Angew. Chem. Int. Ed.* **2021**, *60*, 25013–25019. [CrossRef]
33. Liu, Q.; Hu, Z.; Zou, C.; Jin, H.; Wang, S.; Li, L. Structural engineering of electrode materials to boost high-performance sodium-ion batteries. *Cell Rep. Phys. Sci.* **2021**, *2*, 100551. [CrossRef]
34. Wang, P.-F.; Yao, H.-R.; Liu, X.-Y.; Zhang, J.-N.; Gu, L.; Yu, X.-Q.; Yin, Y.-X.; Guo, Y.-G. Ti-substituted $NaNi_{0.5}Mn_{0.5-x}Ti_xO_2$ cathodes with reversible O3−P3 phase transition for high-performance sodium-ion batteries. *Adv. Mater.* **2017**, *29*, 1700210. [CrossRef]
35. Yao, H.-R.; Wang, P.-F.; Wang, Y.; Yu, X.; Yin, Y.-X.; Guo, Y.-G. Excellent comprehensive performance of Na-based layered oxide benefiting from the synergetic contributions of multimetal ions. *Adv. Energy Mater.* **2017**, *7*, 1700189. [CrossRef]
36. Wei, T.-T.; Zhang, N.; Zhao, Y.-S.; Zhu, Y.-R.; Yi, T.-F. Sodium-deficient O3-$Na_{0.75}Fe_{0.5-x}Cu_xMn_{0.5}O_2$ as high-performance cathode materials of sodium-ion batteries. *Compos. Part B Eng.* **2022**, *238*, 109912. [CrossRef]
37. Wei, T.-T.; Liu, X.; Yang, S.-J.; Wang, P.-F.; Yi, T.-F. Regulating the electrochemical activity of Fe-Mn-Cu-based layer oxides as cathode materials for high-performance Na-ion battery. *J. Energy Chem.* **2023**, *80*, 603–613. [CrossRef]
38. Bredar, A.R.C.; Chown, A.L.; Burton, A.R.; Farnum, B.H. Electrochemical impedance spectroscopy of metal oxide electrodes for energy applications. *ACS Appl. Energy Mater.* **2020**, *3*, 66–98. [CrossRef]
39. Costard, J.; Joos, J.; Schmidt, A.; Ivers-Tiffée, E. Charge transfer parameters of $Ni_xMn_yCo_{1-x-y}$ cathodes evaluated by a transmission line modeling approach. *Energy Technol.* **2021**, *9*, 2000866. [CrossRef]
40. Ma, X.; Chen, H.; Ceder, G. Electrochemical properties of monoclinic $NaMnO_2$. *J. Electrochem. Soc.* **2011**, *158*, A1307. [CrossRef]
41. Zhou, D.; Huang, W.; Lv, X.; Zhao, F. A novel P2/O3 biphase $Na_{0.67}Fe_{0.425}Mn_{0.425}Mg_{0.15}O_2$ as cathode for high-performance sodium-ion batteries. *J. Power Sources* **2019**, *421*, 147–155. [CrossRef]
42. Yan, Z.; Tang, L.; Huang, Y.; Hua, W.; Wang, Y.; Liu, R.; Gu, Q.; Indris, S.; Chou, S.-L.; Huang, Y.; et al. A hydrostable cathode material based on the layered P2@P3 composite that shows redox behavior for copper in high-rate and long-cycling sodium-ion batteries. *Angew. Chem.* **2019**, *131*, 1426–1430. [CrossRef]
43. Liang, X.; Yu, T.-Y.; Ryu, H.-H.; Sun, Y.-K. Hierarchical O3/P2 heterostructured cathode materials for advanced sodium-ion batteries. *Energy Storage Mater.* **2022**, *47*, 515–525. [CrossRef]
44. Peng, B.; Chen, Y.; Wang, F.; Sun, Z.; Zhao, L.; Zhang, X.; Wang, W.; Zhang, G. Unusual Site-Selective Doping in Layered Cathode Strengthens Electrostatic Cohesion of Alkali-Metal Layer for Practicable Sodium-Ion Full Cell. *Adv. Mater.* **2022**, *34*, 2103210. [CrossRef] [PubMed]
45. Wang, H.; Xiao, Y.; Sun, C.; Lai, C.; Ai, X. A type of sodium-ion full-cell with a layered $NaNi_{0.5}Ti_{0.5}O_2$ cathode and a pre-sodiated hard carbon anode. *RSC Adv.* **2015**, *5*, 106519–106522. [CrossRef]
46. Jiang, M.; Qian, G.; Liao, X.-Z.; Ren, Z.; Dong, Q.; Meng, D.; Cui, G.; Yuan, S.; Lee, S.-J.; Qin, T.; et al. Revisiting the capacity-fading mechanism of P2-type sodium layered oxide cathode materials during high-voltage cycling. *J. Energy Chem.* **2022**, *69*, 16–25. [CrossRef]
47. Li, Z.-Y.; Zhang, J.; Gao, R.; Zhang, H.; Hu, Z.; Liu, X. Unveiling the role of co in improving the high-rate capability and cycling performance of layered $Na_{0.7}Mn_{0.7}Ni_{0.3-x}Co_xO_2$ cathode materials for sodium-ion batteries. *ACS Appl. Mater. Interfaces* **2016**, *8*, 15439–15448. [CrossRef]
48. Fu, C.C.; Wang, J.; Li, Y.; Liu, G.; Deng, T. Explore the effect of Co doping on P2-$Na_{0.67}MnO_2$ prepared by hydrothermal method as cathode materials for sodium ion batteries. *J. Alloys. Compd.* **2022**, *918*, 165569. [CrossRef]
49. Hou, P.; Li, F.; Wang, Y.; Yin, J.; Xu, X. Mitigating the P2–O2 phase transition of high-voltage P2-$Na_{2/3}[Ni_{1/3}Mn_{2/3}]O_2$ cathodes by cobalt gradient substitution for high-rate sodium-ion batteries. *J. Mater. Chem. A* **2019**, *7*, 4705–4713. [CrossRef]
50. Rahman, M.M.; Mao, J.; Kan, W.H.; Sun, C.-J.; Li, L.; Zhang, Y.; Avdeev, M.; Du, X.-W.; Lin, F. An ordered P2/P3 composite layered oxide cathode with long cycle life in sodium-ion batteries. *ACS Mater. Lett.* **2019**, *1*, 573–581. [CrossRef]
51. Zhou, Y.-N.; Wang, P.-F.; Niu, Y.-B.; Li, Q.; Yu, X.; Yin, Y.-X.; Xu, S.; Guo, Y.-G. A P2/P3 composite layered cathode for high-performance Na-ion full batteries. *Nano Energy* **2019**, *55*, 143–150. [CrossRef]
52. Zhang, Y.; Kim, S.; Feng, G.; Wang, Y.; Liu, L.; Ceder, G.; Li, X. The interaction between Cu and Fe in P2-type Na_xTMO_2 cathodes for advanced battery performance. *J. Electrochem. Soc.* **2018**, *165*, A1184–A1192. [CrossRef]
53. Mortemard de Boisse, B.; Carlier, D.; Guignard, M.; Bourgeois, L.; Delmas, C. P2-$Na_xMn_{1/2}Fe_{1/2}O_2$ phase used as positive electrode in na batteries: Structural changes induced by the electrochemical (de)intercalation process. *Inorg. Chem.* **2014**, *53*, 11197–11205. [CrossRef]

54. Singh, G.; Tapia-Ruiz, N.; Lopez del Amo, J.M.; Maitra, U.; Somerville, J.W.; Armstrong, A.R.; Martinez de Ilarduya, J.; Rojo, T.; Bruce, P.G. High voltage Mg-doped $Na_{0.67}Ni_{0.3-x}Mg_xMn_{0.7}O_2$ (x = 0.05, 0.1) Na-ion cathodes with enhanced stability and rate capability. *Chem. Mater.* **2016**, *28*, 5087–5094. [CrossRef]
55. Or, T.; Kaliyappan, K.; Bai, Z.; Chen, Z. High voltage stability and characterization of P2-$Na_{0.66}Mn_{1-y}Mg_yO_2$ cathode for sodium-on batteries. *ChemElectroChem* **2020**, *7*, 3284–3290. [CrossRef]
56. Risthaus, T.; Chen, L.; Wang, J.; Li, J.; Zhou, D.; Zhang, L.; Ning, D.; Cao, X.; Zhang, X.; Schumacher, G.; et al. P3 $Na_{0.9}Ni_{0.5}Mn_{0.5}O_2$ cathode material for sodium ion batteries. *Chem. Mater.* **2019**, *31*, 5376–5383. [CrossRef]

 *energies*

MDPI

Article

Effect of Glucose and Methylene Blue in Microbial Fuel Cells Using *E. coli*

Carolina Montoya-Vallejo, Jorge Omar Gil Posada and Juan Carlos Quintero-Díaz *

Grupo de Bioprocesos, Departamento de Ingeniería Química, Universidad de Antioquia, Medellín 050010, Colombia; carolina.montoya1@udea.edu.co (C.M.-V.); jomar.gil@udea.edu.co (J.O.G.P.)
* Correspondence: carlos.quintero@udea.edu.co

Abstract: Microbial fuel cells could be used as an alternative for wastewater treatment and electricity generation. *Escherichia coli* is a representative bacterium that has been widely studied as a model in laboratory assays despite its limited ability to transfer electrons. Although previous studies have employed glucose and methylene blue in electricity production using *E. coli*, there remains a lack of understanding on how current generation would impact the production of metabolites and what the most appropriate conditions for current production might be. To shed light on those issues, this manuscript used a 3^2 factorial design to evaluate the effect of the concentration of organic matter (glucose) and the concentration of the mediator methylene blue (MB) using *E. coli* DH5α as an anodic microorganism. It was found that as the concentration of glucose was increased, the production of electricity increased and at the same time, its degradation percentage decreased. Similarly, a 17-fold increase in current production was observed with an elevation in methylene blue concentration from 0 to 0.3 mM, though inhibition became apparent at higher concentrations. The maximum power generated by the cell was 204.5 μW m^{-2}, achieving a current density of 1.434 mA m^{-2} at concentrations of 5 g L^{-1} of glucose and 0.3 mM of MB. Reductions in the production of ethanol, lactate, and acetate were observed due to the deviation of electrons to the anode.

Keywords: electricity production; carbon source; redox mediator; anodic electrofermentation; *E. coli*

Citation: Montoya-Vallejo, C.; Gil Posada, J.O.; Quintero-Díaz, J.C. Effect of Glucose and Methylene Blue in Microbial Fuel Cells Using *E. coli*. *Energies* **2023**, *16*, 7901. https://doi.org/10.3390/en16237901

Academic Editor: Jun Li

Received: 26 September 2023
Revised: 23 October 2023
Accepted: 28 October 2023
Published: 4 December 2023

1. Introduction

Accelerated industrial development and demographic growth are constantly creating challenges to human society, some of those challenges include a scarcity of fresh water, increasing the demand of energy in the order of 28,500 TWh by the year 2021 [1], the generation of large amounts of waste currently estimated at 360 km^3 year^{-1} [2,3], and environmental deterioration [4]. Therefore, it is important to develop and implement energy-efficient processes that add value to the organic matter present in different types of wastewaters, this will in turn provide a means to achieve water purification and reuse [5].

Microbial fuel cells (MFCs) are bioelectrochemical devices engineered to convert chemical energy into bioelectricity through the redox processes occurring within living microorganisms. Microbial fuel cells can provide ecofriendly solutions to energy scarcity and water pollution [6]. A typical MFC comprises an anode chamber where a microbial culture oxidizes organic matter into electrons and protons thus producing a flow of electrons that is diverted through an external circuit and a flow of protons that travels across the electrolyte, both flows converging to the cathode where oxidized species are finally reduced [7].

Recent advancements have demonstrated that the MFC power density can be substantially improved through alterations in electrode materials and construction, the customization of bacterial cultures to boost MFC performance, and the optimization of MFC geometry and design. Nonetheless, the relatively low power density of MFCs remains a significant challenge that hinders their widespread adoption for large-scale applications [8,9].

Commonly found in the lower intestines of warm-blooded animals, *Escherichia coli*, is a well-known facultative anaerobic bacterium that belongs to the *Enterobacteriacea* family.

Due to its clear genetic background, convenience to be genetically modified, low nutritional requirements, and rapid growth, *E. coli* is one of the most frequently used bacterial models for the electrochemical oxidation of carbon sources [10–12].

Within the respiratory chain of *E. coli*, a significant number of primary dehydrogenases are present to oxidize electron donors, and terminal reductases and quinones are present to reduce electron acceptors. These components are activated depending on the availability of final electron acceptors [13,14]. The transfer of electrons to the anode by *E. coli* has been tested, and although very low, it is not negligible. Since *E. coli* lacks nanotubes for direct electron transfer [15], this transfer appears to be promoted by endogenous redox mediators such as hydroquinones, as well as other soluble molecules that could act as electron carriers [15–17]. *E. coli* DH5α is a widely used strain for maintaining and amplifying plasmid DNA [18]. Its effective utilization in microbial cells can facilitate initiatives to expand its application through improvements aimed at achieving a higher efficiency in electron transfer, as has been done recently [19,20].

Given that the redox potential (E^0) of the NADH/NAD+ pair is −320 mV, the redox potential of terminal reductases under anaerobic conditions such as nitrite reductase (E^0 = +360 mV), DMSO reductase (E^0 = +160 mV), TMAO reductase (E^0 = +130 mV), or fumarate reductase (E^0 = +30 mV), allows for the flow of electrons until they reach the final acceptor. Electron transfer can also be achieved by using exogenous mediators that compete with the natural final acceptors; these species must be capable of penetrating the cellular membrane to receive electron charges from the terminal reductases of the cell, then leave the cell to transfer electrons to the anode. To enable fast electrode reaction kinetics, exogenous mediators should ideally exhibit not only a low toxicity to microorganisms but also a high solubility and stability [21]. Exogenous mediators, such as methylene blue (MB), methyl viologen (MV), neutral red (NR), anthraquinone-2,6-disulfonate (AQDS), 2-hydroxy-1,4-naphthoquinone, and resazurin, have been used to enhance electron transfer when electrically inactive microorganisms like *E. coli* are used with microbial fuel cells [22–26]. Among these redox compounds, methylene blue (E^0 = +110 mV) is highly attractive due to its high redox potential [24]; however, optimization studies of its concentration in relation to the concentration of the carbon source as an electron donor have not been conducted. Figure 1 not only provides a schematic description of a microbial fuel cell but also represents the mediation mechanism of MB. Recent developments have demonstrated the utility of *E. coli* in MFC technology, achieving power densities on the order of 11.7 mW m^{-2} when using *E. coli* K12 and anodes based on carbon [27]; the use of *E. coli* DH5 α and anodes based on carbon nanotubes has rendered even larger power densities, reaching 2740 mW m^{-2} [28,29]. When dealing with *E. coli*, differences in MFC electron transfer measurements can be explained by electrode materials [15], genetic modifications [30], the use of exogen mediators [20], and cocultures [31,32].

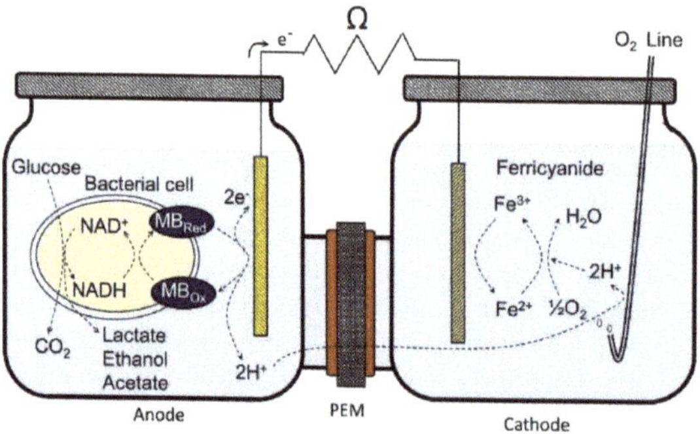

Figure 1. Diagram of electron transfer mediated by methylene blue in a microbial fuel cell.

This manuscript aims at evaluating and improving the production of electric current obtained when *E. coli* degrades glucose by means of an anodic fermentation within a microbial fuel cell, and by using methylene blue as exogenous mediator.

2. Materials and Methods

2.1. Reactor Design and Operation

Escherichia coli DH5α was used as the anodic microorganism and was cultured in LB (Luria-Bertani) medium that contained 5.0 g L^{-1} of NaCl, 10 g L^{-1} of tryptone, and 5.0 g L^{-1} of yeast extract.

A dual-chamber H-type microbial fuel cell was used to conduct experiments. The basic setup consisted of two glass chambers, each of them with a total capacity of 250 mL, separated by a Zirfon$^{®}$ proton exchange membrane with a diameter of 1.5 cm. Graphite brush electrodes were used for both anodic and cathodic chambers (2.5 cm in outer diameter and 2.5 cm long, an average fiber diameter of 0.72 μm, a total area of 0.22 m^2, MILL ROUSE). The anodic chamber was loaded with an *E. coli* DH5α suspension in LB medium that was supplemented with glucose as an electron donor at different concentrations (1.0, 5.0, 10.0) g L^{-1} and methylene blue as a redox mediator also at three concentrations (0, 0.3, 3.0) mM. The cathodic chamber was filled with a solution of 20 mM K$_3$[Fe(CN)$_6$], and oxygen was supplied through air bubbling using an aquarium pump at a constant flow. An external resistance of 1 kΩ was part of the external circuit that was used to connect anode and cathode. The MFC operating temperature was held constant at 35 °C at a constant speed of 150 rpm. Anaerobic conditions within the anodic chamber were reached by minimizing the head space, and by bubbling nitrogen for at least 10 min, prior to each culture.

Data acquisition and electrochemical measurements (polarization curves and voltage profiles) were performed by using a multichannel potentiostat with FRA capabilities (MultiPalmSense4, Palmsens - Houten. Netherlands).

The performance of the MFC was assessed by measuring the net charge (Q_e) generated during each treatment using the current vs. time (*I* vs. *t*) profile, as described in the following expression (Equation (1)):

$$Q_e = \int I\, dt \tag{1}$$

The construction of the power curves (P ▪ V I) (from the polarization curve V vs. I) allowed for the determination of the maximum power generated by the cell (Pmax), and for the determination of the internal resistance which determines a relationship between the maximum power generated by the cell and the square of the intensity of the current (Equation (2)) [33].

$$R_{int} = \frac{P_{mx}}{I^2} \tag{2}$$

The anodic electrofermentation efficiency (η_{EF}) is defined by Equation (3) [34]:

$$\eta_{EF} = \frac{Q_e}{\sum \Delta Q_{pi}} \tag{3}$$

where the number of electrons released to the anode and transferred through the external circuit (Q_e) are calculated by using Equation (1), while the term $\sum \Delta Q_{pi}$ represents the total electron charge due to the formation of products during the open-circuit fermentation (control treatment), minus the electron charge due to product formation during the anodic electrofermentation. To calculate this parameter, it is required to evaluate the number of electrons (N_{Pi}) per mol of product *i* by means of Equation (4).

$$N_{(C_w N_x O_y H_z)} = 4w - 3x - 2y + z \tag{4}$$

If (n_{Pi}) is the number of moles of product i formed and F is the Faraday constant (96,458 C mol^{-1}), then:

$$Q_{pi} = n_{Pi} N_{Pi} F \tag{5}$$

2.2. Analytical Methods

Glucose was measured according to the glucose oxidase method [35]. Biomass was determined by measuring the optical density of a culture sample at 600 nm by using a spectrophotometer. Bacteria metabolites of *E. coli* DH5α (lactate, acetate, and ethanol) were determined by using an HPLC system (Agilent, Tokyo, Japan) equipped with a refractive index detector. Separations were carried out on an ICSEP COREGEL- 87H3 column (Transgenomic, Omaha, NE, USA). The mobile phase consisted of an aqueous solution of sulfuric acid 0.01 N, at a flow rate of 0.6 mL min^{-1} [36].

2.3. Experimental Design and Statistical Analysis

A factorial design 3^2 (duplicate) was performed to evaluate the effect of organic matter concentration as an electron donor (glucose 1, 5 and 10 g L^{-1}) and methylene blue (MB) concentration as an electron carrier (0, 0.3 and 3.0 mM), on the performance of the MFC. Statistical analysis was performed by an ANOVA, a comparation of means, and by the response surface methodology using software Statgraphics Centurion V 19.1.2.

3. Results

As previously explained, a full 3^2 factorial design was used to evaluate the incidence of the organic load (glucose concentration) and redox mediator (methylene blue) on the MFC performance. Experiments were conducted under batch mode and using *E. coli* DH5 α.

As shown in Figure 2A, time-domain voltage profiles for treatments using three different concentrations of mediator and glucose at a constant concentration of 5.0 g L^{-1} were compared. While no significant potential difference across the cell was developed for the culture without mediator, a rapid increase in the MFC voltage profile was detected for the culture when the concentration of methylene blue was increased from 0.0 mM to 0.3 mM, as the maximum voltage drop across the MFC went up from negligible values to almost 0.31 V; once this value was reached, it remained almost constant for the rest of the treatment.

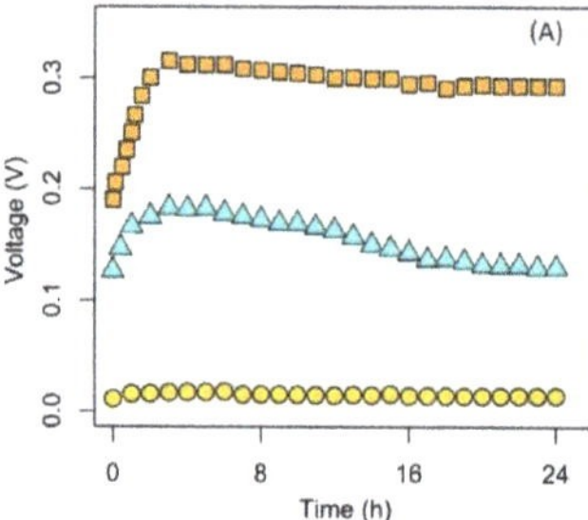

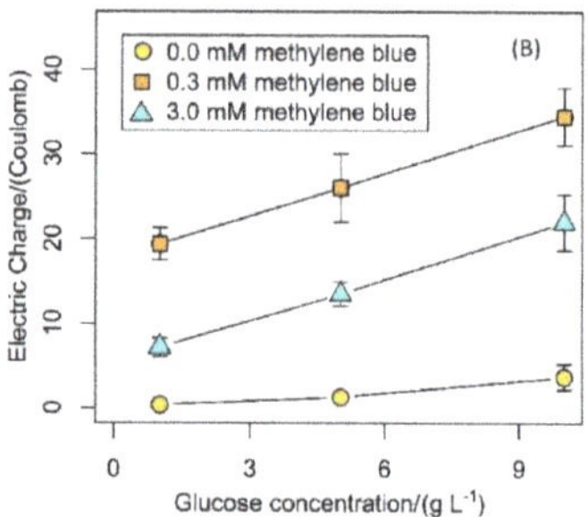

Figure 2. Electricity production in a dual-chamber microbial fuel cell by using *E. coli*. (**A**) MFC voltage profile at 5 g L^{-1} of glucose and three concentrations of methylene blue. (**B**) Electric charge produced vs. glucose and methylene blue concentration.

However, when the culture was conducted at a concentration of methylene blue of 3.0 mM, the maximum voltage drop generated across the MFC went down to 0.19 V. After each treatment was completed, the overall electric charge was calculated as the area under the curve on a current versus time plot using Equation (1). As can be seen from Figure 2B, in the absence of a redox mediator, *E. coli* DH5 α renders neither a significant current nor an appreciable voltage drop across the MFC. However, when methylene blue is added at a concentration of 0.3 mM, there is a significant increase in both current and voltage drops. For example, the treatments conducted at 5.0 g L^{-1} of glucose revealed that when the concentration of methylene blue was increased from 0 to 0.3 mM, the charge increased from 1.3 coulomb to 25.6 coulomb. However, when the methylene blue concentration was further increased to 3.0 mM, the cell performance decreased, and only 13.5 coulombs of charge was produced. Similar results have been reported elsewhere; for example, by using an H-type MFC, Taskan and coworkers [37] evaluated the concentration effect of selected mediators such as methylene blue (MB), 2-hydroxy-1,4-naphthoquinone (HNQ), and neutral red (NR) on cell performance using domestic wastewater as substrate. It was found that for every mediator, there was a concentration threshold above which any further concentration increase would produce a decrease in cell current; that was 50 µM for HNQ and 300 µM for both MB and NR. This reduction was associated with an increase in the internal resistance of the cell, so it was concluded that large concentrations of mediator species would foment the adsorption of mediator molecules on the surface of the electrode, which would in turn affect the overall internal resistance of the cell [37]. Similarly, Rahimnejad and coworkers found that current would decrease when increasing the concentration of methylene blue beyond 3.0 mM; this could be attributed to the formation of methylene blue aggregates that might occur at large concentration [26,38]. On the other hand, methylene blue at concentrations beyond 1.5 mM inhibits microbial growth; therefore, this molecule has been used to treat *E. coli* infections [39–41]. What all of this suggests is that cell performance is adversely affected by large concentrations of methylene blue not only for its antimicrobial properties but also for its absorption on the surface of the electrode (thus increasing internal resistance) and the formation of dimers which might hinder electron transport, and thus any redox process involving the methylene blue itself. Moreover, it has also been reported that for a culture of *E. coli* with methylene blue at concentrations beyond 0.3 mM, current production was at least 10 times larger than the ones corresponding to cultures without a mediator [15]. Finally, no inhibitory effect was detected beyond 10 mM of methylene blue (antimicrobial properties). These findings have been explained in terms of mediator depletion due to adsorption on the walls of the reactor [15].

The presence of glucose was found to have a positive effect on the generation of current only when methylene blue was used. In the absence of methylene blue, current production was marginal at best, as reported in Table 1, which indicates that *E. coli* requires the use of an external redox mediator to generate significant MFC currents. Similar conclusions were drawn with cultures of *E. coli* with neutral red, where currents in the order of 0.1 mA were achieved when no mediator was used; however, in the presence of neutral red, currents in the order of 1.2 mA were reached [42]. On the other hand, cocultures of *E. coli* with *P. aeruginosa* (which is well known for producing several redox mediators) rendered larger current values than its pure-culture counterparts [31,43].

Table 1. Current density (I), electric charge produced (Qe), and remotion efficiency of glucose (%R) for different glucose (G) and methylene blue (M) concentrations.

G (g L^{-1})	M (mM)	I (mA m^{-2})	Qe (C)	%R
1.0	0.0	0.048 ± 0.03	0.36	98.2 ± 1.4
1.0	0.3	0.974 ± 0.04	19.37	93.3 ± 1.0
1.0	3.0	0.432 ± 0.02	7.22	92.8 ± 2.7
5.0	0.0	0.081 ± 0.02	1.28	61.8 ± 7.1
5.0	0.3	1.434 ± 0.18	25.57	40.6 ± 2.8
5.0	3.0	0.832 ± 0.02	13.49	32.1 ± 6.8
10.0	0.0	0.058 ± 0.02	3.68	75.4 ± 6.6
10.0	0.3	2.037 ± 0.13	34.44	44.2 ± 2.8
10.0	3.0	1.201 ± 0.08	21.96	34.2 ± 3.1

In line with the ANOVA and Pareto diagram (Figure 3A), the concentrations of glucose and methylene blue on electric charge were both significant ($p < 0.05$). In the case of methylene blue, a significant second-order model was determined. Neither the glucose quadratic effect nor the mixed term between glucose and methylene blue were significant. The overall model, given by Equation (6), had a regression coefficient (r^2) of 0.973 with an adjusted r^2 of 0.928. From this equation, it followed the linear terms in glucose and methylene blue had a positive effect on the electric charge; however, the methylene blue quadratic term was negative, which indicated inhibition. The quadratic effects of glucose and the interaction were not significant for the model.

$$Qe = -2.93615 + 0.70788 \times G + 89.64 \times M + 0.02224 \times G^2 + 0.2496{*}G \times M - 28.9403 \times M^2 \qquad (6)$$

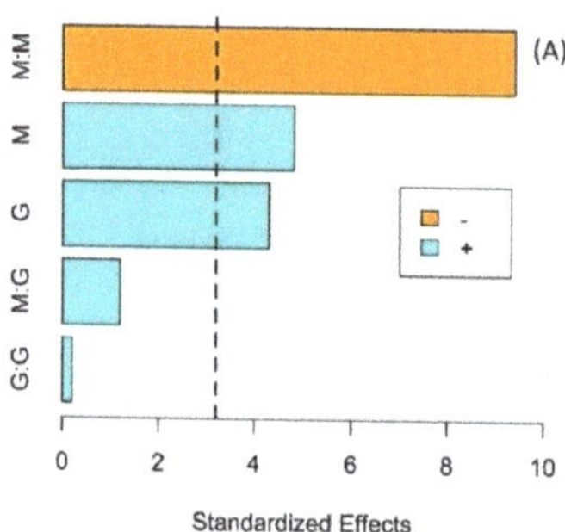
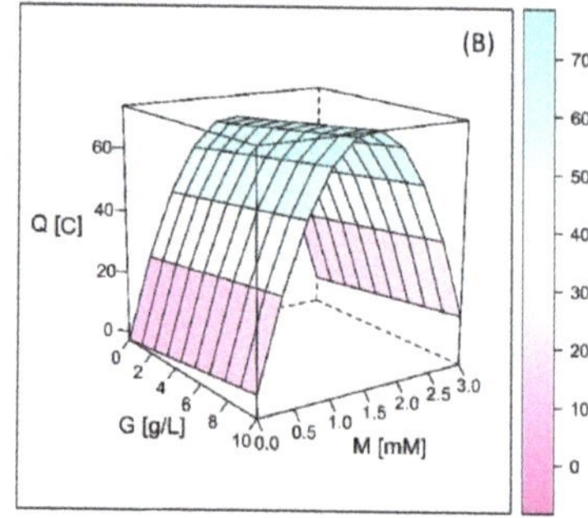

Figure 3. Effect of glucose and methylene blue on electric charge produced in a dual-chamber microbial fuel cell. (**A**) Pareto diagram of standardized effects. (**B**) Surface response of electrical charge as a function of the initial concentrations of glucose (G) and of methylene blue (M).

The surface response model, plotted in Figure 3B, indicates an optimal value for the electric charge of 79.7 C at 1.6 mM of methylene blue and 10.0 g L^{-1} of glucose. It is worth noting that the methylene blue inhibitory effect that occurs at concentrations beyond 1.5 mM. as mentioned in [40], and other aspects that might adversely affect the electric charge (already considered) might limit the usefulness of the model.

As can be seen from Table 1, glucose utilization efficiency values ranged from 93% for an initial concentration of glucose of 1.0 g L^{-1} to nearly 40% for larger initial concentrations of glucose (5–10 g L^{-1}) and all of this at 0.3 mM of methylene blue. As can be seen, initial concentrations of glucose of 1.0 g L^{-1} rendered the largest values of glucose utilization efficiency. In any case, the carbon source was never completely depleted. This behavior has been previously explained by considering that the use of high organic loads will increase the production of organic acids, thus increasing the acidity of the anolyte, which will in turn reduce microbial activity and COD removal [44]. Similarly, the accumulation of fermentation products such as acetate, lactate, and ethanol will inhibit microbial growth when using *E. coli* [45,46]. It has been shown that lactic acid at a concentration of 5.0 g L^{-1} would completely inhibit the growth of *E. coli* [47], Similarly, it has been reported that any additions of acetate species at concentrations beyond 0.45 g L^{-1} would reduce the rate of *E. coli* growth by almost 50% [48].

As can be seen from Table 1, the incidence of methylene blue on glucose degradation was very significant. For example, for an initial concentration of glucose of 5.0 g L^{-1}, the percentage of degradation dropped from 40.6% to 32.1% when the methylene blue concentration was increased from 0.3 mM to 3.0 mM. These results are in line with the potential inhibitory effect the methylene blue has been reported to exert on microbial growth. According to the experimental results, the largest variation on the remotion of organic matter can be mostly explained by the presence of methylene blue.

The largest current density value of 2.0 mA m^{-2} was achieved when using methylene blue at a concentration of 0.3 mM and glucose at 10 g L^{-1}; however, when the concentration of glucose was reduced to 5.0 g L^{-1}, the current density decreased to 1.4 mA m^{-2}; a further reduction in the concentration of glucose to 1.0 g L^{-1} rendered an even smaller current density, in the order of 0.97 mA m^{-2}, but in that case, one of the largest glucose remotion efficiency values (93.3%) was achieved (corresponding to a cell voltage of 210 mV). By taking these observations into consideration, the polarization and power curves shown in Figure 4 reveal the maximum power generated by the cell was in the order of 204.5 μW m^{-2}, and the internal resistance of the cell was in the order of 288.8 ohms. Other authors have claimed that when using dual MFC devices with carbon anodes and *E. coli*, current density values ranging from between 300 and 810 mA m^{-2} were achieved at power densities ranging from between 78 and 350 mW m^{-2}, at cell voltages in the order of 240–250 mV and electrode areas from 1.7 to 8.0 m^2 [49,50]. As can be seen, those experimental findings were observed at similar voltage values to the ones reported within this manuscript; however, there is a significant mismatch in the electroactive area of the electrodes, so it is not possible at this point to conduct a fair comparison.

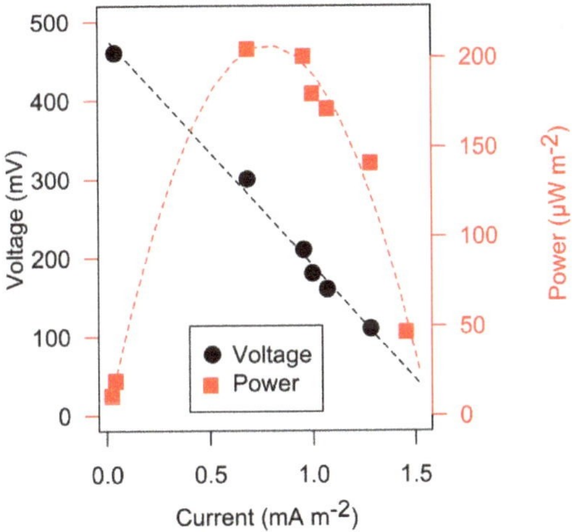

Figure 4. Polarization and power curves for MFC with *E. coli* using 0.3 mM of methylene blue and 5.0 g L^{-1} of glucose.

Regarding the ANOVA and Pareto diagram (Figure 5A) for the concentrations of glucose and methylene blue on the remotion of glucose in the microbial fuel cell, it was observed that both factors were significant with a very significant second-order glucose term ($p < 0.05$). The quadratic effect on the methylene blue as well as the mixed interaction terms were nonsignificant. The overall model, given by Equation (7), has a regression coefficient (r^2) of 0.972 with an adjusted r^2 of 0.926. From this equation, it follows that the linear terms in glucose and methylene blue have a negative effect on glucose degradation; however, the glucose quadratic term is positive. The surface response model, plotted on Figure 5B, indicates an optimal value for the remotion of glucose close to 100% after 24 h at 0.0 mM of methylene blue and 1.0 g L^{-1} of glucose. These values confirm that in the absence of bacterial inhibitors such as methylene blue at low concentrations of glucose, where low numbers of organic acid species are accumulated, the consumption of organic matter is maximized. However, considering the interest in the production of electric current in the MFC, the optimal values for current generation should be reconciled with the optimal values for the degradation of organic matter.

$$\%R = 124.567 - 20.587 \times G - 67.2915 \times M + 1.5174 \times G^2 - 0.9007{*}G \times M + 21.2181 \times M^2 \tag{7}$$

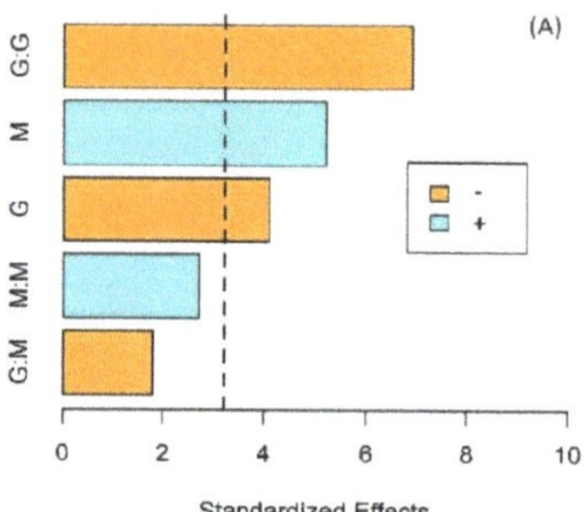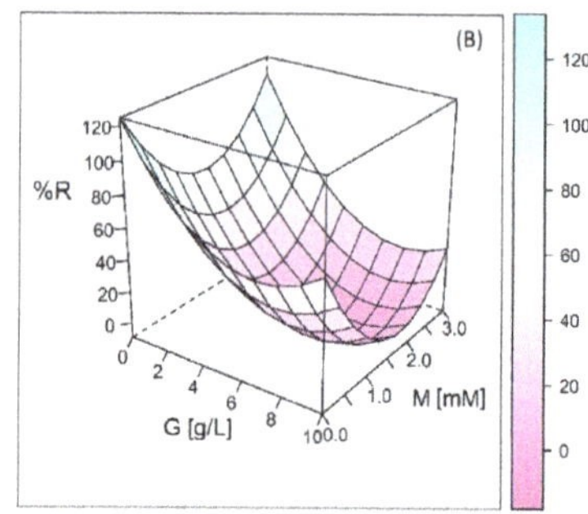

Figure 5. Effect of glucose and methylene blue on organic matter remotion in a dual-chamber microbial fuel cell. (**A**) Pareto diagram of standardized effects. (**B**) Surface response of organic matter remotion (%R) as a function of the initial concentrations of glucose (G) and of methylene blue (M).

The production of electric current from a microbial fuel cell is due to the deviation of the electrons that are produced by the oxidation of organic matter towards the electrode to the detriment of the production of metabolites which are the natural reservoirs of the produced electrons. To evaluate this effect, a set of three experiments were conducted at a constant initial concentration of glucose (5.0 g L^{-1}), and the production of metabolites were evaluated after a 24 h treatment: (i) an anodic fermentation under open-circuit conditions, (ii) an anodic fermentation without methylene blue under closed-circuit conditions, and (iii) an anodic fermentation with 0.3 mM of methylene blue under closed-circuit conditions. As can be seen from Figure 6, for each treatment, the metabolites lactate, acetate, and ethanol were measured. When conducting fermentations within the anodic compartment, the anode receives a fraction of the electrons generated during substrate oxidation. Because of this, a reduction in the levels of intracellular NADH takes place, which, in turn, adversely affects the production of reduced species such as intermediary metabolites [51]. In line with these ideas, Figure 6 reveals significant differences in the production of metabolites when compared against the anodic fermentation treatment that uses methylene blue. It is worth mentioning that no significant differences were found for the treatments lacking methylene blue (conventional fermentation in open-circuit conditions vs. anodic fermentation), where average concentration values of 2.5 g L^{-1} of lactate, 1.5 g L^{-1} of acetate, and 2.0 g L^{-1} of ethanol were found. These results indicate that *E. coli* would have a limited capacity for the transference of electrons to the anode. The treatment with methylene blue was characterized by a net decrement in the production of metabolites, i.e., a 35% reduction in lactate, 98.4% reduction in acetate and 75.5% reduction in ethanol. This could be explained in terms of the capacity of methylene blue to facilitate the diversion to the electrode of a portion of the electrons that resulted from the oxidation of glucose. It was reported that during the production of electricity by means of anodic fermentation with *E. coli*, the concentration of lactate was reduced from 3.3 mM to 2.2 mM and the concentration of acetate was reduced from 4.4 mM to 2.2 mM when using neutral red as the redox mediator [42].

Likewise, the production of ethanol by means of anodic fermentation with *Saccharomyces cerevisiae* was reduced from 1.7% to 0.3% when methylene blue was added [52].

Based on the production of lactate ($C_3H_6O_3$), acetate ($C_2H_4O_2$), and ethanol (C_2H_6O) as presented in Figure 6, the total number of electrons incorporated into these metabolic products during open-circuit fermentation (OCV) was 367.6 mmol; while in anodic electro-fermentation (0.3 mM of MB), 105.0 mmol of electrons were incorporated. The difference between these two values (262.5 mmol) corresponds to the reduction in metabolite production (anodic electrofermentation vs. open-circuit fermentation). On the other hand, the transference of electrons through the external circuit during anodic electrofermentation, calculated from Figure 2A (glucose at 5.0 g L^{-1} and MB at 0.3 mM) by means of Equation (1), would amount to 0.26 mmol (25.57 coulombs), which means an electrofermentation efficiency of 0.1%. It can be observed that the electron flow through the external circuit alone cannot account for the reduction in lactate, acetate, and ethanol during the anodic electrofermentation. Similar results have been found for cathodic electrofermentation processes, where cells consume electrons and increase the production of metabolites, as seen for the acetone–butanol production by *Clostridium*. In this scenario, the flow of electrons through the external circuit, measured by chronoamperometry, is much lower than the increase in electrons incorporated into acetone and butanol, resulting in electrofermentation efficiencies ranging from between 0.2% and 1.0% [53–56]. This electron imbalance could be attributed to the diffusion of oxygen from the cathodic chamber, which has been reported as one of the primary causes of low efficiency in microbial fuel cells [57]. Furthermore, the most significant factor considered was the poor glucose degradation rate, which, under the experimental conditions considered here, was in the order of 40.6% (Table 1), whereas under open-circuit fermentation conditions, glucose degradation reached 76%. The limited glucose degradation hindered the release of electrons that could otherwise be diverted to the anode, thus increasing the electrofermentation efficiency. Additionally, since acetate is not involved in the regeneration of NADH during the *E. coli* metabolism, the low glucose degradation led to a reduced acetate production, which is the metabolic compound that drives ATP production through substrate-level phosphorylation. This diminished ATP production, which in turn resulted in a low biomass yield. Reductions in biomass yield up to 56% have been observed with *E. coli* cultures under closed-circuit conditions compared to their open-circuit counterparts [42]. This study demonstrated the ability of the *E. coli* DH5α strain to generate electricity in a microbial fuel cell using exogenous redox mediators, despite its limited ability to transfer electrons to the anode itself. The experimental evidence gathered here opens the doors to explore new applications of *E. coli* through genetic engineering. For instance, enhancing its electrogenic capacity through the production of endogenous redox mediators [58], direct electron transfer [59], or its ability to form a dense biofilm on the anode surface [20] are strategies that have recently begun to be evaluated. These strategies could further improve the performance of the MFC to achieve large-scale utilization of these devices.

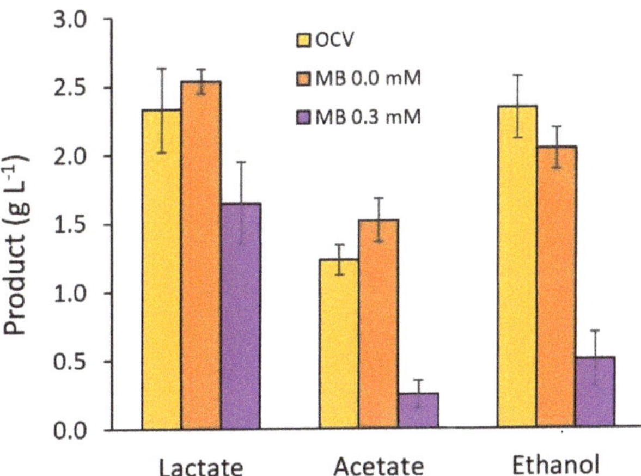

Figure 6. Fermentation products by *E. coli* with 5.0 g L^{-1} of glucose. Conventional fermentation under open-circuit conditions (gold bar). Anodic fermentation without methylene blue (orange bar). Anodic fermentation with 0.3 mM of methylene blue (purple bar).

4. Conclusions

Microbial fuel cells are devices capable of degrading organic matter while simultaneously generating electricity. The most relevant results of this study demonstrate that the highest substrate degradation percentage is achieved at lower concentrations (1.0 g L^{-1}),

Energies **2023**, *16*, 7901

while substrate degradation decreases at higher concentrations. On the other hand, to attain a higher current production, high substrate concentrations are required as the source of electrons (10.0 g L^{-1}), thus necessitating a trade-off relationship for the initial substrate concentration between electricity production and organic matter degradation. It was observed that *E. coli* produces only a very marginal amount of current in the absence of methylene blue (MB) or any redox mediator, indicating its limited capacity for electron transfer to the anode, either directly or through the production of endogenous mediators. The use of MB significantly increases current production, reaching a maximum current of 1.434 mA m^{-2} at 0.3 mM of MB. Nevertheless, at a concentration of 3.0 mM of MB, there is an observed inhibition in current density, resulting in a 47.2% reduction in electricity production. Based on these findings, the experimental conditions considered here, which included 5.0 g L^{-1} of glucose and 0.3 mM of MB, are deemed suitable for the cultivation of microbial fuel cells. Finally, the experimental evidence gathered here would support that anodic electrofermentations carried out in MFC (microbial fuel cell) devices induce the redirection of electrons generated during substrate oxidation towards the anode, thereby reducing the synthesis of fermentation products (lactate and ethanol) traditionally used as intracellular electron acceptors, which are subsequently excreted into the culture medium.

Author Contributions: Conceptualization, C.M.-V. and J.C.Q.-D.; methodology, C.M.-V. and J.C.Q.-D.; software, J.C.Q.-D. and J.O.G.P.; validation, C.M.-V., J.C.Q.-D. and J.O.G.P.; formal analysis, C.M.-V.; investigation, C.M.-V.; resources, J.C.Q.-D.; data curation, C.M.-V., J.C.Q.-D. and J.O.G.P.; writing—original draft preparation, C.M.-V., J.C.Q.-D. and J.O.G.P.; writing—review and editing, C.M.-V., J.C.Q.-D. and J.O.G.P.; visualization, J.C.Q.-D. and J.O.G.P.; supervision, J.C.Q.-D.; project administration, J.C.Q.-D.; funding acquisition, J.C.Q.-D. All authors have read and agreed to the published version of the manuscript.

Funding: This research was funded by the Ministerio de Ciencia, Tecnología e Innovación of Colombia, grant number 80740-177.2019, and was also supported by the Comité para el Desarrollo de la Ciencia y la Tecnología, CODI de la Universidad de Antioquia (Colombia), grant number PROG 2015-7343.

Data Availability Statement: Not applicable.

Conflicts of Interest: The authors declare no conflict of interest.

References

1. Yolcan, O.O. World Energy Outlook and State of Renewable Energy: 10-Year Evaluation. *Innov. Green Dev.* **2023**, *2*, 100070. [CrossRef]
2. Han, J.-H.; Bae, J.; Lim, J.; Jwa, E.; Nam, J.-Y.; Hwang, K.S.; Jeong, N.; Choi, J.; Kim, H.; Jeung, Y.-C. Acidification-Based Direct Electrolysis of Treated Wastewater for Hydrogen Production and Water Reuse. *Heliyon* **2023**, *9*, e20629. [CrossRef] [PubMed]
3. Paraschiv, S.; Paraschiv, L.S.; Serban, A. An Overview of Energy Intensity of Drinking Water Production and Wastewater Treatment. *Energy Rep.* **2023**, *9*, 118–123. [CrossRef]
4. Rocha-Meneses, L.; Luna-delRisco, M.; González, C.A.; Moncada, S.V.; Moreno, A.; Sierra-Del Rio, J.; Castillo-Meza, L.E. An Overview of the Socio-Economic, Technological, and Environmental Opportunities and Challenges for Renewable Energy Generation from Residual Biomass: A Case Study of Biogas Production in Colombia. *Energies* **2023**, *16*, 5901. [CrossRef]
5. Tow, E.W.; Letcher, A.; Jaworowski, A.; Zucker, I.; Kum, S.; Azadiaghdam, M.; Blatchley, E.R.; Achilli, A.; Gu, H.; Melike, G.; et al. Modeling the Energy Consumption of Potable Water Reuse Schemes. *Water Res. X* **2021**, *13*, 100126. [CrossRef]
6. Malik, S.; Kishore, S.; Dhasmana, A.; Kumari, P.; Mitra, T.; Chaudhary, V.; Kumari, R.; Bora, J.; Ranjan, A.; Minkina, T.; et al. A Perspective Review on Microbial Fuel Cells in Treatment and Product Recovery from Wastewater. *Water* **2023**, *15*, 316. [CrossRef]
7. Li, X.; Abu-reesh, I.M.; He, Z. Development of Bioelectrochemical Systems to Promote Sustainable Agriculture. *Agriculture* **2015**, *5*, 367–388. [CrossRef]
8. Ghangrekar, M.M.; Chatterjee, P. A Systematic Review on Bioelectrochemical Systems Research. *Curr. Pollut. Rep.* **2017**, *3*, 281–288. [CrossRef]
9. Santoro, C.; Arbizzani, C.; Erable, B.; Ieropoulos, I. Microbial Fuel Cells: From Fundamentals to Applications. A Review. *J. Power Sources* **2017**, *356*, 225–244. [CrossRef]
10. Goenka, R.; Mukherji, S.; Ghosh, P.C. Bioresource Technology Reports Characterization of Electrochemical Behaviour of *Escherichia coli* MTCC 1610 in a Microbial Fuel Cell. *Bioresour. Technol. Rep.* **2018**, *3*, 67–74. [CrossRef]

11. Sturm-Richter, K.; Golitsch, F.; Sturm, G.; Kipf, E.; Dittrich, A.; Beblawy, S.; Kerzenmacher, S.; Gescher, J. Unbalanced Fermentation of Glycerol in *Escherichia coli* via Heterologous Production of an Electron Transport Chain and Electrode Interaction in Microbial Electrochemical Cells. *Bioresour. Technol.* **2015**, *186*, 89–96. [CrossRef] [PubMed]
12. Feng, J.; Lu, Q.; Li, K.; Xu, S.; Wang, X.; Chen, K.; Ouyang, P. Construction of an Electron Transfer Mediator Pathway for Bioelectrosynthesis by *Escherichia coli*. *Front. Bioeng. Biotechnol.* **2020**, *8*, 590667. [CrossRef] [PubMed]
13. Kracke, F.; Vassilev, I.; Krömer, J.O. Microbial Electron Transport and Energy Conservation—The Foundation for Optimizing Bioelectrochemical Systems. *Front. Microbiol.* **2015**, *6*, 575. [CrossRef] [PubMed]
14. Price, C.E.; Driessen, A.J.M. Biogenesis of Membrane Bound Respiratory Complexes in *Escherichia coli*. *Biochim. Biophys. Acta BBA Mol. Cell Res.* **2010**, *1803*, 748–766. [CrossRef]
15. Sugnaux, M.; Mermoud, S.; Ferreira, A.; Happe, M.; Fischer, F. Bioresource Technology Probing Electron Transfer with *Escherichia coli*: A Method to Examine Exoelectronics in Microbial Fuel Cell Type Systems. *Bioresour. Technol.* **2013**, *148*, 567–573. [CrossRef]
16. Hoang, U.; Nguyen, P.; Grekov, D.; Geiselmann, J.; Stambouli, V.; Weidenhaupt, M.; Delabouglise, D. Electrochimica Acta Anodic Deposit from Respiration Metabolic Pathway of *Escherichia coli*. *Electrochimica Acta* **2014**, *130*, 200–205. [CrossRef]
17. Zhang, T.; Cui, C.; Chen, S.; Yang, H.; Shen, P. The Direct Electrocatalysis of *Escherichia coli* through Electroactivated Excretion in Microbial Fuel Cell. *Electrochem. Commun.* **2008**, *10*, 293–297. [CrossRef]
18. Kostylev, M.; Otwell, A.E.; Richardson, R.E.; Suzuki, Y. Cloning Should Be Simple: *Escherichia coli* DH5α-Mediated Assembly of Multiple DNA Fragments with Short End Homologies. *PLoS ONE* **2015**, *10*, e0137466. [CrossRef]
19. Jahnke, J.P.; Sarkes, D.A.; Liba, J.L.; Sumner, J.J.; Stratis-Cullum, D.N. Improved Microbial Fuel Cell Performance by Engineering *E. coli* for Enhanced Affinity to Gold. *Energies* **2021**, *14*, 5389. [CrossRef]
20. Nguyen, D.-T.; Tamura, T.; Tobe, R.; Mihara, H.; Taguchi, K. Microbial Fuel Cell Performance Improvement Based on FliC-Deficient *E. coli* Strain. *Energy Rep.* **2020**, *6*, 763–767. [CrossRef]
21. Li, T.; Yang, X.; Chen, Q.; Song, H.; He, Z.; Yang, Y. Enhanced Performance of Microbial Fuel Cells with Electron Mediators from Anthraquinone/Polyphenol-Abundant Herbal Plants. *ACS Sustain. Chem. Eng.* **2020**, *8*, 11263–11275. [CrossRef]
22. Nawaz, A.; Hafeez, A.; Abbas, S.Z.; Haq, I.; Rafatullah, M. A State of the Art Review on Electron Transfer Mechanisms, Characteristics, Applications and Recent Advancements in Microbial Fuel Cells Technology. *Green Chem. Lett. Rev.* **2020**, *13*, 365–381. [CrossRef]
23. Ng, I.S.; Hsueh, C.C.; Chen, B.Y. Electron Transport Phenomena of Electroactive Bacteria in Microbial Fuel Cells: A Review of Proteus Hauseri. *Bioresour. Bioprocess.* **2017**, *4*, 53. [CrossRef]
24. Gemünde, A.; Lai, B.; Pause, L.; Krömer, J.; Holtmann, D. Redox Mediators in Microbial Electrochemical Systems. *ChemElectroChem* **2022**, *9*, e202200216. [CrossRef]
25. Martinez, C.M.; Alvarez, L.H. Application of Redox Mediators in Bioelectrochemical Systems. *Biotechnol. Adv.* **2018**, *36*, 1412–1423. [CrossRef]
26. Rahimnejad, M.; Najafpour, G.D.; Ghoreyshi, A.A.; Shakeri, M.; Zare, H. Methylene Blue as Electron Promoters in Microbial Fuel Cell. *Int. J. Hydrogen Energy* **2011**, *36*, 13335–13341. [CrossRef]
27. Reiche, A.; Kirkwood, K.M. Bioresource Technology Comparison of *Escherichia coli* and Anaerobic Consortia Derived from Compost as Anodic Biocatalysts in a Glycerol-Oxidizing Microbial Fuel Cell. *Bioresour. Technol.* **2012**, *123*, 318–323. [CrossRef]
28. Sharma, T.; Mohana, A.L.; Chandra, T.S.; Ramaprabhu, S. Development of Carbon Nanotubes and Nanofluids Based Microbial Fuel Cell. *Int. J. Hydrogen Energy* **2008**, *33*, 6749–6754. [CrossRef]
29. Ojima, Y.; Kawaguchi, T.; Fukui, S.; Kikuchi, R.; Terao, K.; Koma, D. Promoted Performance of Microbial Fuel Cells Using *Escherichia coli* Cells with Multiple—Knockout of Central Metabolism Genes. *Bioprocess Biosyst. Eng.* **2019**, *43*, 323–332. [CrossRef]
30. Popov, A.L.; Kim, J.R.; Dinsdale, R.M.; Esteves, S.R.; Guwy, A.J.; Premier, G.C. The Effect of Physico-Chemically Immobilized Methylene Blue and Neutral Red on the Anode of Microbial Fuel Cell. *Biotechnol. Bioprocess Eng.* **2012**, *370*, 361–370. [CrossRef]
31. Aiyer, K.S. Synergistic Effects in a Microbial Fuel Cell between Co-Cultures and a Photosynthetic Alga Chlorella Vulgaris Improve Performance. *Heliyon* **2021**, *7*, e05935. [CrossRef] [PubMed]
32. Lee, J.; Xu, H.; Xiao, Y.; Calhoun, M.C.; Crisostomo, D.; Cliffel, D.E. The Utilization of Escericia Coli and Shewanella Oneidensis for Microbial Fuel Cell. In *IOP Conference Series: Materials Science and Engineering, Proceedings of the 3rd International Conference on Chemical Engineering Sciences and Applications 2017 (3rd ICChESA 2017), Banda Aceh, Indonesia, 20–21 September 2017*; IOP Publishing Ltd.: Bristol, UK, 2018; Volume 334, pp. 1–6. [CrossRef]
33. del Campo, A.G.; Canizares, P.; Lobato, J.; Rodrigo, M.; Fernandez Morales, F.J. Effects of External Resistance on Microbial Fuel Cell's Performance. In *Environment, Energy and Climate Change II. The Handbook of Environmental Chemistry*; Lefebvre, G., Jiménez, E., Cabañas, B., Eds.; Springer: Cham, Switzerland, 2014; pp. 41–53. ISBN 1433-6863r978-3-642-03970-6.
34. Moscoviz, R.; Toledo-Alarcón, J.; Trably, E.; Bernet, N. Electro-Fermentation: How to Drive Fermentation Using Electrochemical Systems. *Trends Biotechnol.* **2016**, *34*, 856–865. [CrossRef] [PubMed]
35. Zaitoun, M.A. A Kinetic-Spectrophotometric Method for the Determination of Glucose in Solutions. *J. Anal. Chem.* **2006**, *61*, 1010–1014. [CrossRef]
36. Leon-Fernandez, L.F.; Rodrigo, M.A.; Villaseñor, J.; Fernandez-Morales, F.J. Electrocatalytic Dechlorination of 2,4-Dichlorophenol in Bioelectrochemical Systems. *J. Electroanal. Chem.* **2020**, *876*, 114731. [CrossRef]
37. Taskan, E.; Ozkaya, B.; Hasar, H. Effect of Different Mediator Concentrations on Power Generation in MFC Using Ti-TiO$_2$ Electrode. *Int. J. Energy Sci.* **2014**, *4*, 9–11. [CrossRef]

38. Chen, Y.-A.; Yang, H.; Ouyang, D.; Liu, T.; Liu, D.; Zhao, X. Construction of Electron Transfer Chains with Methylene Blue and Ferric Ions for Direct Conversion of Lignocellulosic Biomass to Electricity in a Wide pH Range. *Appl. Catal. B Environ.* **2020**, *265*, 118578. [CrossRef]
39. Gomes, T.F.; Pedrosa, M.M.; de Toledo, A.C.L.; Arnoni, V.W.; dos Santos Monteiro, M.; Piai, D.C.; Sylvestre, S.H.Z.; Ferreira, B. Bactericide Effect of Methylene Blue Associated with Low-Level Laser Therapy in *Escherichia coli* Bacteria Isolated from Pressure Ulcers. *Lasers Med. Sci.* **2018**, *33*, 1723–1731. [CrossRef]
40. Nicu, A.I.; Pirvu, L.; Vamanu, A.; Stoian, G. The European Beech Leaves Extract Has an Antibacterial Effect by Inducing Oxidative Stress. *Romanian Biotechnol. Lett.* **2016**, *22*, 12071–12080.
41. Ash, S.R.; Steczko, J.; Brewer, L.B.; Winger, R.K. Microbial Inactivation Properties of Methylene Blue—Citrate Solution. *ASAIO J.* **2006**, *52*, 17A. [CrossRef]
42. Park, D.H.; Zeikus, J.G. Electricity Generation in Microbial Fuel Cells Using Neutral Red as an Electronophore. *Appl. Environ. Microbiol.* **2000**, *66*, 1292–1297. [CrossRef]
43. Schmitz, S.; Rosenbaum, M.A. Boosting Mediated Electron Transfer in Bioelectrochemical Systems with Tailored Defined Microbial Cocultures. *Biotechnol. Bioeng.* **2018**, *115*, 2183–2193. [CrossRef] [PubMed]
44. Rahmani, A.R.; Navidjouy, N.; Rahimnejad, M.; Alizadeh, S.; Samarghandi, M.R.; Nematollahi, D. Effect of Different Concentrations of Substrate in Microbial Fuel Cells toward Bioenergy Recovery and Simultaneous Wastewater Treatment. *Environ. Technol.* **2022**, *43*, 1–9. [CrossRef] [PubMed]
45. Förster, A.H.; Gescher, J. Metabolic Engineering of *Escherichia coli* for Production of Mixed-Acid Fermentation End Products. *Front. Bioeng. Biotechnol.* **2014**, *2*, 1–12. [CrossRef]
46. Pinhal, S.; Ropers, D.; Geiselmann, J.; de Jong, H. Acetate Metabolism and the Inhibition of Bacterial Growth by Acetate. *J. Bacteriol.* **2019**, *201*, 1–19. [CrossRef]
47. Wang, C.; Chang, T.; Yang, H.; Cui, M. Antibacterial Mechanism of Lactic Acid on Physiological and Morphological Properties of Salmonella Enteritidis, *Escherichia coli* and Listeria Monocytogenes. *Food Control* **2015**, *47*, 231–236. [CrossRef]
48. Roe, A.J.; O'Byrne, C.; McLaggan, D.; Booth, I.R. Inhibition of *Escherichia coli* Growth by Acetic Acid: A Problem with Methionine Biosynthesis and Homocysteine Toxicity. *Microbiology* **2002**, *148*, 2215–2222. [CrossRef]
49. Xiang, K.; Qiao, Y.; Ching, C.B.; Li, C.M. GldA Overexpressing-Engineered *E. coli* as Superior Electrocatalyst for Microbial Fuel Cells. *Electrochem. Commun.* **2009**, *11*, 1593–1595. [CrossRef]
50. Ren, J.; Li, N.; Du, M.; Zhang, Y.; Hao, C.; Hu, R. Study on the Effect of Synergy Effect between the Mixed Cultures on the Power Generation of Microbial Fuel Cells. *Bioengineered* **2021**, *12*, 844–854. [CrossRef]
51. Vassilev, I.; Averesch, N.J.H.; Ledezma, P.K.M. Anodic Electro-Fermentation: Empowering Anaerobic Production Processes via Anodic Respiration. *Biotechnol. Adv.* **2021**, *48*, 107728.
52. Rossi, R.; Fedrigucci, A.; Setti, L. Characterization of Electron Mediated Microbial Fuel Cell by Saccharomyces Cerevisiae. *Chem. Eng. Trans.* **2015**, *43*, 337–342. [CrossRef]
53. Alberto García Mogollón, C.; Carlos Quintero Díaz, J.; Omar Gil Posada, J. Production of Acetone, Butanol, and Ethanol by Electro-Fermentation with *Clostridium Saccharoperbutylacetonicum* N1-4. *Bioelectrochemistry* **2023**, *152*, 108414. [CrossRef]
54. Engel, M.; Holtmann, D.; Ulber, R.; Tippkötter, N. Increased Biobutanol Production by Mediator-Less Electro-Fermentation. *Biotechnol. J.* **2019**, *14*, 1800514. [CrossRef] [PubMed]
55. Choi, O.; Kim, T.; Woo, H.M.; Um, Y. Electricity-Driven Metabolic Shift through Direct Electron Uptake by Electroactive Heterotroph Clostridiumpasteurianum. *Sci. Rep.* **2014**, *4*, 6961. [CrossRef] [PubMed]
56. Moscoviz, R.; Trably, E.; Bernet, N. Electro-Fermentation Triggering Population Selection in Mixed-Culture Glycerol Fermentation. *Microb. Biotechnol.* **2018**, *11*, 74–83. [CrossRef] [PubMed]
57. Wang, C.; Shen, J.; Chen, Q.; Ma, D.; Zhang, G.; Cui, C.; Xin, Y.; Zhao, Y.; Hu, C. The Inhibiting Effect of Oxygen Diffusion on the Electricity Generation of Three-Chamber Microbial Fuel Cells. *J. Power Sources* **2020**, *453*, 227889. [CrossRef]
58. Simoska, O.; Cummings, D.A., Jr.; Gaffney, E.M.; Langue, C.; Primo, T.G.; Weber, C.J.; Witt, C.E.; Minteer, S.D. Enhancing the Performance of Microbial Fuel Cells via Metabolic Engineering of *Escherichia coli* for Phenazine Production. *ACS Sustain. Chem. Eng.* **2023**, *11*, 11855–11866. [CrossRef]
59. Jensen, H.M.; TerAvest, M.A.; Kokish, M.G.; Ajo-Franklin, C.M. CymA and Exogenous Flavins Improve Extracellular Electron Transfer and Couple It to Cell Growth in Mtr-Expressing *Escherichia coli*. *ACS Synth. Biol.* **2016**, *5*, 679–688. [CrossRef]

Article

Design and Control of Extractive Dividing Wall Column for Separating Dipropyl Ether/1-Propyl Alcohol Mixture

Qiliang Ye [1,*], Yule Wang [1], Hui Pan [2], Wenyong Zhou [1] and Peiqing Yuan [1]

[1] School of Chemical Engineering, East China University of Science and Technology, Shanghai 200237, China; y30190958@mail.ecust.edu.cn (Y.W.); wyzhou@ecust.edu.cn (W.Z.); pqyuan@ecust.edu.cn (P.Y.)
[2] College of Environmental and Chemical Engineering, Shanghai University of Electric Power, Shanghai 200090, China; fiona_panhui@shiep.edu.cn
* Correspondence: yql@ecust.edu.cn

Abstract: The focus of this work is the study of the extractive dividing wall column (EDWC) for separating the azeotropic mixture of dipropyl ether and 1-propyl alcohol with N, N-dimethylacetamide (DMAC) as the entrainer. Three separation sequences are investigated, including a conventional extractive distillation sequence (CEDS), EDWC and a pressure swing distillation sequence (PSDS). The static simulation results showed that the EDWC with DMAC as the entrainer is more economically attractive than CEDS and PSDS. Subsequently, a control structure CS1 based on a three-temperature control loop and a control structure CS2 with the vapor split ratio as the manipulated variable are investigated for the EDWC. Their dynamic control performances are evaluated by facing large feed flow rates and composition disturbances. The results showed that the CS1 can deal with feed flow rate disturbance effectively. However, the transient deviation is large and the settling time is too long when facing feed flow composition disturbances. The CS2 can quickly and effectively deal with feed flow rate and composition disturbances, and it can maintain the two products at high purity.

Keywords: extractive dividing wall column; dipropyl ether; 1-propyl alcohol; energy savings

1. Introduction

Dipropyl ether (DPE) is a flammable, colorless liquid with a boiling point of 90.08 °C. It is used mainly as a gasoline additive, stabilizing agent and industrial solvent [1]. 1-propyl alcohol (PA) is a highly flammable, colorless liquid with a boiling point of 97.1 °C. 1-propyl may be used in several applications such as personal care products, cosmetics, pharmaceuticals and solvent, as well as an intermediate for the production of methyl isobutyl ketone, amines, acetone and glycerol [2]. Dipropyl ether can be synthesized by the dehydration of 1-propyl alcohol with sulfuric acid, but the mixture of dipropyl ether and 1-propyl alcohol is an azeotropic mixture and it cannot be completely separated by a simple distillation process.

Much research has been focused on the selection of a separation process for the azeotropic mixture of dipropyl ether and 1-propyl alcohol. Lladosa et al. [3] designed a pressure swing distillation process and an extractive distillation process with 2-ethoxyethanol as an entrainer. Their results showed that pressure swing distillation can reduce the total annual cost by 29.46% compared with that of the extractive distillation process. Mangili [4] further proposed two improved processes based on vapor recompression and direct heat integration, and the results showed that the two improved processes can be increased by 68.69% and 39.54%, making them more economically attractive than the pressure swing distillation process. Luo et al. [5] designed an extractive distillation with 2-methoxyethanol as an entrainer and a heat-integrated pressure swing distillation to separate isopropyl ether/isopropanol azeotropic mixture. Their results revealed that the heat-integrated pressure swing distillation process could reduce the total annual cost by 5.75% compared with that of the extractive distillation process. An extractive distillation under lower pressure

Citation: Ye, Q.; Wang, Y.; Pan, H.; Zhou, W.; Yuan, P. Design and Control of Extractive Dividing Wall Column for Separating Dipropyl Ether/1-Propyl Alcohol Mixture. *Processes* **2022**, *10*, 665. https://doi.org/10.3390/pr10040665

Academic Editor: Maria Gavrilescu

Received: 3 March 2022
Accepted: 26 March 2022
Published: 29 March 2022

was redesigned by You et al. [6]. It was found that the extractive distillation process under lower pressure could reduce the total annual cost by 7.6%.

Dividing wall columns (DWCs) have attracted wide attention because they can greatly reduce the energy consumption of the distillation process. The internal space of the DWC is divided by a dividing wall to realize the separation of the ternary mixture in one column. Compared with the conventional distillation sequence, DWC reduces the number of condensers and reboilers, and thermodynamically reduces the effect of remixing in the column. Generally, DWC can save about 30% of capital costs and energy compared with conventional separation sequence [7–10].

Applying the DWC technology to the extractive distillation process is the extractive dividing wall column (EDWC, see Figure 1a). A dividing wall is added to divide the column into three parts. The feed and the entrainer is fed into the extractive distillation section C1. The entrainer S and components B enter the public stripping section C3, which are finally separated in the entrainer recovery section C2. The AB components are withdrawn from both sides of the column top, respectively. The entrainer is recycled from the bottom of the column. In order to balance the loss of entrainer caused by distillates, it is necessary to add a makeup stream. Figure 1b is the equivalent two columns model commonly used to simulate the EDWC. The entrainer recovery section is simulated by a column C2 without reboiler. The extractive distillation section and the public stripping section are combined into a column C1 for simulation.

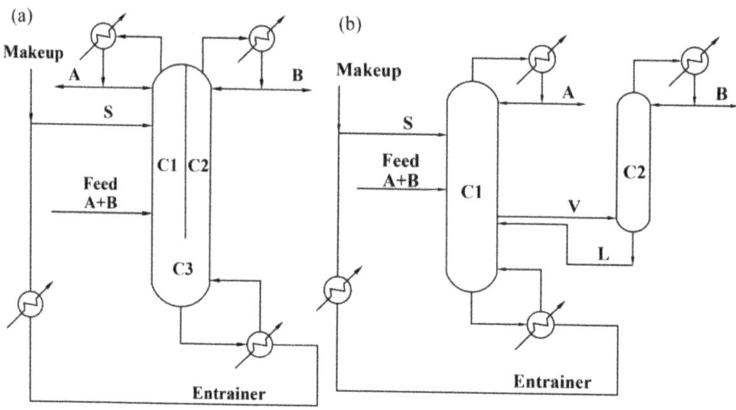

Figure 1. (**a**) Process flowsheet of EDWC and (**b**) equivalent two column model of EDWC.

Compared with the conventional extractive distillation sequence, EDWC has one reboiler, and the structure of the dividing wall also increases the complexity of the structure in the column. Therefore, the control structure of the EDWC is the key factor for its industrially implementation. Wu et al. [11] investigated the acetone/methanol system and designed a three-temperature control loop based on the conventional extractive distillation sequence, two condensers and one reboiler control three parts of the temperature control tray to achieve the stability of product purity. Xia et al. [12,13] investigated the methylal/methanol system and designed two control structures. The first control structure includes four composition controllers and an adjustable vapor split ratio, however, when faced with a decrease in the feed flow rate, the purity of the methanol product decreased relatively quickly. On this basis, the author proposed an improved control structure based on the differential temperature control, which effectively suppresses the offsets of methanol product purity. Zhang et al. [14] investigated the ethyl acetate/isopropanol system, and used a three column model to simulate the EDWC, and optimized the column structure with the goal of minimal total annual cost. Subsequently, three control structures were proposed, in addition to the three-temperature control loop structure, the two improved structures with the entrainer flow rate and the vapor split ratio as the control variables were investigated. The dynamic simulation results showed that the two improved structures have excellent dynamic performance.

Although the control structure for controlling the vapor split ratio has achieved good dynamic performance in simulation, it is difficult to achieve in industrial application. Recently, many researchers [15–20] have designed a special internal column to directly control the vapor split ratio of DWC in laboratory devices. Luyben [21] proposed changing the vapor split ratio by varying the pressure on both sides of the EDWC. Jing et al. [22] added a pressure compensation, based on Luyben's research, to further improve the dynamic performance. Zhu et al. [23] realized the control of the vapor split ratio by adding an intermediate reboiler in the extractive distillation section.

In addition, a variety of new control systems were used in the control of the EDWC, such as artificial neural network [24,25], model prediction control [26–29], which showed good dynamic performance when controlling the EDWC.

In this paper, the conventional extractive distillation sequence (CEDS), the EDWC, and the pressure swing distillation sequence (PSDS) were designed and optimized using N, N-dimethylacetamide (DMAC) as the entrainer. Then, the three processes were compared through economic analysis. Finally, a control structure CS1 based on a three temperature control loop and a control structure CS2 with vapor split ratio as the manipulated variable are investigated to control the EDWC.

2. Steady State Design

2.1. Materials and Methods

The feed flow rate was 100 kmol/h, and the feed composition was 50/50 mol% DPE/PA. The production specifications of DPE and PA reached 99.9 mol%. The condenser pressure was set to 101.325 kPa. The pressure drop of each tray was 0.69 kPa.

The steady-state and dynamic simulations were implemented by Aspen Plus V9.0 and Aspen Dynamics V9.0 commercial software, the DSTWU module for shortcut calculations and RadFrac module for rigorous simulations. According to thermodynamic behaviors of ternary system of DPE + PA + N, N-dimethylformamide, the UNIQUAC model was suitable to describe the ternary system [30]. Thus, the UNIQUAC model was applied to predict the vapor–liquid equilibrium in the simulations.

In this work, the minimal total annual cost (TAC) was the optimization goal, which is defined as the sum of the annual operation costs and capital investments divided by the three-year investment payback period. The capital investments are mainly composed of the column vessels, two condensers, and one reboiler. Here, it is assumed that the heat transfer coefficients of the condenser and the reboiler are $0.852 \text{ kW}/(\text{K·m}^2)$ and $0.568 \text{ kW}/(\text{K·m}^2)$. Some items are usually not considered because the cost of trays, reflux drums, pipelines, pumps, valves is much less than the cost of the column vessels and heat exchanger. For the above basis of the assumption and more calculation details, please refer to Luyben's book [31].

The mainly optimized design variables include the number of stages N_T, the feed stage N_F, the entrainer feed stage N_{FE} when design CEDS and EDWC, and the reflux ratios of the columns.

The optimization strategies mainly include: varying the reflux ratio to meet the production specifications, adjusting the feed stage to minimize the reboiler duty and varying the number of stages to minimize the TAC.

2.2. Steady State Design for CEDS

For the extractive distillation, it was necessary to select the entrainer and its flow rate. *N,N*-dimethylformamide is the best entrainer for separating DPE and PA [3]. However, it is not an environmentally friendly solvent or entrainer. Hence, in this work, DMAC was selected as the entrainer, which shows similar properties as N, N-dimethylformamide was selected as the entrainer.

The effect of entrainer flow rate S and reflux ratio RR1 on the concentration of light component DPE was determined by sensitivity analysis. As illustrated in Figure 2, the increase of S resulted in the improvement of the purity of DPE product. Additionally, the purity of DPE product first improved and then reduced with the increase of RR1. Obviously, the increase of RR1 can prevent more entrainer entering the column top, thereby improving the purity of DPE product. However, too large RR1 also leads to the dilution of the concentration of DMAC by the surplus DPE, so that more component PA enters the top of the column. In order to achieve the production specifications, the entrainer flow rate was selected as 120 kmol/h, and the reflux ratio of the extractive distillation column C1 was selected as 1.2. Figure 3 shows the final optimization results and detailed information of each steam, heat duties and the size of the column.

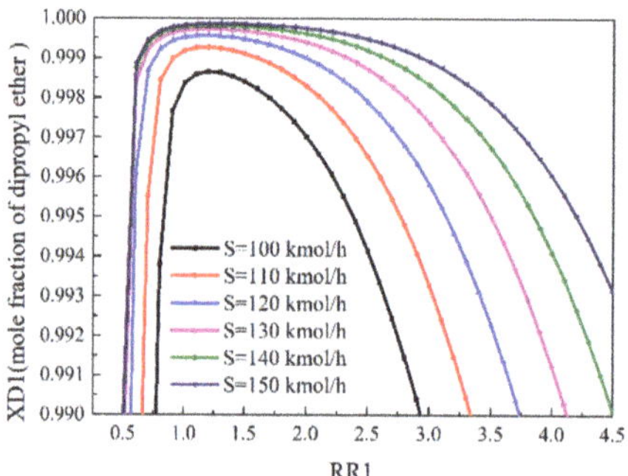

Figure 2. Effect of entrainer flow rate S and reflux ratio RR1 on the purity of DPE.

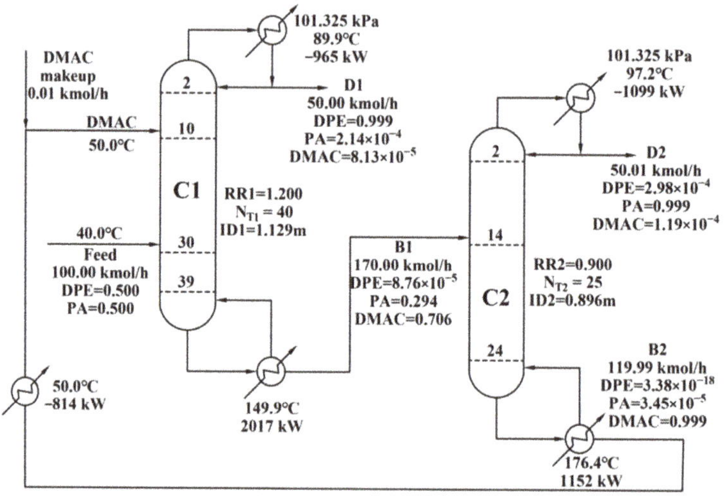

Figure 3. Process flow diagram of the CEDS.

2.3. Steady State Design for EDWC

According to the design results of the CEDS, the initial values of the structural parameters of the two column model (as shown in Figure 1b) were obtained. First, the entrainer recovery column in the CEDS was separated from the feed in two parts, the rectification section parameter of the entrainer recovery column was simulated with a column C2 without reboiler. After the reboiler was removed from the extractive distillation column, the structural parameters of the stripping section of the entrainer recovery column were simulated by column C1. Then, the number of stages and the total pressure drop of the C2 column were increased to be the same as that of the dividing wall of the C1 column. Finally, by adjusting the vapor flow rate V2 between the two columns and the reflux ratio of the two columns, the purity of the top products meets the production specifications [11].

The optimized design variables include the number of stages N_{T1} of the C1 column, the mixture feed stage N_F, the entrainer feed stage N_{FE}, C2 column feed location N_{F2} (also the position of the dividing wall) and the number of stages N_{T2} ($N_{T2} = N_{F2} - 1$), the vapor flow rate V2 between the two columns, and the reflux ratios RR1 and RR2 of the two columns. Figure 4 illustrates the optimization procedure of the EDWC.

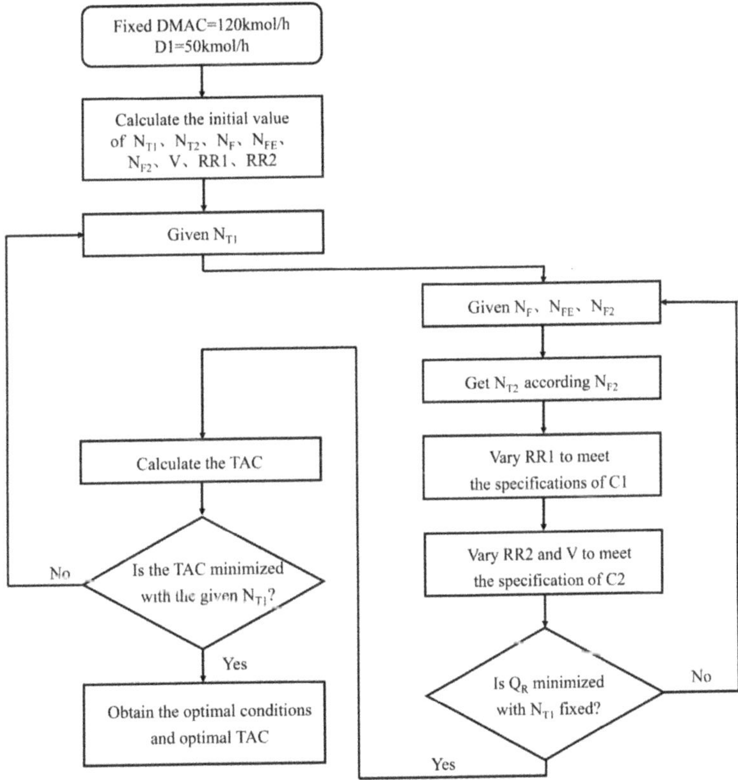

Figure 4. Sequential iterative optimization procedure for EDWC.

The influence of the number of stages N_{T1} of the C1 column and the number of stages N_{T2} of the C2 column on the TAC of the EDWC is depicted in Figure 5. Obviously, the optimized N_{T1} and N_{T2} were 52 and 41, respectively. Figure 6 shows the final optimization results of the EDWC and the detailed information of each stream, heat duties, and the size of the column. As shown in Figure 6, the feed stage N_F was 30, and the entrainer feed stage N_{FE} was 10. The vapor stream flow rate V2 to the C2 column was 60 kmol/h (the vapor flow rate V1 to the C1 column was 158.5 kmol/h, thus the vapor split ratio α = V1/(V1 + V2) = 0.7254). The reflux ratio RR1 was 1.101, and RR2 was 0.273. The three parts of the EDWC share a column vessel, and the upper section is divided into two sections with a dividing wall. To determine the actual diameter, the concept of equivalent diameter needs to be introduced [10]. The diameter of the C1 section was 1.117 m, and the diameter of the C2 section was 0.633 m. The equivalent diameter De can be calculated to be 1.284 m. Because the diameter of the public stripping section C3 was 1.360 m larger than the equivalent diameter De, the diameter of the EDWC was determined to be 1.360 m.

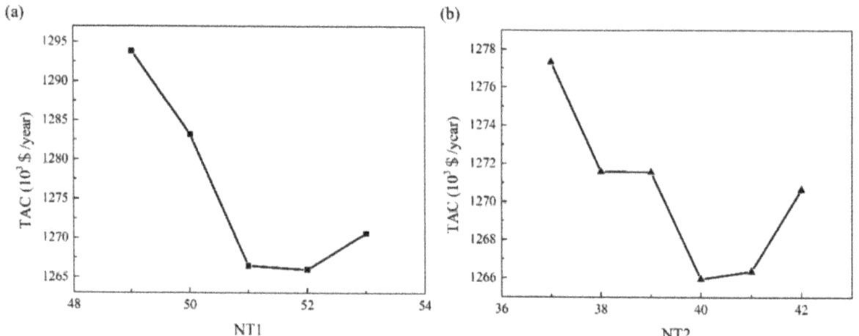

Figure 5. (**a**) Effect of NT1 on the TAC (**b**) effect of NT2 on the TAC.

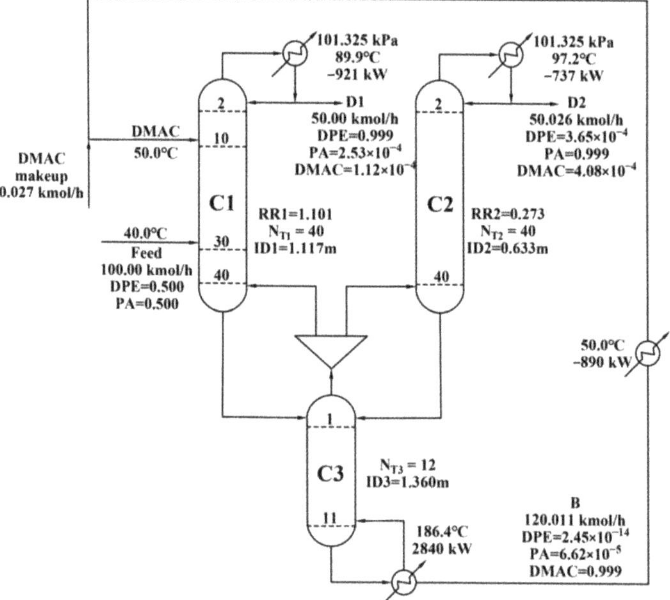

Figure 6. Optimal process flow diagram of the EDWC.

2.4. Steady State Design for PSDS

According to the pressure swing distillation designed by Lladosa et al. [3], the pressure of the low-pressure column is 30 kPa, the composition of DPE at the top of the column is 0.765. The pressure of the high-pressure column is 101.325 kPa, the composition of DPE at the top of the column is 0.674.

First, mass balance calculations were carried out based on the new feed flow rate and production specifications. The calculated distillate rate of the low-pressure column C1 was 178.57 kmol/h, and the distillate rate of the high-pressure column C2 was 128.57 kmol/h. The flow rates of PA and DPE products were 50.00 kmol/h. Then according to the results of the material balance, the structural parameters of the low-pressure column C1 and the high-pressure column C2 were optimized. Figure 7 shows the variation of stage number and heat duty as a function of the reflux ratio for the C1 (a) and C2 (b), it can be determined that the number of stages in the low-pressure column C1 is 16, the feed location is 9, the number of stages in the high-pressure column C2 is 17, and the feed location is 6. Figure 8 shows the final optimization results of the PSDS and the detailed information of each stream, heat duties, and the size of the column.

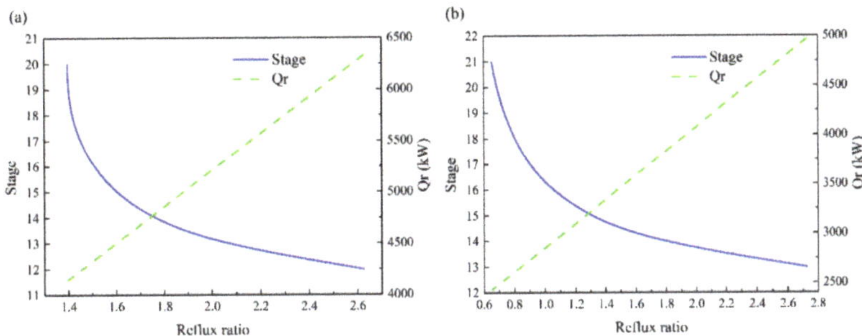

Figure 7. Variation of stage number and heat duty as a function of the reflux ratio for the C1 (**a**) and C2 (**b**).

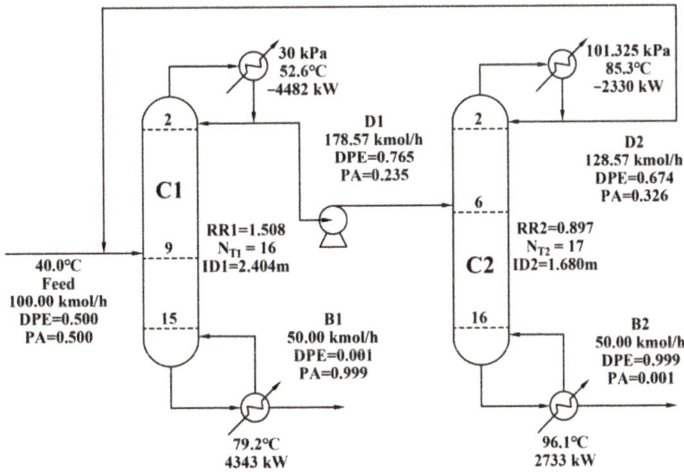

Figure 8. Process flow diagram of the PSDS.

2.5. Comparisons and Analysis of CEDS, EDWC and PSDS

Table 1 summarizes the optimization results of CEDS, EDWC and PSDS. Compared with the CEDS, the EDWC can save 12.45% of capital investments, 10.38% of operation costs, and TAC reduction of 11.01%. Compared with the PSDS, the EDWC can save 41.49% of capital investments, 47.07% of operation costs, and 45.50% of TAC. The results show that the EDWC with DMAC as the entrainer is more economically attractive than the CEDS and the PSDS.

Table 1. Optimization results of CEDS, EDWC and PSDS.

Parameter	CEDS	EDWC	PSDS
N_{T1}	40	52	16
N_{T2}	25	40	17
RR1	1.200	1.101	1.508
RR2	0.900	0.273	0.897
ID1 (m)	1.129	1.360	2.404
ID2 (m)	0.896	-	1.680
Total condenser duty (kW)	2064	1658	6812
Total reboiler duty (kW)	3169	2840	7076
Capital investment (10^3/yr)	1305.739	1143.239	1953.945
Operation cost (10^3/yr)	987.304	884.874	1671.689
TAC (10^3/yr)	1422.551	1265.955	2323.004

3. Control Structure for EDWC

Two control structures CS1 and CS2 are proposed for the dynamic control of EDWC. A three-temperature control loop CS1 was developed, RR1, RR2 and Q_R were used as the manipulated variables to control the reference stages from C1, C2, and C3, respectively. Subsequently, an improved control structure CS2 with combining active vapor split ratio α into CS1 was developed.

3.1. Selecting Temperature Control Trays

For the selection of the temperature control tray, there are usually methods such as slope criterion, sensitivity criterion, and singular value decomposition (SVD) criterion. The singular value decomposition method has the advantages of simplicity and effectiveness, and is widely used in the selection of the temperature of the control tray in the distillation process [32–34]. In this work, the SVD method was used to select the temperature of the control tray. A small change (+0.1%) was applied to one of the three independent variables (RR1, RR2, and Q_R) while keeping the other two variables constant. The steady-state gain of all manipulated variables can be achieved by dividing the change in each stage temperature by the change in the manipulated variable. As a result, a gain matrix K with N_T rows and two columns (the number of manipulated variables) was obtained. The gain matrix K was decomposed into three matrices by the SVD function in Matlab: $K = U\sigma V^T$. The tray corresponding to the largest magnitude in the U vector is the control tray.

Figure 9a shows the steady-state gains of the C1, C2 and C3 column when RR1, RR2, and Q_R were used as manipulated variables, the left and right of the vertical dashed line represent different columns of the EDWC. Figure 9b shows the corresponding values of the U from singular value decomposition analysis, where the U1 largest magnitude represented by the solid line appears on stage 32, the corresponding manipulated variable was RR1. The largest magnitude of the solid line U_3 appeared at stage 5, the corresponding manipulated variable is RR2. The U_3 and U_4 largest magnitudes indicated by the dashed line both appear at stage 43 (stage 3 of the C3 column), the corresponding manipulated variable was Q_R. Finally, the reflux ratio RR1 of the C1 column controlled the temperature of stage 32, the reflux ratio RR2 of the C2 column controlled the temperature of stage 5, and the heat duty Q_R controlled the temperature of stage 3 of the C3 column.

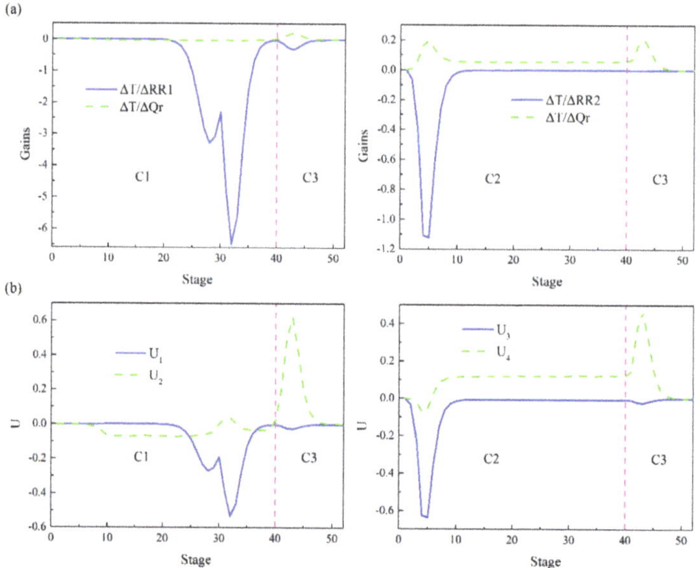

Figure 9. (**a**) Steady-state gains (**b**) SVD analysis for CS1.

3.2. Control Structure CS1

After completing the selection of the temperature control tray, the control structure CS1 was proposed. Figure 10 shows the control structure CS1. The control loops are as follows:

1. The feed is controlled by flow (reverse acting).
2. The flow rate of the entrainer DMAC is controlled by flow (reverse acting), which is cascaded with a fixed ratio of the feed.
3. The pressure at the top is controlled by the condenser heat removal rate (reverse acting).
4. The reflux drum liquid level of the C1 and C2 columns is held by the withdraw flow rate at the top of the column (direct acting).
5. Base level in the C3 column is held by entrainer makeup flow rate (reverse acting).
6. The temperature of stage 32 in the C1 column and the temperature of stage 5 of the C2 column are controlled by manipulating the corresponding reflux ratios RR1 and RR2 (direct acting).
7. The temperature of stage 4 in the C3 column is controlled by manipulating the reboiler duty Q_R (reverse acting).
8. Add 1 min deadtime to all temperature control loops.

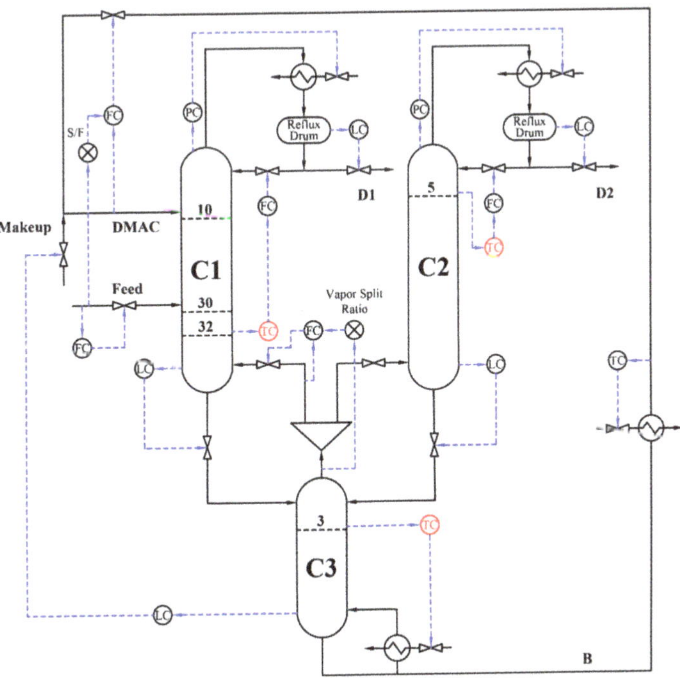

Figure 10. Control structure CS1.

Proportional integral (PI) control was applied to all control loops. All liquid level control loops use proportional control within gain Kc = 2. The pressure controller uses proportional integral control, gain Kc = 20, integral time τ = 12 min. The gain of the flow controller was Kc = 0.5, and the integral time τ = 0.3 min. The temperature control loops with dead time obtained the corresponding control parameters through relay-feedback and Tyreus–Luyben tuning. The final tuning parameters are summarized in Table 2.

The dynamic performance of the control structure CS1 was evaluated by introducing feed flow rate and composition disturbances. Figure 11a illustrates the dynamic responses of EDWC due to $\pm20\%$ disturbances in the feed. Apparently, the product purities (XD1 of the DPE and XD2 of the PA product) recover to their design values after approximately 4 h.

Table 2. Tuning parameters of temperature controllers in CS1.

Parameter	TC1	TC2	TC3
Controlled variable	$C1\text{-}T_{32}$	$C2\text{-}T_5$	$C3\text{-}T_3$
Manipulated variable	RR1	RR2	Q_R
Gain, Kc	1.394	5.197	1.302
Integral time/min	40.92	13.20	13.20

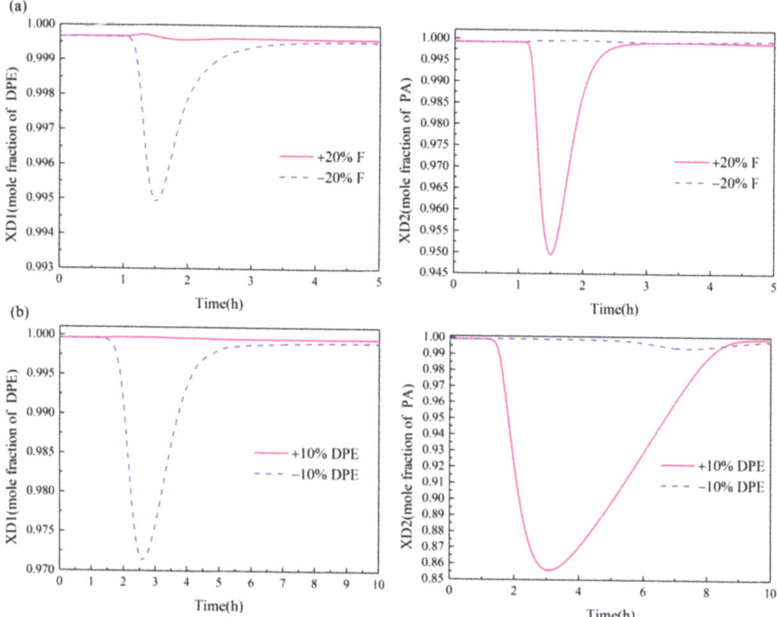

Figure 11. Dynamics responses of CS1 (**a**) 20% feed flow rate disturbances, (**b**) 10% feed composition disturbances.

Figure 11b describes the dynamic responses of EDWC due to ±10% disturbances in the feed composition, with a proportional adjustment in the other components. Obviously, the purity of DPE and PA is stable and qualified at 9 h, it takes a long time to reach the new steady state. The reason is that there will be liquid hydraulic lags (3~6 s/tray) when the liquid flows through the tray, and the temperature control tray of the C1 column is stage 32 at the bottom of the column. When the temperature of the control tray is adjusted by the reflux ratio, it will produce a significant delay time, a large amount of DPE enters the C3 column, and is finally withdrawn from the top of C2 column. This leads to a large decrease in the purity of the PA product in a short time. The control structure CS1 needs to be further optimized to obtain a better dynamic performance.

3.3. Control Structure CS2

The C1 column does not have a reboiler, but there is a vapor stream V1 returning from the C3 column, and the temperature control tray of the C1 column is at the bottom. In order to reduce the impact of liquid hydraulic lags caused by the reflux ratio control of the C1 column on the dynamic control performance, adjusting the temperature control tray of the C1 column was considered, by controlling the return vapor flow rate V1 to improve the dynamic control performance.

Figure 12a shows the corresponding steady-state gains of the C1 and C3 columns when the vapor split ratio α and the heat duty Q_R were used as manipulated variables. Figure 12b depicts the corresponding values of the U from singular value decomposition analysis, where the U5 largest magnitude represented by the solid line appeared at stage 43 and the corresponding manipulated variable was the vapor split ratio α. The C3 column is controlled by the heat duty of the reboiler, and the vapor split ratio replaces the reflux ratio RR1 to control C1 column, so stage 33 was chosen as the control tray. The U6 largest magnitude represented by the dashed line appeared at stage 43, and the corresponding manipulated variable was Q_R. Therefore, the final vapor split ratio α controls the temperature of stage 33 of the C1 column, and the reboiler heat duty Q_R controls the temperature of stage 3 of the C3 column. The control tray location and manipulated variable of the C2 column were the same as the CS1.

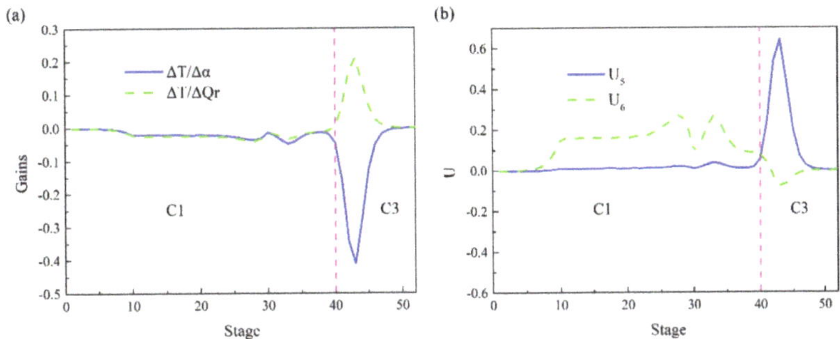

Figure 12. (**a**) Steady-state gains (**b**) SVD analysis for CS2.

The control structure CS2 is depicted in in Figure 13. The temperature control tray of the C1 column is adjusted by controlling the valve opening of the vapor stream V1. The reflux ratio RR1 uses a fixed ratio, and the temperature control loops with dead time obtain corresponding new control parameters through relay-feedback and Tyreus–Luyben tuning. The final tuning parameters are summarized in Table 3.

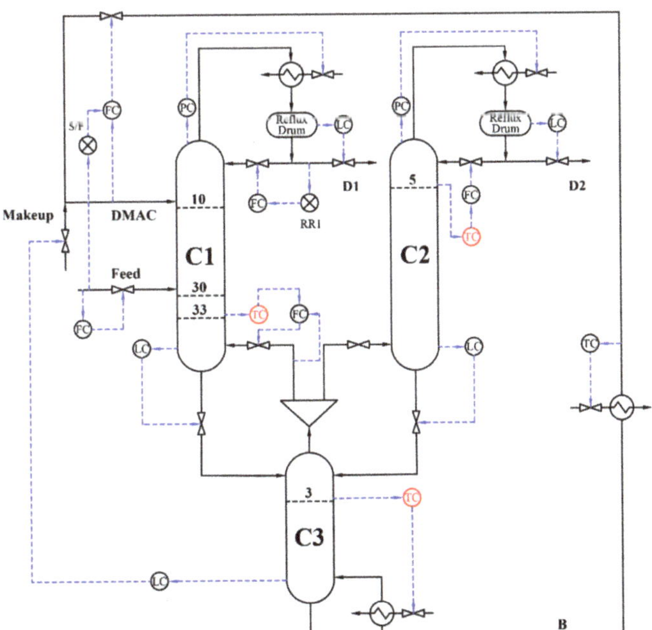

Figure 13. Control structure CS2.

Table 3. Tuning parameters of temperature controllers in CS2.

Parameter	TC1	TC2	TC3
Controlled variable	C1-T_{33}	C2-T_5	C3-T_3
Manipulated variable	V1	RR2	Q_R
Gain, Kc	6.013	7.023	1.316
Integral time/min	9.24	11.88	11.88

Figure 14a shows the dynamic responses of EDWC when facing ±20% disturbances in feed flow rate. The purity of the DPE product returned to its steady states and recovered its designed values about 2.5 h after feed flow rates changed. The purity of the PA product also stabilized at 2.5 h. The PA product purity stable value was 0.9988 only when the feed flow rate increased by 20%, which was slightly lower than the set value in a new steady state.

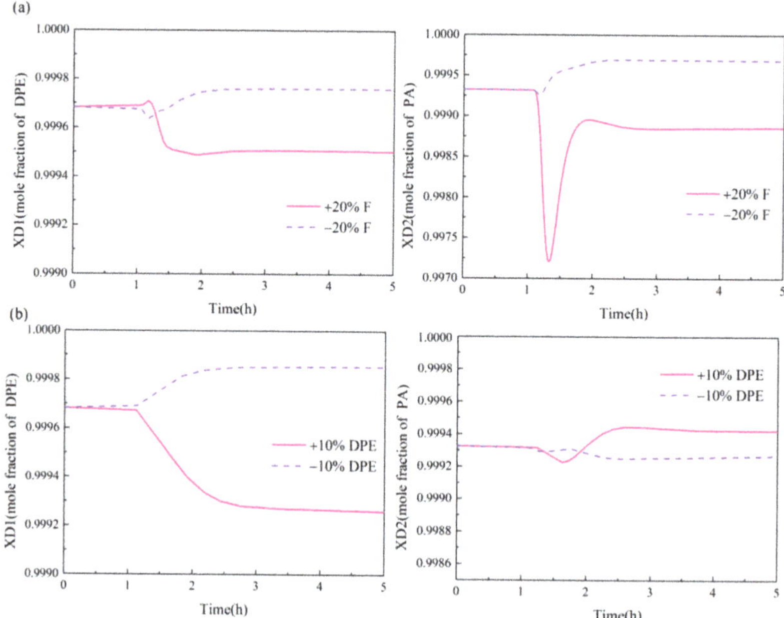

Figure 14. Dynamic responses of CS2 (**a**) 20% feed flow rate disturbances, (**b**) 10% feed composition disturbances.

Figure 14b shows the dynamic responses of EDWC due to ±10% disturbances of the concentration of DPE in the feed composition. Apparently, the product purity of DPE and PA recover their design values. Furthermore, less transient oscillations and a smaller offset were observed with shorter settling times. Obviously, the dynamic control performance of CS2 was superior to that of CS1 in terms of purity control of DPE and PA under feed composition disturbances. This is mainly because when the feed composition changes, it first affects the C1 column, then affects the C3 column, and finally affects the C2 column. When the vapor split ratio is used to control the temperature of the control tray in the C1 column, the response time is more rapid, hence, the purity of the DPE product quickly stabilizes. Most of the DPE is withdrawn from the top of the C1 column, only a small amount of DPE enters the C3 and C2 columns, so the product of PA purity fluctuation is smaller and the settling time is shorter.

4. Conclusions

In this work, DMAC was selected as the entrainer to separate the dipropyl ether/1-propyl alcohol azeotropic mixture. Three sequences including CEDS, EDWC and PSDS were designed and optimized. The results show that the EDWC with DMAC as the entrainer is more economically attractive than the CEDS and the PSDS. Compared with the CEDS, the EDWC can save 12.45% of the capital investments, 10.38% of the operating costs, and 11.01% of TAC. Compared with the PSDS, the EDWC can save 41.49% of the capital investments, 47.07% of operating costs, 45.50% of TAC. Therefore, it is more economical to separate dipropyl ether/1-propyl alcohol azeotropic mixture by EDWC. This study provides technical support for the separation design of such azeotropes.

Consequently, the control structure CS1 based on a three-temperature control loop and the control structure CS2 with vapor split ratio as the manipulated variable are proposed for EDWC. The dynamic control performance of the two control structures was investigated with feed flow rate and composition disturbances. The results show that CS1 has good performance, but the transient deviation is relatively large and the settling time is too long when facing the feed flow composition disturbance. CS2 with the vapor split ratio as the manipulated variable can more effectively deal with the feed flow rate and composition disturbances. It can quickly and effectively maintain the two products at high purity. Therefore, the CS1 can be used in industry to achieve good performance. As research continues on vapor split ratio in the chemical industry, the CS2 could further improve the dynamic performance of EDWC in the future.

Author Contributions: Conceptualization, Q.Y.; methodology, Y.W.; software, Y.W.; validation, Q.Y. and W.Z.; writing—original draft preparation, Y.W.; writing—review and editing, Q.Y. and H.P.; visualization, Y.W.; supervision, P.Y. All authors have read and agreed to the published version of the manuscript.

Funding: This research received no external funding.

Institutional Review Board Statement: Not applicable.

Informed Consent Statement: Not applicable.

Data Availability Statement: Not applicable.

Acknowledgments: This work was financially supported by the National Natural Science Foundation of China [Nos. 22178113].

Conflicts of Interest: The authors declare no conflict of interest.

References

1. Sakuth, M.; Mensing, T.; Schuler, J.; Heitmann, W.; Strehlke, G.; Mayer, D. *Ethers, Aliphatic*; Wiley-VCH Verlag GmbH & Co. KGaA: Weinheim, Germany, 2010.
2. Walther, T.; Francois, J.M. Microbial production of propanol. *Biotechnol. Adv.* **2016**, *34*, 984–996. [CrossRef] [PubMed]
3. Lladosa, E.; Montón, J.B.; Burguet, M. Separation of di-n-propyl ether and n-propyl alcohol by extractive distillation and pressure-swing distillation: Computer simulation and economic optimization. *Chem. Eng. Process.* **2011**, *50*, 1266–1274. [CrossRef]
4. Mangili, P.V. Thermoeconomic and environmental assessment of pressure-swing distillation schemes for the separation of di-n-propyl ether and n-propyl alcohol. *Chem. Eng. Process.* **2020**, *148*, 107816. [CrossRef]
5. Luo, H.; Liang, K.; Li, W.; Li, Y.; Xia, M.; Xu, C. Comparison of pressure-swing distillation and extractive distillation methods for isopropyl alcohol/diisopropyl ether separation. *Ind. Eng. Chem. Res.* **2014**, *53*, 15167–15182. [CrossRef]
6. You, X.; Rodriguez-Donis, I.; Gerbaud, V. Low pressure design for reducing energy cost of extractive distillation for separating diisopropyl ether and isopropyl alcohol. *Chem. Eng. Res. Des.* **2016**, *109*, 540–552. [CrossRef]
7. Yildirim, Ö.; Kiss, A.A.; Kenig, E.Y. Dividing wall columns in chemical process: A review on current activities. *Sep. Purif. Technol.* **2011**, *80*, 403–417. [CrossRef]
8. Dejanović, I.; Matijašević, L.; Olujić, Ž. Dividing wall column-A breakthrough towards sustainable distilling. *Chem. Eng. Process.* **2010**, *49*, 559–580. [CrossRef]
9. Buitimea-Cerón, G.A.; Hahn, J.; Medina-Herrera, N.; Jiménez-Gutiérrez, A.; Loredo-Medrano, J.A.; Tututi-Avila, S. Dividing-Wall Column Design: Analysis of Methodologies Tailored to Process Simulators. *Processes* **2021**, *9*, 1189. [CrossRef]
10. Chen, Z.; Agrawal, R. Classification and Comparison of Dividing Walls for Distillation Columns. *Processes* **2020**, *8*, 699. [CrossRef]

11. Wu, Y.C.; Hsu PH, C.; Chien, I.L. Critical assessment of the energy-saving potential of an extractive dividing-wall column. *Ind. Eng. Chem. Res.* **2013**, *52*, 5384–5399. [CrossRef]
12. Xia, M.; Yu, B.; Wang, Q.; Jiao, H.; Xu, C. Design and control of extractive diving-wall column for separating methylal-methanol mixture. *Ind. Eng. Chem. Res.* **2012**, *51*, 16016–16033. [CrossRef]
13. Xia, M.; Xin, Y.; Luo, J.; Li, W.; Shi, L.; Min, Y.; Xu, C. Temperature control for extractive diving-wall column with an adjustable vapor split: Methylal/methanol azeotrope separation. *Ind. Eng. Chem. Res.* **2013**, *52*, 17996–18013. [CrossRef]
14. Zhang, H.; Ye, Q.; Qin, J.; Xu, H.; Li, N. Design and control of extractive dividing-wall column for separating ethyl acetate-isopropyl alcohol mixture. *Ind. Eng. Chem. Res.* **2014**, *53*, 1189–1205. [CrossRef]
15. Dwivedi, D.; Strandberg, J.P.; Halvorsen, I.J.; Preisig, H.A.; Skogestad, S. Active vapor split control for dividing-wall columns. *Ind. Eng. Chem. Res.* **2012**, *51*, 15176–15183. [CrossRef]
16. Huaqiang, G.; Xiangwu, C.; Nan, C.; Wenyi, C. Experimental study on vapor splitter in packed divided wall column. *J. Chem. Technol. Biot.* **2016**, *91*, 449–455. [CrossRef]
17. Li, C.; Li, J.; Li, D.; Ma, S.; Li, H. Experimental study and CFD numerical simulation of an innovative vapor splitter in dividing wall column. *AIChE J.* **2020**, *66*, e16266. [CrossRef]
18. Harvianto, G.R.; Kim, K.H.; Kang, K.J.; Lee, M. Optimal operation of a dividing wall column using an enhanced active vapor distributor. *Chem. Eng. Res. Des.* **2019**, *144*, 512–519. [CrossRef]
19. Kang, K.J.; Harvianto, G.R.; Lee, M. Hydraulic driven active vapor distributor for enhancing operability of dividing wall column. *Ind. Eng. Chem. Res.* **2017**, *56*, 6493–6498. [CrossRef]
20. Hu, Y.; Chen, S.; Li, C. Numerical research on vapor splitter in divided wall column. *Chem. Eng. Res. Des.* **2018**, *138*, 519–532. [CrossRef]
21. Luyben, W.L. Vapor split manipulation in extractive divided-wall distillation columns. *Chem. Eng. Process.* **2018**, *126*, 132–140. [CrossRef]
22. Jing, C.; Zhu, J.; Dang, L.; Wei, H. Extractive dividing-wall column distillation with a novel control structure integrating pressure swing and pressure compensation. *Ind. Eng. Chem. Res.* **2021**, *60*, 1274–1289. [CrossRef]
23. Zhu, J.; Jing, C.; Hao, L.; Wei, H. Insight into controllability and operation of extractive dividing-wall column. *Sep. Purif. Technol.* **2021**, *263*, 118362. [CrossRef]
24. Araújo Neto, A.P.; Farias Neto, G.W.; Neves, T.G.; Ramos, W.B.; Brito, K.D.; Brito, R.P. Changing product specification in extractive distillation process using intelligent control system. *Neural Comput. Appl.* **2020**, *32*, 13255–13266. [CrossRef]
25. De Araújo Neto, A.P.; Sales, F.A.; Brito, R.P. Controllability comparison for extractive dividing-wall columns: ANN-based intelligent control system versus conventional control system. *Chem. Eng. Process.* **2021**, *160*, 108271. [CrossRef]
26. Rewagad, R.R.; Kiss, A.A. Dynamic optimization of a dividing-wall column using model predictive control. *Chem. Eng. Sci.* **2012**, *68*, 132–142. [CrossRef]
27. Rodriguez, M.; Li, P.Z.; Diaz, I. A control strategy for extractive and reactive dividing wall columns. *Chem. Eng. Process.* **2017**, *113*, 14–19. [CrossRef]
28. Feng, Z.; Shen, W.; Rangaiah, G.P.; Dong, L. Proportional-integral control and model predictive control of extractive dividing-wall column based on temperature differences. *Ind. Eng. Chem. Res.* **2018**, *57*, 10572–10590. [CrossRef]
29. Feng, Z.; Shen, W.; Rangaiah, G.P.; Dong, L. Closed-loop identification and model predictive control of extractive dividing-wall column. *Chem. Eng. Process.* **2019**, *142*, 107552. [CrossRef]
30. Lladosa, E.; Montón, J.B.; Burguet, M.C.; Munoz, R. Phase equilibria involved in extractive distillation of dipropyl ether + 1-propyl alcohol using N, N-dimethylformamide as entrainer. *J. Chem. Eng. Data* **2007**, *52*, 532–537. [CrossRef]
31. Luyben, W.L. *Distillation Design and Control Using Aspen Simulation*; John Wiley & Sons: Hoboken, NJ, USA, 2006.
32. Luyben, W.L. Evaluation of criteria for selecting temperature control trays in distillation columns. *J. Process Control* **2006**, *16*, 115–134. [CrossRef]
33. Qian, X.; Liu, R.; Huang, K.; Chen, H.; Yuan, Y.; Zhang, L.; Wang, S. Comparison of Temperature Control and Temperature Difference Control for a Kaibel Dividing Wall Column. *Processes* **2019**, *7*, 773. [CrossRef]
34. Ling, H.; Luyben, W.L. Temperature control of the BTX divided-wall column. *Ind. Eng. Chem. Res.* **2010**, *49*, 189–203. [CrossRef]

processes

Article

Residence Time Distribution of Non-Spherical Particles in a Continuous Rotary Drum

Saeed Mahdavy [1], Hamid Reza Norouzi [1,*], Christian Jordan [2], Bahram Haddadi [2,*] and Michael Harasek [2]

[1] Center of Engineering and Multiscale Modeling of Fluid Flow (CEMF), Department of Chemical Engineering, Amirkabir University of Technology (Tehran Polytechnic), No. 350, Hafez, Tehran 15875-4413, Iran; saeedmahdavy@aut.ac.ir

[2] Technische Universität Wien, Institute of Chemical, Environmental and Bioscience Engineering, Getreidemarkt 9/166, 1060 Vienna, Austria; christian.jordan@tuwien.ac.at (C.J.); michael.harasek@tuwien.ac.at (M.H.)

[*] Correspondence: h.norouzi@aut.ac.ir (H.R.N.); bahram.haddadi.sisakht@tuwien.ac.at (B.H.); Tel.: +98-21-6454-3157 (H.R.N.); +43-1-58801-166252 (B.H.)

Abstract: The motion of non-spherical particles with sharp edges, as they are commonly involved in practice, was characterized by residence time distribution (RTD) measurement in a continuous drum. Particles with two sizes, 6 and 10 mm, and two densities, 750 and 2085 kg/m^3, were used in the experiments. The effects of rotation speed (3–11 rpm), incline angle (2–4°), feed rate, and mixture composition were investigated and compared to the results of other researchers on particles without sharp edges. We also fitted the RTD with an axial dispersion model to obtain a better insight into the flow behavior. MRT of non-spherical particles with sharp edges depends on $\omega^{-\alpha}$ similar to other shapes, while the value of alpha is higher for particles with sharp edges ($0.9 < \alpha < 1.24$), especially at high incline angles. The MRT depends on incline angle, β^{-b}, where b varies between 0.81 (at low ω) and 1.34 (at high ω), while it is close to 1 for other shapes. Feed rate has a slight effect on the MRT of particles with sharp edges and the effect of particle size diminishes when rotation speed increases. The MRT linearly increases with volume fraction of light particles in a mixture of light and heavy particles (from pure heavy to pure light particles).

Keywords: rotary drum; non-spherical particles; residence time distribution; mean residence time; biomass

Citation: Mahdavy, S.; Norouzi, H.R.; Jordan, C.; Haddadi, B.; Harasek, M. Residence Time Distribution of Non-Spherical Particles in a Continuous Rotary Drum. *Processes* **2022**, *10*, 1069. https://doi.org/10.3390/pr10061069

Academic Editor: Jianhong Xu

Received: 5 May 2022
Accepted: 25 May 2022
Published: 26 May 2022

1. Introduction

Rotary drums are used in many industries such as chemical, pharmaceutical, metallurgical, and food processing [1]. They provide good conditions for mixing, drying, coating, granulation, chemical reaction [2], clinkering of cement, regeneration of spent catalyst, and char activation [3]. They can process a wide range of particle sizes and shapes. High construction cost, non-uniform temperature profile, and limited reaction rate between solid and gas are their disadvantages [3].

Rotary drums can be used in batch or continuous processes [4]. In continuous mode, solids are fed into the drum through a feeder. They move forward and mix by a combined action of rotation and gravitation due to inclination of the drum. The time that particles stay in the drum is defined as residence time [5]. Residence time is one of the most important parameters that can be used to evaluate heat and mass transfer conditions and reaction extent in the rotary drum [6]. For instance, if heat transfer derives the drying or reaction of particles, the residence time of particles must be longer than time that is required to accomplish desired drying or reaction [7]. Thus, residence time distribution (RTD) can affect the reaction progress or the pyrolysis performance in the kiln [8].

Plug flow and completely mixed flow are two different ideal models for flow patterns. In reality, solid flow differs from these ideal flow models. Thus, the RTD of particles deviates from the RTD of plug flow and mixed flow [9]. Many factors affect the RTD, such as the rotation speed of the drum, incline angle, feed rate (as operation variables) [10,11], internal geometry [12], and particle shape [13].

Most of the studies on the RTD measurement in rotary drums are dedicated to spherical particles as reviewed, e.g., by Lu et al. [14], while in industrial application, particles are mostly non-spherical [15,16]. Some other researchers study non-spherical shape in rotary drums via the means of experiments and simulation, such as cylindrical particles (such as broken rice or wood pellet) [10,11,17,18], cuboid particles (such as wood chips or alumina) [5,15,17,19], pharmaceutical tablets [1], ellipsoid particles [20], polyhedral particles [21], and cubic particles [16,18]. In addition, even spherical particles may undergo shape change due to cohesion, breakage, or reaction [7,22]. Non-spherical particles' flow behavior is more complex than spherical particles [23]. For instance, correlations and criteria used for characterizing flow regime for spherical particles may not be used for non-spherical particles [2,20,24].

The axial dispersion model (ADM) is used to describe non-ideal solid flow [25]. RTD of solids can be defined as a function of Peclet number. Peclet number includes the combined effect of dispersion coefficient and mean residence time (MRT). Thus, a change in the axial dispersion coefficient or the MRT can lead to a change in the RTD. Variables in the ADM are meaningful and can be compared to other phenomena such as reaction and heat transfer [16,25]. Gao et al. measured RTD of spherical, cylindrical and quadrilobe particles. Their results showed that non-spherical particles have less axial dispersion than spheres [5]. Lu et al. studied axial dispersion coefficient for cubic particles. Their results showed that axial dispersion coefficient depends on rotation speed, acceleration of gravity and volume equivalent diameter [16].

In addition to experiments, discrete element (DEM) simulations answer many unknowns concerning granular flow through accurate modeling of inter-particle interactions. DEM is a robust and mature modeling approach for granular flow. The sub-models of mechanical contact interactions and techniques for representing particle surface are well established for spherical particles, while sub-models and shape representations of non-spherical particles still need improvement [26]. One fundamental step in utilizing DEM for granular flow is to validate simulation results against reliable experimental measurements. Therefore, providing detailed and vast experimental data on the flow behavior of granular flows (especially for non-spherical particles) for DEM-based research is another aim of this research.

Among non-spherical particle with regular and irregular shape, cubic and cuboid particles can be obtained by six cuts, and their dimensions can be easily specified. Thus, we used cubic particles as the model particle for non-spherical particles with sharp edges. Additionally, wooden and ceramic cuboid particles were used in a mixture to find the effect of density variation. This mixture resembles the conditions that the solid flow contains particles with various density (such as wood pyrolysis in rotary kiln with ceramic as a heat carrier [27]). RTD was measured at different rotation speeds, incline angles, feed rates, particle sizes, and volume fraction of wooden particles, and then fitted to the ADM. In all cases, we tried to compare our results with similar results on spherical and non-spherical particles without sharp edge to find the similarities and the differences between their flow behavior. All the experimental results on RTD were also provided as Supplementary Data to be used by other researchers who are aiming to validate their DEM simulation results.

2. Material and Procedure

Figure 1 shows the shape and properties of particles that we used in this study. The dimensions of cubic wooden particles were $6 \times 6 \times 6$ mm^3 and $10 \times 10 \times 10$ mm^3. We also tested the variation in the density with cuboid ceramic (heavy) and wooden (light) particles whose dimensions were $4 \times 10 \times 10$ mm^3. The density variation is important, since, in reality, ceramic particles can be used as heat carriers in the system [27].

Material	Wood	Wood	Wood	Ceramic
Shape	Cubic	Cubic	Cuboid	Cuboid
Size (mm³)	6 × 6 × 6	10 × 10 × 10	4 × 10 × 10	4 × 10 × 10
Density $(\frac{kg}{m^3})$	660	660	750	2085

Figure 1. Particle shapes and their properties.

All experiments were performed in a rotary drum with length 50.0 cm, inside diameter 8.4 cm, and outside diameter 9.0 cm and at room temperature. The drum was made of plexiglas with very smooth surface. This smooth surface could not resemble the actual roughness of the rotary kilns in practice. Therefore, a thin sandpaper P1200 was affixed to the inside surface of the drum to achieve the desirable roughness in the surface. Figure 2 shows the experimental setup including parts that were used in this research.

Figure 2. Rotary drum setup and its parts: (**a**) vibrator (**b**) particle feeder (**c**) inlet of particles into drum (**d**) rotary drum (**e**) outlet of drum (**f**) motor (**g**) setup support.

A three-phase AC motor (400 V) was used with a frequency inverter for rotation speed control. The feeder was vibrated using a Retsch DR 100 vibrator (Retsch, Hann, Germany) to achieve the desired flow rate of particles. The incline angle of the support of the kiln was changed by a jack. Table 1 lists experimental conditions. Rotation speed varied between 3 rpm to 11 rpm; incline angle, between 2° to 4°; and feed rate, between 0.67 cm^3/s and 1.08 cm^3/s for cubic particles. Cuboid particles (wood and ceramic mixture) were used to determine the effect of density at 7 rpm and 2°, whereas the volume fraction of wood was varied in the mixture of wood and ceramic particles.

We used colored particles as the tracers. When rotary drum operation reached steady state condition, these colored tracers were fed into the drum. A Samsung HMX-F90 camera, located near the outlet of drum, recorded the particles' movement and their exit moments. Entrance and exit times of each tracer give us the residence time. Recorded videos were processed by Kdenlive 19.12.3. Each experiment was repeated three times. Figure 3 demonstrates cross view of drum at different rotation speed, incline angle and feed rate for 6 mm and 10 mm cubes. Tracers are particles with dark color.

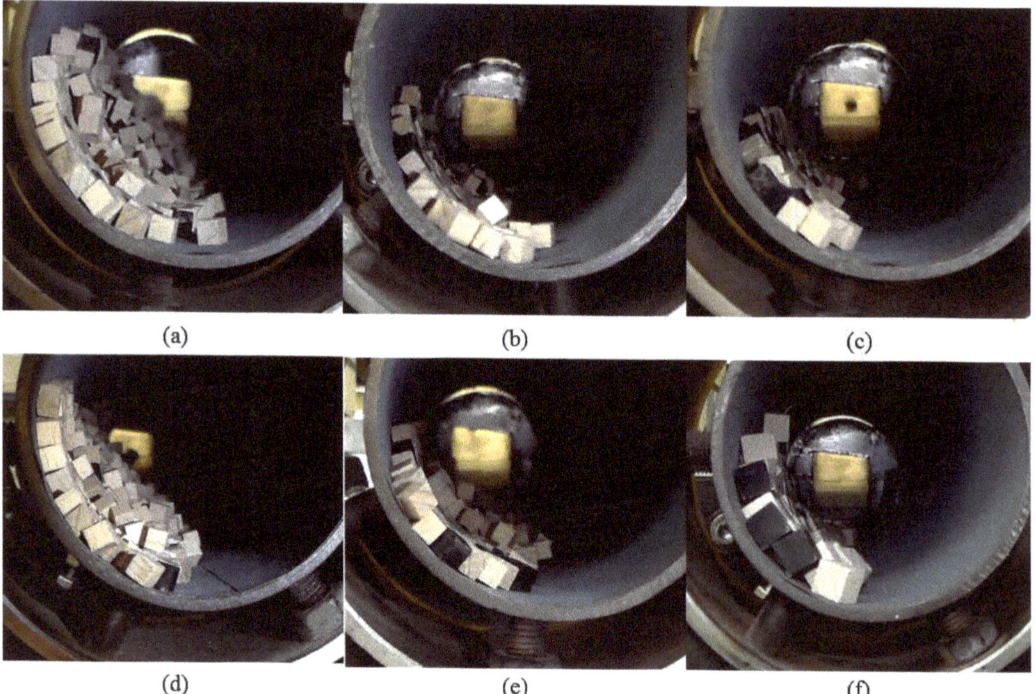

Figure 3. Cross view of drum at (**a**) 3 rpm, 3°, and 1.08 cm^3/s for 6 mm cubes (**b**) 7 rpm, 3°, and 1.08 cm^3/s for 6 mm cubes (**c**) 11 rpm, 3°, and 1.08 cm^3/s for 6 mm cubes (**d**) 3 rpm, 2°, and 1.08 cm^3/s for 6 mm cubes (**e**) 3 rpm, 3°, and 0.67 cm^3/s for 6 mm cubes (**f**) 3 rpm, 3°, and 1.08 cm^3/s for 10 mm cubes (black particles are tracer).

Table 1. Run conditions.

Run. No	Rotation Speed (rpm)	Incline Angle (°)	Feed Rate (cm³/s)	Size (mm³)	Shape	Number of Tracer (-)	Wood Volume Fraction (-)
1	3						
2	5						
3	7	2					
4	9						
5	11						
6	3						
7	5						
8	7	3					
9	9						
10	11		1.08				
11	3			$6 \times 6 \times 6$		50	
12	5						
13	7	4					
14	9						
15	11						
16	5	2.5					
17	11						
18	5	3.5					
19	11						
20	3				cube		1
21	7	3	0.67				
22	11						
23	3						
24	5						
25	7	2					
26	9						
27	11						
28	3						
29	5						
30	7	3					
31	9						
32	11		1.08	$10 \times 10 \times 10$		30	
33	3						
34	5						
35	7	4					
36	9						
37	11						
38	5	2.5					
39	11						
40	5	3.5					
41	11						
42			0.28			50	0
43			0.31				0.25
44	7	2	0.34	$10 \times 10 \times 4$	cuboid	60	0.5
45			0.37				0.75
46			0.40			50	1

3. Residence Time Distribution (RTD)

The system reaches a steady-states condition when the numbers of entering and exiting particles are the same. After that, tracers are entered into the drum. Figure 4 shows the method used for introducing the tracers. We used 30–60 tracer particles in each experiment.

Sixty tracers were used in the run numbers 42–46 (effect of volume fraction of the wood). For example, when the volume fraction of wood is 0.25, the number of tracers for wood is 25% of all tracers, hence 15 black tracers are wood, and 45 red tracers are ceramic (see Figure 4b).

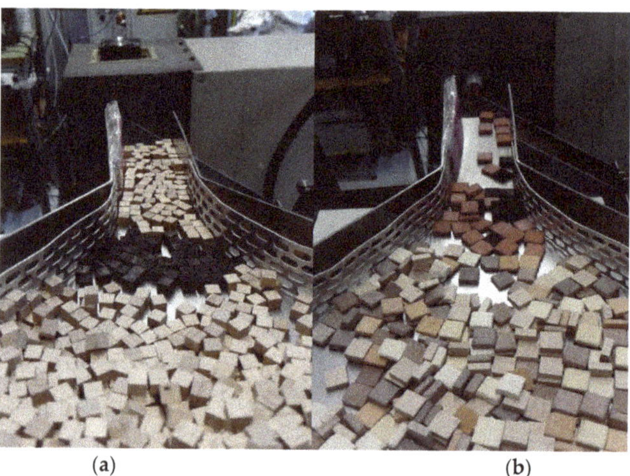

(a)   (b)

Figure 4. Introduction of tracers into the feeder for single component and two components, (**a**) black tracer for 6 mm cubes (**b**) black wood tracer and red ceramic tracer for wood-ceramic mixture (x_{wood} = 0.25).

Entrance and exit moments of tracer in the drum were recorded by camera. Residence time of a tracer is the difference between these two moments. The cumulative distribution function, $F(t)$, demonstrates the fraction of tracers that have left the drum until t. $F(t)$ is computed by [28]:

$$F(t) = \frac{N(t)}{N_0} \tag{1}$$

where $N(t)$ is number of tracer particles that left the drum until t, and N_0 is total number of tracer particles. The mean residence time (MRT) is obtained by:

$$\text{MRT} = \frac{1}{N_0} \sum_{i=1}^{n} t_i \tag{2}$$

where t_i is residence time of tracer.

4. The Axial Dispersion Model (ADM)

A one-dimensional axial dispersion model (ADM) was used for fitting experimental data. This model describes convective-dispersive transport of particles at unsteady state condition without reaction [28,29].

$$\frac{L}{u_z} \frac{\partial C(z,t)}{\partial t} = \frac{1}{Pe} \frac{\partial^2 C(z,t)}{\partial z^2} - \frac{\partial C(z,t)}{\partial z} \tag{3}$$

where $C(z,t)$ shows dimensionless tracer concentration, z is dimensionless length ($z = l/L$), l is position of tracer in the axial direction, t is time, u_z is the mean axial velocity, L is the drum length, and Pe is the Peclet number.

Pe number indicates the ratio of convective mass transport to dispersive mass transport. If the Pe number is high, the convective mass transport is dominant; while if Pe number is low, the dispersion is dominant. A narrow time distribution is obtained when Pe number is high, while a wide distribution is obtained when Pe number is low.

$$Pe = \frac{u_z.L}{D_{ax}} \tag{4}$$

where D_{ax} is axial dispersion coefficient, as a measure for erratic movement [10].

All dispersion is assumed to happen in the drum; which means that the tracer concentration gradient begins from the drum inlet by dispersion and convection (Equation (5)) and ends at the drum outlet (Equation (6)) [28,29].

$$C_{in} = C_{(0+,t)} - \frac{1}{Pe}\frac{\partial C}{\partial z}\Big|_{z=0+} \tag{5}$$

$$\frac{\partial C}{\partial z}\Big|_{z=1} = 0 \tag{6}$$

ADM equation with above boundary conditions is solved numerically [28,29], though the analytical solution can be found for $Pe > 50$ [30]. An implicit finite difference method was used to solve ADM according to the above boundary conditions. Solution of ADM gives the tracer concentration at all positions and times. $C(z = 1, t)$ shows the concentration of tracer at the drum outlet that depends on Pe. Time distribution changes due to changes in Pe. Average absolute square error (AASE) was defined (Equation (7)), to find the best fitted curve of ADM on experimental data. The nearest curve is obtained based on the minimum value of AASE. MRT is obtained from the experiment based on Equation (2). Consequently, the axial dispersion coefficient is calculated by Equation (4). fminbnd function is used in MATLAB R2018b to minimize AASE.

$$\text{Average Absolute Square Error (AASE)} = \frac{\sum_{i=1}^{n}\left(F_{experiment} - F_{ADM}\right)^2}{N_0} \tag{7}$$

5. Result and Discussion

Figure 5 shows the cumulative residence time distributions obtained from experiments and fitted with the ADM. The cumulative residence time distribution, $F(t)$, shows the fraction of tracers that has left the drum. It includes RTD results of three rotation speeds (3, 7, and 11 rpm), three incline angles (2, 3 and 4°), and two feed rates (0.67 and 1.08 cm^3/s) for 6 mm cubes. The ADM fitted curves can describe one-dimensional flow behavior in the drum very well. The difference between the first and the last tracer residence time for 3 rpm (Figure 5a) is approximately 550 s. For the two other cases (7 rpm and 11 rpm), these differences are approximately 300 s and 80 s; therefore, when rotation speed increases, tracers leave the drum faster, and RTD becomes narrower. Figure 5d compares cumulative residence time distribution at three different rotation speed. A narrower distribution is obtained due to higher rotation speed. Other researchers have also shown that narrower RTD are obtained when rotation speed increases [5,7,10,24,31–33]. Figure 5e shows the effect of the incline angle on the RTD. When the incline angle increases, the residence time of particles decreases and RTD becomes narrower. Previous studies similarly showed that RTD becomes narrower when incline angle increases [7,10,24,33]. Figure 5f compares the RTD results of 6 mm at two different flow rates. According to Figure 5f, an increase in the feed causes a slight shift to higher values of RTD. The shapes of RTD curves at two flow rates are similar. Njeng et al. [33] showed that feed rate had a slight effect on RTD curves for cuboid particles.

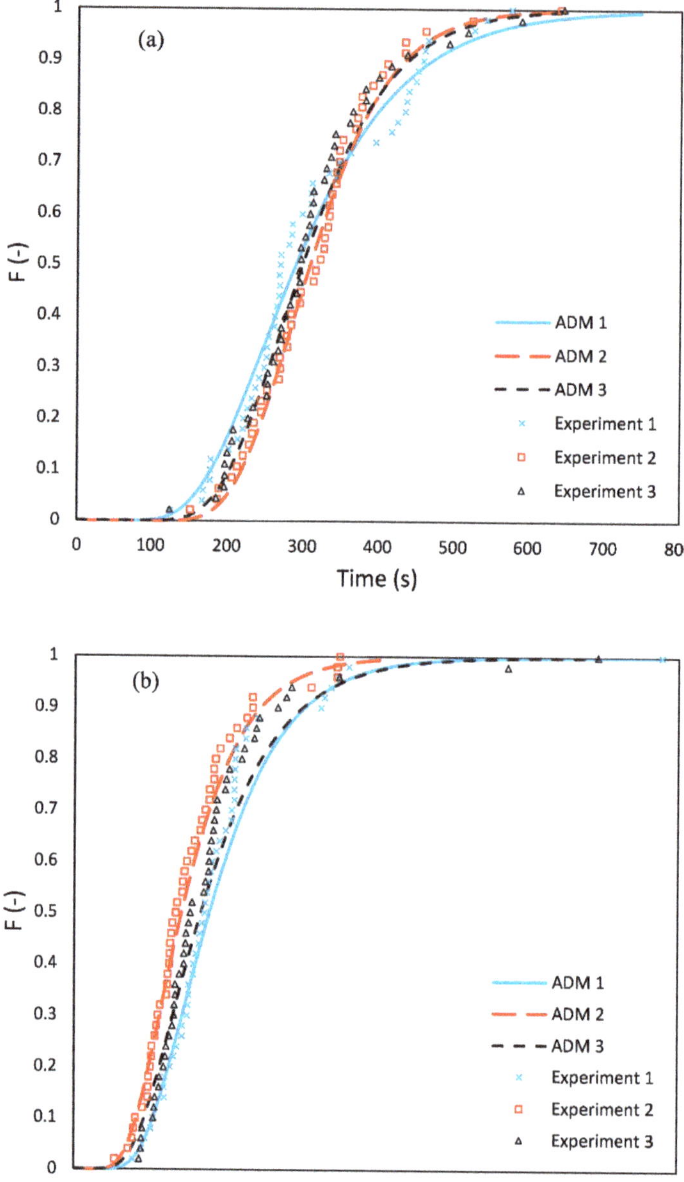

Figure 5. *Cont.*

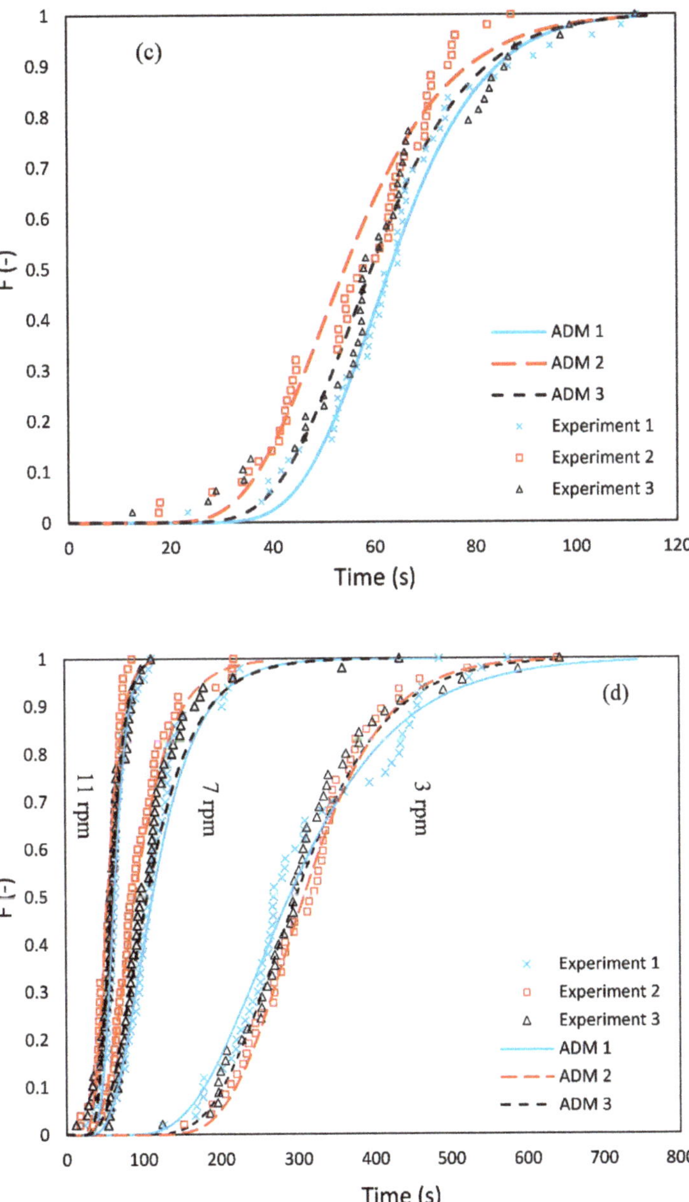

Figure 5. *Cont.*

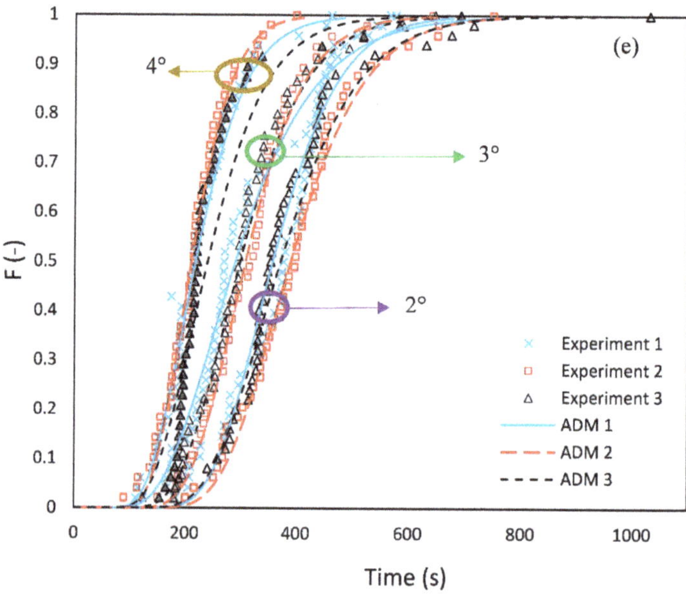

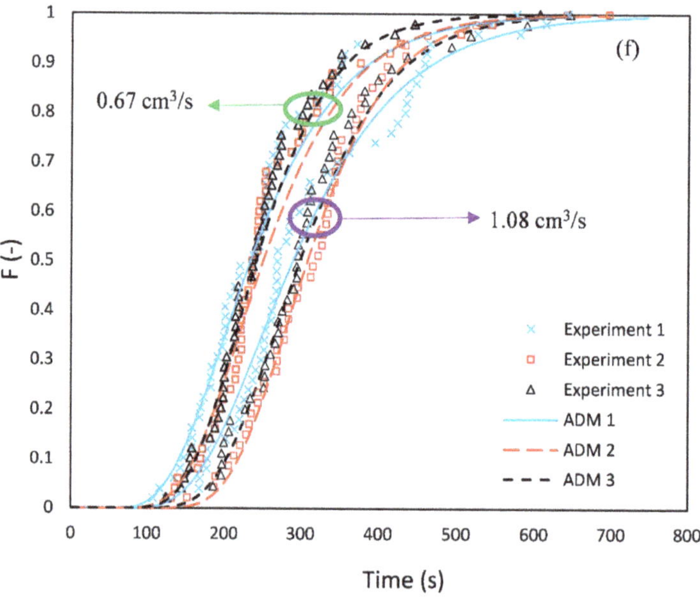

Figure 5. Cumulative distribution as a function of outlet time in experiments and fitted with ADM (**a**) 3 rpm, 3°, and 1.08 cm^3/s for 6 mm cubes; (**b**) 7 rpm, 3°, and 1.08 cm^3/s for 6 mm cubes; (**c**) 11 rpm, 3°, and 1.08 cm^3/s for 6 mm cubes; (**d**) 3–11 rpm, 3° and 1.08 cm^3/s for 6 mm cubes; (**e**) 3 rpm, 2–4° and 1.08 cm^3/s for 6 mm cubes; (**f**) 3 rpm, 3° and 0.67 and 1.08 cm^3/s for 6 mm cubes. (Each experiment repeated 3 times and ADM model was fitted on the three sets of data.)

Table 2 shows the results of Figure 5a–c. MRTs of each row are obtained from the experiment based on Equation (2). *Pe*, AASE, and D_{ax} are obtained from ADM. An increase in the rotation speed led to a decrease in the MRT. The MRT value changes from 313 s to 61 s when the rotation speed is increased from 3 to 11 rpm. Axial dispersion coefficient increases when rotation speed changes from 3 rpm to 7 rpm; however, it decreases when rotation speed changes from 7 rpm to 11 rpm.

Table 2. Results of experiment and ADM at 3° and 1.08 cm^3/s for 6 mm cubes. (Each experiment was repeated 3 times and ADM model was fitted on the three sets of data.)

Rotation Speed		MRT [s]		Pe. No [-]	AASE [-]	D_{ax} [cm^2/s]
3 rpm	Experiment 1	310.0	ADM 1	12.4	2.1×10^{-3}	0.651
	Experiment 2	318.8	ADM 2	26.0	9.0×10^{-4}	0.290
	Experiment 3	311.5	ADM 3	20.7	1.5×10^{-3}	0.386
	Average	313.4		19.7	1.5×10^{-3}	0.442
	Standard deviation	3.9		5.6		0.153
7 rpm	Experiment 1	122.2	ADM 1	10.2	3.6×10^{-3}	1.997
	Experiment 2	98.1	ADM 2	9.5	1.1×10^{-3}	2.664
	Experiment 3	116.9	ADM 3	8.0	3.4×10^{-3}	2.660
	Average	112.4		9.2	2.7×10^{-3}	2.440
	Standard deviation	10.4		0.9		0.313
11 rpm	Experiment 1	64.7	ADM 1	38.5	1.5×10^{-3}	1.002
	Experiment 2	56.5	ADM 2	22.2	4.1×10^{-3}	1.987
	Experiment 3	61.4	ADM 3	29.7	2.5×10^{-3}	1.368
	Average	60.9		30.1	2.7×10^{-3}	1.452
	Standard deviation	3.4		6.7		0.407

In the following sections, we present the effects of rotation speed, incline angle, feed rate, particle size and density on the MRT and the axial dispersion coefficient. All the RTD experimental data and ADM analysis results can be found in the Supplementary Data that accompanies this article. The data are stored in the spreadsheet (Microsoft Excel) format. The data can be used for validation DEM simulations for non-spherical particles.

5.1. Effect of Rotation Speed

Figure 6a shows the effect of rotation speed on MRT for 6 mm cubes in which rotation speed changes from 3 rpm to 11 rpm at three different incline angles while feed rate is fixed at 1.08 cm^3/s. The error bars also show the interval change in mean with the confidence level of 95%. An exponential model was used to find the relationship between rotation speed and MRT. Regressed equations are presented in the figure. On average, particles remain shorter in the drum when rotation speed increases. At low rotation speeds (3–5 rpm), the rotation speed has a greater effect on MRT than high rotation speeds. MRT decreases when incline angle increases. The figure indicates an overall 69 % decrease in the MRT when the rotation speed changes from 3 rpm to 11 rpm at 2° incline angle. This change in MRT is more notable for higher incline angle values.

When the incline angle increases, the magnitude of exponents of rotation speed (ω) in the regressed equations in the figure also increases ($|-0.899| < |-1.227| < |-1.242|$), which shows that rotation speed has a more pronounced effect on the MRT at higher incline angles. Previous studies showed that MRT $\propto \omega^{-1}$, MRT $\propto \omega^{-0.98}$, and MRT $\propto \omega^{-0.88}$ for spherical particles [7,34,35]. Njeng et al. [33] established a relationship for spherical and non-spherical particles (broken rice and beech chips) which shows MRT $\propto \omega^{-0.884}$. Gao et al. [5] showed that the effect of rotation speed on the MRT of spherical particles is not same as non-spherical particles (cylindrical and cuboid). Our results show that the magnitude of the exponent for cubic particles with sharp edges is higher than that for spherical and non-spherical particles without sharp edge (see Table 3 to compare the operating conditions in this research with others).

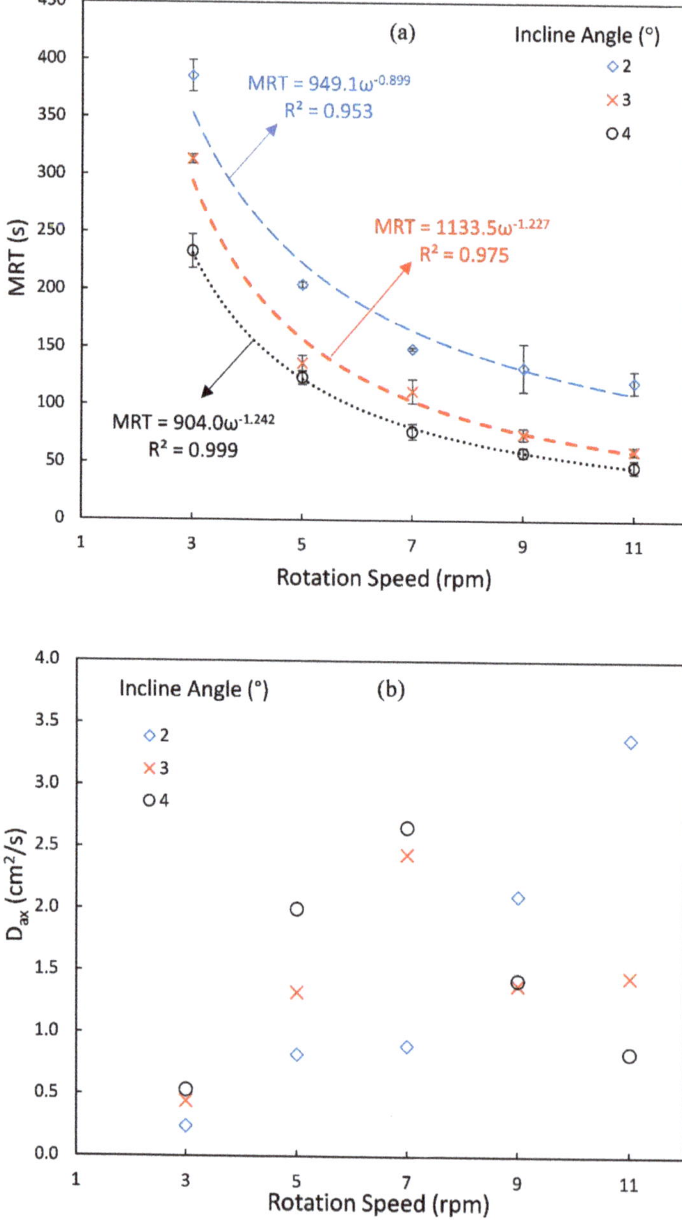

Figure 6. Effect of rotation speed on (**a**) MRT (**b**) axial dispersion coefficient at 1.08 cm^3/s for 6 mm cube.

Figure 6b shows the effect of rotation speed on the axial dispersion coefficient at three incline angles. At 2° of incline angle, the axial dispersion coefficient increases with rotation speed. This behavior can also be seen for other incline angles (3° and 4°) up to 7 rpm, while after 7 rpm, the axial dispersion coefficient decreases. When the rotation speed increases, particles rise to a higher height on the drum wall, detach and avalanche from that height. This means that avalanching particles on the surface have more chance of having random motions on the surface (contrary to the particles in layers below the surface). It should be noted that the axial dispersion coefficient shows random displacement of particles [10]. Generally, other studies showed that axial dispersion coefficient increases with rotation speed for cohesive powder and catalyst [7,31], spherical particles [10,11,34], cylindrical particles [10,10,36], cuboid particles [33] and cubic particles [16].

Table 3. Comparison between operating conditions, drum geometry and particle shape in the literature with this research.

	L (cm)	D (cm)	Shape	dp * (mm)	ω (rpm)	β (°)	dp/D (-)	L/D (-)	$\omega^2.R/g$ (-)
Sai et al. [34]	590	14.7	sphere	1.1	1–3	0.78–1.37	0.0075	40.14	8.2×10^{-5}–7.4×10^{-4}
Sherritt et al. [37]	90	20	sphere	3	5–25	0	0.015	4.5	2.8×10^{-3}–7.0×10^{-2}
Third et al. [38]	15.2	10	sphere	1–3	2.5–30	0	0.01–0.03	1.52	6.9×10^{-5}–3.3×10^{-1}
Lu et al. [16]	15.2	10	cube	1.692–3.384 (2–4)	2.5–30	0	0.02–0.04	1.52	6.9×10^{-5}–4.9×10^{-1}
Njeng et al. [33]	420	21	cuboid	10 × 4.5 × 2 (5.56)	2–6	1–3	0.026	20	4.7×10^{-4}–4.2×10^{-3}
Chen et al. [31]	323.23 and 600.46	15.24 and 33.02	sphere	0.070	1–13	1.1–3	0.00046 & 0.00021	18.18 & 21.21	8.5×10^{-5}–3.1×10^{-2}
This research	50	8.4	Cube and cuboid	6–10 (7.44–12.41)	3–11	2–4	0.089–0.148	5.95	4.2×10^{-4}–5.7×10^{-3}

* Diameter of equivalent sphere was reported in the parenthesis for non-spherical particles.

Reviewing the captured movies in the experiments revealed that a change in flow behavior occurs at 9 rpm when the incline angle is 3° and 4° and the bed flow behavior is similar to slumping regime. Sherritt et al. [37] indicated that for slumping regime, the axial dispersion coefficient decreased when rotation speed increased. Our results beyond 7 rpm at 3° and 4° incline angle show a similar effect. Thus, increasing rotation speed has an inverse effect on axial dispersion coefficient.

5.2. Effect of Incline Angle

Figure 7 illustrates the effect of incline angle on the MRT and the axial dispersion coefficient. Figure 7a shows the effect of incline angle on the MRT at two different rotation speeds and feed rate 1.08 cm^3/s for 6 mm cubes. In addition, the power law functions were regressed on the data and equations are presented in the figure. MRT decreases as the incline angle increases. MRT does not change too much between 3° to 4°, while this effect is considerable from 2° to 3°. An increase in the rotation speed at each incline angle causes a shift of MRT to a lower value (almost 80 s in each case). The effect of incline angle on the MRT is similar to the effect of rotation speed, so we can write MRT $\propto \beta^{-a}$ (where β indicates incline angle). The relations MRT $\propto \beta^{-1.054}$ and MRT $\propto \beta^{-1.02}$ can be found for spherical particles [7,34,35], while Njeng et al. [33] showed that MRT $\propto \beta^{-0.928}$ for spherical and non-spherical particles. Our results show a wider span of exponent for non-spherical particles with sharp edges. This indicates that the incline angle has a stronger dependency on the MRT than it does on spherical and non-spherical particles without sharp edges.

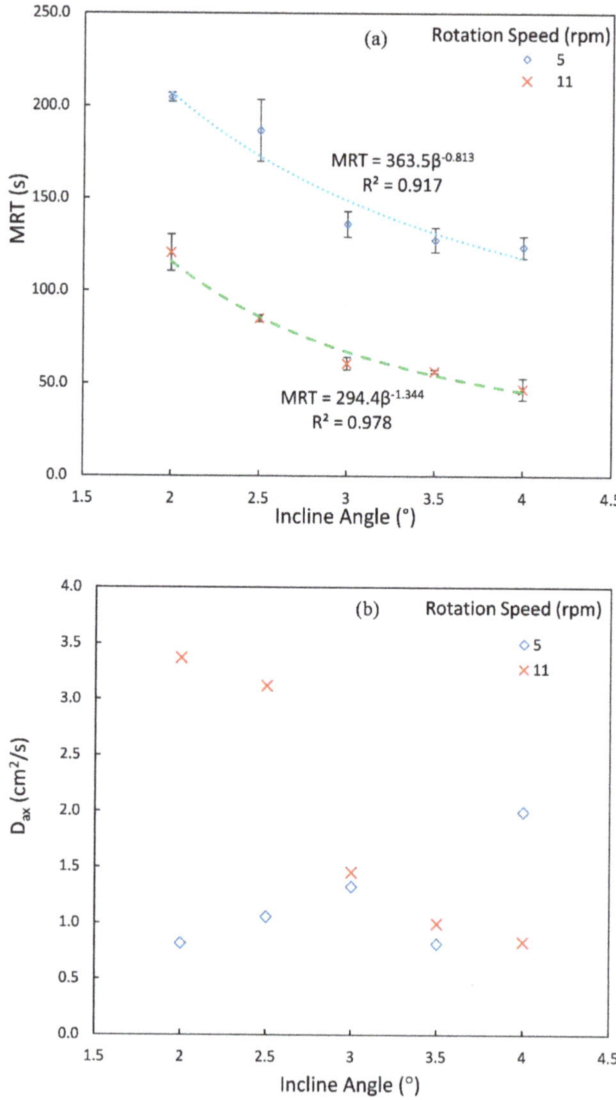

Figure 7. Effect of incline angle on (**a**) MRT (**b**) axial dispersion coefficient at 1.08 cm^3/s for a 6 mm cube.

In Figure 7b, when incline angle increases, axial dispersion coefficient increases (except at 3.5°) at 5 rpm. The standard deviation of axial dispersion coefficient at 3°, 3.5° and 4° are, respectively, equal to 0.058, 0. 376 and 0.803 cm^2/s, which clearly shows that the reduction in the average value at 3.5° is not statistically significant. The axial dispersion coefficient decreases as the incline angle increases at 11 rpm. When the rotation speed is high, the number of particles in the drum is low and only one or, at most, two layers of particles are formed, which means that there is directed motion of particles rather than random motion of particles during avalanching. Consequently, the axial dispersion coefficient decreases.

5.3. Effect of Feed Rate

The effects of feed rate on the MRT and the axial dispersion coefficient are shown in Figure 8 at 3° incline angle for 6 mm cubes. When the feed rate decreases, the number of particles and formed layers decrease, and hence, residence time decreases. Figure 3a,e illustrate that a decrease in feed rate leads to a decrease in the fill level in the drum. On average, particles remain for a shorter time in the drum when the feed rate decreases. This fact is more considerable at low rotation speeds. When rotation speed increases from 3 rpm to 7 rpm, the flowability of particles in the drum enhances and, hence, the difference between the MRTs of different feed rate decreases. An 11 rpm feed rate does not have a significant effect on the MRT. Other studies also confirm slight effect of feed rate on the MRT. For instance, Gao et al. [5] observed that the MRT was weakly influenced by feed rate for sphere, cylinder or cuboid. Similar results were observed for cuboid particles by Njeng et al. [33] and for cylindrical particles by Sudah et al. [36]. It is noteworthy that changing feed rate can slightly change the MRT, which was reported in previous research [5,7,11,33–35,39,40]. Our results have good qualitative agreement with other studies [5,7,11,33–35,39,40].

Figure 8b shows that the feed rate does not have a pronounced effect on the axial dispersion coefficient. At 3 rpm and 11 rpm, the values of axial dispersion coefficient are close together, whereas at 7 rpm they are far from together. In other studies, usually, when the feed rate increased, the axial dispersion coefficient decreased [10,33]. This shows that the randomness of particle movement decreases when the feed rate decreases. However, there were some cases in which feed rate had a direct effect on axial dispersion coefficient. Sudah et al. [36] observed that for cylindrical particles, the axial dispersion coefficient decreased from 0.1 cm^2/s to 0.034 m^2/s when the feed rate increased form 0.46 cm^3/s to 1.83 cm^3/s; and then it gradually increased to 0.042 cm^2/s when the feed rate further increased to 4.34 cm^3/s. Gao et al. [5] showed that the axial dispersion coefficient of cuboid particles did not change as a function of feed rate, but feed rate had a small effect on the axial dispersion coefficient of cylindrical particles, and had a more noticeable effect on the axial dispersion coefficient of spherical particles. Our results at 7 rpm do not follow these trends from other literature.

5.4. Effect of Particle Size

Figure 9 shows the effect of particle size on the MRT and the axial dispersion coefficient. To understand the effect of size, 6 mm and 10 mm cubes at 3° were compared at the same volumetric flow rate (1.08 cm^3/s). At these conditions, we observed a lower number of large particles in the drum (see Figure 3a,f). When size increases, the number of particles decreases and then particle–particle and particle–wall interactions decrease. Then, on average, the MRT decreases with an increase in the size. Other studies also confirm these results [41,42]. When rotation speed increases, the flowability of particles in the drum increases, and then the difference between MRT for 6 mm cubes and 10 mm decreases. Nafsun et al. [43] showed a similar effect of particle size on the thermal mixing time for changes in rotational speed (thermal mixing time is directly proportional to the MRT). They discovered that at a high rotation speed effect of particle size diminishes, but at low rotation speeds, particle size had an significant effect on the thermal mixing time [43].

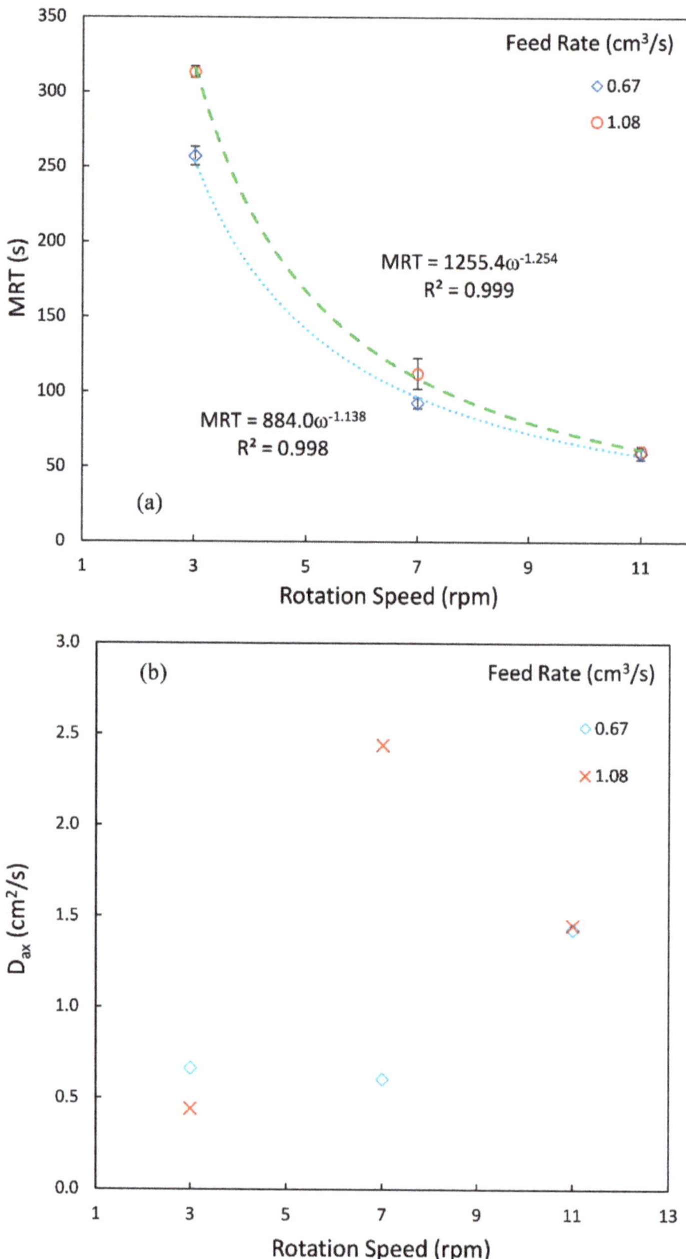

Figure 8. Effect of feed rate on (**a**) MRT (**b**) axial dispersion coefficient at 3° for 6 mm cube.

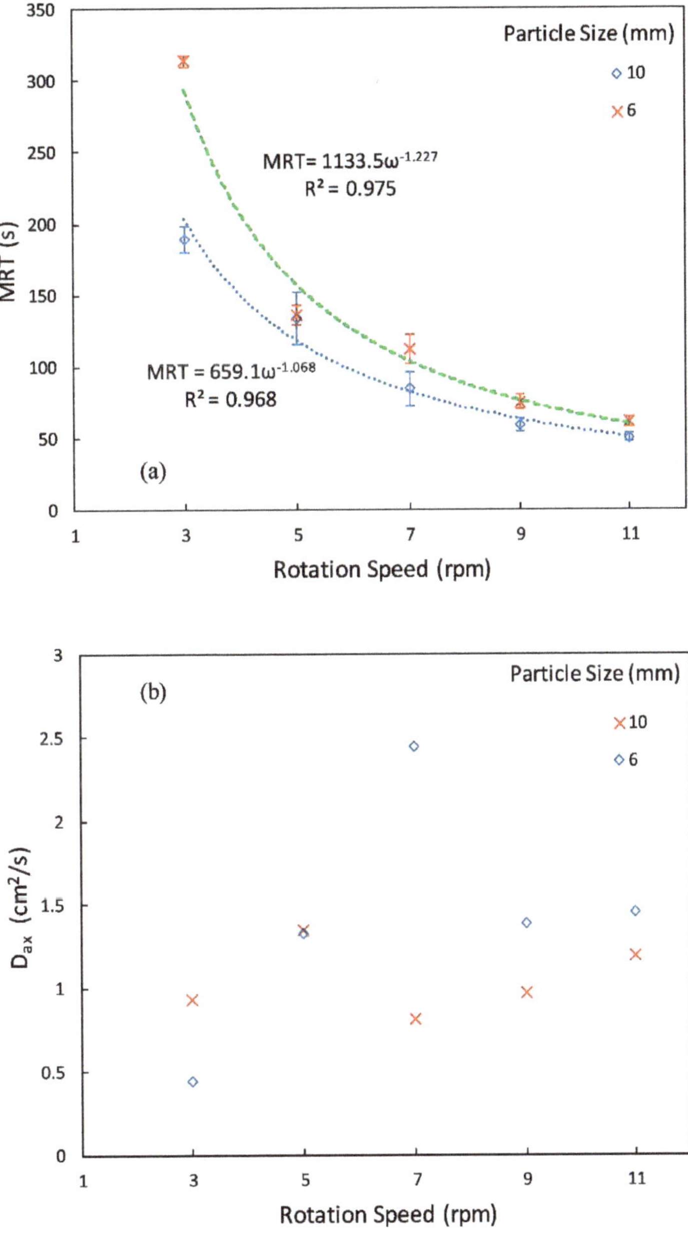

Figure 9. Effect of particle size on (**a**) MRT (**b**) axial dispersion coefficient at 3° and 0.72 cm^3/s for 6 mm and 10 mm cube.

The axial dispersion coefficient for larger particles is lower than that for smaller particles (see Figure 9b). This shows that smaller particles are dispersed better. Rotation speed does not significantly affect the axial dispersion coefficient for 10 mm cubes, while rotation speed has a notable effect on the axial dispersion coefficient for 6 mm cubes (see Figure 9b). Bensmann et al. [42] showed that when particle size is increased from 0.55 mm to 1.5 mm, the axial dispersion coefficient is increased from 0.019 cm^2/s to 0.240 cm^2/s, and then decreased to 0.035 cm^2/s.

5.5. Effect of Mixture Composition in the Feed

Ceramic and wooden particles of the same size ($10 \times 10 \times 4$ mm^3) were entered into the drum to study the effect of volume fraction of heavy and light particles. Rotation speed and incline angle were fixed at 7 rpm, and 2°, while wood and ceramic volume fractions were changed. The number of tracers of each kind (wooden and ceramic) was proportional to the volume fraction of each particle in the mixture. We could not fix the feed rate at the various volume fractions due to the changes in the density. Therefore, in this series of experiments, feeder vibration was kept constant and we let the feed rate change with volume fraction. The feed rate increased when the wood volume fraction increased in the experiments. For example, for pure ceramic (wood volume fraction = 0) the feed rate was 0.28 cm^3/s, while for pure wood, it was 0.40 cm^3/s. In Figure 8a, we showed that the feed rate does not significantly change the MRT of particles in the drum. Based on this observation, we can still draw some conclusions about the pure effect of wood volume fraction on the MRT (without considering the fact that feed rate changes with wood volume fraction).

Figure 10a shows the effect of wood volume fraction in the feed on the MRT (the ceramic volume fraction is 1-wood volume fraction). Since we could distinguish between wood and ceramic particles in the outlet, we calculated separate MRTs for wooden and ceramic particles. Then, the mixture MRT is calculated by Equation (8).

Wood particles stay for a longer time in the drum than ceramic particles. This is mainly related to the higher weight of the wood particles relative to the ceramic particles. Reviewing the captured movies in the experiment, we observed that the mixture sticks to the wall of the drum and rise with the wall. At a certain point, they detach from the wall and avalanche down. During avalanching, ceramic particles travel a longer axial distance (the drum is inclined). This process is repeated a number of times until the particles exit the drum. Thus, ceramic particles stay a shorted time in the drum. The MRT of the mixture is also shown in this figure. We provided a linear regression for the MRT of the mixture as a function of wood volume fraction. It clearly shows that the MRT of the wood and ceramic mixture (particles with different densities) can be linearly interpolated by using the MRTs of pure particles. The regressed line of the MRT of the mixture in Figure 10a shows that the MRT of mixture is a linear function of wood volume fraction:

$$\text{MRT}_{\text{mixture}} = \sum_{i=1}^{2} x_i \times \text{MRT}_i \tag{8}$$

Figure 10b shows the effect of wood volume fraction on the axial dispersion coefficient. When wood volume fraction changes, the axial dispersion coefficient does not change notably. Therefore, a change in the wood volume fraction in the mixture does not have a significant effect on the axial dispersion coefficient in this range of operating conditions. We could not find a clear trend for axial dispersion coefficient; however, the maximum of axial dispersion values was observed at the wood volume fraction of 0.25.

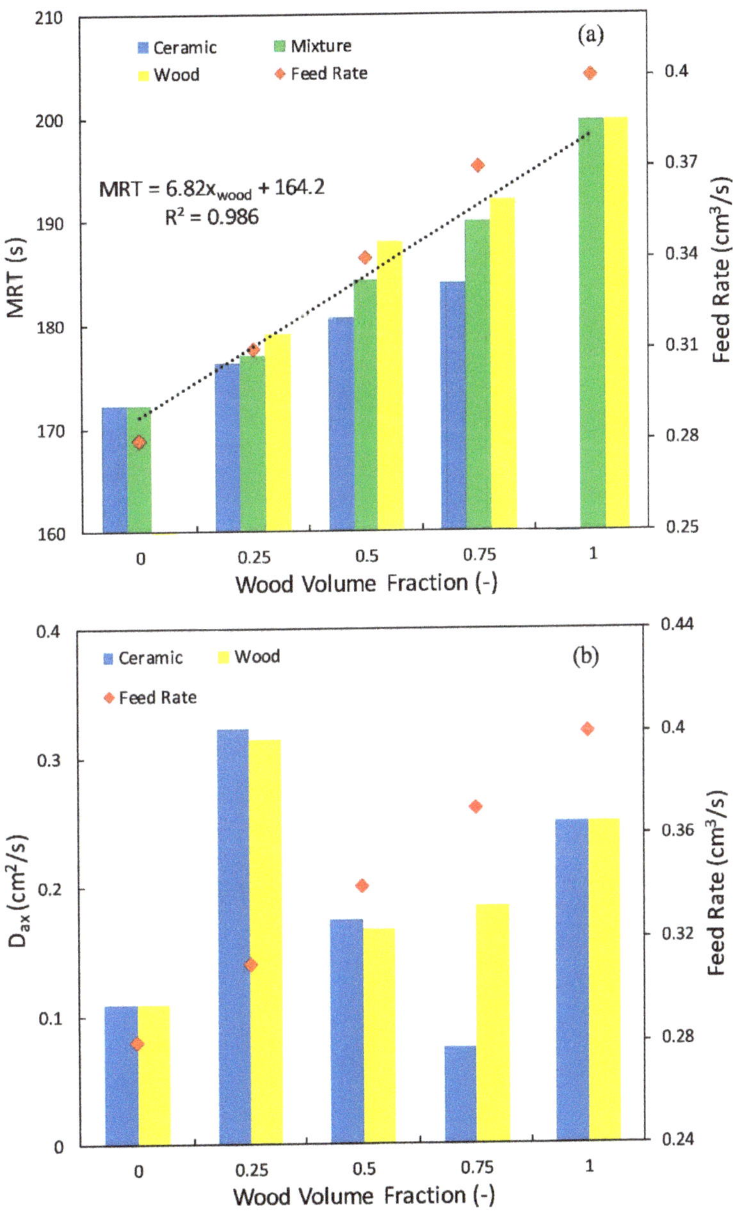

Figure 10. Effect of volume fraction of light (wood) and heavy (ceramic) cuboid particles on (**a**) the MRT (**b**) the axial dispersion coefficient at 7 rpm and 2°.

6. Conclusions

Understanding the motion pattern of non-spherical particles in a continuous rotary drum is of great importance for design and scaleup of pyrolysis processes, food manufacturing processes, etc. The motion pattern of spherical particles is characterized by RTD measurements that reflect the cumulative effect of motion of individual particles in the drum. RTD was measured in a continuous rotary drum with the length 50.0 cm and internal diameter 8.4 cm, using colored particles as tracers. Non-spherical particles with sharp edges were selected in the experiments, since in practice we are facing this class of shapes rather than spherical or non-spherical particles without sharp edges.

Particles with two sizes, 6 and 10 mm, and two densities, 750 and 2085 kg/m^3, were used in experiments. We included the particles with two densities in our experiments to resemble the condition in which wooden particles (as pyrolysis media) are processed along with ceramic materials (as heat carrier media). In the experiments, rotation speeds varied between 3 and to 11 rpm; incline angle, between 2 and 4°; and feed rate, between 0.28 cm^3/s to 1.08 cm^3/s. The effect of mixture composition (a mixture of low- and high-density particles) on the RTD was also investigated. We also fitted the RTD data with the axial dispersion model to obtain a better insight on the flow behavior of particles in the drum. The results were compared with those from similar studies on spherical and non-spherical particles without sharp edges to better understand the similarities and differences between the two (particles with sharp edges and without sharp edges). All the experimental results on RTD are also provided as supplementary data to be used by other researchers who are aiming to validate their DEM simulation results.

Our results showed that the MRT of non-spherical particles with sharp edges depends on $\omega^{-\alpha}$ similar to other shapes (spherical and non-spherical particles without sharp edge), but the exponent is higher in our experiments ($0.9 < \alpha < 1.24$), especially at high incline angles. At rotation speeds between 3 and 7 rpm, the axial dispersion coefficient increased with rotation speed, while this trend was not observed with further increase in the rotation speed due to change in the flow regime in the drum. The MRT also depends on incline angle MRT $\propto \beta^{-b}$, where b varies between 0.81 (at low rotation speed) to 1.34 (high rotation speed). However, the exponent b is closer to 1 for other particles with other shapes, suggesting the more pronounced role of incline angle on the particle motion for particles with sharp edges. The axial dispersion coefficient increases with incline angle at 5 rpm and it decreases with increasing the incline angle at 11 rpm. When feed rate increases, fill level increases, and consequently the MRT increases, especially at low rotation speed. Feed rate does not affect the axial dispersion coefficient in our experiments. Smaller particle has higher MRT than bigger particle. We could show that the value of α in the relation MRT $\propto \omega^{-\alpha}$ is higher for smaller particles and the difference between MRT of large and small particles diminishes as the rotation speed increases. Rotation speed has a significant effect on the axial dispersion coefficient of 6 mm cubes, while rotation speed has a small effect on the axial dispersion coefficient of 10 mm cubes. Wooden particles (low density) have higher MRT than ceramic particles (high density). For a mixture of wooden and ceramic particles, the MRT linearly increases with volume fraction of wooden particles in a mixture. The MRT of mixture scales with the MRT of pure wooden and ceramic particles, with volume fraction as the scaling factor.

Supplementary Materials: The following supporting information can be downloaded at: https://www.mdpi.com/article/10.3390/pr10061069/s1, Supplementary data includes a Microsoft excel file named data.xlsx. The first sheet of this file lists experimental conditions table (similar to Table 1 of the article). That sheet has hyperlinks to the corresponding experimental data. The rest of the sheets contain the residence time of tracers and cumulative distribution of residence time (F) in experiments and fitted curves obtained from the ADM. In addition, the calculate MRT, Pe and D_{ax} of particles are available in each sheet.

Author Contributions: Conceptualization, H.R.N., B.H. and M.H.; methodology, S.M. and B.H.; software, S.M.; validation, S.M. and B.H.; formal analysis, S.M.; investigation, S.M.; resources, C.J.; data curation, S.M., B.H. and C.J.; writing—original draft preparation, S.M.; writing—review and editing, H.R.N., C.J. and B.H.; visualization, S.M.; supervision, H.R.N., B.H. and C.J.; project administration, M.H.; funding acquisition, M.H. All authors have read and agreed to the published version of the manuscript.

Funding: Open Access Funding by TU Wien.

Institutional Review Board Statement: Not applicable.

Informed Consent Statement: Not applicable.

Data Availability Statement: Not applicable.

Acknowledgments: The authors would like to thank Mario Pichler, Florian Wesenauer and Franz Winter who helped us in this research.

Conflicts of Interest: The authors declare no conflict of interest.

Nomenclature

symbols

$C(t)$	Dimensionless concentration of tracers at the drum outlet as a function of time, (-)
C_0	Initial dimensionless concentration, (-)
d	Particle diameter, (m)
D_{ax}	Axial dispersion coefficient, (cm^2/s)
$F(t)$	Cumulative distribution as a function of time, (-)
l	Position of tracer in the axis direction of drum, (-)
L	Drum length, (m)
N_0	Total number of tracer particles, (-)
$N(t)$	Number of tracer particles leave the drum until t as a function of time, (-)
Pe	Peclet number, (-)
t	Time, (s)
u_z	Linear velocity, (m/s)
x	Volumetric fraction in the mixture, (-)
z	Dimensionless length ($\frac{l}{L}$), (-)
β	Incline angle of drum, (°)
ω	Rotation speed of drum, (rpm; 1/min)

abbreviation

ADM	Axial dispersion model
MRT	Mean residence time, (s)
RTD	Residence time distribution

References

1. Dubé, O.; Alizadeh, E.; Chaouki, J.; Bertrand, F. Dynamics of Non-Spherical Particles in a Rotating Drum. *Chem. Eng. Sci.* **2013**, *101*, 486–502. [CrossRef]
2. Norouzi, H.R.; Zarghami, R.; Mostoufi, N. Insights into the Granular Flow in Rotating Drums. *Chem. Eng. Res. Des.* **2015**, *102*, 12–25. [CrossRef]
3. Kunii, D.; Chisaki, T. *Rotary Reactor Engineering*; Elsevier: Amsterdam, The Netherlands, 2007; ISBN 0080553338.
4. Boateng, A.A. *Rotary Kilns: Transport Phenomena and Transport Processes*; Butterworth-Heinemann: Oxford, UK, 2015; ISBN 0128038535.
5. Gao, Y.; Glasser, B.J.; Ierapetritou, M.G.; Cuitino, A.; Muzzio, F.J.; Beeckman, J.W.; Fassbender, N.A.; Borghard, W.G. Measurement of Residence Time Distribution in a Rotary Calciner. *AIChE J.* **2013**, *59*, 4068–4076. [CrossRef]
6. Hamawand, I.; Yusaf, T. Particles Motion in a Cascading Rotary Drum Dryer. *Can. J. Chem. Eng.* **2014**, *92*, 648–662. [CrossRef]
7. Paredes, I.J.; Yohannes, B.; Emady, H.N.; Muzzio, F.J.; Maglio, A.; Borghard, W.G.; Glasser, B.J.; Cuitiño, A.M. Measurement of the Residence Time Distribution of a Cohesive Powder in a Flighted Rotary Kiln. *Chem. Eng. Sci.* **2018**, *191*, 56–66. [CrossRef]
8. Pichler, M.; Haddadi, B.; Jordan, C.; Norouzi, H.; Harasek, M. Influence of Particle Residence Time Distribution on the Biomass Pyrolysis in a Rotary Kiln. *J. Anal. Appl. Pyrolysis* **2021**, *158*, 105171. [CrossRef]
9. Levenspiel, O. *Chemical Reaction Engineering*; John Wiley & Sons: Hoboken, NJ, USA, 1999; ISBN 047125424X.
10. Bongo Njeng, A.S.; Vitu, S.; Clausse, M.; Dirion, J.L.; Debacq, M. Effect of Lifter Shape and Operating Parameters on the Flow of Materials in a Pilot Rotary Kiln: Part I. Experimental RTD and Axial Dispersion Study. *Powder Technol.* **2015**, *269*, 554–565. [CrossRef]
11. Bongo Njeng, A.S.; Vitu, S.; Clausse, M.; Dirion, J.L.; Debacq, M. Effect of Lifter Shape and Operating Parameters on the Flow of Materials in a Pilot Rotary Kiln: Part II. Experimental Hold-up and Mean Residence Time Modeling. *Powder Technol.* **2015**, *269*, 566–576. [CrossRef]
12. Priessen, J.; Kreutzer, T.; Irgat, G.; Behrens, M.; Schultz, H.J. Solid Flow in Rotary Drums with Sectional Internals: An Experimental Investigation. *Chem. Eng. Technol.* **2021**, *44*, 300–309. [CrossRef]
13. Santos, D.A.; Barrozo, M.A.S.; Duarte, C.R.; Weigler, F.; Mellmann, J. Investigation of Particle Dynamics in a Rotary Drum by Means of Experiments and Numerical Simulations Using DEM. *Adv. Powder Technol.* **2016**, *27*, 692–703. [CrossRef]

14. Lu, G.; Third, J.R.; Müller, C.R. Discrete Element Models for Non-Spherical Particle Systems: From Theoretical Developments to Applications. *Chem. Eng. Sci.* **2015**, *127*, 425–465. [CrossRef]
15. Gu, C.; Fan, J.; Pan, D.; Yao, S.; Dai, L.; Guan, L.; Wu, K.; Yuan, Z. Effect of Baffle Structure on the Dynamic Transportation Behavior of S-Liked Biomass Fuels in a Rotating Drum. *Energy Sci. Eng.* **2020**, *9*, 743–756. [CrossRef]
16. Lu, G.; Third, J.R.; Müller, C.R. The Parameters Governing the Coefficient of Dispersion of Cubes in Rotating Cylinders. *Granul. Matter* **2017**, *19*, 12. [CrossRef]
17. Fantozzi, F.; Colantoni, S.; Bartocci, P.; Desideri, U. Rotary Kiln Slow Pyrolysis for Syngas and Char Production from Biomass and Waste—Part I: Working Envelope of the Reactor. *J. Eng. Gas Turbines Power* **2007**, *129*, 901–907. [CrossRef]
18. Hlosta, J.; Jezerská, L.; Rozbroj, J.; Žurovec, D.; Nečas, J.; Zegzulka, J. DEM Investigation of the Influence of Particulate Properties and Operating Conditions on the Mixing Process in Rotary Drums: Part 2-Process Validation and Experimental Study. *Processes* **2020**, *8*, 184. [CrossRef]
19. Maione, R.; Kiesgen De Richter, S.; Mauviel, G.; Wild, G. DEM Investigation of Granular Flow and Binary Mixture Segregation in a Rotating Tumbler: Influence of Particle Shape and Internal Baffles. *Powder Technol.* **2015**, *286*, 732–739. [CrossRef]
20. He, S.Y.; Gan, J.Q.; Pinson, D.; Yu, A.B.; Zhou, Z.Y. Flow Regimes of Cohesionless Ellipsoidal Particles in a Rotating Drum. *Powder Technol.* **2019**, *354*, 174–187. [CrossRef]
21. Xie, C.; Song, T.; Zhao, Y. Discrete Element Modeling and Simulation of Non-Spherical Particles Using Polyhedrons and Super-Ellipsoids. *Powder Technol.* **2020**, *368*, 253–267. [CrossRef]
22. Ahmadian, H.; Hassanpour, A.; Ghadiri, M. Analysis of Granule Breakage in a Rotary Mixing Drum: Experimental Study and Distinct Element Analysis. *Powder Technol.* **2011**, *210*, 175–180. [CrossRef]
23. Cleary, P.W. DEM Prediction of Industrial and Geophysical Particle Flows. *Particuology* **2010**, *8*, 106–118. [CrossRef]
24. Colin, B.; Dirion, J.L.; Arlabosse, P.; Salvador, S. Wood Chips Flow in a Rotary Kiln: Experiments and Modeling. *Chem. Eng. Res. Des.* **2015**, *98*, 179–187. [CrossRef]
25. Gao, Y.; Muzzio, F.J.; Ierapetritou, M.G. A Review of the Residence Time Distribution (RTD) Applications in Solid Unit Operations. *Powder Technol.* **2012**, *228*, 416–423. [CrossRef]
26. Norouzi, H.R.; Zarghami, R.; Sotudeh-Gharebagh, R.; Mostoufi, N. *Coupled CFD-DEM Modeling: Formulation, Implementation and Application to Multiphase Flows*; John Wiley & Sons: Hoboken, NJ, USA, 2016; ISBN 1119005132.
27. Fu, P.; Bai, X.; Yi, W.; Li, Z.; Li, Y. Fast Pyrolysis of Wheat Straw in a Dual Concentric Rotary Cylinder Reactor with Ceramic Balls as Recirculated Heat Carrier. *Energy Convers. Manag.* **2018**, *171*, 855–862. [CrossRef]
28. Fogler, H.S. *Elements of Chemical Reaction Engineering*, 5th ed.; Prentice Hall: Upper Saddle River, NJ, USA, 2016.
29. Bérard, A.; Blais, B.; Patience, G.S. Experimental Methods in Chemical Engineering: Residence Time Distribution—RTD. *Can. J. Chem. Eng.* **2020**, *98*, 848–867. [CrossRef]
30. Cangialosi, F.; Di Canio, F.; Intini, G.; Notarnicola, M.; Liberti, L.; Belz, G.; Caramuscio, P. Experimental and Theoretical Investigation on Unburned Coal Char Burnout in a Pilot-Scale Rotary Kiln. *Fuel* **2006**, *85*, 2294–2300. [CrossRef]
31. Chen, I.Y.; Navodia, S.; Yohannes, B.; Nordeck, L.; Machado, B.; Ardalani, E.; Borghard, W.G.; Glasser, B.J.; Cuitiño, A.M. Flow of a Moderately Cohesive FCC Catalyst in Two Pilot-Scale Rotary Calciners: Residence Time Distribution and Bed Depth Measurements with and without Dams. *Chem. Eng. Sci.* **2021**, *230*, 116211. [CrossRef]
32. Sai, P.S.T. Drying of Solids in a Rotary Dryer. *Dry. Technol.* **2013**, *31*, 213–223. [CrossRef]
33. Bongo Njeng, A.S.; Vitu, S.; Clausse, M.; Dirion, J.L.; Debacq, M. Effect of Lifter Shape and Operating Parameters on the Flow of Materials in a Pilot Rotary Kiln: Part III. Up-Scaling Considerations and Segregation Analysis. *Powder Technol.* **2016**, *297*, 415–428. [CrossRef]
34. Sai, P.S.T.; Surender, G.D.; Damodaran, A.D.; Suresh, V.; Philip, Z.G.; Sankaran, K. Residence Time Distribution and Material Flow Studies in a Rotary Kiln. *Metall. Trans. B* **1990**, *21*, 1005–1011. [CrossRef]
35. Chatterjee, A.; Sathe, A.V.; Mukhopadhyay, P.K. Flow of Materials in Rotary Kilns Used for Sponge Iron Manufacture: Part II. Effect of Kiln Geometry. *Metall. Trans. B* **1983**, *14*, 383–392. [CrossRef]
36. Sudah, O.S.; Chester, A.W.; Kowalski, J.A.; Beeckman, J.W.; Muzzio, F.J. Quantitative Characterization of Mixing Processes in Rotary Calciners. *Powder Technol.* **2002**, *126*, 166–173. [CrossRef]
37. Sherritt, R.G.; Chaouki, J.; Mehrotra, A.K.; Behie, L.A. Axial Dispersion in the Three-Dimensional Mixing of Particles in a Rotating Drum Reactor. *Chem. Eng. Sci.* **2003**, *58*, 401–415. [CrossRef]
38. Third, J.R.; Scott, D.; Scott, S.A. Axial Dispersion of Granular Material in Horizontal Rotating Cylinders. *Powder Technol.* **2010**, *203*, 510–517. [CrossRef]
39. Gu, C.; Li, P.; Yuan, Z.; Yan, Y.; Luo, D.; Li, B.; Lu, D. A New Corrected Formula to Predict Mean Residence Time of Flexible Filamentous Particles in Rotary Dryers. *Powder Technol.* **2016**, *303*, 168–175. [CrossRef]
40. Liu, X.Y.; Specht, E. Mean Residence Time and Hold-up of Solids in Rotary Kilns. *Chem. Eng. Sci.* **2006**, *61*, 5176–5181. [CrossRef]
41. Hatzilyberis, K.S.; Androutsopoulos, G.P. An RTD Study for the Flow of Lignite Particles through a Pilot Rotary Dryer Part i; Bare Drum Case. *Dry. Technol.* **1999**, *17*, 745–757. [CrossRef]
42. Bensmann, S.; Subagyo, A.; Walzel, P. Residence Time Distribution of Segregating Sand Particles in a Rotary Drum. *Part. Sci. Technol.* **2010**, *28*, 319–331. [CrossRef]
43. Nafsun, A.I.; Herz, F.; Liu, X. Influence of Material Thermal Properties and Dispersity on Thermal Bed Mixing in Rotary Drums. *Powder Technol.* **2018**, *331*, 121–128. [CrossRef]

chemengineering

Article

Comprehensive Modeling of Vacuum Systems Using Process Simulation Software

Eduard Vladislavovich Osipov *, Daniel Bugembe, Sergey Ivanovich Ponikarov and Artem Sergeevich Ponikarov

Machines and Apparatus in Chemical Production Department, Institute of Chemical and Petroleum Engineering, Kazan National Research Technological University, Karl Marx Street, 68, 420015 Kazan, Russia; dbugembe@kstu.ru (D.B.); ponikarovsi@corp.knrtu.ru (S.I.P.); ponikarovas@corp.knrtu.ru (A.S.P.)
* Correspondence: eduardvosipov@gmail.com; Tel.: +7-(917)-293-66-00

Abstract: Traditional vacuum system designs often rely on a 100% reserve, lacking precision for accurate petrochemical computations under vacuum. This study addresses this gap by proposing an innovative modeling methodology through the deconstruction of a typical vacuum-enabled process. Emphasizing non-prescriptive pressure assignment, the approach ensures optimal alignment within the vacuum system. Utilizing process simulation software, each component was systematically evaluated following a proposed algorithm. The methodology was applied to simulate vacuum-driven separation in phenol and acetone production. Quantifying the vacuum system's load involved constructing mathematical models in Unisim Design R451 to determine the mixture's volume flow rate entering the vacuum pump. A standard-sized vacuum pump was then selected with a 40% performance margin. Post-reconstruction, the outcomes revealed a 22.5 mm Hg suction pressure within the liquid-ring vacuum pump, validating the efficacy of the devised design at a designated residual pressure of 40 mm Hg. This study enhances precision in vacuum system design, offering insights that are applicable to diverse petrochemical processes.

Keywords: vacuum system; phenol and acetone waste recycling; liquid-ring vacuum pump; reconstruction of vacuum blocks; process simulation software; Unisim Design R451

Citation: Osipov, E.V.; Bugembe, D.; Ponikarov, S.I.; Ponikarov, A.S. Comprehensive Modeling of Vacuum Systems Using Process Simulation Software. *ChemEngineering* **2024**, *8*, 31. https://doi.org/10.3390/chemengineering8020031

Academic Editor: Roumiana Petrova Stateva

Received: 19 January 2024
Revised: 18 February 2024
Accepted: 23 February 2024
Published: 6 March 2024

1. Introduction

Modeling vacuum units in petrochemical installations is challenging due to the uncertainties in defining the key parameters. This arises from the complexity of characterizing all the contributing factors, resulting in a task lacking a clear structure. Established principles governing processes under atmospheric pressure undergo changes in a vacuum, revealing new patterns and characteristics.

The computational methodologies guiding the design of vacuum apparatus were explained in previous works [1–4]. The assessment of any vacuum system involves two crucial stages: the design phase and subsequent validation. During the design phase, decisions regarding the types of vacuum pumps are made, and the initial geometric parameters for the interconnecting pipelines among the key system elements are estimated [1]. The configuration of the vacuum system resembles analogous setups with a similar purpose, while the selection of the primary vacuum pump depends on the specified pressure thresholds, capacity requirements, and evacuation timelines.

The scientific literature contains numerous examples illustrating the calculation of technological processes that operate under vacuum conditions. For instance, in reference [5], factors influencing fuel–oil separation processes were outlined, while reference [6] explored ways to reduce the workload on the vacuum system based on the computational results. Reference [7] argued for the necessity of installing a fore-vacuum pump when modeling a membrane separation block, and reference [8] quantified the energy consumption of a vacuum pump during the purification process of butanol using hybrid membranes.

Reference [9] investigated methods to reduce the boiling point differential in vacuum distillation, while reference [10] discussed the complexities of designing a Vacuum Distillation Unit (VDU).

The above-referenced articles heavily relied on process simulation software such as Aspen HYSYS (https://www.aspentech.com/en/products/engineering/aspen-hysys), Aspen Plus (https://www.aspentech.com/en/products/engineering/aspen-plus), and ChemCad (https://www.chemstations.com/CHEMCAD/) for conducting thermal and mass balance computations. However, when it came to the vacuum-based technological processes, the influence of the vacuum system was often overlooked, potentially compromising the accuracy of the modeling results. In reference [11], a specialized user module was incorporated into HYSYS to address this gap by computing steam ejector pumps, which are crucial in vacuum systems. Additionally, [12] emphasized that neglecting the intricacies of the technological process when modeling vacuum mixture condensation using such programs could introduce errors.

Traditionally, designers of vacuum systems and operational staff rely on industry experience when selecting the Vacuum Overhead System (VOS). However, during the design phase, there is often a lack of consideration for the specific requirements of the vacuum installation, leading to difficulties in achieving the desired residual pressure in the equipment.

The challenge of selecting the optimal VOS becomes apparent during comprehensive upgrades of technological installations due to the changes in the characteristics of the vacuumed object, rendering existing VOS specifications inadequate.

Therefore, designing a technological process under a vacuum requires simultaneous computations of the primary process within the technological object, the condensation block preceding the VOS, and the VOS itself. The complexity of designing and computing vacuum-based technological installations in chemical and petrochemical complexes lies at the intersection of vacuum technology and chemical engineering. Hence, this study aimed to devise a specialized methodology for modeling vacuum systems utilizing the capabilities of process simulation software. The developed methodology was intended for validation on a standard petrochemical installation, specifically targeting the waste separation block in phenol and acetone production.

2. The Conventional Approach to Designing Vacuum Systems

The traditional approach to designing vacuum generation systems involves calculating the overall gas flow entering the system and determining the effective pumping speed using Formula (1). Subsequently, the capacity of the vacuum pump is computed using Formula (2).

$$V_0 = \frac{Q_{sum}}{p_0} \tag{1}$$

$$V_n = V_0 \cdot v \tag{2}$$

In Equations (1) and (2), p_0—inlet pressure, Pa; Q_{sum}—summary leak rate, m$^3\cdot$Pa/s; V_0—required load on the VOS, m^3/s; and V_n—efficiency of the vacuum pump. The coefficient v's magnitude varies depending on the vacuum pump type and is referenced in specialized literature. For example, volumetric pumps hold a coefficient of 1.4, whereas jet pumps hold a coefficient of 2, indicating capacity margins of 40% and 100%, respectively [1–4].

The gas flow entering the evacuated object is calculated using Formula (3). In Equation (3), Q_{pr}—represents the flow due to the permeability of the vacuum chamber walls, measured in m$^3\cdot$Pa/s; Q_{in}—denotes the flow of atmospheric air entering the system, measured in m$^3\cdot$Pa/s; Q_{diff} signifies the gas flow resulting from the diffusion gas emission from the depth of structural materials and processed products, measured in m$^3\cdot$Pa/s; Q_{surf} indicates the gas flow from the surface of the working chamber and its components, measured in m$^3\cdot$Pa/s; and Q_{proc} represents the gas flow generated during the technological process, measured in m$^3\cdot$Pa/s.

Industrial vacuum systems typically operate within pressure ranges of 0.5 to 500 mm Hg and exhibit temperatures ranging from 10 to 150 degrees Celsius. The gas release effects from constituent materials in these systems minimally impact the vacuum creation and maintenance processes in chemical-technological operations. Therefore, Q_{pr}, Q_{diff}, and Q_{surf} can be neglected. Studies [12,13] demonstrated that leakage gas flows are not influenced by pressure but are mainly determined by temperature or equipment volume. Therefore, in Equation (3), Q (leak rate, m³·Pa/s) can be replaced by V (m³/s), resulting in Equation (4):

$$Q_{sum} = Q_{pr} + Q_{in} + Q_{diff} + Q_{surf} + Q_{proc} \tag{3}$$

$$V_0 = V_{in} + V_{proc} \tag{4}$$

To estimate V_{in}, the empirical relationships outlined in [12,13] prove beneficial, whereas determining V_{proc} necessitates the construction of a computational model for both the block and the vacuum system (VS).

Current modeling and design tools for vacuum systems lack the requisite functionality to accurately compute V_{proc}. The methodologies embedded within vacuum Computer-Aided Design (CAD) systems are primarily devised for evacuating pristine air. Consequently, to attain the desired pressure, vacuum pump performance is often overstated by introducing performance safety margins.

Present-day advancements in process simulation software enable the calculation of V_{proc}, allowing for the potential abandonment of performance safety margins (or their reduction to an acceptable level). However, it is imperative to disassemble the vacuum block into its constituents and employ these software tools in a manner that accommodates the intricate interactions between the individual elements within the vacuum block.

3. Methodology for Coupled Modeling of Vacuum Blocks

The technological apparatus functioning under a vacuum presents a complex assembly, interlinking various apparatuses interconnected by technological streams. Processes involving chemical, petrochemical, or crude oil raw material processing under a vacuum can be deconstructed into key blocks:

- Main technological unit (or group of units) operating under vacuum;
- Vacuum condensers;
- Vacuum overhead system (VOS).

The vacuum block consists of diverse equipment, requiring separate energy resources (cooling water, steam, electrical power, etc.). Gases from leaks (atmospheric air) enter the system, and during the technological process, decomposition gases may be formed. Therefore, a typical vacuum chemical-technological installation can be represented as blocks shown in Figure 1 (solid lines depict the flows entering the system components).

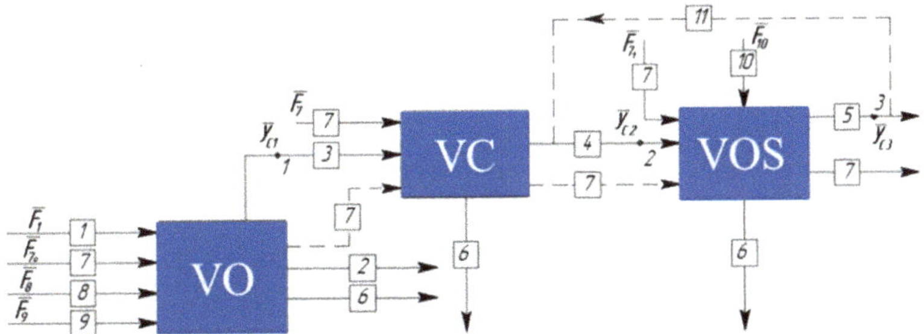

Figure 1. Decomposition of a typical chemical-technological process conducted under vacuum.

In Figure 1, the following notations for streams are adopted: 1—raw material inlet; 2—product outlet; 3—non-condensable gases; 4—flow directed to the Vacuum Overhead

System (VOS); 5—exhaust; 6—condensate; 7—energy resources; 8—decomposition gases; 9—ingress gases; 10—energy resource; 11—bypass flow facilitating VOS property regulation. Notably, points 1, 2, and 3 represent pivotal conjugation points.

The interconnected components of the examined VOS establish direct linkages via the gas pipelines housing the conjugation points: 1 (technological object pressure); 2 (VOS inlet pressure); and 3 (VOS exhaust). These junctures ensure equilibrium between the state parameters of the outgoing flows from the antecedent blocks and the incoming flows into the subsequent segments. If necessary, the recirculation of the energy resources or the flows between blocks can be organized. For example, the heat transfer fluid used from the previous block may be utilized in the subsequent one (stream 7), while a portion of the exhaust gas is returned to suction for regulating the performance of the VOS (stream 11). These connections are not always implemented in vacuum blocks; hence, they are depicted with dashed lines in Figures 1–3.

The transformation of the input variable vector $\overline{F}$ into output estimate vectors $\overline{y}_c$ can be represented as a vector function $\overline{\varphi}$. In this study, $\overline{\varphi}$ stands for the series of analytical or numerical (approximate) solutions to the system of equations that describe the chemical and technological processes within the equipment elements. $\overline{a}$ refers to the coefficients within this system of equations $\overline{\varphi}$.

Then, the vector functions for the decomposition scheme in Figure 1 are expressed as:

$$\overline{y}_{c1} = \overline{\varphi}_1\left(\overline{F}_1, \overline{F}_{7_0}, \overline{F}_8, \overline{F}_9, \overline{a}_1\right) \tag{5}$$

$$\overline{y}_{c2} = \overline{\varphi}_2\left(\overline{y}_{c1}, \overline{F}_7, \overline{a}_2\right) \tag{6}$$

$$\overline{y}_{c3} = \overline{\varphi}_3\left(\overline{y}_{c2}, \overline{F}_{7_1}, \overline{F}_{10}, \overline{a}_3\right) \tag{7}$$

In Equations (5)–(7), the indices to $\overline{y}_c$ and $\overline{\varphi}$ correspond to the points indicated in Figure 1. The indices for $\overline{F}$ relate to the streams specified in Figure 1 and $\overline{a}$ represents the numerical values of the equations describing the chemical and technological processes within the equipment elements.

Given that the choice of vacuum pump relies on the volumetric flow rate of the mixture during suction, $\overline{y}_c$ is established based on the dependence of the mixture's volumetric flow rate on the intake pressure.

The Vacuum Overhead System (VOS) of an industrial chemical-technological process also consists of various equipment, allowing this block to be divided into a series of subsystems at one hierarchical level. If the VOS consists of three stages, the decomposition can be represented as shown in Figure 2.

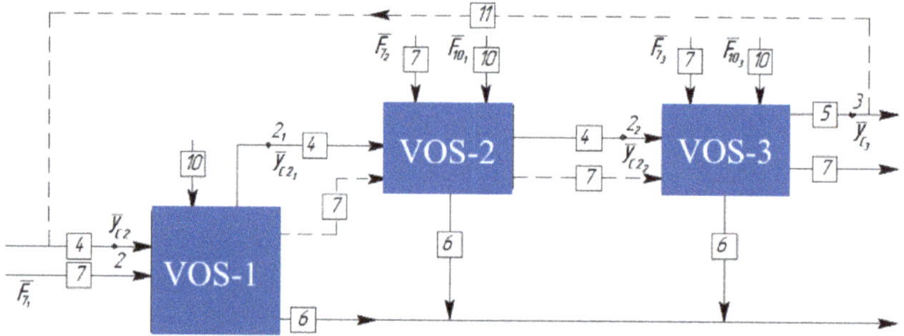

Figure 2. Decomposition of the Vacuum Overhead System (VOS) consisting of three blocks.

The additional conjugation points, identified as 2_1 and 2_2, arise from the interconnections among the constituent blocks, matching the number of linkages between the elements.

Consequently, the vector functions governing the intricate chemical-technological system operating under vacuum, comprising a VOS with three blocks, can be formulated as follows:

$$\overline{y}_{c2_1} = \overline{\varphi}_{2_1}\left(\overline{y}_{c2}, \overline{F}_{7_1}, \overline{F}_{10_1}, \overline{a}_{2_1}\right) \tag{8}$$

$$\overline{y}_{c2_2} = \overline{\varphi}_{2_2}\left(\overline{y}_{c2_1}, \overline{F}_{7_2}, \overline{F}_{10_2}, \overline{a}_{2_2}\right) \tag{9}$$

If the VOS consists of n (n = 1, ..., N) blocks, then the decomposition of the system will take the form shown in Figure 3.

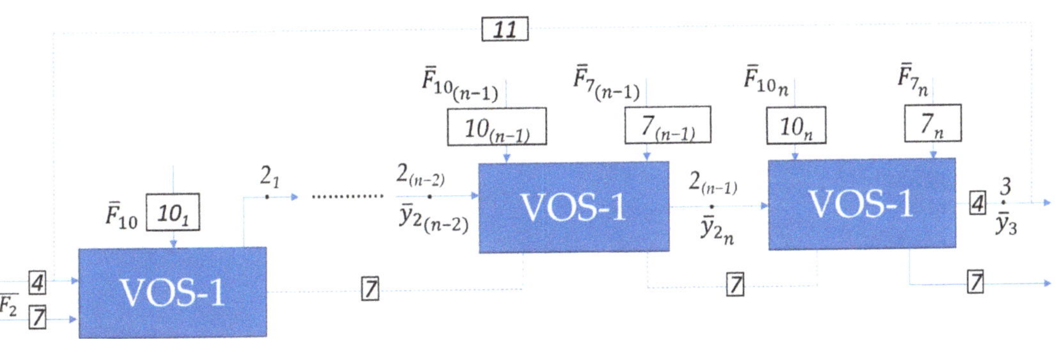

Figure 3. Decomposition of a VOS consisting of n blocks. The vector functions of the constituent elements of a vacuum overhead system consisting of n elements will be written as follows:

$$\overline{y}_{c2_{(n-1)}} = \varphi_{2_{(n-1)}}\left(\overline{y}_{c2_{(n-2)}}, \overline{F}_{7_{(n-1)}}, \overline{F}_{10_{(n-1)}}, \overline{a}_{2_{(n-1)}}\right) \tag{10}$$

$$\overline{y}_{c3} = \varphi_3\left(\overline{y}_{c2_{(n-1)}}, \overline{F}_{7_n}, \overline{F}_{10_n}, \overline{a}_3\right) \tag{11}$$

In this context, the vector function representation for the first element (stage) of the VOS would resemble Equation (1).

Characterizing the components of the CVCTS (Complex Vacuum Chemical Technology System) facilitates the calculation of the thermal and mass balances of the installation and addresses synthesis and optimization tasks.

The alignment of properties between the technological object and the vacuum creation system implies the coordination of their fundamental characteristics. By fundamental characteristics, we mean the key parameters describing the operating conditions of these blocks as a unified whole.

Interconnecting the elements within the vacuum block is envisaged based on the block diagram depicted in Figure 4. Opposite each symbol in the flowchart is the step number, and the description is provided below.

1. Introduce the minimum and maximum possible pressures in the object (p_1 and p_2, respectively), the minimum and maximum temperatures in the condenser (t1 and t2, respectively), the allowable pressure drop Δp, and the heat exchange surface margin e ($0 < e \leq 5\%$);
2. Set the initial value of the pressure p in the vacuumized technological object;
3. If the condition $p_1 \leq p \leq p_2$ is met, proceed to step 4;
4. Calculate the vacuumized technological object and determine the parameters of the flow entering the condenser (or group of condensers);
5. Introduce the initial approximation of the cooling temperature of the mixture, t;
6. If the condition $t1 \leq t \leq t2$ is met, proceed to step 7. If the condition is not met, assign a new value of pressure to the technological object (step 2);
7. Calculate the condenser and determine the load on the VOS;

8. If the heat exchange surface margin and pressure drop do not exceed the specified values, proceed to the next step. Otherwise, assign a new value to t (repeat step 5);
9. Assign the pressure at the VOS inlet, which is less than the pressure in the technological object by the pressure drop in the condenser Δp;
10. Calculate VOS;
11. If the recorded performance of VOS is greater than the volumetric flow rate at the condenser outlet (V1) and does not exceed the maximum permissible performance (V2), proceed to step 12. If this condition is not met, assign a new pressure to the technological object (step 2);
12. Fix the pressure in the technological object and conclude the calculation;
13. If, in the end, the pressure in the technological object satisfies all the specified conditions, the calculation is completed, and it is concluded that the specified VOS is not suitable for the investigated process.

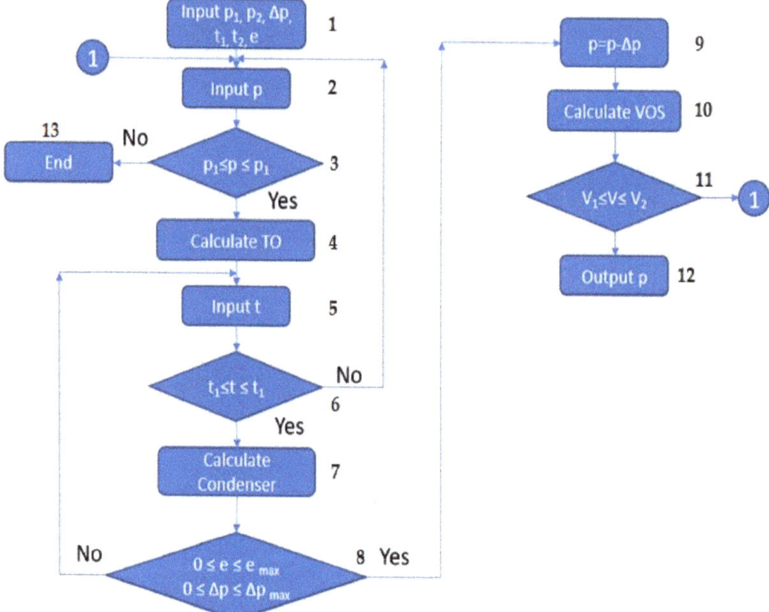

Figure 4. Flowchart of the coupled modeling methodology for vacuum blocks.

The introduced maximum and minimum values of the pressures and temperatures, pressure drop, and heat exchange surface margin at the first stage depend on the characteristics of the technological equipment and the type of chemical-technological processes taking place in it. For instance, the cooling temperature of the mixture (t) depends on the initial and final temperatures of the coolant. If the temperature of the refrigerant at the condenser inlet is 28 °C, and at the outlet is 38 °C, then the temperature t cannot be lower than 39 °C [5,6]. Meanwhile, the maximum temperature (t2) is approximately 50–60 °C since at higher temperatures, the condensation of the mixture would be insufficient [10,11]. Additionally, if the rear end head types are L or M (according to standards [14]), the allowable temperature difference between the wall and the jacket cannot exceed 40 °C.

Maximum and minimum pressures are determined by the norms of the technological regime. For rectification processes, the pressure is chosen to eliminate product decomposition in cubic sections (or in the furnaces of the preheating raw materials) [5] while reducing energy input. When conducting vacuum condensation, the pressure drop should be minimal [5,6].

The proposed approach for modeling vacuum blocks is only feasible with an accurate calculation of the technological object, condensation block, and VOS. Present vacuum system design tools [15,16] lack the functionality to compute the technological components of petrochemical installations. Therefore, process simulation software is recommended for this purpose.

In process simulation software, a mathematical model of blocks is constructed, comprising interconnected modules selected from the program's database. Assigning temperatures and pressures is executed through a developed external control program. This program includes an algorithm, depicted in Figure 4, utilizing calculated heat and physical parameters of flows as input data from the process simulation software and providing the temperature within the condensation block and the pressure in the object as output data.

The program is designed to load the required flow parameters from the calculated Process Flow Diagram (PFD) of the process:

- Pressures in the technological object;
- Temperatures in the condenser;
- Volumetric flow rate of the mixture at the condenser outlet;
- Pressures in the condenser;
- Heat exchange surface margin of the condenser;
- VOS performance.

The program checks the conditions of steps 3, 6, 8, and 11. If any of the conditions in the flowchart (Figure 4) are not met, the user inputs a new temperature or pressure value into the program, which is then loaded into the corresponding modules of the process simulation software. The PFD is initiated for calculation, and the calculated values of e and Δp are loaded back into the program. The execution of steps 12 and 13 is determined by the user and depends on the required temperature and pressure ranges. The program is written in VBA using recommendations [17,18].

The recommended process simulation software for this purpose is the Unisim Design R451 software, widely employed for solving similar tasks [19–22].

4. Description of the Research Object

The research subject chosen was a vacuum block for processing phenol and acetone waste, representing a typical petrochemical installation operating under vacuum conditions. The vacuum overhead system in the distillation unit for the separation of phenol and acetone products is typically equipped with steam ejection units [23,24]. Despite their simplicity of design and ease of operation [5], steam ejection units exhibit several significant drawbacks, the most prominent of which is their low thermodynamic efficiency and consequent high energy consumption. Moreover, when the working agent condenses in the inter-stage barometric condensers of the steam ejection units, the distillate components from the pumped gas co-condense, leading to a mixture of distillate products and condensate from the working water vapor. Consequently, the condensate from vacuum overhead systems is unsuitable for reuse in power plants and must be treated as chemically contaminated water, incurring additional costs.

High-potential superheated water vapor acts as a working agent in steam ejection units, with the residual pressure reached by the pump determined by its temperature, pressure, and flow rate. It is worth noting that the parameters of the technological approach, including the operating pressure sustained in the vacuum columns, deviate considerably from those specified in the project. Therefore, the steam ejector pumps operate below their calculated modes, leading to a further decrease in their efficiency.

Replacing steam ejector vacuum pumps with an environmentally friendly hydro-ejection Vacuum Overhead System (VOS) is a crucial task to reduce energy consumption and minimize pollutant emissions.

The waste processing rectification unit in the main phenol–acetone production incorporates a hydrocarbon fraction processing unit aimed at the extraction of isopropylbenzene

(IPB) and alpha-methylstyrene (AMS). This unit consists of six distillation columns, with five of them operating under vacuum conditions.

The Complex Chemical Technological System (CCTS) comprises several key components, including vacuum distillation columns (VC), condensation units (CU), vacuum overhead systems (VOS), and transfer pipelines (TP) connecting the VC, CU, and VOS. The efficiency of the system is heavily influenced by the performance of these pipelines, which affect the vacuum pressure attained within the rectification column and thus the properties of the entire CCTS. Given the system's integrative nature, it is impossible to analyze the individual components in isolation from the conjugate elements. Modeling tools such as Unisim Design R451, designed to simulate a broad range of chemical processes and devices including rectification systems, provide an effective means of studying such complex systems.

Schematically, the plant for processing waste from the production of phenol and acetone is shown in Figure 5 [24].

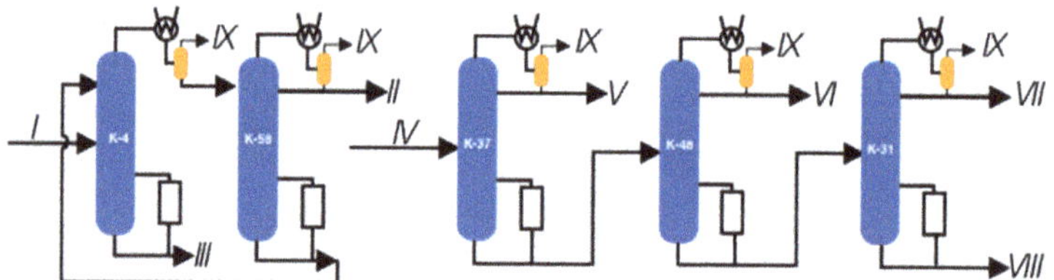

Figure 5. Vacuum columns of a plant for processing phenol and acetone production waste. I—K-4 column feed; II—phenolic fraction; III—phenolic tar; IV—supply of K-37; V—fraction of isopropylbenzene; VI, VII—fractions of alphamethyl styrene; VIII—residue, IX—flow on the VOS.

In the initial phase, an examination of the current operating conditions of the distillation columns, namely K-4, K-31, K-37, K-48, and K-58, was conducted. The primary aim of this survey was to establish the technical mode parameters of the process, which would be employed as inputs for the mathematical model of the block. Particular attention was paid to measuring the temperatures and pressures of the pumped gases along the path from the distillation column to the vacuum overhead system. These data were deemed critical for the successful completion of the survey.

Sampling points were chosen at each installation to measure pressure as close as possible to the outlet of the steam–gas mixture from the distillation column (helmet line), the first and second stages of dephlegmation, and the entrance to the suction pipe of the Steam Ejector Pump (SEP), where feasible. The measurement locations of the analyzed parameters were coordinated with the technical service department of the plant. Pressure measurements were carried out using a model vacuum meter that was verified metrologically and met the required measurement accuracy. The measured parameters were compared with the data of technological regulations and indications of stationary measuring devices in the production environment. The results of the survey are presented in the table.

Upon analyzing the results presented in Table 1, it can be inferred that the vacuum overhead system (VOS) of the K-31 and K-48 columns is operating in the overload region. This is attributed to the high condensation temperature of the condensation unit [25]. Additionally, the maximum residual pressure developed by the VOS depends on the gases that are not condensed in the condensation unit. These uncondensed vapor–gas mixtures (VGMs) are formed due to the thermal decomposition of the cubic product and the entry of "flow gasses" like atmospheric air into the system, as any evacuated system will receive an external environment to some extent through micro-densities such as welds, gaskets, pump seals, etc. Although the amount of leakage gases is relatively small, it cannot be ignored in this case, as they significantly influence the load on the evacuation unit (VOS).

Table 1. Survey results.

Column №	Pressure in the Helmet Line, mmHg.	Condensation Temperature, °C
4	14	40
58	22	35
37	65	45
48	100	55
31	Over 100	70

5. Description and Validation of Simulation Model of the Block

According to recommendations [17,26,27], for calculating VLE, one can use the Wilson model, UNIQUAC, and NRTL.

The Wilson equation is more complex than the van Laar or Margules equations and requires more processor time [17]. However, it satisfactorily represents nearly all non-ideal liquid solutions, except for electrolytes and solutions demonstrating limited miscibility. The Wilson equation provides similar results to the Margules and van Laar equations for weak non-ideal systems but consistently outperforms them for increasingly non-ideal systems.

NRTL is an extension of the Wilson equation [28]. It uses statistical mechanics and liquid cell theory to represent the liquid structure. It is recommended for highly non-ideal chemical systems and can be used for VLE and LLE applications.

UNIQUAC uses statistical mechanics and the quasi-chemical theory of Guggenheim to represent the liquid structure [29]. The equation is capable of representing LLE, VLE, and VLLE with accuracy comparable to the NRTL equation, but without the need for a non-randomness factor.

During the production of phenol and acetone using the cumene method, phenol tar is formed [30]. This complex material comprises various components, including phenol, acetophenone (AP), α, α-dimethylbenzylalcohol (DMBA), dimers of α-methylstyrene (AMSD), o, p-cumylphenols (CP), unidentified components, and a small number of salts (mainly Na_2SO_4). The exact composition of phenol tar depends on the specific phenol production technology and can vary widely. The composition of phenol tar is not determined in the considered production process, which complicates the characterization of the feed stream and PFD setup.

Since the phenol tar component is not available in the program's database, it was created based on the p-cumylphenol component.

To use the NRTL model, the parameter c_{ij} needs to be inputted, and the recommended values depend on the type of mixture. Due to the presence of the phenol tar component, it was challenging to select this value, so the decision was made to try using the UNIQUAC model. However, during the calculation of columns K-4 and K-58, convergence was not achieved. These columns were only calculated using the Wilson model, where the parameter a_{ij} had to be adjusted to match the production data in terms of the flow rate, composition, and temperature of cube K-4. The thermodynamic model used for the vapor phase is the Ideal Gas law.

The model is based on the Equations (12)–(14):

$$y_i = \gamma_i \cdot \frac{p_i^0}{p} \cdot x_i \tag{12}$$

$$ln\gamma_i = 1 - ln\left(\sum_j A_{ij} \cdot x_j\right) - \sum_{k=1}^{n} \frac{x_k \cdot A_{kj}}{\sum_{j=1}^{n} x_j \cdot A_{kj}} \tag{13}$$

$$A_{ij} = \frac{V_j}{V_i} exp\left[-\frac{a_{ij} + b_{ij}T}{RT}\right] \tag{14}$$

Once the fluid package and the component list are defined, the a_{ij} parameters used are the ones set as default by the software database, as presented in Table 2. It is to be stressed that such values have not been modified, i.e., Equations (12)–(14) are solved with the numbers in Table 2. The parameter b_{ij} was assumed to be equal to 0.

Table 2. Parameters α_{ij} of components in the Wilson equation.

	IPB	AMS	Acetophenone	Air	Phenol	Cumylphenol	DMPC	Phenol Tar
IPB	-	57.60	335.69	829.40	1324.01	869.43	1013.17	18.02
AMS	−51.14	-	273.39	1008.22	761.73	599.15	749.16	101.65
Acetophenone	112.84	60.59	-	-	3175.79	−4.76	120.53	-
Air	1445.89	2145.26	-	-	-	-	-	-
Phenol	−27.95	25.48	−1335.68	-	-	−929.53	−776.69	-
Cumylphenol	76.48	68.27	121.00	3172.93	1499.71	646.19	814.84	270.26
DMPC	−2.67	88.21	211.48	-	−388.44	−250.98	-	-
Phenol tar	61.17	101.47	-	-	-	-	-	-

Further, calculation models of vacuum columns were synthesized in the Unisim Design R451 software package, in which the data of the technological survey were set as specifications. The calculation schemas of the block are shown in Figures 6 and 7.

In Figures 6 and 7, the designations are identical to Figure 5, except for stream X, which simulates the intake of atmospheric air into the evacuated objects.

The results of comparing the calculated data with the results of the industrial survey are presented in the tables below.

The notation of streams in Tables 3 and 4 is similar to Figure 5.

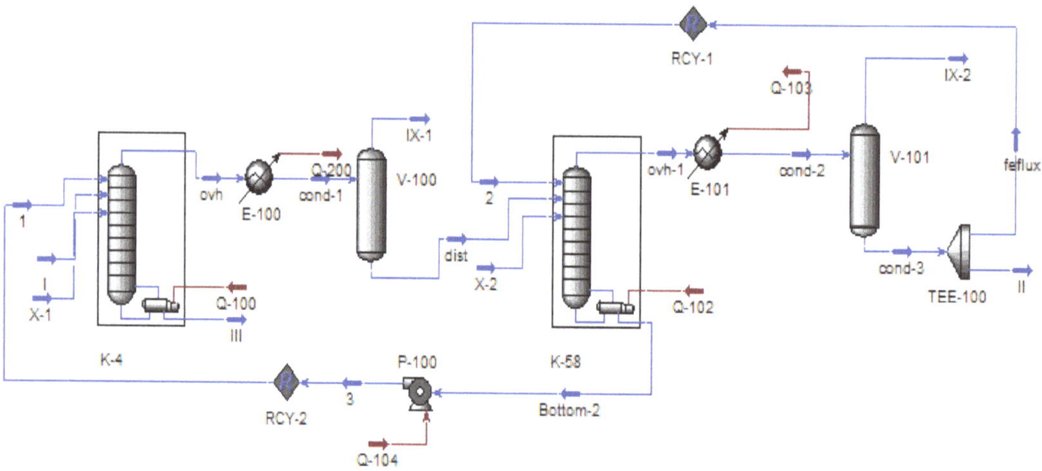

Figure 6. Process flow diagram of columns K-4 and K-58. I—K-4 column feed; II—phenolic fraction; III—phenolic tar; IX—flow on the VOS; X—air leak rate.

For a better assessment of Tables 3 and 4, calculated overall and component balances based on the Process Flow Diagrams of Figures 6 and 7 are presented in the Appendix A.

In Tables 3 and 4, the values of temperatures and flow rates calculated by the models are in good agreement with the production data. The contents of the main (target) components in the flows also match the technological regulations (deviation is not more than 10%). The content of auxiliary components (non-target) may deviate up to 100% but their total content in the flows does not exceed five (mass%), which does not significantly affect the accuracy of the calculation. Based on Tables 3 and 4, it can be concluded that the calculation results for the streams denoted with an asterisk (*) are in close agreement with the corresponding data obtained from the survey (streams without an asterisk). This observation indicates that the mathematical model developed for the process is sufficiently accurate and reflects the actual technological process being studied.

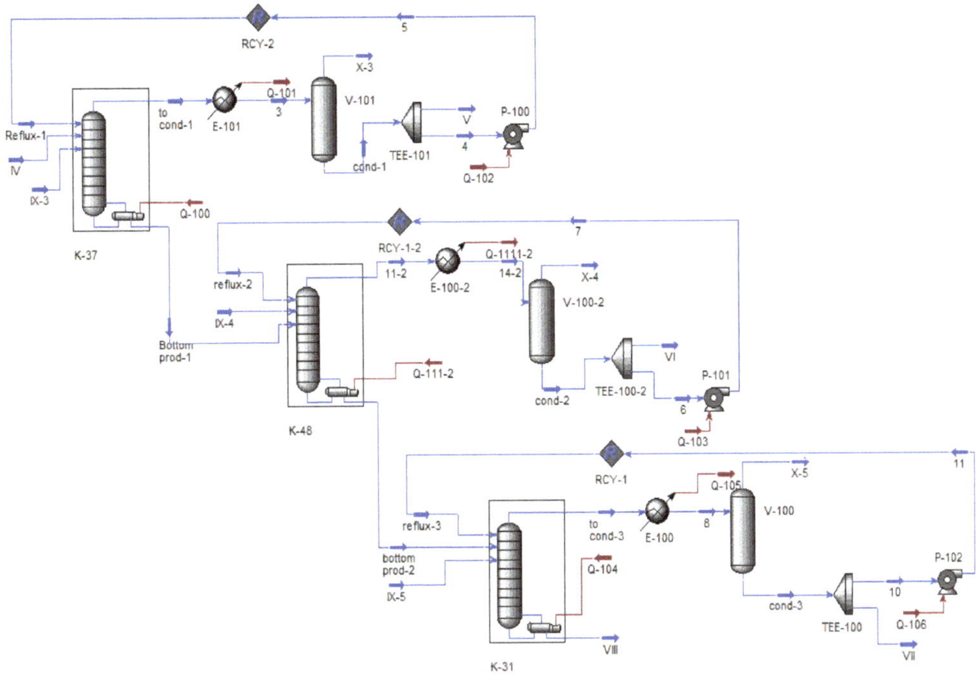

Figure 7. Process flow diagram of columns K-37, K-48, and K-58. IV—supply of K-37; V—fraction of isopropylbenzene; VI, VII—fractions of alphamethyl styrene; VIII—residue; IX—air leak rate; X—flow on the VOS.

Table 3. Technological parameters of the column and calculation data for columns K-3 and K-58.

Stream Name	Composition, Mass Fraction.					Temperature, °C	Flow Rate, kg/hr
	Phenol	Acetophenone	DMPC	Cumylphenol	Phenol Tar		
I	0.48	0.08	0.012	0.2	0.22	<150	2294
I *	0.48	0.08	0.012	0.2	0.22	140	2294
Δ, %	0	0	0	0	0	6	0
II	0.97	0.02	0.01	-	-	30–40	1362
II *	0.84	0.14	0.02	0.01	-	35	1354
Δ, %	14%	86%	47%	100%	-	0%	1%
III	0.08	0.13	0.03	0.36	0.4	122	932
III *	0.01	0.05	-	0.4	0.54	180	940
Δ, %	88%	62%	-	11%	26%	48%	1%

Stream name with (*) represents the calculation data.

The temperature of stream I set in the program (140 °C) differs from the industrial one because the conditions are maintained at the plant to ensure that the inlet stream I does not overheat above 150 °C. Typically, the operating personnel maintain the temperature at around 140 °C; therefore, this value was chosen as the input parameter.

Table 4. Technological parameters of the column and calculation data for columns K-37, 48, and 31.

Stream Name	Composition, Mass Fraction.			Temperature, °C	Flow Rate, kg/hr
	IPB	AMS	Acetophenone		
IV	0.22	0.76	0.02	<170	1305.3
IV *	0.221	0.76	0.019	170	1305
Δ, %	0.5%	0.0%	5.3%	0.0%	0.0%
V	0.98	0.015	0.05	30–40	242.5
V *	0.98	0.02	-	40	239.5
Δ, %	0%	25%	-	0%	1%
VI	0.081	0.92	0.001	30–40	350
VI *	0.14	0.85	0.01	40	349
Δ, %	42.1%	8.2%	90.0%	0.0%	0.3%
VII	0.02	0.96	0.02	30–40	568
VII *	0.01	0.98	0.01	40	563
Δ, %	50.0%	2.1%	50.0%	0.0%	0.9%
VIII	-	0.84	0.16	122	143.5
VIII *	-	0.83	0.17	127	143.5
Δ, %	-	1.2%	5.9%	3.9%	0.0%

Stream name with (*) represents the calculation data.

6. Selection of a Vacuum Overhead System

The circumstances outlined led to the proposal to revamp the vacuum overhead system (VOS) across all the distillation columns within the unit. The aim was to replace the existing steam ejector pumps (SEP) with a new generation of energy-efficient and environmentally friendly VOS. This transition seeks to reduce operational expenses related to establishing and maintaining vacuum levels while also minimizing the production of chemically contaminated effluents.

Innovative hydro-circulatory VOS designs have been developed and implemented in the industry, utilizing either single-stage liquid ejectors (LE) or liquid-ring vacuum pumps (LRVP) as operational mediums, with distillates from the distillation columns being used in both cases. Previous analyses demonstrated that a VOS relying on LRVP demonstrates better cost-effectiveness in terms of both SEP and VOS operational expenses within the vacuum range of 50 mm Hg and higher compared to a VOS relying on LE. This observation also extends to capital investments, which is crucial given the need to ensure VOS redundancy for the secure operation of these facilities, considering the intricacies of vacuum rectification technology.

A conceptual proposal for the reconstruction involves the implementation of a hydro-circulatory single vacuum-generating station employing a liquid-ring vacuum pump (LRVP). Figure 8 illustrates the block diagram depicting the envisioned reconstruction plan for the vacuum block.

Figure 8 depicts the block diagram of the proposed reconstruction of the vacuum block, where the stream designations are similar to those in Figure 5, except for stream XI, which represents the condensate stream formed in a condenser installed at the outlet of the uncondensed gases from columns K-37, 48, and 31. As recommended in [31–35], the vacuum-generating system must be selected based on the required load. A LRVP with a capacity of 610 m^3/hour was chosen as the VOS. The parameters and composition of the pumped mixture were calculated according to the models of Figures 2 and 3 and are presented in Table 5.

The pairing of the characteristics of a group of coupled columns and the standard characteristics of the LRVP are presented in Figure 9.

It was observed that the characteristics of these elements intersect at point A, which corresponds to a pressure of 33 mm Hg (43.89 mbar). Due to the limited accuracy of modeling the hydraulic resistance of the gas path and condensing units, the system pressure was adjusted to 40 mm Hg (53.2 mbar) using anti-surge protection of the pump at point B.

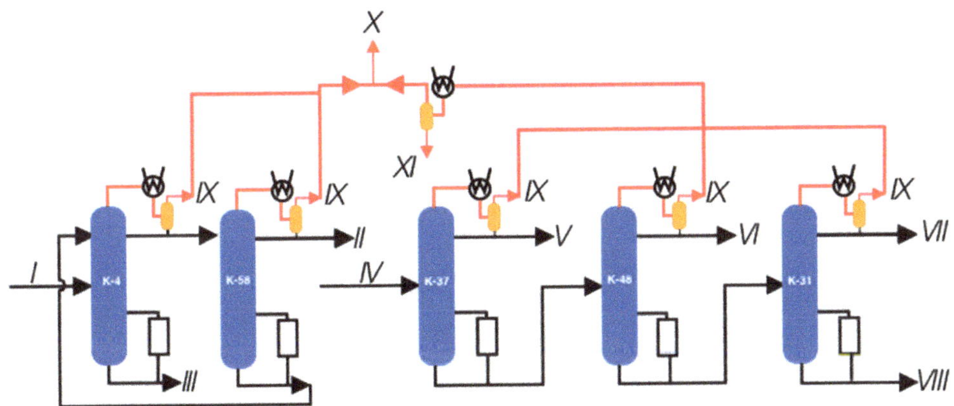

Figure 8. Block diagram of the proposed reconstruction. I—K-4 column feed; II—phenolic fraction; III—phenolic tar; IV—supply of K-37; V—fraction of isopropylbenzene; VI, VII—fractions of alphamethyl styrene; VIII—residue, IX—flow on the VOS; X—summary flow to VOS; XI—condensate.

Table 5. Parameters of the pumped mixture.

Parameter	Value
Pressure, mm Hg	40
Temperature, °C	35
Consumption, kg/hour	25
Composition, (mass fraction)	
IPB	0.16
AMS	0.04
Acetophenone	0.006
Air	0.794
Phenol	-
Cumylphenol	-
DMPC	-

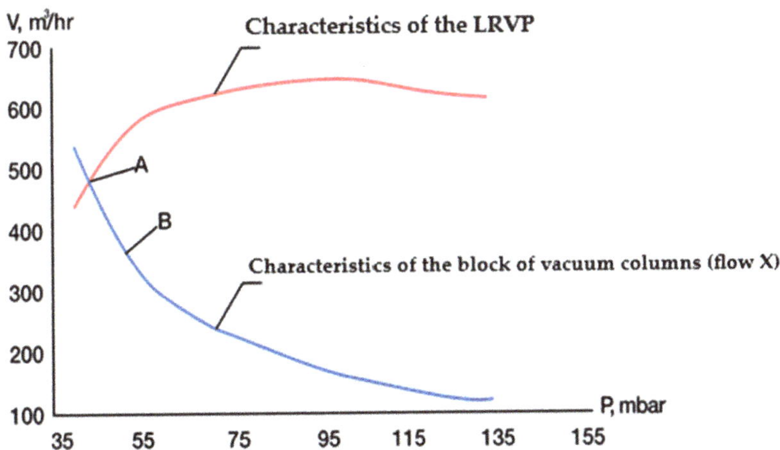

Figure 9. Coupling of the characteristics of the pipeline path of the evacuated object and the housing pump.

In light of the results obtained, it can be concluded that the LRVP utilized in this CCTS operates with a 40% productivity margin at the working point. This margin serves to mitigate the potential impact of temperature increases in the condensation units that may occur during the summer operating period of the installation.

7. Connection of the VOS to the Reconstructed Unit

Upon completion of the reconstruction, the start-up of the vacuum system was conducted in a specific sequence:

- Initially, all column sections were transitioned to operate at the recommended mode with a pressure of 40 mm Hg, using regular SEPs to establish the vacuum;
- Subsequently, the LRVP P2L 65327 Y 4B was launched via a separately mounted communication in parallel with the existing SEP. As the anti-surge system automatically activated, the vacuum pressure in the system remained largely unaffected;
- Following this, at intervals of 10–15 min, the SEPs were turned off for all columns, except for the K-31 column, while maintaining the pressure at the prescribed level. The entire system restart process did not exceed 1 h.

The K-31 column experienced a pressure of 55 mm Hg, surpassing the design value of 40 mm Hg. This deviation likely stemmed from the inadequate performance of the column's condensation unit, resulting in a condensation temperature exceeding 60 °C, along with significant hydraulic resistance in the condenser. As a result, the column failed to achieve the required separation efficiency. To rectify this issue, a standard steam ejector pump with a barometric condenser was installed downstream of K-48, effectively reducing the column's pressure to the target value of 40 mm Hg. It is important to note that phenolic water with a phenol concentration of up to 4% by weight was utilized as the working fluid, necessitating recalibration of the pump's characteristics based on the methodology outlined in [36,37], or the model presented in [38]. To mitigate potential calculation inaccuracies, a 40% performance margin was incorporated to enhance the modeling of the system's bottlenecks. This approach proved effective overall, except for the K-31 column, where the issue of high condensation temperature in the condensation unit persisted.

8. Coupling of the Characteristics of the VOS and the Vacuum Unit, Taking into Account the Operating Conditions

The utilization of the methodology and thinned models of an LRVP [38] can provide a means to manage the selected stock coefficients and enhance the effectiveness of the proposed design solutions. Therefore, recalculating the block of distillation columns to suit new production conditions and coupling their characteristics with the recalculated characteristics of the selected LRVP would be of significant scientific and practical interest.

Subsequently, a technological survey was conducted to gather temperature data at the top and bottom of vacuum columns, as well as reflux rate costs, after the completion of the reconstruction project of the vacuum-generating system for the separation of the rectification of phenol and acetone production waste. These data were crucial for establishing a mathematical model and verifying its adequacy.

It should be noted that during the reconstruction and replacement of the vacuum overhead system for improved vacuum creation, various modifications were implemented, including the installation of an additional condenser before the LRVP, the use of salted brine (a mixture of water with 5% mass phenol) instead of water, and changes in the pipelines connecting the column condensers with the vacuum overhead system. As a result, not only did the operating parameters of the system change, but the technological topology was also altered.

Consequently, the previously developed models, as presented in Figures 6 and 7, became unsuitable and required additional modifications. To this end, the models in Figures 6 and 7 were updated with the addition of blocks for mixing the streams of uncondensed gases that were suctioned at the VOS, and a block for cooling and condensing the streams of gases leaving the K-37, 48, and K-31 columns was also included.

The modification involved adding a model for the first stage SEP to column K-48 (Figure 10) following the recommendations [39]. The parameters for the stage were adopted based on the data from [40]. As a result of these modifications, the models of the columns and the condensation unit had to be updated, which can be observed in the modified versions presented in Figures 10–12.

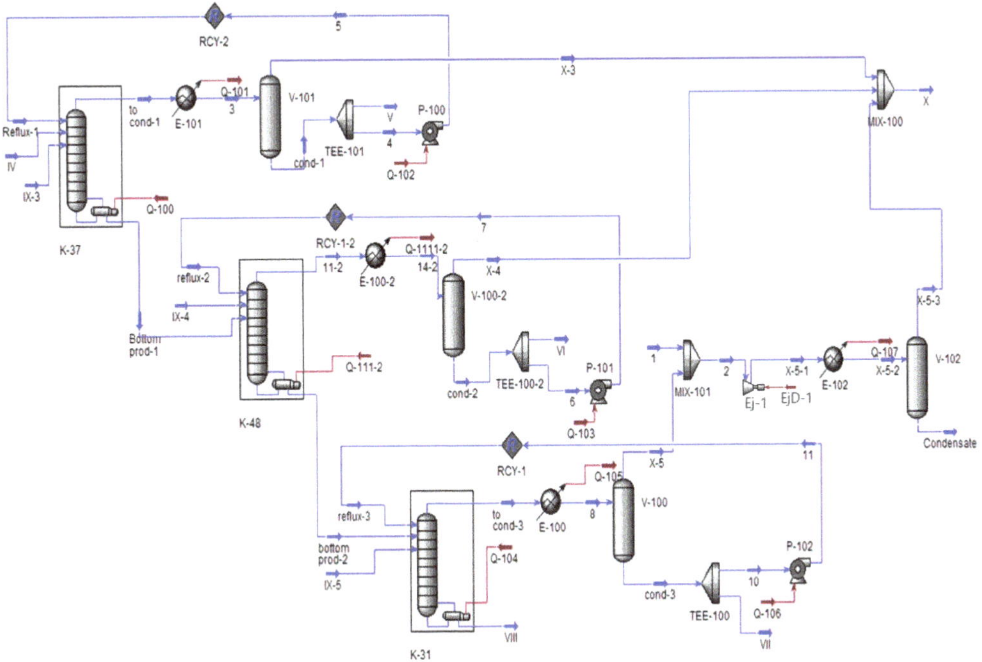

Figure 10. Modified process flow diagram of columns K-37, 48, and 31 in Unisim Design R45. IV—supply of K-37; V—fraction of isopropylbenzene; VI, VII—fractions of alphamethyl styrene; VIII—residue; IX—air leak rate; X—flow on the VOS.

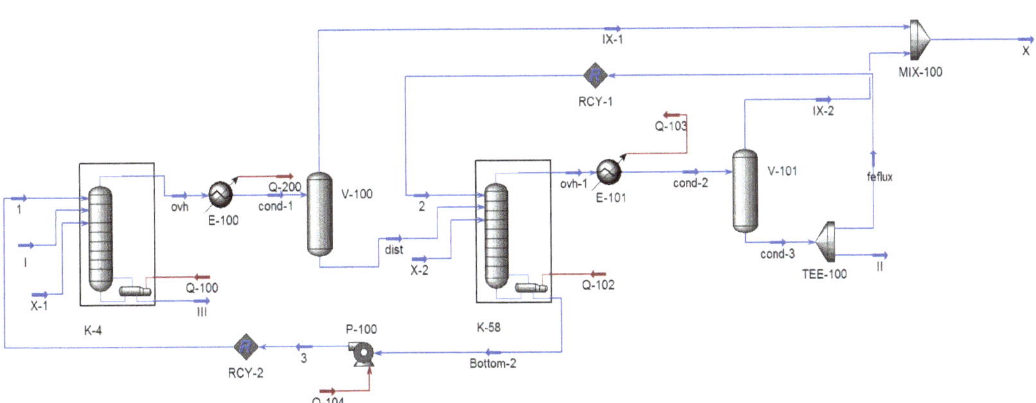

Figure 11. Modified process flow diagram of columns K-4 and K-58 in Unisim Design R45. I—K-4 column feed; II—phenolic fraction; III—phenolic tar; IX—air leak rate; X—flow on the VOS.

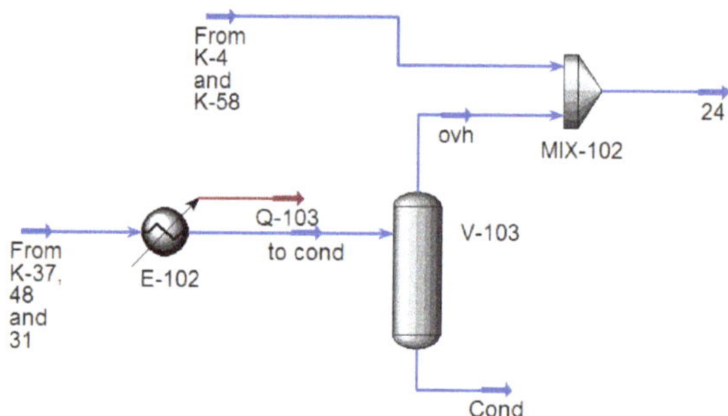

Figure 12. Process flow diagram for calculating the load on the VOS in Unisim Design R45.

The temperature in the inter-tube space has a significant impact on the point of conjugation between the characteristics of the vacuum column and the vacuum overhead system. To account for this, temperature data were incorporated into the specifications of the cooler modules in Unisim Design R45. The temperature values were determined through a technological survey and used to obtain accurate results.

As noted previously, a characteristic feature of volume flow is its dependence on pressure. The characteristics of the vacuum distillation column unit are defined by the output of the uncondensed vapor phase in the condensation units, whereas for the vacuum overhead system, the corresponding characteristic is the input stream at the inlet of the vacuum pump. The points where these characteristics intersect can be referred to as conjugate points.

Simulation flowchart VO and VOS are shown in Figure 13. The conjugate points of the block of rectification columns and the VOS were determined using the following method:

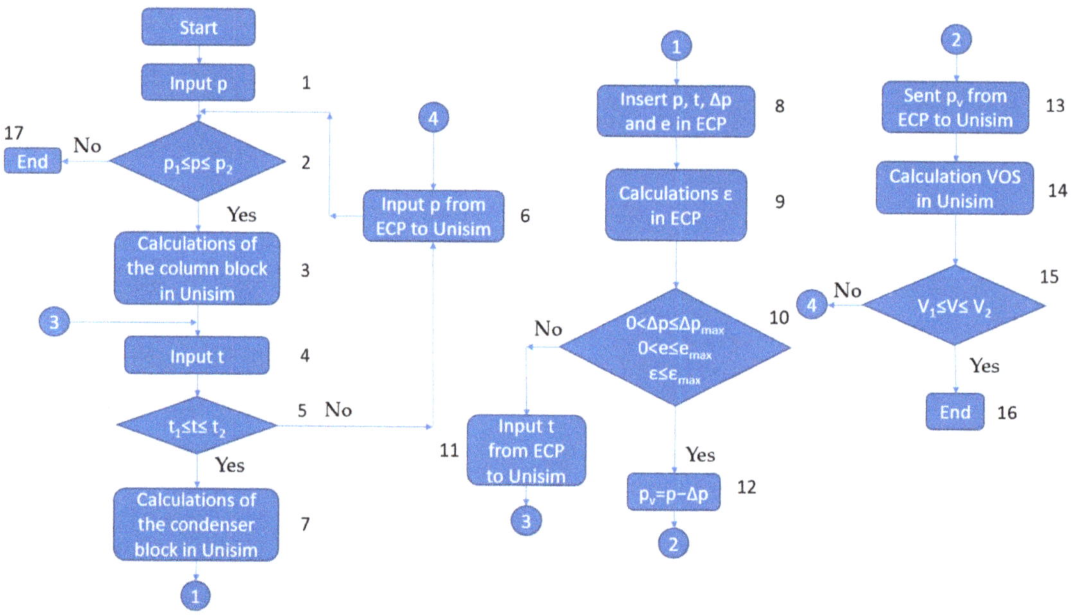

Figure 13. Simulation flowchart. VO and VOS.

1. At the initial stage, an initial pressure approximation is introduced into the columns;

2. If the pressure falls within the specified range, the calculation proceeds. If not, the calculation is terminated and it is considered that the VOS is not operational in this production;
3. The columns are calculated in the Unisim software package;
4. The initial temperature value in the condensers is introduced;
5. The condition that the temperature value must lie within the range specified in Table 6 is checked. If the condition is met, the calculation continues (step 7);

Table 6. Ranges of values.

Parameters	Value	Unit of Measurement
p_1	15	mbar
p_2	60	mbar
t_1	20	°C
t_2	40	°C
e	5	%
V_2	$1.1 \cdot V_1$	-
Δp	$0.4 \cdot p$	-

6. If the condition is not met, a new pressure value is introduced via the specially developed ECP, and the calculation starts again from step 1;
7. The condensation block is calculated in the Unisim software package;
8. The values of p, t, Δp, and e are exported from the Unisim software package to the ECP;
9. The heat exchange surface reserve ε is calculated in the EPC;
10. The condition that p, t, Δp, and ε must lie within the range specified in Table 6 is checked;
11. If the values of p, t, Δp, and ε are outside the range specified in Table 6, a new temperature value is introduced and the calculation starts again from step 4;
12. If the condition is met, the pressure p is reduced by the amount of Δp, which corresponds to the pressure at the VOS inlet;
13. The pressure is imported into Unisim;
14. The performance of the VOS is calculated;
15. The condition that the V performance must be within the limits specified in Table 6 is checked. If the condition is not met, a new pressure value is introduced and the calculation starts again from step 6;
16. If the condition is met, the calculation is completed and the pressure is fixed in the block;
17. If it is not possible to adjust the pressure, the calculation is terminated and a conclusion is made that the VOS cannot meet the specified values of the technological process.

Following the proposed approach, a numerical analysis was conducted to determine the conjugate point between the vacuum overhead system (VOS) and the block of vacuum columns.

The analysis was performed using the following method: pressure values were assigned to key nodes of the system, including the columns and the condensers, and cooling temperatures were introduced. Subsequently, the columns and condensers were calculated and if the surface heat exchange margin was within the range of 0–5%, the calculation was considered complete; otherwise, new temperature values were entered into the model and the calculation was repeated. In accordance with the load on the VOS, the performance of the Liquid-Ring Vacuum Pump (LRVP) was computed.

If the calculated LRVP efficiency was higher than the load by more than 10%, the pressure in the column was decreased, while if the load on the VOS was greater than the LRVP efficiency, then the pressure in the column was increased.

The results of the analysis are presented in Figure 14.

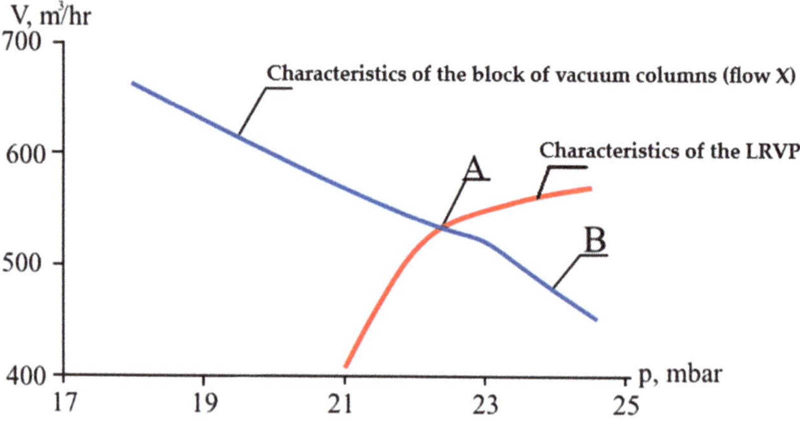

Figure 14. Coupling of the characteristics of the vacuum column block and the VOS.

The intersection of the characteristics of the VOS and the block of vacuum columns was observed at point A, corresponding to a pressure of 22.5 mm Hg, while different values of top pressure were established in the columns. Point B in Figure 14 corresponded to the pressure at the outlet of the load calculation unit for the VOS, i.e., the pressure at the inlet to the pipeline connecting the columns and the LRVP, which was found to be 24.4 mm Hg.

A comparison of the remaining technological parameters of the columns with the calculation results is given in Table 7.

Table 7. Comparison of technological parameters of columns with calculation results.

Parameter	Survey Data	Calculation Data	Unit of Measurement	Δ, %
K-4				
Top pressure	23	26	mm Hg	13%
Top temperature K-4	93	103	°C	11%
Bottom temperature	122	180	°C	48%
K-58				
Top pressure	31.5	32	mm Hg	2%
Top temperature	106.8	101.2	°C	5%
Bottom temperature	123.9	130	°C	5%
Reflux rate	1203	1200	kg/hr	0%
K-37				
Top pressure	45.75	45	mm Hg	2%
Top temperature	63.3	69.26	°C	9%
Bottom temperature	111	124	°C	12%
Reflux rate	3400	3400	kg/hr	0%
K-48				
Top pressure	58.5	53.25	mm Hg	9%
Top temperature	78.7	87.1	°C	11%
Bottom temperature	111	124	°C	12%
Reflux rate	4900	5100	kg/hr	4%
K-31				
Top pressure	30.25	33	mm Hg	9%
Top temperature	78.5	74.2	°C	5%
Bottom temperature	131.5	127.1	°C	3%
Reflux rate	4000	3900	kg/hr	3%
VOS				
Suction pressure	-	22.5	mm Hg	-
Temperature of the service liquid at the inlet	5	5.5	°C	10%
The temperature of the service fluid at the outlet	11.8	11.6	°C	2%

Table 7 demonstrates that the calculated data align with the results obtained during the technological inspection of the installation, with deviations within 15%, except for the bottom temperature of column K-4. This column is used for distilling phenol tar from the phenol stream and is linked with K-57 through a recycling stream. In the literature [41–49], there are no data provided on the composition of the substance phenol tar, but a separate component is included in the material balance according to production data. Section 5 describes the method of introducing phenol tar into the scheme. As a result of the calculation, the deviations of the calculated results from the production data on composition (Table 3) are as follows: phenol tar—26% and Cumylphenol—11%. Deviations in the content of phenol and acetophenone were 88% and 62%, respectively; however, the final calculated and production values were of the same order of magnitude. There was also a significant deviation in the temperature of column K-4—48%. However, the aim of this article is to determine the load on VOS, and the temperature of the column and the content of distillate components (phenol and acetophenone) have little influence on the calculation of the final load on VOS. Therefore, it can be concluded that the calculation data are in good agreement with the industrial data, indicating the adequacy of the model of the investigated setup.

Figure 15 illustrates the pressure distribution of the top of the vacuum columns and the vacuum overhead system (VOS). The results indicate that even though the pressure at the inlet to the liquid-ring vacuum pump (LRVP) is nearly half of the required pressure (40 mm Hg), the pressure at the top of the columns remains close to the design parameter, suggesting that the pipelines exhibit high hydraulic resistances.

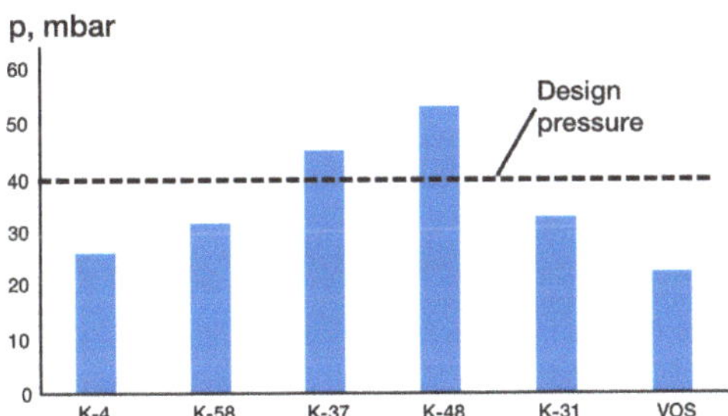

Figure 15. Pressure distribution across columns and VOS.

The adoption of an increased performance margin has resulted in the successful attainment of the objectives of the reconstruction project, leading to significant cost savings. Notably, had a lower VOS efficiency been chosen, the reconstruction tasks would have been incomplete. Therefore, the decision to raise the performance margin was critical in ensuring the project's success.

9. Evaluation of the Effectiveness of the Vacuum Overhead System

An efficiency assessment can be made by determining the operating costs for the functioning of the VOS. To do this, according to Formula (15), the costs of the consumed resources are determined:

$$Cost_R = Pr_R \cdot TC_R \cdot h \tag{15}$$

Further, all costs are summed up and the total operating costs are determined:

$$Cost_{TR} = \sum_R Cost_R \tag{16}$$

Further, according to Formula (17), the economic effect of the reconstruction of the existing VSS is determined.

$$EE = Cost_{TR1} - Cost_{TR2} \tag{17}$$

After that, with the accepted capital expenditures, the payback period of the project can be determined by:

$$PP = \frac{CE}{EE} \tag{18}$$

The key outcome of the calculation based on the above-described methodology is the payback period of the project, which serves as a determinant for the viability of a particular VOS layout option. Moreover, this approach is considerably sensitive to the prevailing energy resource prices, which are shaped by the production specifics and location. The studies in [36,37] proposed prices for fundamental energy resources that were employed to estimate the economic impact. Prices for energy resources are provided in Table 8.

Table 8. Energy prices.

Resource Name	Value	Unit of Measurement
Water vapor	15.28	USD/Gcal
Recycled water	0.021	USD/m^3
Electricity	0.04	USD/kW
Cleaning the Chemical Contaminated Condensate	0.16	USD/m^3

The above prices were converted using RUB/USD exchange rate of 28 August 2023.

It is worth noting that the production of electricity and water vapor is typically achieved through the combustion of various types of fuels. Therefore, an alternative method for assessing energy efficiency in the operation of a vacuum overhead system is to convert the energy consumed into units of conventional fuel, which enables the estimation of pollutant emissions (e.g., CO_2). Conventional fuel is a standard unit for accounting for organic fuel in calculations, used to determine the useful effects of various types of fuels in their overall accounting. It is assumed that the combustion of 1 kg of solid (liquid) or 1 cubic meter of gas releases 29,300 kJ (7000 kcal) of energy. Thus, to calculate the amount of conventional fuel consumed in producing G kilograms per hour of water vapor, the following formula can be used:

$$G_{fe} = \frac{G \cdot \Delta H}{29300 \cdot \eta_b} \tag{19}$$

If the task is to determine how much conventional fuel needs to be spent to produce N kW of electricity, then you can use the following formula:

$$G_{fe} = \frac{N \cdot 3600}{\eta_t}, \tag{20}$$

In order to estimate the cost of conventional fuel for recycled water based on flow rate, temperature, and pressure, Formula (20) can be utilized to calculate the necessary drive power of feed pumps. Then, Formula (19) can be applied to determine the cost of conventional fuel for producing electricity.

$$N = L \cdot \rho \cdot g \cdot P_N / (1000 \cdot 3600 \cdot \eta_p) \tag{21}$$

With a known consumption of conventional fuel, it is possible to determine the CO_2 output:

$$G_{CO_2} = C_1 \cdot K \cdot G_{fe} \tag{22}$$

In order to evaluate the environmental impact of a vacuum overhead system (VOS), it is important to calculate the emissions of carbon dioxide (CO_2) that occur during its operation. The coefficient C_1, which corresponds to the formation of CO_2 per unit of conventional fuel consumption, can be obtained from reference data. It should be noted that the carbon oxidation coefficient is assumed to be 1.

The calculation of CO_2 emissions is a crucial factor in determining the environmental sustainability of a particular type of VOS. In addition to CO_2 emissions, the operating costs and conventional fuel costs must also be considered. These values are presented in the Table 9.

Table 9. Comparison of the current and reconstructed VOS.

№	Parameter	Unit	Type of VOS	
			Current	LRVP
1	Steam consumption	kg/hr	300	0
2	Steam consumption	Gcal/hr	0.192	0
3	Recycled water	m^3/hr	10	2
4	consumption Electricity consumption	kW	1.09	18
5	Conditional fuel consumption	t/hr	33.196	6.914095
6	CO_2 emissions	kg/hr	0.033	0.008988
7	Resource prices			
8	Water vapor	USD/Gcal	14.079	
9	Recycled water	USD/m^3	0.020	
	Cleaning of Chemical Contaminated Condensate	USD/m^4	0.153	
10	Electricity	USD/kW	0.039	
11	Costs (8000 h per year)			
12	Water vapor	USD/fe	21,682.71	0
13	Recycled water	USD/fe	1578.94	315.79
14	Electricity	USD/fe	344.21	5684.21
	Cleaning of Chemical Contaminated Condensate	USD/fe	366.31	0
15	Total	USD/fe	23,972.18	6000
16	% of the baseline	%	100%	25%

The implementation of a hydro-circulatory vacuum overhead system (VOS) presents an opportunity to eliminate the need for costly water usage. Furthermore, based on an annual operating time of 8000 h for the system, the associated costs of the proposed VOS are only a quarter of those for the existing vapor steam stripping (VOS) system. In terms of conventional fuel efficiency, the hydro-circulatory VOS would result in a reduction in the costs of conventional fuel and CO_2 emissions by more than threefold. In addition, the hydro-circulatory VOS would eliminate the formation of Chemical Contaminated Condensate (CCC), which would positively impact the environmental conditions at the facility and alleviate the burden on the environment.

10. Conclusions

The problems of designing technological systems for creating a vacuum are relevant and are not widely discussed in scientific literature. This is because, in general, researchers solve local problems by designing a specific process at a specific plant. Furthermore, no methodology exists for designing and calculating technological systems for creating a vacuum, depending on the type of process and equipment used.

Due to the uncertainty in assigning technological parameters for the Vacuum Overhead System (VOS) of an industrial chemical-technological process and the difficulty in determining the load, designers tend to overestimate the size of the selected vacuum pump. This leads to increased consumption of valuable energy resources and the generation of chemically contaminated wastewater. The methodology proposed in the article allows for the utilization of process simulation software capabilities. In this approach, the pressure in the system is not pre-determined but selected to harmonize the characteristics of the main elements of the vacuum block. This eliminates the need for introducing large safety factors since the system is calculated as a whole, considering the mutual influence of its components.

The capabilities of process simulation software allow for the rapid and highly accurate multifactor optimization of vacuum systems, enabling the identification of system bottlenecks. However, the proposed design methodology has significant drawbacks, including:

1. The initial models of vacuum blocks and the Vacuum Overhead System (VOS) must be accurate and qualitatively describe the behavior of the technological object over a wide range of input data;
2. The primary elements of the model must be tuned to ensure the convergence of calculations across a broad range of input parameters;
3. Specialized software is required for managing the process simulation software since it involves comparing different data and assigning initial values based on this comparison;
4. Challenges in automating the input of pressure (p) and temperature (t) values.

While these drawbacks are significant, they are not critical, and the proposed methodology can still be applied in the design of vacuum blocks for chemical-technological processes.

The process of selecting a vacuum overhead system for the cumulus method of processing phenol–acetone production waste was described in this paper; mathematical models of the process itself (distillation column), condensation units, and VOS were created for this purpose. The dependencies (volume flow and pressure) were utilized to connect these various elements. During the initial design, a vacuum pump (LRVP) with a performance margin of 40% was chosen. Following the reconstruction, it was discovered that the pressure at the inlet to the vacuum-overhead system was nearly twice as low as in the design.

To determine the reasons for this, modified models of the process and the vacuum overhead system were used based on the results of the post-modernization inspection of the installation, and numerical experiments were performed on the coupling of the characteristics of the block and the vacuum-creating system using the provisions in [36–38]. It was discovered that using cooled brine in the condensation unit and phenolic solution as the working fluid of the vacuum overhead system allowed for a significant reduction in the load on the vacuum pump while increasing its productivity. At the same time, the pressure in the individual columns is nearly twice that of the pressure at the system's inlet.

The example under consideration demonstrated that assigning technological parameters in key nodes of the system a priori can result in both a significant decrease in pressure in the block and failure to achieve the required vacuum value, which necessitates additional research.

Author Contributions: Conceptualization E.V.O. and S.I.P.; methodology, E.V.O. and A.S.P.; software, D.B.; validation E.V.O., S.I.P. and D.B.; formal analysis, E.V.O.; investigation, E.V.O. and A.S.P.; resources, S.I.P.; data curation, A.S.P.; writing—original draft preparation, E.V.O.; writing—review and editing, D.B.; visualization, D.B.; supervision, S.I.P. and E.V.O.; project administration, S.I.P.; funding acquisition, S.I.P. All authors have read and agreed to the published version of the manuscript.

Funding: This research was funded by the Ministry of Science and Higher Education of the Russian Federation grant number 075-01261-22-00 «Energy saving processes of liquid mixtures separation for the recovery of industrial solvents» and the APC was funded by the authors.

Data Availability Statement: The data that supports the findings of this study are available within the article.

Abbreviations

AMS	alpha-myethyl styrene;
DMFC	Dimethylphenylcarbinol
ACP	alpha-myethyl styrene
C	CO_2—formation coefficient;
SEP	steam ejector vacuum pump
IPB	isopropylbenzene
G	mass flow, kg/hr
CCTS	complex chemical and technological systems
DC	distillation columns
CD	contact devices

VO	object under vacuum;
VC	vacuum condenser;
VOS	vacuum overhead systems
L	liquid volume flow, m^3/hr;
LLE	liquid–liquid equilibrium
LE	liquid ejectors
LRVP	liquid-ring vacuum pumps
TP	communication pipelines
K-4, K-31, K-37	distillation columns
K	carbon oxidation coefficient;
SGM	steam—gas mixture
ECP	external control program;
VLE	vapor–liquid equilibrium;
VLLE	vapor–liquid–liquid equilibrium;
A	coefficients of equations describing physicochemical processes;
a_{ij}	non-temperature dependent energy parameter between components i and j (cal/gmol);
b_{ij}	temperature dependent energy parameter between components i and j (cal/gmol-K);
e	surface heat exchange margin, %;
$\bar{y}_c$	output variables vector;
y	molar fraction of component i in vapor phase;
x	molar fraction of component i in liquid phase;
p_o	pressure of inflated vapors, mbar;
p	pressure, mbar
p_i^0	vapor pressure of component i, mbar;
P_N	pump power, kW;
$\Delta P \cdot \Delta p$	pressure drop, mbar
$\bar{F}$	Input variables vector;
G	mass flow, kg/hr
g	acceleration due to gravity, m^2/s;
T	temperature, K;
t	temperature, °C
V_i	molar volume of pure liquid component i in $m^3/kgmol$ (litres/gmol)
V	volume flow rate, $m^3/hour$
V_1	volume flow rate at the outlet, $m^3/hour$
V_2	volume flow rate at the inlet, $m^3/hour$
Q	leak rate, $m^3 \cdot Pa/s$;
Greek	
Δ	error, %
γ	activity coefficient;
φ	vector function;
Subscriptions	
$diff$	diffusion;
pr	permeability;
Min	minimum values;
max	maximum value;
in	air leak
fe	fuel equivalent;
t	turbine;
b	boiler;
p	pump;
proc	process;
$surf$	surface;
i, j and k	component number
1	parameters related to the output;
2	parameters related to the input.

Appendix A

Table A1. Calculated values for the flows indicated in Figure 6.

Name	I	1	III	ovh	cond-1
Vapor fraction	0.00	0.00	0.00	1.00	0.00
Temperature (°C)	140.00 **	130.52 **	180.62	103.34	40.00 **
Pressure (mm Hg)	760.04 *	1500.12 *	116.56	23.00	22.62
Molar flow (kgmole/h)	18.16	36.27	4.54	50.00	50.00
Mass flow (kg/h)	2294.00 **	4000.05 **	949.92	5333.53	5333.53
Liquid Volume flow (m³/h)	2.22	3.89	0.96	5.16	5.16
Heat flow (KJ/h)	−2,770,225	−4,952,186.71	−803,012.11	−4,554,105.46	−7,663,623.96
Name	IX-1	dist	2	ovh-1	Bottom-2
Vapor fraction	1.00	0.00	0.00	0.99	0.00
Temperature (°C)	40.00	40.00	35.00 *	101.19	130.39
Pressure (mm Hg)	22.62	22.62	27.75 **	31.50	83.26
Molar flow (kgmole/h)	0.12	49.89	12.28	26.05	36.28
Mass flow (kg/h)	3.68	5329.85	1200.00 **	2534.40	4000.05
Liquid Volume flow (m³/h)	0.00	5.16	1.14	2.41	3.89
Heat flow (KJ/h)	−389.06	−7,663,234.90	−1,842,412.81	−2,260,965.91	−4,952,067.72
Name	cond-2	IX-2	cond-3	II	reflux
Vapor fraction	0.0	1.00	0.00	0.00	0.00
Temperature (°C)	35.00 **	35.00	35.00	35.00	35.00
Pressure (mm Hg)	27.75	27.75	27.75	27.75	27.75
Molar flow (kgmole/h)	26.05	0.16	25.89	13.60	12.28
Mass flow (kg/h)	2534.40	5.13	2529.27	1329.27	1200.00 *
Liquid Volume flow (m³/h)	2.41	0.01	2.41	1.27	1.14
Heat flow (KJ/h)	−3,883,588	−486.80	−3,883,101.49	−2,040,783.96	−1,842,317.52
Name	3	X-1	X-2	-	-
Vapor fraction	0.00	1.00	1.00	-	-
Temperature (°C)	130.52	20.00 *	20.00 *	-	-
Pressure (mm Hg)	1500.12 **	760.04 *	760.04 **	-	-
Molar flow (kgmole/h)	36.28	0.11	0.16	-	-
Mass flow (kg/h)	4000.05	3.21 **	4.60 **	-	-
Liquid Volume flow (m³/h)	3.89	0.00	0.01	-	-
Heat flow (KJ/h)	−4,950,997	−15.89	−22.77	-	-

Stream name with (*) represents the calculation data. Values marked with ** denote inputs that were not calculated by the software (except for reflux streams, as they are recycled and were calculated using a special module).

Table A2. Calculated values for the streams indicated in Figure 7.

Name	IX-3	IV	VI	Reflux-2	11-2
Vapor fraction	1.0	0.0	0.0	0.0	1.0
Temperature (°C)	20.00 *	160.00 *	55.00	55.16 *	82.71
Pressure (mm Hg)	760.04 **	760.04 **	49.50	1500 **	53.25
Molar flow (kgmole/h)	0.14	11.00	2.92	43.05	46.08

Table A2. *Cont.*

Name	IX-3	IV	VI	Reflux-2	11-2
Mass flow (kg/h)	3.95 **	1305.30 **	345.90	5100.00 **	5452.19
Liquid Volume flow (m^3/h)	0.00	1.46	0.39	5.70	6.09
Heat flow (KJ/h)	-19.55	821,035.87	182,340.96	2,692,787.32	5,085,198.27
Name	14-2	IX-4	Reflux-1	to cond-1	3
Vapor fraction	0.00	1.00	0.00	1.00	0.01
Temperature (°C)	55 **	20 **	45.17 *	69.26	45 **
Pressure (mm Hg)	49.50	760.0 *	1500.12 *	45.75	42.00
Molar flow (kgmole/h)	46.08	7.53E-02	28.30	30.46	30.46
Mass flow (kg/h)	5452.19	2.18 **	3400.00 **	3647.25	3647.25
Liquid Volume flow (m^3/h)	6.09	0.00	3.93	4.21	4.21
Heat flow (KJ/h)	2,874,260.95	-10.79	$-968,792.90$	412,735.44	$-1,033,177.47$
Name	X-3	V	4	cond-1	5
Vapor fraction	1.00	0.00	0.00	0.00	0.00
Temperature (°C)	45.00	45.00	45.00	45.00	45.17
Pressure (mm Hg)	42.00	42.00	42.00	42.00	1500.12 *
Molar flow (kgmole/h)	0.21	1.96	28.30	30.25	28.30
Mass flow (kg/h)	12.35	234.90	3400.00 *	3634.90	3400.00
Liquid Volume flow (m^3/h)	0.01	0.27	3.93	4.20	3.93
Heat flow (KJ/h)	675.42	$-66,812.24$	$-967,040.65$	$-1,033,852.89$	$-965,994.01$
Name	IX-5	reflux-3	Bottom prod-1	X-4	bottom prod-2
Vapor fraction	1.00	0.00	0.00	1.00	0.00
Temperature (°C)	20.00 **	60.17 **	124.85	55.00	127.32
Pressure (mm Hg)	760.04	1500.12 *	236.27	49.50	247.52
Molar flow (kgmole/h)	0.09	33.00	8.98	0.11	6.02
Mass flow (kg/h)	2.58 **	3900.00 **	1062.00	6.29	711.99
Liquid Volume flow (m^3/h)	0.00	4.33	1.18	0.01	0.79
Heat flow (KJ/h)	-12.77	2,616,102.33	751,196.24	3458.36	526,180.91
Name	cond-2	6	7	to cond-3	VIII
Vapor fraction	0.00	0.00	0.00	1.00	0.00
Temperature (°C)	55.00	55.00	55.16	74.20	127.11
Pressure (mm Hg)	49.50	49.50	1500.12 *	35.00	228.77
Molar flow (kgmole/h)	45.97	43.05	43.05	37.90	1.21
Mass flow (kg/h)	5445.90	5100.00 **	5100.00	4471.07	143.49
Liquid Volume flow (m^3/h)	6.09	5.70	5.70	4.97	0.16
Heat flow (KJ/h)	2,870,802.59	2,688,461.63	2,689,971.44	4,748,832.61	71,468.00
Name	8	X-5	cond-3	10	VII
Vapor fraction	0.01	1.00	0.00	0.00	0.00
Temperature (°C)	60.00 **	60.00	60.00	60.00	60.00
Pressure (mm Hg)	31.25	31.25	31.25	31.25	31.25
Molar flow (kgmole/h)	37.90	0.21	37.69	33.00	4.69
Mass flow (kg/h)	4471.07	17.22	4453.85	3900.00 **	553.85
Liquid Volume flow (m^3/h)	4.97	0.02	4.95	4.33	0.62
Heat flow (KJ/h)	3,001,246.98	15,374.97	2,985,872.01	2,614,568.91	371,303.10
Name	11	-	-	-	-
Vapor fraction	0.00	-	-	-	-

Table A2. *Cont.*

Name	IX-3	IV	VI	Reflux-2	11-2
Temperature (°C)	60.17	-	-	-	-
Pressure (mm Hg)	1500.12 **	-	-	-	-
Molar flow (kgmole/h)	33.00	-	-	-	-
Mass flow (kg/h)	3900.00	-	-	-	-
Liquid Volume flow (m^3/h)	4.33	-	-	-	-
Heat flow (KJ/h)	2,615,734.78	-	-	-	-

Stream name with (*) represents the calculation data. Values marked with ** denote inputs that were not calculated by the software (except for reflux streams, as they are recycled and were calculated using a special module).

Table A3. Composition of the flows indicated in Figure 6.

Name	I	1	III	ovh	cond-1
Mass fraction (Phenol)	0.488 **	0.429 **	0.010	0.530	0.530
Mass fraction (Acetophenone)	0.084 **	0.367 **	0.050	0.311	0.311
Mass fraction (DMPC)	0.013 **	0.186 **	0.000	0.145	0.145
Mass fraction (Cumylphenol)	0.220 **	0.000 **	0.400	0.000	0.000
Mass fraction (Phenol tar)	0.220 *	0.017 *	0.540	0.012	0.012
Mass fraction (Airl)	0.000 *	0.000 *	0.000	0.001	0.001
Name	IX-1	dist	2	ovh-1	Bottom-2
Mass fraction (Phenol)	0.09	0.53	0.84 **	0.84	0.43
Mass fraction (Acetophenone)	0.04	0.31	0.14 **	0.14	0.37
Mass fraction (DMPC)	0.01	0.15	0.02 **	0.02	0.19
Mass fraction (Cumylphenol)	0.00	0.00	0.00 **	0.00	0.00
Mass fraction (Phenol tar)	0.00	0.01	0.00 *	0.00	0.02
Mass fraction (Airl)	0.87	0.00	0.00 *	0.00	0.00
Name	cond-2	IX-2	cond-3	II	reflux
Mass fraction (Phenol)	0.84	0.10	0.84	0.84	0.84
Mass fraction (Acetophenone)	0.14	0.00	0.14	0.14	0.14
Mass fraction (DMPC)	0.02	0.00	0.02	0.02	0.02
Mass fraction (Cumylphenol)	0.00	0.00	0.00	0.00	0.00
Mass fraction (Phenol tar)	0.00	0.00	0.00	0.00	0.00
Mass fraction (Airl)	0.00	0.90	0.00	0.00	0.00
Name	3	X-1	X-2	-	-
Mass fraction (Phenol)	0.43	0.00 **	0.00 **	-	-
Mass fraction (Acetophenone)	0.37	0.00 *	0.00 **	-	-
Mass fraction (DMPC)	0.19	0.00 **	0.00 **	-	-
Mass fraction (Cumylphenol)	0.00	0.00 **	0.00 **	-	-
Mass fraction (Phenol tar)	0.02	0.00 **	0.00 **	-	-
Mass fraction (Airl)	0.00	1.00 **	1.00 **	-	-

Stream name with (*) represents the calculation data. Values marked with ** denote inputs that were not calculated by the software (except for reflux streams, as they are recycled and were calculated using a special module).

Table A4. Composition of the flows indicated in Figure 7.

Name	IX-3	IV	VI	Reflux-2	11-2
Mass fraction (IPB)	0.000 **	0.222 **	0.144	0.143 **	0.144
Mass fraction (AMS)	0.000 **	0.759 **	0.856	0.857 **	0.856
Mass fraction (Acetophenone)	0.000 **	0.019 **	0.000	0.000 **	0.000
Mass fraction (Air)	1.000 **	0.000 **	0.000	0.000 *	0.000
Name	14-2	IX-4	Reflux-1	to cond-1	3
Mass fraction (IPB)	0.144	0.000 **	0.980 *	0.978	0.978
Mass fraction (AMS)	0.856	0.000 **	0.020 *	0.021	0.021
Mass fraction (Acetophenone)	0.000	0.000 **	0.000 *	0.000	0.000
Mass fraction (Air)	0.000	1.000 **	0.000 *	0.001	0.001
Name	X-3	V	4	cond-1	5
Mass fraction (IPB)	0.672	0.979	0.979	0.979	0.979
Mass fraction (AMS)	0.008	0.021	0.021	0.021	0.021
Mass fraction (Acetophenone)	0.000	0.000	0.000	0.000	0.000
Mass fraction (Air)	0.320	0.000	0.000	0.000	0.000
Name	IX-5	reflux-3	Bottom prod-1	X-4	bottom prod-2
Mass fraction (IPB)	0.000 **	0.001 **	0.051	0.144	0.001
Mass fraction (AMS)	0.000 **	0.998 **	0.926	0.509	0.964
Mass fraction (Acetophenone)	0.000 **	0.001 **	0.024	0.000	0.035
Mass fraction (Air)	1.000 **	0.000 **	0.000	0.347	0.000
Name	cond-2	6	7	to cond-3	VIII
Mass fraction (IPB)	0.144	0.144	0.144	0.001	0.000
Mass fraction (AMS)	0.856	0.856	0.856	0.998	0.831
Mass fraction (Acetophenone)	0.000	0.000	0.000	0.001	0.169
Mass fraction (Air)	0.000	0.000	0.000	0.001	0.000
Name	8	X-5	cond-3	10	VII
Mass fraction (IPB)	0.001	0.001	0.001	0.001	0.001
Mass fraction (AMS)	0.998	0.849	0.998	0.998	0.998
Mass fraction (Acetophenone)	0.001	0.000	0.001	0.001	0.001
Mass fraction (Air)	0.001	0.150	0.000	0.000	0.000
Name	11	-	-	-	-
Mass fraction (IPB)	0.001	-	-	-	-
Mass fraction (AMS)	0.998	-	-	-	-
Mass fraction (Acetophenone)	0.001	-	-	-	-
Mass fraction (Air)	0.000	-	-	-	-

Stream name with (*) represents the calculation data. Values marked with ** denote inputs that were not calculated by the software (except for reflux streams, as they are recycled and were calculated using a special module).

References

1. Hoffman, D.M. *Handbook of Vacuum Science and Technology*; Academic Press: Cambridge, MA, USA, 1997.
2. Umrath, W. *Fundamentals of Vacuum Technology*; Oerlikon Leybold Vacuum Gmbh: Cologne, Germany, 2007.
3. Martin, L.H.; Hill, R.D. *A Manual of Vacuum Practice*; University of Melbourne Press: Melbourne, Australia, 1947.
4. Danilin, V.S. *Vacuum Pumps and Systems*; Tocenergoisdat: Moscow, Russia, 1957.

5.	Martin, G.R.; Lines, J.R.; Golden, S.W. Understanding Vacuum-System Fundamentals. Hydrocarbon Processing 1994; pp. 1–7. Available online: https://graham-mfg.com/wp-content/uploads/2023/02/Understand-Vacuum-System-Fundamentals.pdf (accessed on 22 February 2024).

6.	Gaikwad, W.; Warade, A.R.; Bhagat, S.L.; Bhasarkar, J.B. Optimization and Simulation of Refinery Vacuum Column with an Overhead Condenser. *Mater. Today Proc.* **2022**, *57*, 1593–1597. [CrossRef]

7.	Rom, A.; Miltner, A.; Wukovits, W.; Friedl, A. Energy Saving Potential of Hybrid Membrane and Distillation Process in Butanol Purification: Experiments, Modelling and Simulation. *Chem. Eng. Process. Process Intensif.* **2016**, *104*, 201–211. [CrossRef]

8.	Kancherla, R.; Nazia, S.; Kalyani, S.; Sridhar, S. Modeling and Simulation for Design and Analysis of Membrane-Based Separation Processes. *Comput. Chem. Eng.* **2021**, *148*, 107258. [CrossRef]

9.	Li, X.; Cui, C.; Sun, J. Enhanced Product Quality in Lubricant Type Vacuum Distillation Unit by Implementing Dividing Wall Column. *Chem. Eng. Process.—Process Intensif.* **2018**, *123*, 1–11. [CrossRef]

10.	Habibullah, A. Crude and Vacuum Unit Design Challenges. *J. Chem. Eng.* **2017**. Available online: https://www.researchgate.net/publication/318405793_CRUDE_AND_VACUUM_UNIT_DESIGN_CHALLENGES (accessed on 22 February 2024).

11.	Sohrabali, G.; Shahryar, J.N. Ejector Modeling and Examining the Possibility of Replacing Liquid Vacuum Pump in Vacuum Production Systems. *Int. J. Chem. Eng. Appl.* **2011**, *2*, 91–97.

12.	Coker, A.K. *Ludwig's Applied Process Design for Chemical and Petrochemical Plants*; Elsevier Inc.: Amsterdam, The Netherlands, 2007.

13.	Hicks, T.G. *Handbook of Mechanical Engineering Calculations*; McGRAW-HILL Professional: New York, NY, USA, 2006.

14.	*Standards of the Tubular Exchanger Manufacturers Association*, 10th ed.; TEMA, Inc.: Tarrytown, NY, USA, 2019; Section 5:1.1.1.

15.	Leybold GmbH. Vacuum Calculations. Available online: https://calc.leybold.com/en/ (accessed on 22 February 2024).

16.	VacTran. Vacuum Technology Software. Available online: https://www.lesker.com/newweb/technical_info/vactran.cfm (accessed on 22 February 2024).

17.	Unisim Design. In *Tutorials and Applications*; Honeywell: Mississauga, ON, Canada, 2017.

18.	August, D.; Chang, J.; Girbal, S.; Gracia-Perez, D.; Mouchard, G.; Penry, D.A.; Temam, O.; Vachharajani, N. UNISIM: An Open Simulation Environment and Library for Complex Architecture Design and Collaborative Development. *IEEE Comput. Arch. Lett.* **2007**, *6*, 45–48. [CrossRef]

19.	Bartolome, P.S.; Van Gerven, T. A comparative study on Aspen Hysys interconnection methodologies. *Comput. Chem. Eng.* **2022**, *162*, 107785. [CrossRef]

20.	Miranda, T.d.C.R.D.d.; Figueiredo, F.R.; de Souza, T.A.; Ahón, V.R.R.; Prata, D.M. Eco-efficiency analysis and intensification of cryogenic extractive distillation process for separating CO_2–C_2H_6 azeotrope through vapor recompression strategy. *Chem. Eng. Process.—Process Intensif.* **2024**, *196*, 109636. [CrossRef]

21.	Paiva, M.; Vieira, A.; Gomes, H.T.; Brito, P. Simulation of a Downdraft Gasifier for Production of Syngas from Different Biomass Feedstocks. *ChemEngineering* **2021**, *5*, 20. [CrossRef]

22.	Vaccari, M.; Pannocchia, G.; Tognotti, L.; Paci, M. Rigorous simulation of geothermal power plants to evaluate environmental performance of alternative configurations. *Renew. Energy* **2023**, *207*, 471–483. [CrossRef]

23.	Baynazarov, I.Z.; Lavrenteva, Y.S.; Akhmetov, I.V.; Gubaydullin, I.M. Mathematical Model of the Process of Production of Phenol and Acetone from Cumene Hydroperoxide. *J. Phys. Conf. Ser.* **2018**, *1096*, 012197. [CrossRef]

24.	Osipov, E.V.; Ponikarov, S.I.; Telyakov, E.S.; Sadykov, K.S. Reconstruction of Vacuum Overhead Systems of the Department of Waste Processing of Phenol-Acetone Production. *Bull. Kazan Technol. Univ.* **2011**, *18*, 193–201.

25.	Powers, R.B. *Steam Jet Ejectors for the Process Industries*; McGraw-Hill: New York, NY, USA, 1994.

26.	Edwards, J.E. *Chemical Engineering in Practice: Design, Simulation and Implementation*; P & I Design Ltd.: Stockton-on-Tees, UK, 2011.

27.	Wilson, G.M. Vapor-Liquid Equilibrium. XI. A New Expression for the Excess Free Energy of Mixing. *J. Am. Chem. Soc.* **1964**, *86*, 127. [CrossRef]

28.	Renon, H.; Prausnitz, J.M. Local Compositions in Thermodynamic Excess Functions for Liquid Mixtures. *AIChE J.* **1968**, *14*, 135–144. [CrossRef]

29.	Abrams, D.S.; Prausnitz, J.M. Statistical Thermodynamics of liquid mixtures: A new expression for the Excess Gibbs Energy of Partly or Completely Miscible Systems. *AIChE J.* **1975**, *21*, 116–128. [CrossRef]

30.	Dyckman, A.S.; Boyarsky, V.P.; Malinovskii, A.S.; Petrov, Y.I.; Krasnov, L.M.; Zinenkov, A.V.; Gorovits, B.I.; Chernukhim, S.N.; Sorokin, A.D.; Fulmer, J.W. Phenol Tar Processing Method. U.S. Patent 5,672,774A, 30 September 1997.

31.	Reddy, C.C.S.; Rangaiah, G.P. Retrofit of Vacuum Systems in Process Industries. In *Chemical Process Retrofitting and Revamping: Techniques and Applications*; John Wiley & Sons, Ltd.: Hoboken, NJ, USA, 2016; pp. 317–346. [CrossRef]

32.	Govoni, P. An Overview of Vacuum System Design. *Chem. Eng.* **2017**, *124*, 52–60.

33.	Golden, S.; Barletta, T.; White, S. Vacuum unit performance. GEA Wiegand. *Pet. Technol. Q.* **2012**, *17*, 131.

34.	Sterling Fluid Systems Group. Liquid Ring Vacuum Pumps & Compressors: Technical Details & Fields of Application. 2017. Available online: https://www.scribd.com/document/487161329/ZHidkostno-Koltsevyie-Nasosyi-i-Kompressoryi-Harakteristiki-i-Primenenie-Liquid-Ring-Vacuum-Pumps-Compressors-Technical-Details-Fields-of-Applicatio (accessed on 22 February 2024).

35.	Aliasso, J. How to Make Sure You Select the Right Dry Vacuum Pump. *World Pumps* **2000**, *2000*, 26–27. [CrossRef]

36.	Osipov, E.V.; Telyakov, E.S.; Ponikarov, S.; Bugembe, D.; Ponikarov, A. Mini-Refinery Vacuum Unit: Functional Analysis and Improvement of Vacuum Overhead System. *Processes* **2021**, *9*, 1865. [CrossRef]

37. Osipov, E.V.; Telyakov, E.S.; Ponikarov, S.I. Coupled Simulation of a Vacuum Creation System and a Rectification Column Block. *Processes* **2020**, *8*, 1333. [CrossRef]
38. Osipov, É.V.; Telyakov, É.S.; Latyipov, R.M.; Bugembe, D. Influence of Heat and Mass Exchange in a Liquid Ring Vacuum Pump on Its Working Characteristics. *J. Eng. Phys. Thermophys.* **2019**, *92*, 1055–1063. [CrossRef]
39. Karmanov, E.; Lebedev, Y.; Chekmenev, V.; Aleksandrov, I. Ejector Systems. *Chem. Technol. Fuels Oils* **2004**, *40*, 80–83. [CrossRef]
40. Giproneftemash, M. (Ed.) *Steam Jet Vacuum Pumps*; Schutte & Koerting: Moscow, Russia, 1965.
41. Zou, Y.; Jiang, H.; Liu, Y.; Gao, H.; Xing, W.; Chen, R. Highly Efficient Synthesis of Cumene via Benzene Isopropylation over Nano-sized Beta Zeolite in a Submerged Ceramic Membrane Reactor. *Sep. Purif. Technol.* **2016**, *170*, 49–56. [CrossRef]
42. Schmidt, R.J. Industrial Catalytic Processes—Phenol Production. *Appl. Catal. A Gen.* **2005**, *280*, 89–103. [CrossRef]
43. Verma, R.K. IHS Markit | PEP Review 2020-02 Phenol Production by ExxonMobil 3-Step Process. 2020. Available online: https://cdn.ihsmarkit.com/www/pdf/0420/RW2020-02_toc.pdf (accessed on 22 February 2024).
44. Zou, B.; Hu, Y.; Huang, X.; Zhou, X.; Huang, H. Manufacturing and New Technology Research of Phenol. *Shiyou Huagong/Petrochem. Technol.* **2009**, *38*, 575–580.
45. Zakoshansky, V.M. Alternative Technologies for Obtaining Phenol. *Russ. Chem. J.* **2008**, *52*, 53–71.
46. Zakoshansky, V.M. Mechanism and Kinetics of Acid-Catalyzed Decomposition of Cumene Hydroperoxide. *Catal. Chem. Petrochem. Ind.* **2004**, *4*, 3–15.
47. Kruzhalov, B.D.; Golovanenko, B.N. *Joint Production of Phenol and Acetone*; Goskhimizdat, M., Ed.; Goskhimizdat: Moscow, Russia, 1963; 191p.
48. Zakoshansky, V. Actual Performance of Key Stages of the Phenol Process: Present State and Expected Future. *Russ. J. Appl. Chem.* **2013**, *86*, 1118–1140. [CrossRef]
49. Perego, C.; Ingallina, P. Recent Advances in the Industrial Alkylation of Aromatics: New Catalysts and New Processes. *Catal. Today* **2002**, *73*, 3–22. [CrossRef]

MDPI

St. Alban-Anlage 66

4052 Basel

Switzerland

www.mdpi.com

MDPI Books Editorial Office

E-mail: books@mdpi.com

www.mdpi.com/books